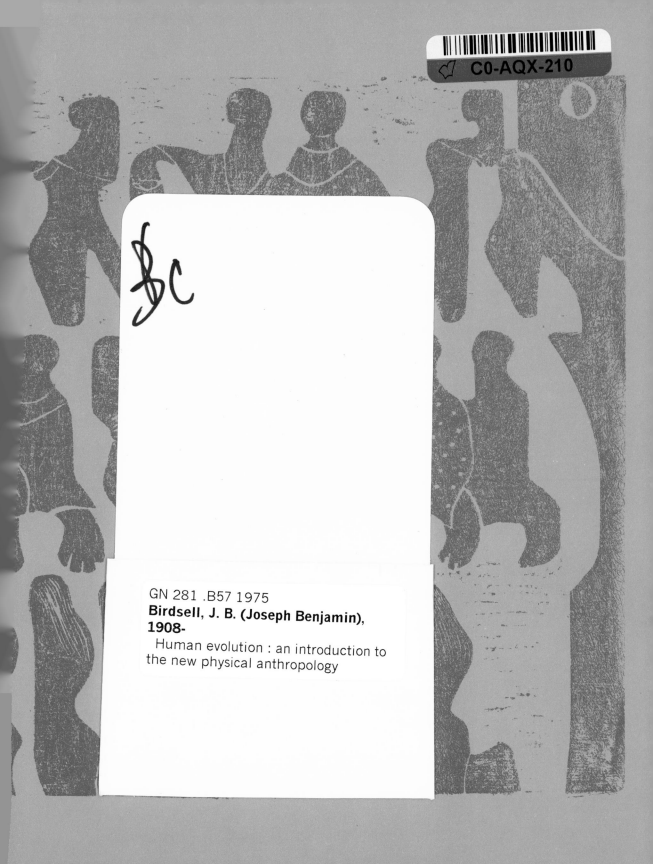

Human Evolution

Human Evolution

SECOND
EDITION

An Introduction to the New Physical Anthropology

J. B. BIRDSELL

RAND McNALLY COLLEGE PUBLISHING COMPANY / Chicago

RAND MCNALLY
ANTHROPOLOGY SERIES

Human Evolution:
An Introduction
to the New Physical
Anthropology
Second Edition

J. B. BIRDSELL

Handbook of
Social and Cultural
Anthropology

JOHN J. HONIGMANN

Make Men of Them:
Introductory Readings
for Cultural Anthropology

CHARLES C. HUGHES

Current printing (last digit) 15 14 13 12 11 10 9 8 7 6 5 4 3 2 1

To those students who see the need for change
and who care enough to work for it.

CONTENTS

PLATES AND CREDITS

FIGURES

TABLES

Classic physical anthropology primarily concerned itself with the description and classification of man. Through the use of anatomical measurements and other data, it attempted a classification of human races, and also tried to relate fossil forms to living populations. It had little sense of process, and it lacked comprehension of general evolution as this applies to the ascent of man. There has developed—in recent decades—a new physical anthropology concerned with processes, the formulation and testing of hypotheses, and the application of neo-Darwinian evolution to the human species. This new approach is heavily involved with human ecology, primate and human behavior, and the nature and structure of human populations, all framed in terms of population genetics. When possible, problems are investigated by laboratory experiments; in other situations, what we may call experiments in nature are examined. Classification has become an obsolete exercise and the main game involves the analysis of evolutionary processes as they apply to us and to our ancestors.

The purpose of an introductory text is to effectively present the materials necessary to introduce the reader to a new intellectual area, and it presumes no prior exposure. The organization of such books varies with the authors' preferences. In my case, having taught this course by choice for more years than I care to confess, I choose to emphasize the broad issues. If my own students can recall a dozen of the major issues presented once they leave the university, the course can be judged a success. My emphasis is upon relationships, particularly those that provide guiding principles in the understanding of general organic evolution as well as the specific evolution of man. There are exceptions to all laws in evolution; however, a significant number of principles guides our thoughts through the maze of complicated data and the missing pages of the record which must be bridged.

A number of authorities begin their presentation with the primates and lead quickly into human evolution. This approach has the advantage of directness, but it suffers from a lack of breadth. Most of the processes of evolution are best illustrated by materials drawn from outside our own taxonomic order. General evolution is much richer in examples of convergence or divergence, in rates of evolution that may approach zero or that may involve rapid change, and indeed in the basic principles which are needed to interpret our own record. A great many scientists have contributed to the overall investigation of evolution, and their names are specifically noted only where their research or point of view is directly useful to the reader or worth pursuing outside of the course. Naturally, the ideas of many other evolutionists have greatly contributed to the subject matter.

INTRODUCTION TO THE PROPER USE OF THIS TEXT

I have incorporated the technical or taxonomic name of many animals in the evolutionary background of man. These are unimportant except in cases where the reader may wish independently to confirm the exact genus and species of a given creature, or to independently investigate the situation in which it appears.

The record of evolution has long been taught by means of bones and, more especially, teeth, which are so well preserved in the rocks. Today throughout the natural sciences there is a wholesome trend to clothe the bones with flesh, to place the animal or man in the proper environment, to attempt to visualize its relations with the rest of the natural community, and where possible to reconstruct the major aspects of behavior. It is now recognized that the form and structure of bone are influenced by the way an animal lives and the manner in which it behaves. Some behavioral interpretations are purely speculative, but others rest upon the firmer foundation of behavior observed among living animals today. Many extinct forms of life have no living analogues, and so their behavior remains conjectural.

My style of lecturing contains many digressions and asides to illustrate points considered worthwhile; however, this text is so organized as to let the major ideas proceed unimpeded. Supplements following many of the chapters serve to expand certain areas, and also to provide further biological guidance. In some cases the supplements offer greater depth of information than is needed in the text itself. Instructors may ignore the supplements where time is short, but in my own view they are valuable.

Examples based upon the Aborigines of Australia are presented in the later chapters. I have had the good fortune to spend four years in field work among these people who perhaps better than any other now living provide reasonable and generalized models for the lives of men in the last half million years of the Pleistocene. They are not perfect models, for they live in a continental environment which does not show all of the variations found in Eurasia and Africa. But they do provide examples of ongoing evolutionary processes which have never been observed elsewhere. And, of course, men make better models for our ancestors than do monkeys.

Currently, the controversy between religion and science is moderating. There is no conflict between the natural sciences and religion, for natural sciences are based on orthodox materialism, which does not concern itself with the supernatural. If the student understands that this book is materialistically oriented in an explicit way, he will recognize that he has no obligation to accept such a position, nor to allow it to alter his personal views. Today the area of disagreement between religion and science focuses on a single central point, the origin of life. Most natural scientists ultimately explain this in materialistic terms, and as a consequence of the very nature of matter and the planetary environment. Those who may reject this view for reasons of faith may take some consolation in the idea that life has not yet been produced in the laboratory, and even if it had, the creation of living molecules would not necessarily prove that life had been created in the same fashion four billion years ago. For while laboratory experiments may be analogous to the real act, and certainly will throw some light upon the event, they cannot duplicate an occurrence so far back in time. The reader of faith may conclude that the depth of this intervening time makes the exact origin of life on this planet essentially unknowable. The scientific optimist on the other hand will look foward to the day, quite possibly in the next few decades, when living molecules will have been produced in laboratories. In either case, the sweep of organic evolution has a grandeur which compels the individual to some sense of personal humility. Its perspectives do offer all of us some guidelines in our living.

A new edition of a text must reflect the changing world about it. As knowledge has grown, so new data must be included. Where new discoveries alter previous theories, the necessary revisions must be made. The input from teachers who have used a text and the students who have read it should be incorporated. General comments have indicated that the format and style of the first edition of this book needed no drastic revision, but every chapter has been brought up to date. An effort has been made to reduce the number of pages without losing important evidence and principles. Thus, Chapter 4 of the first edition, which was devoted to the progressive evolution of functional systems, has been omitted. Instead some examples taken from it have been incorporated in the general discussion of the fossil record and in two new supplements. The general response to the supplements has been favorable, and their use is continued here. Even where time in a course is pressing, it is hoped that students will find the supplements interesting enough to read, for they contain important perspectives.

Knowledge about human evolution usually increases at a uniform rate in most areas, but fossil man stands as a marked exception to this rule. The constant recurring changes in dating bring about dramatic reinterpretations of the fossils themselves. New finds, such as the very early fossil ER-1470 from East Rudolf in Africa, have required the entire reworking of the chapters on fossil man in the first edition. Certainly new discoveries lie just ahead, and professional gossip has it that East Rudolf will provide more in the near future. Human evolution is complex, and students are warned that with new discoveries it will become even more complicated.

A few readers have commented that too many of the examples used in the book are derived from the author's Australian data and experiences. This is not due to egotism, but follows from the fact that the Australian Aborigines have provided better data than other hunters for many areas in human microevolution.

I receive letters from readers, and their advice and criticisms are considered. I would appreciate a continuing flow of comments about this edition.

Students have often remarked that they find some difficulty picking out the important theories from the illustrative examples in the book. To solve this problem it is suggested that a red marking pen be used to underline important principles and a yellow one for indicating the most useful examples.

INTRODUCTION TO THE SECOND EDITION

Human Evolution

CHAPTER 1

THE
UNIVERSE
AND OUR
PLACE IN IT

This book is about evolution. As an essential part of the framework of the natural sciences, it is important to understand the meaning of evolution. It is here used in the broad sense to include changes in our universe over time, ideas about the very origin of life, the changing nature of our own solar system and the planet Earth—all as a background for the study of how man came into being. It is useful to know the language of science for it contains ideas of different levels of reliability. It does not pretend to absolute truth, for science continues to grow with new knowledge and concepts.

Our Frame of Reference

As the contents of the book will indicate, the emphasis is upon principles rather than details. This frame of reference does not mean that either general organic evolution, or the evolution of man himself, is totally understood, or that its guiding principles have all been defined and tested. The frame of reference does suggest that emphasis is placed upon underlying processes and such regularities as can be detected in organic evolution at this time. This approach is meant to be suggestive rather than pretentious.

The full sweep of the story of organic evolution provides needed examples for comprehension of the rise of man. In part this is the result of the vastly longer time scale with which we deal, for the origin of life is now estimated to have begun four billion years ago; in part it results from the greater variety provided by general evolution in contrast to primate evolution with its limited time scale and condensed patterns of variation.

The boundary between living (organic) substances and nonliving (inorganic) material is blurred and connected by an ambiguous bridge through the viruses. Further, on our earth, in our solar system, and throughout the universe, there are regular patterns of change in time which must be considered evolutionary in the broad sense. At a time when man has set foot on the moon and proposes to do the same upon Mars, there is every reason to consider evolution in the broadest possible terms. Thus we begin with current concepts about our universe and follow with some consideration of the planet Earth and its position in our solar system. These are not viewed as digressions but as a part of the necessary framework of understanding organic evolution, and even of the ascent of man within that framework.

On Our Use of the Language of Science

Understanding requires that communication be both simple and clear. Scientific regularity forms a kind of hierarchy with validity rising in the higher levels. Since its language is frequently misused, this is a good point at which to define the scientific vocabulary used in this book. A *proposition,* assertion, or conjecture is nothing more than a proposal which needs confirmation. Students of human evolution are particularly prone to build structures by piling one proposition upon another without submitting any of them to proper testing for validation. There are even writers of considerable talent who conclude that if a phrasing sounds good it therefore must be true. Propositions have their place as preliminary formulations of ideas, but they should carry little weight if they have not been tested, or if they are by nature untestable, an all too frequent occurrence.

The word *hypothesis* has been so misused that it requires clarification. Derived from the same word in Greek, it means groundwork or foundation. It consists of a proposition formulated to explain certain observed facts. In science a hypothesis preferably refers to the relations between two or more variable factors and so it deals with relationships. A good hypothesis should be stated in such a way that it is testable, for otherwise it is a useless assertion. In another sense, a hypothesis is an unproved theory that in time may be proved true or false.

The word *theory* is also derived from the Greek and originally meant a mental viewing or contemplation of a situation. Today the meaning is changed to a formulation of apparent relationships or underlying principles of certain observed phenomena. The use of the word theory implies that some verification has been achieved. An unproved theory should mean that some confirmation has been obtained but that further testing is needed before its general acceptance as a good theory. So the concept covers a wide range in terms of degrees of verification. To become good theory in the world of science the verifications must be repeated by a number of workers in different laboratories or field situations. Good experiments are the tests of hypotheses. Thus, the growth from hypothesis to good theory may involve many experiments by different people.

Beyond the level of well-tested theory are a series of simple terms which are helpful in indicating the degree of regularity found in natural phenomena. A *rule* is a statement of what normally or usually happens. It clearly implies that there are exceptions to it, which may or may not be completely explained. Most of the regularities in the natural sciences do not rise above the level of rules, even though they may be given names implying greater rigidity. A *law* involves a sequence of events in nature that has been observed to occur with unvarying uniformity as long as the conditions are constant. Many of the so-called laws of nature fall short of this definition. A *principle* is a fundamental law upon which other laws are based. This is another overworked word and many evolutionary principles fall short of the rigor of this definition.

An *axiom* is a statement which is universally accepted as true and so seldom applies to the fields of the natural sciences. It is more properly used in mathematics. A term that I am prone to use is *dogma,* derived from the same Greek word which refers to what one thinks to be true. In this book it will be used to indicate a consensus among interested scientists without any necessary implication that it represents a final statement. Dogma frequently proves either to be untrue or to need considerable modification as scientific knowledge evolves. Nonetheless, it is a useful term to indicate a current state of scientific opinion.

A term coming into more general use in the contemporary world is *model.* Originally referring to a miniature replica of a real object, its use has now been extended considerably. Intellectual models can be constructed to simulate complex processes in the world about us. Since they are designed to aid the investigator in comprehending what goes on in very complex situations, it follows that good models are always simpler than the natural process. Whether or not they are useful depends upon the skill of the model builder and whether his simplified construct has properly focused on the important variables and ignored the unimportant ones. Models always occur in the mind of the investigator and are never found lying around loose. A cosmologist may construct a model to explain his ideas of the nature and origins of the universe. A biologist may construct a model to explain the set of biochemical processes occurring in the normal activity of a single cell. A physical anthropologist constructs models of various kinds. They might range from a simple plastic replica of the lower jaw of man intended to be subjected to stresses in a testing machine to determine how much its form is dependent upon function, to purely intellectual models designed to relate a complex variety of processes to one another. The latter types become necessary when dealing with evolutionary processes spanning a number of generations, or in the tracing of population variables in ways which differ from the methods of demographers. Or models may be constructed to try to understand fluctuations in the gene pool of a population as evolutionary processes vary in intensity. These complex intellectual models are frequently designed as computer simulations, for they contain too many operations to be done by the old pencil and paper method. Even these models which are submitted to the computer for successful solution must be simpler than the real phenomena found in natural populations. Yet they do enable investigators to ask questions which were not even conceived of prior to the advent of the computer age.

One may ask: Where does absolute truth fit into the scientific system? And the clear answer is that it does not. It properly belongs in theology and areas of inquiry involving the supernatural. Both the natural and the physical sciences are materialistic and seek answers to their problems in the nature of the material world observed. Investigators in both fields are content to see a good hypothesis tested sufficiently to become a good theory. When this is later revised in the light of further information, or even totally superseded by a better theory, the scientists show a little reluctance, shed a sentimental tear or two, and recognize that this is the normal way in which a science grows, and in which scientific knowledge evolves. The theoretical dogma of today inevitably becomes obsolete tomorrow and ultimately will be discarded as newer information develops. At no time does scientific knowledge become absolute. Even the well-tested law of the conservation of matter may apply only to our planet rather than to the entirety of the universe.

On the Meaning of Evolution

A book devoted to the discussion of the principles of evolution must early give a proper definition of evolution. The Latin root from which the word is derived means an unrolling, an opening out, or a working out, especially in the process of development of growth. In the broadest senses in which it is used today *evolution has come to mean simply changing characteristics in time.* For organic evolution this means a change in the genetic contents of the *pool which represents the total inheritable properties of a population.* The great breadth of the basic definition allows it to be properly applied not only to organic life, but to human behavior, man's culture, his technology, and the very growth of science itself. In this form evolution can further be applied to those regular changes which have occurred in our planet, our solar system, and which are believed to affect the total universe.

Let us return for the moment to organic evolution. In Darwin's day evolutionary concepts were not cast in genetic terms, for the gene was not yet known. Darwinian evolution was based upon the general idea that the action of natural selection upon the variability found in nature produced populations better fitted for survival in their environment. This has been recast into genetic terms and elaborated through the ideas of mathematical evolutionists.

Macroevolution

It is wise to distinguish the differences in organic evolution between long-term, or *macroevolution*, and short-term, or *microevolution.* As viewed through the fossil records of hundreds of millions of years, organic evolution appears to be primarily directional in its progress, although admittedly complicated as a process. The fossil record reveals that our present living animals are a product of progressive evolutionary changes which have affected most of their major structures and systems. Even though macroevolution must primarily be read through the limited evidence of bones and teeth, nonetheless it tells its story very effectively. Many early students of evolution believed that evolutionary trends were linear in nature. From such ideas came the concept of *orthogenetic evolution*, which consisted of a series of changes proceeding in so direct a fashion as to appear as if the animals were literally seeking a goal, or to perfect their parts. In recent years it has become evident that these straight-line trends were a consequence of a faulty fossil record in which many of the pages were actually missing. It is now known that organic evolution proceeds with populations branching in a variety of directions and producing many populations, some of which ultimately prove less successful as competitors than others. While some random components operate in the macroevolutionary system, they are difficult to identify from fossil remains, and there is a general agreement that macroevolution primarily manifests itself as a series of seemingly oriented or systematic changes.

Microevolution

Microevolution, which is small in scale, involves short time periods and is primarily studied upon living populations. Because of the difference in the time scale and the type of materials involved, microevolution somehow seems to be bitterly debated. The scale of time by which we read microevolutionary changes is so fine and detailed that broad trends are not always evident in it. The overall picture is confused by a bewildering array of populations, each apparently moving in its own direction, guided by processes which we do not completely understand at the observable level. Chance seems to play a fairly important part in microevolutionary change.

The broad definition of evolution—changes in time—allows broad application of its concepts. For example, Herbert Spencer formulated his ideas about social evolution before Darwin published his works on organic evolution. Human societies and social behavior have shown evolutionary changes, and when some of these—pollution, overpopulation, violence, drugs—are projected into the near future, they suggest reason for alarm. Man's ideas have gone through an unfolding process that is evolutionary in nature. Turning to the inorganic world, changes have also proceeded in such regular ways that it is convenient to view them in evolutionary terms. The position, structure, and chemistry of our planet Earth, of our solar system, and of our home galaxy the Milky Way have changed in regular ways with time and so have evolved. Presumably even the universe itself is subject to evolutionary regularities.

In recent years men have pushed back the boundaries of their ignorance and begun to record some of the regularities that seem to govern the infinity of space and time that make up the universe in which our planet is but a pebble. All people, even the most primitive, have invented their own legends to explain the origin of the world as they know it. Many of these start with the statement that in the beginning there was darkness. Into this darkness came supernatural figures who created the first man. After a suitable interval, light was created and was alternated with darkness in a regular pattern of night and day. These primordial figures molded the major land forms and in their later wanderings created the geographical features known to local groups. These myths vary in detail, but all of them picture the world as created out of nothingness and structured by the ancestral figures into its present form. The universal impulse in men to see the world as beginning and ending at some particular point in time seems to arise from the fact that the life of each individual has a specific beginning and goes on to an unavoidable end.

Evolution in the Inorganic Universe

The concept that evolution consists of changes in time allows us to apply it to the nonliving universe. Physical matter on earth contains 92 different kinds of atoms, beginning with hydrogen, the lightest, and ending with uranium, the heaviest. Eleven other elements, all heavier, can be produced inside of the stars under natural circumstances, or artificially here on earth in high energy accelerators. These 103 different elements are arranged in an atomic table and differ systematically from each other in terms of the number of protons contained in the nucleus. Hydrogen, the simplest, has a single proton, while the 103rd element, lawrencium, contains 103 protons in its nucleus. The differences between the known elements are systematic and orderly, and the heavier ones can be produced through energy inputs. In the world of nature, heat is usually the energy involved. Most of the heavier elements are created in the inner furnaces of stars at certain evolutionary stages in their life cycle. There is considerable evidence to suggest that most of the atoms in our present short-lived human bodies have the same content as, and in fact may be, debris cast out by exploding stars billions of years ago.

The inner furnaces of the stars are the only places in the universe known to be capable of creating the 102 elements heavier than hydrogen out of that original material. Hence, in the creed of the materialists, immortality involves atoms rather than spirits or bodies.

Speculations about the nature of the universe properly belong to a certain kind of astrophysicist known as a *cosmologist*. Working at levels higher than that of galaxies, the cosmologist attempts to clarify the overall apparent chaos in the universe so that he can seek simple orderly relationships in terms of density, mass motions, and pressures of the total system. For such a theorist each galaxy is but a single atom in a supergas. Limited by human perceptions, and able to examine the universe only through a small variety of instruments, the cosmologist faces a difficult task.

Currently there are three different models to explain the nature of the universe. They vary in fundamental assumptions concerning the nature of time and space. The first hypothesis conceives of a model of a *steady state universe* in which both time and space are infinite, as shown in Table 1–1. This hypothesis involves some difficulty, for it cannot explain why galaxies seem to be expanding away from each other at speeds approaching that of light.

On the Nature of the Universe

TABLE 1–1

Model	Time	Space
1) Steady State	Infinite	Infinite
2) Big-Bang	Finite	Infinite
3) Oscillating	Infinite	Finite

A totally different model is based upon the dramatic *"big-bang"* hypothesis. In this case the cosmologists hypothesize that the universe began about fifteen billion years ago as an aftermath of an incredibly huge explosion of densely packed primordial matter, that is, essentially hydrogen. This is a rather popular model, since it explains the apparent expansion of the galaxies of the universe, but it has not yet been subjected to critical tests. In this hypothesis time is conceived as finite, with space as infinite.

A third model of the nature of the universe is an *oscillating* one. In this hypothesis the universe expands after a "big-bang" and then in time contracts back to an extreme density, when it explodes again. As originally formulated, the cycle in this model was conceived of being about 80 billion years in duration, with a number of such cycles endless in time. So this hypothesis assumes a finite space and infinite time.

As interesting as these constructions are, they do not rise to the level of good theory, for they have not yet passed a sufficient number of adequate tests. Some scientists doubt that man will ever perceive the universe accurately enough to determine its real nature. Nevertheless, it remains clear that the universe has evolutionary trends of its own and that our concepts about it will evolve in time toward better hypotheses, or even attain the state of a satisfactory theory.

On the Evolution of Galaxies

The universe contains giant clusters of stars called *galaxies,* which may number in the billions. Astronomers have studied 1,000 of the brightest observable galaxies and classified them according to their form. They hypothesize that our own home galaxy originated perhaps ten billion years ago from a pervasive cloud of hydrogen. Gravitational pull caused a gradual shrinkage so that the gaseous ingredients were mixed and a giant sphere began a slow rotation of an orderly sort. With the passage of billions of years this gas sphere diminished in size, and its mass was compressed into smaller and smaller volume. Occasionally there occurred extra dense eddies of gas and some of these were held together by gravitational force, thus giving birth to the first clusters of stars.

Plate 1-1
The Spiral Galaxy in Andromeda, similar in shape to our home galaxy, but somewhat larger. It is the most distant cosmic feature visible to the naked eye and lies 750,000 light years away.

The 1,000 galaxies nearest to us, and hence the most observable, vary from flattened forms with spiral arms, as in our own home galaxy, to beautifully symmetrical elliptical systems that are otherwise featureless. They suggest that the sequence of galactic evolution begins with a rotating, spirally armed form, which changes in structure into a thickening elliptical disc, and as its rate of turning slows, finally approaches an ultimate spherical stage. This concept may be correct, but it cannot yet be tested.

Our Home Galaxy

Our home galaxy, also called the Milky Way, is but one of an unknown number of billions of galaxies in the universe. It contains perhaps a billion luminous and potentially visible stars. Many of these may be surrounded by systems of planets, as is our own sun, but these dark bodies are too far away to be seen through our optical telescopes. Astronomers generally agree that these planetary systems themselves undergo an evolutionary sequence, during a part of which time some of the planets would be suitable for the origin of life. In the enormous distances of our own galaxy there may be hundreds of thousands of planets with conditions suitable for living organisms.

Our Solar System

Our sun, and its retinue of nine planets, are situated about two-thirds of the way out in one of the spiral arms of our galaxy. Our sun is a rather mediocre star, but even so it is impressive. Its diameter is more than one hundred times greater than the Earth, that is, 864,000 miles. It is a thermonuclear furnace consisting largely of hydrogen and helium which every second converts 4,000,000 tons of matter into energy, as the enormous temperatures of its interior combine four molecules of hydrogen into one of helium. It is estimated that in the core of the sun temperatures reach the unbelievable level of 27,000,000° F. Our sun, judging from the variation in the other stars in the visible universe, is about midway in its life cycle, having originated about five billion years ago, and with an expected future life-span of another five billion years. The estimated age of our solar system, approximately 4.6 billion years, is now receiving further confirmation from the absolute dates obtained from rocks brought back by our moon expeditions.

The Planet Earth

The nine planets rotate about the sun in the same direction, in essentially elliptical orbits which approximate a single plane. The outer five planets, most of which are known as gas giants, could host no life as we know it. The innermost planet, little Mercury, about the size of our own moon, surprisingly has been able to hold a very thin atmosphere about it. Its rotation is nearly stopped by the gravitational pull of the sun, rotating once every 58 and one-half days of Earth time, so that the sunlit side reaches temperatures up to 750° F., while the dark side has temperatures approaching absolute zero.

Beyond Mercury lie the three so-called terrestrial planets, Venus, Mars, and Earth. Recent space probes have penetrated the dense atmosphere of Venus and recorded a ground temperature in excess of 900° F. Life cannot exist upon it. Mysterious Mars has a thin atmosphere consisting primarily of carbon dioxide plus a little water vapor and oxygen. It lacks nitrogen, an essential component for life as we know it. The Martian atmosphere is so thin that it offers little protection from violent bombardment of ultraviolet radiation from the sun. Most scientists agree that this is so severe as to prevent the formation and maintenance of organic chemical compounds basic to life.

Our own planet, Earth, has very special attributes which enable it to support life as we know it. Its atmosphere consists primarily of nitrogen, about one-fifth of its content is all-important oxygen, and a small and variable amount of the atmosphere is water vapor. One-half of the atmospheric molecules occur in its lower three and one-half miles of altitude. This constitutes the *biosphere* of our planet, the zone in which life can occur. An important attribute of our planetary atmosphere is that it acts as a greenhouse, trapping much of the solar energy it receives. As a consequence the climate of Earth is a reasonable one which falls below freezing at some times and in some places, but never becomes too hot to maintain organic life. The presence of water upon our planet is of enormous consequence. As L. J. Henderson noted in an interesting book called *The Fitness of the Environment* (1913), water has unusual physical and chemical properties. It is simple in structure, consisting of a molecule containing two atoms of hydrogen and one of oxygen. It has the unusual property of being able to dissolve all of the other elements, even inert gold. Thus it is an ideal medium for the complex transport of other elements in solution and so provides a broad foundation for the processes of life. Water has a relatively high freezing point, and the almost unique property of expanding its volume as it freezes. This is important, for all bodies of water on Earth would freeze solid from bottom to top if ice were not buoyant, and aquatic life could not exist in freezing areas. In going from one gram of ice at 0° C. to water the same temperature, 80 calories of heat are required. The conversion of water at 100° C. into water vapor, through boiling or evaporation, absorbs or requires 540 calories. This latent heat of evaporation provides a most convenient

Plate 1-2
The Planet Earth as viewed from
Space. Its unique weather-making
atmosphere is clearly visible in
cloud formations. The outlines of
North and South America can be
seen.

way of cooling, including the cooling of human bodies in hot
countries. Virtually all forms of life are intimately associated with
water and are dependent upon it.

The solid portion of our planet is hypothesized to consist of an
innermost core of extremely dense material with a radius of about
800 miles. It is about 12 to 18 times as heavy as water, and is
presumed to consist primarily of iron and nickel. Its temperature is
not known, but estimates range upward into the thousands of
degrees Fahrenheit. This inner core is further surrounded by a zone
ranging from 800 to 2,200 miles outward of very dense materials,
again presumably iron and nickel. This second belt seems to be in a
liquid state and hence very hot. A third layer, called the mantle,
comprises the next 1,800 miles of the Earth's radius. It consists
primarily of silicon dioxide, which is the chief chemical building
material in rocks. Finally an outer crust reaches a maximum depth of
50 miles beneath continents and a minimum depth of only three
miles in some regions of the ocean floors. It is largely a veneer of

igneous rocks which have solidified from previous hot and molten conditions. Both the upper granites and lower basalts float as lighter materials upon the mantle proper.

Figure 1-1
Earth Structure showing its inner layers

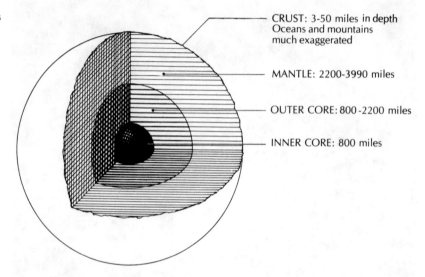

CRUST: 3-50 miles in depth
Oceans and mountains
much exaggerated

MANTLE: 2200-3990 miles

OUTER CORE: 800-2200 miles

INNER CORE: 800 miles

In very recent years it has become evident that the once rejected idea of Alfred Wegener concerning continental drift is in fact correct. Current evidence shows that the crust of the Earth consists of great plates which move slowly as the basalt flows up through great fissures in the ocean floor, causing it to spread. India, Africa, South America and Australia were once all firmly attached to the Antarctic land mass. This explains why some of their fauna show remarkable similarities. But these continental masses began drifting apart several hundred million years ago and so are of little importance in the evolution of the mammals. Supplement No. 6 deals with the idea of continental drift in greater detail.

The Cooling Cycle of Our Planet

In terms of the hypothesis now currently most attractive, our solar system came into being some five billion years ago due to the condensation of matter through gravitational forces acting on the primeval cloud of hydrogen and cosmic dust. Evidence suggests that in its original form our planet was considerably hotter than it is today, for the deeper we drill into the Earth's crust, the higher the temperatures become. Figure 1–2 illustrates the type of cooling curve representative of our planet as it aged. Its initial temperature at birth cannot be specified but it presumably has been gradually cooling since its formation, for it has no substantial inner sources of energy in the form of its own nuclear fuels. The rate of cooling has been moderated by solar energy from the sun. As characterizes most cooling curves, the value drops off rapidly in the early part of the planet's life and then settles down to a very gradual rate of decline over the years of its maturity. In its final phase, which is estimated to be about five billion years in the future, our sun will itself be aging, through the exhaustion of the sources of its nuclear fuels. As the critical point of exhaustion approaches, it will begin to expand into a red giant star which may even engulf the inner planet, little Mercury. In its dying phases the heat radiated will be so extreme that it is calculated that everything on the surface of our planet will be vaporized. Fortunately this disaster lies so far ahead in time that we need not be concerned with it, although it certainly indicates that life on Earth will come to an end.

The graph in Figure 1–2 is important in demonstrating that in the early portion of its life our planet was too hot to support life as we know it, that it entered a favorable range of temperatures lasting over many billion years and that finally it will become too hot to support life. While the basic life supporting molecule of organic nature may form at higher temperatures, such as perhaps exist in the gases of Jupiter today, nonetheless life as we know it finds the temperature of boiling water an effective upper limit. Some types of living bacteria can survive at or slightly above 212° F. but they cannot

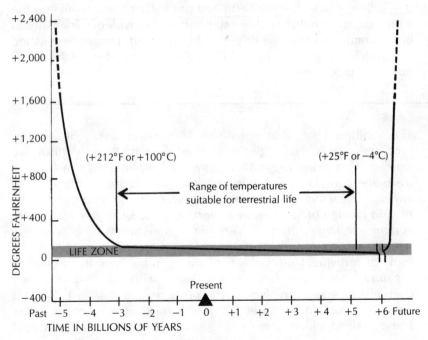

Figure 1-2
Cooling Curve Representative of Earth Temperature

reproduce at this temperature. Simple plants such as red algae are well adapted to survival and reproduction at temperatures as high as 167° F. as in Yellowstone Park's hot springs. Bacteria, which are so adaptable, have intermittently survived temperatures below freezing, but it is probably safe to say that the lower limit on effective life is generally 32° F. or 0° C. Many higher organisms, both invertebrate and vertebrate, are well adapted to survival only a few degrees above freezing in the waters of the Antarctic and Arctic regions. But it is probably fair to conclude that in the Earth's cooling cycle the limit for living forms lies between the boiling and freezing points of water.

A few years ago it was thought that life on Earth might have originated only a billion or so years ago. But with advancing techniques of identification and improved dating methods, we find that life originated much earlier than was formerly thought. The most recent evidence was derived from South Africa in 1968 by Dr. Engel of the Scripps Institute of Oceanography at La Jolla, California, in the form of fossils of one-celled organisms resembling bacteria and algae. The rocks containing them are definitely dated as 3.5 billion years old, plus or minus an error of 100 million years. The fossils have cup-like or rod-like shapes and are so tiny that they are best discerned with an electron microscope at a magnification of 50,000. The largest of these is only 39 millionths of an inch long. Based upon such evidence, the scientific consensus today considers that life on Earth may have originated four billion years ago or only one billion years after the Earth and our solar system came into the early stages of their being. As startling as this evidence is, it should be remembered that the origin of life on Earth cannot be pushed very much further back in time, based on present ideas of the age of our planetary origin.

The Characteristics of Living Creatures

The definition of what constitutes life poses a set of its own problems. Paraphrasing the definition given by Rensch (1960), we may say that *living organisms represent systems, each individualized, and containing proteins and especially nucleoproteins, which are capable of metabolism, energy transfer, reproduction, excitability, and change of form, while maintaining the same specific type of system unchanged through very many generations.* The definition stresses a number of important points. Life comes packaged in the form of individual units. Each of these individuals in the higher organisms has its own beginning and ends with its own death. Their chemical structure involves nucleoproteins and it is in these that their life program is embedded in the form of the genetic code. These individual units are energy-using and so must capture or assimilate materials to provide energy, and, having transformed it to their own uses, then rid themselves of unwanted residue. This is accomplished through diffusion across boundary membranes in the simplest living forms, or through excretion in the more complex ones. Excitability means that the individual unit has the capacity to respond to the varied stimuli received from the external environment which surrounds it. This capability provides the individual animal or plant with signals by which it regulates its own physical or physiological activities. The change in form involves a variety of processes, which may include growth, maturing, and aging. Above all, living individual units must be capable of reproducing. This capability of reproduction is an essential aspect of life that allows for the perpetuating of the same systemic organization through many generations, a necessity for evolution as we know it to operate effectively.

Each of these attributes of life has in itself undergone evolutionary change from the early and simple forms of life to the later and more complex ones. The attributes of life may be briefly summed up as involving the capabilities of reproduction, eating and excreting, and excitability. This concept of life adequately covers most of the living creatures seen in daily life. But as the scale of size descends to microorganisms, considering first those observable with light telescopes and ending with those visible only through the powers of the electron microscope, we approach a point where it is difficult to say whether the entities are living or not, for they contain some but not all of the attributes of living organisms. Viruses once were held as possible models for the earliest life on Earth, but currently this view is not as popular. It may be that their inability to convert energy within their own body is a result of their universal parasitism, a condition which usually leads to the simplification of form and function.

The Scale of Living Creatures

Life ranges from the incredibly minute forms, such as the bacteria, through to the greatest of waterborne vertebrates, such as the blue whale, which exceeds a length of 100 feet and weighs more than 150 tons. All creatures within this enormous range are characterized by the attributes of life, which are obviously more complicated in multicellular forms than in unicellular ones. Although man stands less than six feet tall and weighs an average of 150 pounds, he is to be classed in the overall scheme as a large animal.

Viruses are borderline between the living and nonliving universe since they cannot reproduce themselves in the way which characterizes living forms. But they are of interest to us and are the smallest forms to approach a living condition. The hoof and mouth disease virus, which causes fatalities in cattle, would require 2,500,000 laid end to end to extend an inch. Some viruses reach considerably larger size; the one which causes the parrot disease, psittacosis, would require only 50,000 to extend a linear inch. Thus within the level of organization of viruses the largest are 50 times the length of the smallest.

The lowest among the truly living organisms are *bacteria,* which are single-celled in organization and parasitic in habit. Although they have no chlorophyll, as occurs in the higher plants, they are classed in the plant kingdom. In size the smaller spherical bacteria, such as the staphylococci and streptococci, are so small that it requires from 20,000 to 30,000 to reach a linear inch. Some of the very largest of bacteria run only 200 to the inch. Here again, as among the viruses, there is a great range of variation in size, with the largest being about 150 times the length of the smallest. But as befits parasites, bacteria as a class are very small and one cubic inch could contain 9,000,000,000,000 bacteria of average size if they were efficiently packed into this measured space.

The multicellular animals, or the *metazoans,* include all the higher forms of life. In the animal kingdom let us first look at the *invertebrates,* the animals without backbones. They range in size from some of the tiny mites, whose bulk is smaller than that of a human ovum and less than that of the largest forms of single-celled paramecia, up through the largest of living invertebrates, the giant squid, whose extended length may exceed 50 feet. This is a considerably greater range of variation in size than is found in single-celled forms of animals and reflects the greater potential inherent in the evolution of multicellular organisms.

Man, of course, belongs to the *vertebrates,* those animals with backbones. Their range of size is of more interest to us. The smallest living vertebrates include some very tiny fish, some weighing less than $1/5,000$ of an ounce; the smaller forms of hummingbirds, of which 14 would make an ounce; and among the mammals, shrews, fierce little insect eaters of approximately the same weight. From these midget vertebrates we proceed upward to the whale shark among fish, whose length exceeds 50 feet, and to the great blue whale among the mammals, which grows to a length of 100 feet. The largest of living vertebrates are marine inhabitants, for water as an environment solves many of the structural problems that go with great size. The real weight of such creatures does not impose stresses such as would occur in a land animal, for the volume of water displaced buoys their bulk. Birds, on the other hand, have no such buoyancy to sustain them, for the weight of the displaced air is negligible. Flying is an expensive form of locomotion and among living forms their wingspan seldom exceeds the 10 feet which occurs among some albatrosses and the great condors. Both of these birds

are significantly characterized by gliding flight. Among the vertebrates the range of weight is great, for the blue whale is at least 50
million times as heavy as the smallest of birds and mammals, and
this ratio would be greatly exceeded if we included some of the
miniature fish.

Genetic Complexity Increases with Structural Complexity

While the simplest of viruses may contain no more than half a dozen
genes, large multicellular animals require many more to program
their biochemistry throughout life. Man is a large animal comprised
of a great number of individual cells for which estimates range from
lower values of a few trillions to an upper limit of a few quadrillions.
The remarkable thing is that each of these cells within the matured
human body originates and continues to function through instructions programmed in *DNA,* that is, *deoxyribonucleic acid.* In the
course of evolution the amount of DNA needed in each cell to
program the necessary course of instructions increases in the more
complicated forms. In the obsolete theory of the gene, which was
perfectly good and workable for the purposes of many genetic
predictions, estimates range from 10,000 to 50,000 genes per human
being. This is a very large number of separate instructions in the
code of life, and it makes possible a colossal variety of biochemical
activity. Nevertheless, with the replacement of the theory of the
gene by the DNA code, it is now estimated that man may contain
anywhere from 20,000 to 100,000 sequences of instruction, and that
these would equate with the separate genes under the outmoded
theory. The exact number does not concern us, for the lowest of
either of these estimates provides us with enormous potentials for
individual variation and much greater variation within populations.
populations.

Bibliography

RENSCH, BERNHARD.
   1960   Evolution Above the Species
          Level. Columbia Biological Se-
          ries, No. 19. New York: Colum-
          bia University Press.

The Problem of the Origin of Life

A general principle which operates for all energy transfer, at least for everywhere on our planet, holds that every conversion of energy from one form to another involves some inefficiency and therefore energy losses. For this reason scientists agree that the origin of life was an unlikely event, for the very activities of living organisms require an increase in energy potential. The paradox is apparent in the fact that living organisms have somehow learned to raise their level of molecular organization from a relatively simple structure to fantastically complicated ones. This requires energy input. The problem is determining how previously nonliving molecules evolved into those with the ability to incorporate energy into themselves.

Research in astrophysics has made rapid progress in the early seventies, and it is now known that more than a dozen molecules of the kind considered necessary for the ultimate origins of life do in fact exist in outer space. Such molecules constitute building blocks for amino acids, and these in turn are basic for the creation of proteins, which are essential parts of all living creatures, including man. One of the more complex molecules discovered in outer space is a *porphyrin,* which is a precursor of chlorophyll and is much like it in its basic structure.

Theorists have long held that the primitive planet Earth provided an environment very different from the one we know today. H. C. Urey, who won the Nobel Prize in chemistry in 1933, listed the necessary and reasonable characteristics of the early atmosphere for the origins of life. He concluded that it must contain water (H_2O), hydrogen (H_2), ammonia (NH_3), and methane (CH_4), all in unknown amounts and proportions. Further, he had chemical reasons for suggesting that the abundance of hydrogen must have been roughly equal to that of water. His scheme contained two important points: first, methane and not carbon dioxide furnished the bulk of the carbon for the first organic compounds; second, free hydrogen was available for a long time, acting as a reducing agent during the formative period of life. These necessary molecules are among those identified by astronomers in interstellar clouds of dust.

Upon framing this hypothesis, Urey urged a colleague, Dr. Stanley Miller, to put it through tests in his laboratory. Miller passed electrical discharges through a flask containing a mixture of the above gases at a high temperature, between 80° and 90° C. He obtained spectacular results. The experiment yielded no less than 25 amino acids, more than are essential as the basic building blocks of life. Other acids and more complex compounds were also produced by this energy input experiment. It is significant that the resulting compounds could not have been synthesized in the presence of oxygen.

Scientists generally agree that the primitive atmosphere of the Earth must have contained much free hydrogen and little or no oxygen. Energy was available in the form of ultraviolet rays, ionizing radiation, potential chemical energy due to molecular recombination, and local heating resulting from volcanic activities. In time the atmosphere became generally hydrogen poor and began to contain enough traces of oxygen to begin to buffer the action of ultraviolet light through the formation of ozone (O_3). At this stage organic substances became more diversified and were built into more complex forms; still life as we know it had not yet appeared.

At yet a later stage living molecules appeared, among which a single *clone* (that is, the descendants of a single cell resulting from generations of division) outgrew all of the other competing clones and so came to be the primordial seed for the further rapid expansion of life in the Darwinian sense. This hypothesis allows for a multiple origin of life, in that a number of complex molecules may have crossed the difficult boundary between nonliving and living in the primitive seas of our planet. But after allowing for the creation of life a number of times, the door is then closed in that one type of molecule so prospered that it multiplied and outcompeted all other forms of living molecules. In essence this restores us to the concept of a single origin of life and that all creatures living today are descendants of this one successful clone.

Researchers in the past few years have also filled in some of the missing parts of the story of how life originated here. The oldest rocks on earth so far identified are 3.76 billions of years old and come from West Greenland. They are sedimentary in character, demonstrating that water then existed not only in the atmosphere of our planet, but as oceans in its basins. The existence of such deposits implies both erosion and systematic sedimentation in water. Thus the stage was set for the origin of life this far back in time and perhaps as much as 4 billion years ago. While the age of our solar system is estimated at 4.6 billions of years, rocks as old as this have not yet been found within it, although some moon samples approach that date.

J. William Schopf of the Department of Geology, UCLA, has recently made the interesting discovery that about 3.3 billions of years ago the isotopes carbon 12 and carbon 13 for the first time appeared in approximately the same ratio as they do in modern deposits, apparently indicating the presence of life at that time. This deduction is based upon the fact that plants use C^{12} in preference to C^{13}, because the former moves and reacts faster chemically in their systems. At present it is thought that this indication of organic life probably resulted from bacteria rather than advanced plant life. A little later at 3.1 billion years ago, there appeared the first sporadic deposits of *stromatolites.* These are complex limey dome-like structures apparently resulting from the fixation of calcium carbonates during the life processes of blue-green algae. At the same time there are so-called *microfossils* which appear to be individual organisms but may just as well be inorganic particles in the form of grains of various minerals and the like. These stromatolites certainly are evidence for life, but these "microfossils" which are their contemporaries, seem too heterogeneous themselves to be acceptable as evidence for life. But by 2.3 billions of years ago stromatolites had become common and widespread and were accompanied by a variety of the tiny fossils of the *microbiotas.* Life had clearly become established by this latter time and had reached a considerable stage of differentiation. The presence of plants indicates the existence of *photosynthesis,* which, along with the splitting of the molecule of water, released oxygen into our atmosphere. By 2.0 billions of years ago oxygen was present in the atmosphere as shown by the oxidized iron strata, known as "red beds" which can still be found in various parts of the Earth. Stromatolites disappeared from the oceans of the earth as the algae which formed them could not withstand the predatory multicellular animals which grazed upon them. It is of some interest that there is only one place known on Earth today, Shark Bay, Western Australia, where stro-

matolites still exist. This refuge area apparently exists because the waters of the bay are so alkaline that the predatory animals which eat the algae cannot survive in it.

From the Bitter Springs formation of Central Australia, dated to a billion years ago, Professor Schopf has found evidence of mitotic division of the kind that regularly occurs in the formation of new body cells in single-celled organisms (as well as in multicellular ones). The process is illustrated in Plate 1–3. From the same beds come suggestions of sexual cell divisions in plants in the form of four cells that remained grouped together in the stony strata as though they had just completed a meiotic, or sexual, division. Thus the evidence suggests that both types of fundamental cell division were present a billion years ago.

Up to this point there is no evidence whatsoever for multicellular life. Strata 680 million years old from the Soviet Union showed tracks and traces made in soft mud, and subsequently fossilized, that could only have been produced by multicellular and segmented creatures. A little later, in the Ediacara Hills of South Australia, in what are clearly Precambrian times, occurs an impressively diversified fauna of soft-bodied multicellular animals. These have been well described by Martin F. Glaessner (1961). The animals include various jellyfishes which have been classified into four or more orders taxonomically. There are also coral-like fossils, while the most common specimens, those of *Dickinsonia,* may be related to certain living worms. There are in addition two totally unknown types which are not represented among all present-day living creatures. Perhaps the most spectacular animal of these shallow seas was a bilaterally segmented worm named *Spriggina* after one of the men responsible for the collections on the site. This little animal, about an inch and one-half long, was apparently free-swimming. On the whole the deposit contained a surprising diversity of multicellular forms for so early a time. It is of interest that their soft bodies evidenced the fact that the metabolism to produce limey hard parts had not yet evolved, although it was to become important in the Cambrian era millions of years into the future. Thus the stage was set, as the fossil record shows in these early times, for the great proliferation that was to occur.

As convincing as this empirical evidence is, even scientific optimists realize that today there remains an immense gap in the explanation of how life originated. The successful experiment of Dr. Miller produced amino acids, which are necessary for life, but not life itself. The optimists consider that man will rediscover, one by one, the chemical and physical conditions which once made possible the formation of living molecules. But since this original transformation occurred so far back in time, it is possible that men in their laboratories will never have either the time or the intellectual ingenuity to solve the basic mystery. There is ample room at this stage for a side by side existence of belief in both the creative origin of life and in scientific materialism. Certainly the origin of life itself is a continuing and wondrous mystery.

Excellent sources for the reader are S. L. Miller's article, "A Production of Amino Acids Under Possible Primitive Earth Conditions," *Science*, 117 (1953): 528–529; and Miller's and H. C. Urey's joint article, "Organic Compound Synthesis on the Primitive Earth," *Science*, 130 (1959): 245–251. The fauna from Ediacara is described by Martin F. Glaessner in "Pre-Cambrian Animals," *Scientific American*, 204, March 1961, 72–78.

Plate 1-3
Successive stages of mitotic division in a unicellular green algae from the Precambrian Bitter Springs formation of Central Australia. These billion-year-old fossils are related to living species.

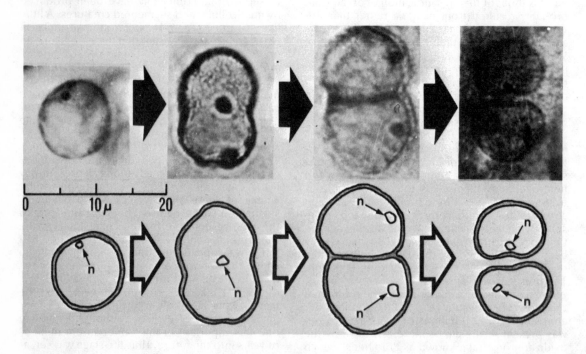

On the Nature of the Viruses

Viruses are submicroscopic forms in the ambiguous position of bridging the gap between nonliving and living worlds. They contain the code of life but do not reproduce themselves in the fashion of really living organisms. They reduplicate themselves quickly and easily by parasitically capturing the materials in the cells of living organisms and converting them to their own use. They are classed as nonliving in that they do not possess metabolic processes within their own bodies, being incapable of capturing and using free energy-producing materials. Viruses do respond to some stimuli in order to go about their own business. Whether they are degenerate parasitic forms of life or forms bridging the inorganic and organic worlds, they are worth our passing attention.

More than 500 different kinds of viruses have been identified within the bodies of men, including more than 60 which cause the common cold. Viruses are clearly important to man as parasites, constituting a considerable threat to health and life, for they are not subject to control through antibiotics. Viral illnesses in general must be allowed to run their course.

The viruses show considerable difference in the complexity of their organization. The tobacco mosaic virus is representative of the lowest viral level of complexity. Its code of life consists of a single strand of *RNA*, that is, *ribonucleic acid,* and seems to contain no more than five or six genes, or separate sets of instructions. This much studied virus is very simple and yet apparently succeeds well with the limited instructions with which it is equipped. It has the interesting attribute of actively replicating itself when placed upon a tobacco leaf, but when removed from this favored environment, it transforms itself into a crystalline form. Obviously there is great difficulty in deciding whether this virus is living or actually inorganic in nature.

The T4 virus which parasitizes the colon bacillus provides a much more complicated example of organization. In this virus the code of life is written in *DNA* and contains several hundred genes or separate sets of instructions. The T4 virus particle is a curious mechanism with a polyhedral head which is connected through a collar and neck to a coiled tail from which protrudes an end plate with six attached angled tail fibers evenly spaced around its perimeter. An electron microscope photograph is given in Plate 1–4.

The genetics and construction of this virus particle have been elegantly investigated. One of the modes of analysis included noting how specific identifiable mutations, that is, changes in the code, block portions of the work of the assembly

Plate 1-4
T4 Virus seen through an electron microscope

in the different lines which are necessary to produce the completed virus structure. Experiments show that the finished virus particles of T4 result from three independent assembly lines which work in close cooperation. These are illustrated in Figure 1–3. Sixteen of the genes are now known to cooperate in the formation of the DNA-containing polyhedral head of the virus. Nineteen other genes cooperate to produce the end plate and its tail spikes. Then the tail sheath surrounding the tail is constructed and the neck is added. Experiments demonstrate that these three assembly lines operate in a purely mechanical way to produce what seems to be a mechanical product.

The life cycle of the T4 virus begins with its attaching itself to specified chemical points on the surface of a colon bacillus. By means of the spikes on its end plate, and with the tail fibers serving to hold it in a perpendicular position on the surface, the tail then contracts like a spring and drives the tubular core of the tail through the wall of the bacterial cell. The implanted tail provides an entry through which the DNA in the head of the virus can pass intact into the bacterial interior.

Once inside, the viral DNA quickly takes over the machinery of that doomed cell. First it breaks down the DNA of the host bacterium, thus stopping the production of bacterial protein. In less than a minute it has begun its own assembly project of manufacturing proteins of the viral type. Five minutes later, the captured cell is making the first proteins in the form of enzymes needed for viral DNA replication. Only three minutes later another set of genes acts to direct the synthesis of the structural proteins which will form the head and tail components. At this stage the process of construction of the complete virus particle begins and at the end of no more than 13 elapsed minutes the first T4 virus comes off the assembly line. The process of synthesis continues for 12 more minutes. Thirty minutes after the initial injection, the cycle ends as a viral enzyme, lysozyme, attacks the cell of the bacterium from the inside, breaks it open, and so liberates the 200 new T4 viral particles. These emerge to quickly start a new round of infection by seeking other bacterial hosts. It is small wonder that a man quickly succumbs to viral illness and most fortunate that our body mechanisms usually can damp them down in time and control them.

The reader who wishes to pursue this subject can turn to William B. Wood and R. S. Edgar, "Building a Bacterial Virus," *Scientific American,* 217 (1967): 60–74.

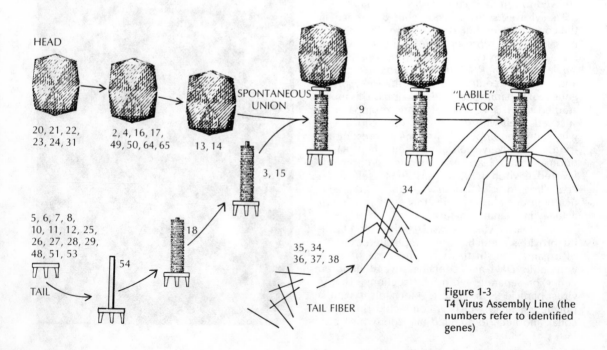

HEAD

20, 21, 22, 23, 24, 31

2, 4, 16, 17, 49, 50, 64, 65

13, 14

SPONTANEOUS UNION

9

"LABILE" FACTOR

3, 15

5, 6, 7, 8, 10, 11, 12, 25, 26, 27, 28, 29, 48, 51, 53

18

34

54

35, 34, 36, 37, 38

TAIL

TAIL FIBER

Figure 1-3
T4 Virus Assembly Line (the numbers refer to identified genes)

CHAPTER 2

DARWIN
AND THE
EVOLUTIONARY
REVOLUTION

It is seldom that one man so profoundly affects the thought of future times as did Charles Darwin. In order to understand his impact upon Western civilization it is necessary to trace the ideas of some of his predecessors. Darwin's contributions to the understanding of evolution are permanent in character, although they do not make up the complete story. His explanation contained errors, and it is important to understand what they were and how he came to make them.

Darwin in Perspective

In the first years of the atomic revolution and the space age it is appropriate to pause to reflect upon the evolutionary revolution caused by that quiet and insecure man, Charles Darwin. He was and he remains one of the truly eminent figures in human history. Darwin demonstrated the inescapable fact that evolution had occurred, he made it understandable as a process, and it became pervasive as a concept. His idea, that natural selection in the environment could transform species over the course of time, gave biology a wholesome jolt, and redirected the other natural sciences toward new goals. But he had a disorganizing impact upon philosophy and religion, from which neither has completely recovered.

Charles Darwin (1809–1882) is an interesting example of a man who suited his times well. As Sir Julian Huxley has expressed it, had Darwin lived in the eighteenth century he might have become a good amateur naturalist. Had his life been delayed until the beginning of the twentieth century, he no doubt would have contributed to ecological studies, tracing relationships between organisms and their total environment. His publication of *On the Origin of Species* in 1859 came at a most suitable time. Geologists had shortly before set the stage for the acceptance of gradual changes on the face of the earth. Linnaeus had provided the basis for the systematic classification of living creatures. Malthus in his essay on population principles had posed the paradox of human numbers outstripping their resources. The times were ripe for the tying together of these progressive ideas through the unifying principles of evolution. And this was Darwin's great contribution.

Men and Ideas Behind Darwin

Attempts to order nature go back into classical and preclassical Greece. Thales of Miletus, who lived in the sixth century before Christ, was the first ancient to leave a record of an orderly approach to the interpretation of natural phenomena. He broke away from mythological explanations and expressed his belief that all life had originated in and rose out of the waters of the seas. This early speculation was not too far from modern truth. Though he left no writings, and the exact dates of his existence are uncertain, his ideas survived through the students that he influenced. One of these, Anaximander (611–546 B.C.) produced a complete but childishly clumsy hypothesis of evolution. He thought that the Earth came into being through a kind of condensation of water. Originally it consisted of mud which floated on the surface of an Earth centered under a circular vault of sky. Life arose out of the primordial mud, first with animals and then plants, and finally human beings came into existence. Men were originally formed like fishes, and lived in the water with them. Later these men-fish cast off their scaly skin and came out upon dry land, where they lived thereafter.

Aristotle.   During the fourth century before Christ, the great Greek philosopher and naturalist, Aristotle (384–322 B.C.), put his stamp upon the intellectual achievements of the classic period. This high point in man's mental life was not to be exceeded during the following two thousand years. He was the first to make a serious attempt to collect and organize the known geological facts of his times. Driven by an inexhaustible energy and gifted with remarkable powers of organization, he wrote an incredible amount in the fields of logic, metaphysics, politics, psychology, art, and biology. Today there remain ten books from his purely biological output. Aristotle classified animals on the basis of anatomy, and much of his knowledge seems to have come from his own firsthand dissections. His primary contribution consisted of a system of thought based upon an evolution which was subject to rigid laws and which reflected a guiding intelligence. His evolutionary changes appropriately proceeded from lower to higher forms. In the Aristotelian system of logic the realities of nature became incorporated in the type (*eidos*). This abstract type was unchanging, fixed, and real. Observable variations within populations he considered to be an illusion or to represent accidental or pathological deviations from the true type. This ancient notion of fixity and unchangeability persisted down through more than two millennia into our own times. It still lies hidden in the thinking of some recent students of man.

Darwin's Eighteenth Century Precursors

Buffon.   Aristotle's belief in fixity in nature still characterized the thinking of most of the great biologists in the century preceding Darwin. The ideas of a number of them make a useful contrast to those held by Darwin. Georges Louis de Buffon (1707–1788), a distinguished French natural philosopher, organized the known phenomena of nature into a rigid system governed by physical laws.

Buffon, who held a law degree, became interested in forestry, particularly the properties of timbers and their improvements on his family estates. He translated Newton's *Fluxions* into French before his own major contribution, which consisted of the 15 volumes of *Histoire Naturelle.* Included in this great work were essays on the theory of the origin of the Earth, and its development into an environment fit for living creatures. He was one of the first to have the courage to reject the biblical account of creation, some interpreters of which suggest a 6,000-year-old Earth, and to postulate a far greater age for our planet. Buffon divided Earth history into seven periods, beginning with its creation and ending with the advent of man. Buffon's greatest contribution came from his realization that both the plant and animal kingdoms changed their characteristics from one time to another. Thus, he helped to set the stage for Charles Darwin.

Lamarck.   France continued to produce men who influenced the natural sciences. One of the most important and yet least understood of these was Chevalier de Lamarck (1744–1829). He was a brilliant scholar and a committed evolutionist, but a man whose rating is low today because his explanations for the processes of evolution are considered both inadequate and incorrect. He stressed that the most fundamental aspect of life is change, and he held that evolution involved change through some sort of striving upward. He arranged living forms in an evolutionary scale in which the mammals naturally occupied the top level. His classification was sufficiently sensitive so that the system by which he distinguished the various invertebrates remains true down to the present time, allowing for one or two alterations. But he had the virtue of believing that the Earth had evolved continuously, and not through a series of catastrophes, as did his contemporary, Cuvier.

We remember Lamarck today chiefly for his false doctrine of *the inheritance of acquired characteristics.* Ironically, Lamarck himself did not consider this an important part of his evolutionary scheme, and he used it only to explain some exceptions which occurred in his evolutionary hierarchies. Unfortunately, Lamarckian evolutionists in the twentieth century lifted this unimportant fragment from his thinking, and Lamarck himself would be amazed at the interpretation some zoologists put on his theories. Charles Darwin was, of course, aware of Lamarck's then respected work and ideas. He rejected Lamarck's primary evolutionary beliefs, although necessarily basing some of his own position upon a belief in the doctrine of the inheritance of acquired characteristics. This error in Darwin's formulation is directly traceable to his lack of knowledge about the real processes of heredity. He still believed that parents transmitted their characteristics to their children as a blended quality rather than as particulate characteristics, as we now know in Mendel's terms.

Cuvier. Not all influential Frenchmen were helpful in preparing the stage for Darwin's appearance. Georges Cuvier (1769–1832) was a comparative anatomist and the first with a relatively modern point of view. His importance in our context results from his studies of fossil animals derived from the rich limestone beds of the Paris district. During his extensive paleontological studies of these animal remains, he faced the questions concerning what changes had taken place in the character of the Earth's surface and how these might explain the differences between fossil faunas and the living forms. Of necessity he became a geologist in addition to his other intellectual fields of specialization. Cuvier discovered that each stratum had its own definite association of fossil species which did not occur in other beds. To explain these discontinuities in time he devised his *theory of catastrophes.* Cuvier attempted to prove that the changes in the character of the animals had been caused by great catastrophes which had repeatedly overwhelmed the Earth's surface in ancient times. These were violent in nature, involving mass volcanic upheavals and resultant floods on the land surfaces. This faulty formulation resulted from incomplete geological evidence and Cuvier's failure to understand its total meaning. He believed in the fixity of each species and thought that after each catastrophe a few animals survived in some isolated and unaffected portion of the world, and that these provided the seed stock for the next formal succession. With his insistence on the fixity of the species and his repeated catastrophes of near extinction, Cuvier stands in sharp contrast to the later ideas of Darwin.

Linnaeus. Karl von Linné (1707–1778), better known today as Linnaeus, was a gifted Swedish botanist who brought order into the system of naming and classifying plants and animals. His first great work, *Systema Naturae,* was published in 1735, when he was only 25 years old. His classification grouped animals and plants into major categories, ranging from the lowest, the species, in an ascending hierarchical order of inclusiveness. Fortunately, he selected Latin or latinized names for the identification of his individual categories, and so through the use of a dead but scholarly language his system was taken out of the vernacular languages of his day, thus achieving an objective precision. His original binomial system in which both the genus and the species are named has sometimes today been extended to include a trinomial form to specify the subspecies or race. Remarkably few changes have been made since Linnaeus formulated his system more than two centuries ago. He made some errors in his detailed classifications, such as including the bats among the primates, but in general his judgment was good. Further details on Linnaeus and his system are given in Supplement No. 3. His importance to Darwin lies in the fact that a systematic scheme of classification had been provided, thus enabling Darwin to properly classify and organize his field collections.

Malthus.   The ideas of a British clergyman and political economist, Thomas Malthus (1766–1834), exerted an important influence on Darwin in his formative years. Malthus's chief work, *An Essay on the Principle of Population,* was widely printed and discussed. Opposing the widely held view that the improvement of man's lot depended upon a fair distribution of wealth, Malthus insisted that human miseries arose from man's fertility and from his frivolous ways of life. He pointed out that natural populations of plants and animals demonstrated that natural reproduction is stronger than all the potentials for maintaining life. In short, *reproductive potential exceeds resources.* Malthus argued correctly that the nature of reproduction produces a geometric increase in the numbers of a given species, whereas its food resources tend to remain constant in nature. Malthus applied this concept to man and showed that the increase in his numbers is indeed geometric in nature, while his food support base, even granting technological improvement, only increases at an arithmetical rate. Therefore, he concluded that men are doomed to increasing misery as starvation becomes inevitable among the poorer social classes. Malthus's own solution would have required men to control themselves procreatively and would have allowed no man to raise a family unless he could provide definite guarantees of its means of subsistence. In the light of present knowledge, this Malthusian doctrine seems perfectly correct even if it was proposed too early in time. The basic concept of Malthus, that *excessive fertility provided large numbers of expendable individuals in balanced nature,* gave Charles Darwin the foundation he needed for his evolutionary theory.

Lyell.   Darwin's intellectual development depended upon finding an antidote to Cuvier's ideas of catastrophism. His support came from a friend and contemporary, Charles Lyell (1798–1875), who was the founding father of modern geology. Lyell took the present form of the Earth's surfaces and studied the changes imposed upon them by various natural forces. He concluded that the same type of forces had always operated with the same intensity as they did in his day. Such simple things as changes in temperature and running water could over long periods of time produce the landscapes as we see them now. He saw mountain building as a slow and continuous

Two Influential Englishmen

process rather than as a dramatic and sudden upheaval, thus denying the theory of catastrophism. He discussed the distribution of animal species in the various geological periods and detailed the course of their development and subsequent extinction in time. Lyell's ideas provided Darwin with an important background of continuing gradual change against which to review his own work.

Darwin's Early Life and Education

The impact of Darwin's ideas upon the world has been so great that it is helpful to look at the man himself. He was born in 1809, on the same day as Abraham Lincoln. He lost his mother early in life and had little to buffer him from the excessively positive views of his father, Robert Darwin, a wealthy physician. Charles was very nearly ruined by the dominance of his gigantic father, and was a three-time academic dropout.

As a boy, Charles enjoyed the out-of-doors and was a collector of pebbles, birds' eggs, insects, and plants. Despite his interest in nature, his academic record at the Shrewsbury Grammar School was poor. He made very little progress in either Greek or Latin composition, a failure which greatly displeased his father. During his middle teen-age years he was still absorbed in bird-shooting and rat-catching, and according to his father coming to no good end. Dr. Darwin sent Charles and his older brother, Erasmus, to the University of Edinburgh to study medicine, for he was determined that they should not grow up as idle sportsmen. There, while visiting the operating theater in which he observed a screaming child being operated upon without anesthetic, Charles Darwin turned and ran from both the room and the field of medicine, never to return to either again.

At about this same time young Charles Darwin learned that he would ultimately inherit enough property to insure his reasonable comfort all his life. Revolted by his studies of medicine and given this new assurance, he dabbled faintheartedly with an education in the field of law. It soon became apparent that he had no talent in this field either. His formidable father then decreed that he should become a clergyman. Always acquiescent to his parent's requirements, Charles at the age of nineteen agreed to accept the creed of the Church of England and went up to Cambridge University to study for the required degree in divinity. He worked hard enough with tutors at the University to obtain respectable grades, but the life he really loved was a very different one. In his free time he ran around with a sporting crowd and shot small game. He continued his varied collecting and read widely in the writings of Lamarck and the personal narratives of Von Humboldt's travels in tropical America. At about this time he reread the works of his grandfather,

Erasmus Darwin, who had had some evolutionary glimmerings showing through his writings.

The stage was now fully set for Charles Darwin to become an obscure clergyman in a country church and to leave no mark upon the life of his century. But through his Cambridge contacts he received an invitation to join the Honor Second Extensive Exploratory Scientific Expedition as the naturalist on the ship *H.M.S. Beagle.* Charles was eager to accept but his father refused both to grant permission and to provide his support during the trip. Fortunately, his mother's family, the Wedgwoods of pottery fame, urgently intervened, and in the end Dr. Darwin finally yielded and agreed to finance the trip for his son.

Late in 1831, as the trim 235-ton brig sailed away from England for five years of surveying and mapping, the career of Charles Darwin was finally safely launched. The young naturalist sailed along both coasts of South America, and dug up fossil animals in Patagonia, including a skull of the enormous notoungulate, *Toxodon.* He visited the relatively unknown Galapagos Islands, where some of his most interesting observations on fauna and flora were made, including the dramatic radiation of the archipelago's finches. The *Beagle* then traveled on to New Zealand and Australia, then finally

Plate 2-1
H.M.S. Beagle at anchor in the Straits of Madgellan at the tip of South America.

returned home to England almost exactly five years after its departure. Darwin, now 27, had become a seasoned naturalist and returned with trunks full of meticulously kept journals and notebooks. The foundations were laid for his further intellectual development, and he was ready to seek a better explanation than supernatural creation to account for the observed and undeniable basic unity of the entire living world.

But Darwin's ideas took a long time to hatch. Still subordinate to his father and not yet certain of his own beliefs and abilities, he needed time to form his evidence into a new system. Never an aggressive individual, Darwin no doubt was further inhibited by the harsh criticism heaped upon some of the early advocates of evolutionary ideas. He went down to London in 1837 to finish his work on the journal of the voyage, and during this period his workbooks give evidence that he had hit upon the essence of his major hypothesis: "The Transmutation of Species." He had certainly set out on his trip believing in a divine creation. He returned from it filled with doubts on this score and convinced that species do change and therefore were not fixed by a creator.

A year after reaching London he read a paper by Malthus on the principles of populations and its ideas struck him so forcibly as to provide him with the missing foundation for his future theory. He suddenly realized that under the circumstances in which reproductive capacities exceed the available food supply that favorable variations in individuals would tend to be preserved, while those less fit would be unlikely to survive. This concept of differential survival in terms of fitness is directly traceable to the stimulus provided by Malthus.

Plate 2-2
Charles Darwin in 1849

It was characteristic of Darwin to undertake the task of amassing all of the data to substantiate his ideas on organic evolution, and in doing so he nearly lost the race to be recognized as its originator. He worked nearly two decades to collect the evidence he needed, and then early in the summer of 1858 he received an essay, "On the Tendency of Varieties to Depart Indefinitely from the Original Type," from Alfred Russel Wallace (1823–1913), a naturalist working in Malaysia. Darwin had been in correspondence with Wallace for some years, so it was natural that the latter should send him his short paper. In those few pages Wallace summarized the main point of the theory on which Darwin had toiled for almost 20 years.

Unlike Darwin, Wallace moved quickly. The ideas for his paper came to him as he lay fever-ridden on the island of Ternate in the Dutch East Indies (Indonesia). His synthesis of ideas came as a matter of slashing insight, and in a few days he had written out the frame for his paper. He had requested Darwin to ask Charles Lyell to look it over. Darwin's response to the crisis was an inclination to suppress his own unpublished work. Lyell proposed a solution for the dilemma. Portions of both papers were read before the Linnaean

Society in 1858. So each of these remarkable men received simultaneous credit for the convincing formulation of the way in which evolution had operated in nature. It is ironic that neither paper excited any attention, and the limited comments were critical in character, for the members of the society were conservative, and the ideas in both papers were revolutionary.

"On the Origin of Species"

Shaken by Wallace's essay, Darwin reduced his proposed four volumes to a single one entitled *On the Origin of Species*. This world-shaking book was published November 24, 1859, and its first edition of 1,205 copies sold out on the first day of its appearance. Its publication made certain that the major credit for the new hypothesis of evolution would go to Charles Darwin. But the brilliant formulation of Wallace should be better remembered than it is today.

This time reaction came very quickly and a sea of criticism arose, the ripples of which still spread in intellectual backwaters. For Darwin, in spite of his unaggressive nature, had felt it necessary to include in his final chapter a simple statement, "Much light will be thrown on the origin of man and his history." Both Darwin the man and his evolutionary ideas were publicly attacked by both churchmen and scientists. Fortunately, he did not have to undertake his own public defense, for he could not have done it effectively. He was saved that effort by Thomas Huxley (1825–1895), a well-known British biologist and skillful writer, who became his dogged defender. A climax was reached in 1860 at a meeting of the British Association for the Advancement of Science held in Oxford. Samuel Wilberforce, the Bishop of Oxford, took to the platform before an audience of 700 listeners and slashed at the absent Darwin and his evolutionary ideas. The Bishop was known familiarly as "Soapy Joe," for he was both glib and empty of content in his declarations. At the end of his tirade he turned to Thomas Huxley and posed his famous sneering question as to whether the latter claimed descent from the apes through his grandfather or his grandmother. Huxley accepted the challenge and retorted that he would rather be descended from a humble monkey than from a man who used his knowledge and eloquence to misrepresent those searching for truth. Huxley thereafter became known as the "bishop-eater."

Darwin's Contributions

In his own time the major contribution of Charles Darwin was the demonstration of the great fact of organic evolution and its documentation with a wealth of careful observations. He also designated the mechanisms by which evolution might proceed. But it was the fact of evolution which had the greatest impact, for this idea forced the restructuring of human thought throughout the world of the natural sciences, the social sciences, and the broad sweep of the humanities. Philosophy and religion were so greatly shaken that even today, more than a century after the first appearance of his book, the raw fact is still unpalatable and unaccommodated in some circles.

The entirety of *On the Origin of Species* was a sustained argument based upon three great observed facts and included two deductions that the author drew from them. First, Darwin pointed out that all living creatures vary. This fact denied the concept of the fixity of the species. His second fact, derived from Malthus, was that all living creatures tend to increase their numbers in a geometric ratio in nature, and the resultant overcrowding requires elimination of large numbers of individuals in each generation. His third fact involved the balance systems in nature, for in spite of this excessive reproductive efficiency, each species tends to maintain the number of individuals at a relatively constant level. From these three great interlocking facts, Darwin drew two crucial deductions. First, it cannot be doubted that there is a struggle for existence in such a system; and second, in that struggle the fittest tend to survive. While his defender Huxley tended to frame survival in bloody and violent physical terms, Darwin himself recognized the variety and subtlety of its form and even that differential fertility alone could be its mechanism. These three facts and two deductions have been little changed by more than a century of continuing research and stand as basic in modern Darwinian or neo-Darwinian evolution.

Darwin's Errors

Although Darwin's facts and deductions concerning evolution have stood firm, his explanation of its processes has become modified in the course of time. The heart of the Darwinian explanation of evolutionary process involved natural selection. He recognized this as his most important contribution, and posterity agrees. It does not detract from his position that earlier workers had foreshadowed the idea, or even that Wallace had achieved it independently. The process of natural selection remains primarily Darwin's own discovery. In his full and final theory Darwin recognized four factors or causes of evolution, in the following descending order of importance: (1) natural selection; (2) the inheritance of acquired characteristics due to the use or disuse of various organs; (3) the inheritance of acquired characteristics due to direct effects of the environment; and (4) suddenly recurring variation in individuals, changes which are now called mutations in the broader sense. His final theory was framed totally in strictly naturalistic or materialistic terms. He rejected a variety of vitalistic or supernatural theories about evolution.

Natural selection continues to be recognized as the most important force in producing evolutionary change. His facts are under no dispute, but his dependence upon the inheritance of acquired characteristics as a part of evolutionary process has long since been discarded. Darwin, of course, unfortunately never knew about Mendel's work in genetics. Based upon Mendel's principle, the geneticists in the twentieth century in intensive experimental exploration have been forced to reject completely the doctrine of the inheritance of acquired characteristics, whether these originate due to use or disuse, or whether they are the result of direct environmental effects. The twentieth century genetical framework provides for the explanation of evolutionary processes.

Darwin's facts and hypotheses have fared well under the assault of both critics and advancing newer forms of scientific investigation. With its modern modification the Darwinian theory of organic evolution stands firmly established. The world owes the quiet naturalist a great intellectual debt, and his name and memory fully deserve the respect given to them.

Bibliography

APPLEMAN, PHILIP.
  1970 Darwin, A Norton critical edition. New York: W. W. Norton and Company, Inc.

BATES, MARSTON and
PHILIP S. HUMPHREY.
  1956 The Darwin Reader. New York: Charles Scribner's Sons.

DARWIN, CHARLES.
  1967 On the Origin of Species. A facsimile of the first edition. New York: Atheneum.

The Basis for Classification in the Natural Sciences

In the eighteenth century Western civilization found itself uniquely prepared for one of the greatest scientific advances up to that time. A mingling of Judeo-Greek ideas with those of the medieval church provided the foundation for the recognition of a history of life and further demonstrated that all of its forms were totally interrelated. But the role of theology lay in setting the intellectual stage and providing a static or unchanging concept of nature. Coming out of earlier times and spread throughout the literature of the seventeenth and eighteenth centuries was the theological doctrine known by various titles, among which were the Chain of Being, *Scala Naturae,* or the Ladder of Perfection. These labels covered the idea of a gradation of living organisms in an orderly fashion, although there was no implication that this represented any arrangement of biological relationships. The scale of nature ran upwards from minerals, through small changes to the lower forms of life, then plants and animals, and even to man himself before extending on to such purely spiritual existences as the souls of men, the angels, and the other occupants of the spiritual universes. Eighteenth century scholars were well aware that the ape stood next to man in the scale of nature but were not dismayed because the doctrine of the *Scala Naturae* proclaims the unchanging character of species. It basically assumes that the entire chain of life had been created in its present order when God through his supernatural acts created the universe out of original chaos.

Into this eighteenth century world, with its unchanging but varied forms of life, came the great taxonomist, Carl von Linné (1707–1778). An eminent naturalist, he held the chair of botany at the University of Uppsala and was a devout Christian believer. He lived at a time when reports from exploring seamen and other travelers were flooding Europe with accounts of new plants, strange animals, unknown apes, and indescribably savage humans. He recognized that the first step in any science is to know one thing from another, and that this involved detailed and significant descriptions. But von Linné proceeded to take the second step, that of imposing order through systematic classification upon the multitude of descriptive details at his disposal. His fundamental book, *Systema Naturae,* was first published in 1735, and successive editions continued to develop in judgment and in accuracy until the tenth edition was issued in 1758.

As the originator of systematization and classification in both zoology and botany, von Linné had the genius to introduce a flexible but consistent nomenclature of binary nature, that is, a system involving two names for identification of any species. Each of the presumably static species of plants and animals was given a primary generic name, and a secondary specific one. While each species was considered immutable, those which seemed most like each other were classified together as a genus. His final system contained seven categories beginning with the kingdom at the top and ending at its most finely divided lower level, with the species. In Table 2–1 these

---

TABLE 2–1

Linnaean System as Originally Created with the Domestic Dog Used as an Example

KINGDOM—*Animalia*
 PHYLUM—*Chordata*
  CLASS—*Mammalia*
   ORDER—*Carnivora*
    FAMILY—*Canidae*
     GENUS—*Canis*
      SPECIES—*Familiaris*

---

seven categories are scaled in a descending order and their uses illustrated by identifying the domestic dog. This animal we know belongs to the *kingdom* of animals, that is, *Animalia*. This is the most inclusive category and represents in a figurative sense the top level of a seven-stepped pyramid. Descending to the next level, the category *phylum* is based upon the basic structural organization and dogs belong to the *phylum Chordata* since they have lengthwise stiffening in a body essentially symmetrical bilaterally. They may also be put in the class *Mammalia,* since the females nurse their young. The mammals may be further divided into a substantial number of *orders,* and as primarily meat-eaters, dogs were placed into the *order Carnivora.* This order includes a number of families, and dogs have been placed in the *family Canidae.* Families may be further subdivided, and dogs have been placed in the *genus Canis.* There are a considerable number of species within the genus *Canis,* and the domestic dog has been given the *species* title of *C. familiaris.* In the final Linnaean system the

---

TABLE 2–2

Current Expansion of Linnaean Hierarchy

KINGDOM
 PHYLUM
  SUBPHYLUM
   SUPERCLASS
    CLASS
     SUBCLASS
      INFRACLASS
       COHORT
        SUPERORDER
         ORDER
          SUBORDER
           INFRAORDER
            SUPERFAMILY
             FAMILY
              SUBFAMILY
               TRIBE
                SUBTRIBE
                 GENUS
                  SUBGENUS
                   SPECIES
                    SUBSPECIES

---

hierarchy had seven steps and no further subdivisions were made. But today this scheme of classification has been expanded so that as it is ordinarily used it contains 21 categories or *taxa.* These are illustrated in Table 2–2. In the expanded classification, of course, the many varieties of the domestic dog, even though created by man under artificial breeding systems, would be classed as subspecies or races.

When first devised, the system served to classify the 4,235 species of animals then known. Using its modern expanded form scientists not only accommodate a third of a million species of

plants, but more than one million species of animals, of which no fewer than 40,000 belong to the vertebrates alone. If considered only as a simple classifying device, it is a great success, for it provides a systematic guide by which scientists can find their way among this enormous number of different living current species.

At the upper end of his classification, von Linné included the order of the *Primates,* or the first order, which contained four genera. The highest was that of *Homo,* which included two primary subdivisions, *sapiens* and *sylvestris,* the latter including the manlike apes. It is interesting that in his day von Linné should place man and the anthropoids together in a single genus. The order included several other families. *Simia* contained the monkeys, and *Lemuria* the lemurs. A final family, *Vespertilio,* consisted of the bats, an error for which we can forgive the first great classifier.

Within the classification *Homo sapiens,* von Linné distinguished a number of subdivisions which we would call races today. He recognized the difference between the peoples of North and South America under the title *americanus,* Europeans under *europeus,* Mongoloids of Eastern Asia under *asiaticus,* Negroes under the title *asser.* He had two further categories—that of *ferus* contained all of the so-called savages, while into the category *monstrous* he thrust all of the abnormal human beings which appeared in the verbal and written accounts of some of the overly imaginative early explorers. To be consistent, Carl von Linné latinized his name to Karolus Linnaeus, a term which we shall use hereafter.

The attitude of Linnaeus underwent some evolution with the passage of time. His final edition of *Systema Naturae* shows more awareness of variation among individuals of a given species, although he never formally renounced his ideas about the static nature of the species. He observed new varieties appearing spontaneously; he saw abnormal plants derived from normal ones and gradually became aware of the varietal confusion and disorder in his time. Linnaeus began to distinguish between the true species of the Creator, and the variation existing at the present time. He became reconciled to the possibility of new species arising through cross-breeding and in his last editions removed the statement that no new species could arise.

In the more than two centuries which have passed since his classification was first published, it has been affected by two revolutionary changes. The first involved the realization that the species are not static, not the result of supernatural creation, but rather that they have evolved one from another throughout the vast periods of life on this planet. The insertion of an evolutionary scheme into the Linnaean taxonomy requires no great changes in its structure, for the same general criteria that Linnaeus used to classify animals also generally serves to describe their evolutionary position with regard to each other. Animals which look very much alike are likely to be closely related in biological descent. This is particularly true if the catalogue of likenesses is based upon really equivalent structures, that is, *homologies.* The principle only fails when applied within the category of the species, a point that will be dealt with later. The second revolution which has changed the nature of the Linnaean system results from the abandonment of the concept of a type and substituting for it the population as the basic unit in nature. The type, an Aristotelian concept, involved viewing a species in terms of an essential abstraction which incorporated all of its reality. This meant suppressing visible variation and focusing on the abstraction. With Darwin's recognition of the reality of populations as opposed to type, scientists became aware of the essential nature of variation within population, thus opening a door to the study of evolution. Indeed today the *taxon* is recognized as a formal unit in a hierarchical classification that is based upon a group of real organisms. Through incorporating these two changes, the Linnaean system continues to function as an important classifying device in the modern natural sciences.

The expansion of his classification from about 4,000 animals to comfortably handle more than a million has produced some temporary difficulty and has required establishing certain rules. Authority is now incorporated in an International Code of Zoological Nomenclature. The Code lists the rules which must be followed if scientists are to be able to communicate accurately among themselves, and if they are to be able to find their way through the maze of categories which the system contains. The Code establishes a law of priority which goes back to the names used in 1758 by Linnaeus. But in spite of the best of intentions there remains considerable terminological confusion. For example, the chimpanzee, which represents a localized and clearly identifiable type of anthropoid, has been given no less than 21 different generic names, and these have been combined with 73 different specific names. This is depressing. And the situation is even worse among fossil men. The discoverer of a new form of fossil hominid usually appears incapable of refusing the temptation to glorify his discovery by giving it a new and unique name. Under the rules of nomenclature this is allowable only if the new discovery differs so greatly from all previously known genera and species with full account taken for variability in age and sex differences, that the new form must be given a new label for proper identification. This perpetual splitting of the concept of the species or the genus has produced a literature on human evolution which is terminologically confused. The same early men are known by a wide number of synonyms resulting from changes in labeling, and the formal binomial Linnaean label bears no relationship to lines of descent, or to the real differences in anatomy. For these sufficient reasons, in this book fossil men are primarily identified by the name of the place where they were discovered, for in most cases this provides a unique label. Such a name carries no improper implications about relationship or level of organization. To simplify the labeling, the living primates will also usually be identified by a common name. On the other hand, fossil primates, in pre-Pleistocene times, are necessarily labeled by their proper genus and species titles. The most recent and comprehensive discussion of classification is that of G. G. Simpson (1961 *Principles of Animal Taxonomy,* Columbia University Press). Readers interested in the problems of taxonomy are advised to turn to it.

It is important to realize that the *taxon* called the species is the only one of the complex hierarchy which really exists in nature. All of the other *taxa* occur only in the minds of human classifiers. Genera are constructs which group together those species which the systematist considers to be closely related species. *Subspecies* and *races* are impossible to define in nature unless isolating factors provide real boundaries. But the species does exist, and it consists of those populations which can interbreed with no loss of fertility in the offspring, when opportunity allows. All modern men belong to a single species, admittedly widespread and highly variable. While the species has reality in nature, there is a complex literature in the new systematics concerning exact definitions and their applications. These problems will not concern us in this book. A comprehensive discussion of the problem is given by Ernst Mayr, *Animal Species and Evolution,* (Cambridge, Mass.: Belknap Press of Harvard University Press, 1966).

CHAPTER 3

MENDEL'S DISCOVERY OF THE GENETIC BASIS FOR VARIATION

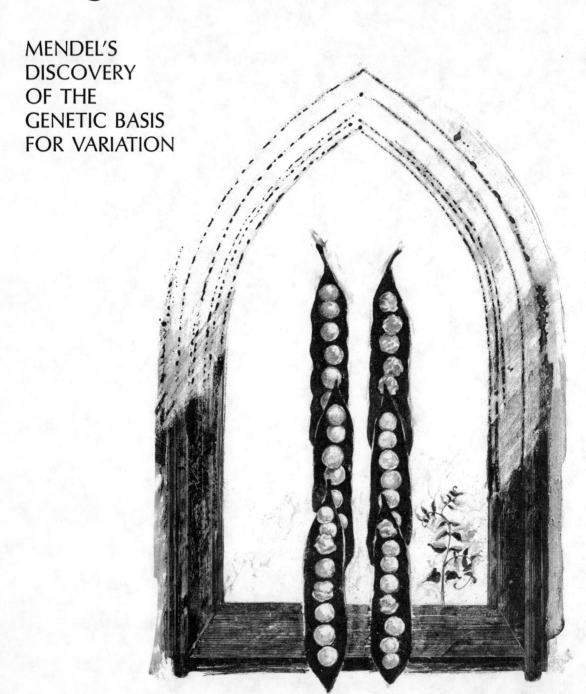

It is ironic that Charles Darwin and Gregor Mendel never knew of each other's work. The two men used very different approaches: Darwin obtained his great results from observation, Mendel by experimentation. Mendel's conclusion that inheritance was particulate in nature laid the foundation for modern genetics and ultimately for understanding the code of life in molecular terms.

Darwin's Failure

Charles Darwin's contributions are great enough, but he also narrowly missed being the father of modern genetics. His own garden experiments in the breeding of hybrid plants produced whole number ratios of three to one in the second generation of crossing. The plants before him held the key with which to unlock the mystery of inheritance through separate particles, but the enormous significance of the small whole numbers in the ratio of the hybrid progeny escaped him. He was a careful observer of the realities of nature, but he lacked the mathematical insight which would lead him to a real understanding of these simple numbers in the special context in which he found them. And so Mendel became the discoverer of the genetic principles of inheritance.

Some Erroneous Concepts of the Nature of Inheritance

The manner in which children inherit their physical characteristics from their parents had been debated for many centuries by Darwin's time. The great evolutionist regarded descent and reproduction as involving blood, as did the people of his day. The manner in which blood might act to convey inheritable characteristics from parents to offspring had been speculated upon by a variety of reputable scientists. Indeed, Darwin himself was guilty of conceiving of minute particles which he called "gemmules," which he suggested were given off by every cell to be scattered throughout the body by the currents of the blood and other body fluids. His "gemmules" could rejoin under the proper circumstances, and so the sexual cells of both parents could contain "gemmules" from all parts of their individual bodies. And Darwin hypothesized that they combined to form the embryo. Needless to say, neither "gemmules" nor any other fanciful agencies are involved in heredity.

But the idea that blood somehow is concerned with inheritance dies hard. There are many old ladies still alive in the Western world proud of their own good blood, who wonder how their daughters could marry men with bad blood. These ancient ones are, of course, confusing their issues. In some cases they equate inheritance with status, especially with economic status. The rich, and in particular the respectable rich whose fortunes date back to conveniently forgotten grandparents, are, of course, always persons of good blood. So they carry with them the capability of good breeding, that is, producing superior children. Necessarily the poor, whose bad blood is reflected in their impoverished status, can be predicted to breed only poor children. These ancient snobberies still persist in most stratified populations, that is, in almost all technologically advanced ones.

The curious practices of the breeders of domestic animals have also served to keep alive the myth that blood is the mechanism by which inherited traits are transmitted from parents to offspring. Even today among animal breeders, who may be striving to "improve the breed" of racing horses, and among dog fanciers striving to meet the latest breed standards, the concept of bloodlines is still supreme. Such breeders talk knowingly of good bloodlines for staying power in horses, for good noses among the hunting dogs. Ancestors do contribute to the characteristics of their descendants but not by bloodline and not in any rigorously predictable fashion. That serious breeders, intelligent men and women with great sums of money invested in their animals, can persist in talking this nonsense seems remarkable. But the explanation is actually rather simple.

Man originally created his domestic breeds of animals without any knowledge of exactly how inheritance operates by the simple device of controlling matings among his animals. Those animals with useful characteristics were mated together to increase the number of their offspring, while animals with bad characteristics were eliminated, or at least not allowed to reproduce. There is no doubt that this simple practice of mating desirable animals to each other has resulted in the creation of more and more useful animals and plants through time. Breeding by eye has achieved results which we must be thankful for. The truth is that even now few practical animal breeders know genetics, or even the way in which characteristics they find desirable in their animals are inherited.

Cattle provide another instance in which the breeder's ignorance of genetics has led to difficulties. Here again the conflict arose because of the breeders' following show standards instead of stressing economic success. Cows of dairy cattle prove their worth in terms of their individual milk production rate. But there is no direct way in which the bulls that sire the herds can do the same. Until recent years, herd bulls were chosen for their external appearance in terms of show standards. In theory, a prize-winning Holstein bull should sire the best heifers. But milk production is determined by a sizable number of specific genes, and these do not show themselves in any visible fashion in the exterior shape of the animal. Breeders frequently found that their prize bulls lowered the milk yield in the cows they sired, rather than improving it. To their further chagrin, the consulting geneticists determined that some very scrubby-looking bulls, whose shape was a disgrace to the breed, nevertheless transmitted improved milk yield capacity to the cows they sired. Sheer economics forced dairy cattle breeders to give up the ancient practice of breeding by eye and to resort to what are now called progeny tests. The new method requires that the bull be judged by the real qualities he transmits to his daughters, rather than by the blue ribbons he may earn in the show ring.

Mendel's Education

The first man to understand the real mode of inheritance was Gregor Johann Mendel (1822–1888). Both his life and his mental qualities stand in interesting contrast to those of Charles Darwin. He was born of peasant parents in a German colony at Heinzendorf in the midst of the Slavic population of the country now known as Czechoslovakia. Little Gregor did so well in his early classes that he was recommended for higher schooling. But his family was too poor to pay his tuition and board so he had to enter the school at Troppau on half rations. Later a two-year philosophy course at Olmutz Institute convinced young Mendel that he would do well to seek a profession which would provide him with certain economic security. At a teacher's suggestion he entered the Augustinian Monastery at Brünn.

Mendel's Experiments

Entering a monastery does not suggest the beginning of a distinguished scientific career; however, Mendel had the mental qualities for such achievements. He had always been a lover of nature, and during his free hours he began the botanical experiments which subsequently led to his discovery of the basic principle of heredity and ultimately the science of genetics. Previous attempts to unravel the mode of inheritance had been both chaotic and lacking in persistence. An orderly thinker, Mendel realized that systematic breeding for generation after generation was necessary, and he further sensed that exact records must be kept of the characteristics appearing in each of his plants. His remarkable successes can be traced to an unusually keen sense of experimental design and the tenacity to carry through all of its vital elements.

Mendel's experimental success resulted from four rather simple procedures. First, he decided that his experiments must be based upon true breeding plants and ones that could be easily protected from all foreign pollen, either insect-borne or wind-carried. Secondly, he decided that the traits to be studied should be easily recognizable and not yield blended forms among their offspring. Next, he decided to concentrate on one trait at a time and thus avoid the hazard of trying to record all traits among the hybrids. Finally, as simple as it sounds, he decided that accurate counts must be kept of the hybrid offspring.

Mendel's experiments were of such a critical nature that some details are worth recounting. He achieved his first experimental requirement by choosing the common garden pea as his primary material. This plant had the advantages of being self-fertilizing and was easy to protect from the intrusion of outside pollen. His second goal was attained by the simple procedure of ordering 34 varieties of pea seeds from commercial sources and by giving them trial plantings over a preparatory two-year period. These preliminary results led him to select 22 varieties as most suitable for his experiments. Being both patient and intellectually well disciplined, Mendel narrowed down his choices to seven characters: (1) in form the ripe seeds were either smoothly round or wrinkled; (2) in color the peas were either yellow or intense green; (3) the seed coats enclosing the peas were either gray or white; (4) the form of the seed pod was either inflated or constricted between the peas it contained; (5) the unripe pods were either green or very yellow in color; (6) the position of the flowers varied dramatically, for they were distributed along the stem or occurred only at the top end; and (7) the pea plant stem was either normal, that is, from seven to eight feet high, or dwarfed, that is, limited to nine to eighteen inches. To satisfy his third requirement Mendel examined each of these seven characters separately and recorded the frequencies one at a time in his various experimental crosses. This was a considerable task, for he recorded 12,980 different observations on individual plants. Thus he fulfilled his fourth requirement. His results from these single-factor crosses came in simple whole number ratios as 1:2:1, 3:1, and 1:1.

An Early Mendelian Experiment

The beauty of Mendel's methods can best be shown by summarizing one of his early experiments. He decided to investigate inheritance by directly crossing plants with wrinkled seeds with those carrying round seeds. Taking 70 plants, he made a total of 287 fertilizations of flowers. When their pods matured he opened them and found that they contained only round peas. Thus their inheritance was not a process of blending the traits. The wrinkled seeds which had occurred in one half of the parents of the hybrids had disappeared as completely in their offspring as if they had never existed. The first generation of hybrids, now know as the F-1, was the first filial generation, and universally had round peas. Mendel called the character that prevailed in the hybrids a *dominant* trait, and the one that disappeared, a *recessive*. Persistent and curious, he then proceeded to the next stage of his experiment by letting the plants which grew from the hybrid round pea fertilize themselves the next spring. When these second generation hybrids, or *F-2s*, had grown to maturity, Mendel opened the pods and there lay both round peas and wrinkled peas nestled side by side. The wrinkling from the grandparents, which had been lost in the first hybrid generation, had reappeared undiminished in the second generation of hybrids. This experiment illustrated one of the laws that he formulated, the *segregation* of inheritable characters, a law that ranks high among those in the biological sciences, and is as true today as it was in his time. Mendel harvested 7,324 peas, of which 5,474 were round, and 1,850 were wrinkled. This exact count allowed him to calculate that 74.7 percent of his peas were round, while 25.3 percent of them were wrinkled. He was mathematician enough to know that these values deviated very slightly from the ratio of 3 to 1 by a magnitude that might be produced by chance sampling errors. He confirmed this ratio for single factor hybrid crossing for a number of his other contrasting characters, and soon made his most important contribution, which was the fact that inheritance was *particulate* in nature as though determined by separate particles, and not a blending of characters as results from a mixing of different fluids.

In this and his other experiments, each of the contrasting traits behaved as though controlled by a single factor or particle. Mendel had no microscope with which to probe the inner structure of the sex cells, but in an operational way he had shown that inheritance operated to produce ratios that could be expressed as simple whole numbers in the offspring. He did not proceed to formulate the idea of the gene as the basic unit in inheritance, but his conclusion would have given close support to such a hypothesis.

Building confidence as he proceeded, in his later experiments Mendel observed crosses involving two of his independent characters. From these he obtained more complex results but they could still be expressed in the form of ratios involving simple whole numbers. In an experiment in which he crossed round yellow peas with wrinkled green ones, he observed both the color differences and the form differences simultaneously. Taking the hybrid offspring and allowing them again to self-fertilize, he produced a second generation of hybrids which yielded 315 round yellow peas, 101 wrinkled yellow peas, 108 round green ones, and 32 wrinkled green ones. The resulting ratio expressed in the same order was almost exact; 9:3:3:1. In this and other similar experiments involving two different hereditary factors, he consistently produced frequencies which approximated this ratio. Thus he was not only able to demonstrate that inheritance was particulate, but further that characters *segregated independently* from each other in his second generation hybrid peas. Thus from two kinds of grandparental plants, he obtained four kinds of second generation hybrid offspring in frequencies which could be predicted in terms of dominance and recessiveness of the characters involved, and their random assortment of the generations of descent. This process will be better understood when the division of the sex cells, *meiosis,* has been described a little later.

Mendel had made a fortunate choice in biasing his experiments by limiting his breeding plants to those which showed large contrasting pairs of traits. The genes producing these readily visible and non-overlapping differences are to this day called Mendelian genes. While he could not anticipate it, this wise selection of readily visible differences allowed him to avoid the complexities which are inherent in characteristics which are determined by a complex of many genes, or *polygenes.* Polygenes are difficult to analyze in any detailed fashion and to date none have been adequately investigated in man. Mendel laid the foundations for modern genetics in a little kitchen garden in the monastery cloister, which

Plate 3-1
The monk Gregor Mendel pollinating peas in his kitchen garden.

only contained a few hundred square feet. His materials were the simplest, while his conclusions remain of great importance.

Mendel's Publication

The good monk wrote the results of his various experiments in February 1865 and read his paper before the Brünn Society for the Study of Natural Science. At the following meeting he explained the implications of the whole number ratios which he had found. His combination of simple mathematics with well-controlled botanical experiments was unprecedented, and further, his results ran completely counter to the general belief that heredity involved the whole individual and was determined by blood. His two presentations before scientific audiences had little impact. Nevertheless, he was invited to publish in the Society's *Proceedings,* and in 1866 his work appeared under the title "Experiments in Plant Hybridization."

In retrospect, it is tempting to think that Gregor Mendel's findings were disregarded during the remaining 34 years of his century simply because people did not learn about a paper which was published so obscurely. But this is not the truth of the matter. Copies of his paper were sent to more than 120 scientific organizations and universities in both Europe and America. These were again received in silence, with neither praise nor criticism for his work. In fact, his paper seems to have received no attention whatsoever. Mendel's discrete units of inheritance could not be related to the known, continuously varying anatomical and physiological units of his day. Hence his great discoveries were buried by rejection.

The quality that made Mendel a great experimenter also caused him to rise in his church order. In 1868 he was elected the abbot of his monastery, and with administrative duties pressing upon him, he dropped his experiments with plant hybridization entirely. When he died in 1884 no one in the world seemed to realize that a great scientist had gone, and even his experimental notes and records disappeared.

The Rediscovery of Mendel's Laws

It is interesting that some men, such as Gregor Mendel, were so far in advance of their time, but it is ironic that other men independently arrived at similar conclusions. Thus in the first year of the twentieth century, no less than three other botanists were led by their own experiments in hybridization to rediscover Mendel's findings. Hugo DeVries, a botanist at the University of Amsterdam in Holland, had long been concerned with variations and modifications in nature. Working primarily with primrose plants, DeVries had noted variations that finally led him to uncover a reference to

Mendel's publication, and in the year 1900 he read it. From his own work DeVries at once recognized the importance of Mendel's findings. Only one month later a German botanist by the name of Karl Erich Correns, who had also been hybridizing peas and corn, published that he had been encountering simple whole number ratios from generation to generation in his hybrid offspring. But the coincidence is yet more complex, for a Viennese botanist with the improbable name of Tschermac von Fursenegg made the same discovery at the same time. He had undertaken to repeat Darwin's earlier experiments with peas and likewise had found constant whole number ratios characteristic of the hybrid offspring. Thus in three months between April and June of 1900, Mendel's laws were rediscovered and recognized for their real worth to the world. Thus the great experimenter finally received the scientific acclaim which had passed him by in his lifetime.

The Rapid Development of Genetics

With the rediscovery of Mendel's principles, genetics developed rapidly. Soon biologists throughout the Western world were specializing in the science of heredity, and there arose a new group called geneticists. By 1902 it was suggested that the *chromosomes,* color-staining bodies, visible by microscope within the nucleus of the cell, might be the units through which heredity was transmitted. A series of elaborate experiments confirmed this hypothesis and as time went on the theory of the gene was gradually developed. The nature of *genetic linkage,* involving inheritance due to genes at different places on the same pair of chromosomes, was established, making possible the mapping of the sites of the hereditary units on the chromosomes of the fruit fly. New discoveries came tumbling in so rapidly that in the future the twentieth century may be remembered as the age of genetics as well as the age of atomic discoveries.

The Theory of the Gene

The theory of the gene grew out of Mendel's work and served satisfactorily to explain the mechanisms of inheritance during the first half of the twentieth century. Since then the unraveling of the genetic code and the deciphering of *DNA, deoxyribonucleic acid,* in which genetic instructions are written, have made the theory of the gene obsolete. Even though it has been superseded, the ideas incorporated in the theory of the gene, and its language, make it a better basis for communication both in familial genetics and in evolution. On the other hand, the genetic code provides the entering wedge into the new field of molecular biology and some day, although not for the present, its language will supersede that embedded in the theory of the gene.

Let us first consider the theory of the gene and some of the more important words in its language. *Chromosomes* are important structures in the nucleus of the cell and they have long been known to carry the units of heredity. With suitable materials, such as the enlarged salivary glands of the fruit fly, it is possible, under the microscope, to see that chromosomes consist of materials arranged in alternating and visibly different bands. It was long thought that the genes, themselves the carriers of inheritance, were arranged along the length of the chromosomes like beads on a necklace. Each gene was conceived of as a molecular entity having its own arrangement of atoms and consisting of hundreds of thousands or even millions of these. *Mutations,* which were conceived of as abrupt changes in inheritance, were thought of as involving a change of atomic arrangement within the molecule or the gene. The chromosomes are duplicated in cell division both within the body and in the sex glands. This act of multiplication was conceived of as somehow going on as though each gene could, out of the nuclear materials available to it, construct an identical image of itself, that is, a molecule of the same atomic arrangement. As a figure of speech it is often explained that the gene used its own structure as a template or pattern by which it could construct another identical molecule. Even at this early stage in exploring the DNA code of life we know that chromosomes are duplicated by a complementary process, not by duplication of self. It is of some philosophical interest that even though the theory of the gene was wrong on a number of counts, it still served as an adequate basis to explain inheritance as we see it occurring both in families and in populations. For this reason we shall use its simple language throughout the remainder of this work, although where it is pertinent, reference will be made to the code of life.

The Code of Life

As an organic compound of considerable complexity, DNA has been known for more than a century, for its presence within the nucleus of the cell was discovered by a German chemist. Only in recent years has DNA been recognized as providing the genetic code for all life, both in terms of the biochemistry of individual cells and in terms of the transmission of inheritable properties. The initial breakthrough came when J. D. Watson and F. H. C. Crick in 1953 published their paper on the structure of deoxyribonucleic acid. Many critical items needed in this construct had been available earlier, but it took the genius of Crick and Watson to complete the picture of how DNA is structured (Figure 3–1). It is now known to consist of a single pair of equidistant and entwined spirals, or helices, which are connected by pairs of chemical bases. The chemical bases are so chosen as to always provide the equal spacing between the spirals. To oversimplify, the structure of the DNA molecule can be visualized as a long ladder constructed of flexible material, twisted so its sides form the spirals while the base pairs form the rungs of the structure. The sides of the ladder consist of alternate sugar and phosphate groups. Each sugar is joined to the sugar on the opposite spiral by a pair of chemical bases consisting of one purine and one pyrimidine joined together. The base pairs are very simple and consist of two kinds of purines and two types of pyrimidines. The two purines are *adenine* and *guanine,* while the pair of pyrimidine bases are *cytosine* and *thymine.* Hereafter they will be represented by their initial letters, A, G, C, and T. The

Figure 3-1
Section of a Model of the
Genetic Code and an
Oversimplified Ladder-
like Visualization

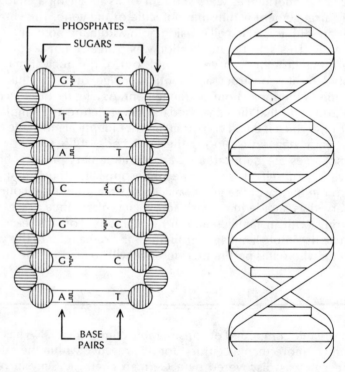

molecular requirement that the pair of bases always be of the same structural length limits their combinations to only two. A is always bonded to T; G is always linked with C. Since the order in terms of one side of the ladder may just as well be T-A as A-T, the two types of paired bases provide us with four basic letters in the alphabet making up the code of life. These bases have the property that one particular purine always combines with one particular pyrimidine, while the other type of base pair consists of the other two similar components. A and T are joined by two chemical bonds, while three connect C and G. This is a rather rigid chemical construction but it serves to make the spaces between the spirals exactly the same.

Since we are not concerned with molecular chemistry for its own sake, let us move on to the way in which the DNA molecule acts to provide the code of life. Its messages are carried in the form of a four-letter alphabet derived from the arrangement of the two orderings each of the two base pairs. They are ordered in a sequence as fixed as that in the written English language, which we always read from left to right. At first sight it might appear that a language limited to only four letters must have a very restricted communication capability. But it will be recalled that the English language, which contains 26 letters, uses them in all varieties of combinations and so creates perhaps as many as 200,000 words. The nearly infinite variety in which these can be arranged with regard to each other gives us enormous linguistic flexibility. The code of life is even more flexible, for while it has only four letters with which to send its message, they are arranged in much longer sequences than found in any English word. In fact the code behaves in a somewhat simpler fashion than this. It has been determined by the molecular biologists that the coding unit, or the message unit in the instructions, also called a *codon,* consists of a sequence of three base pairs. Four different letters taken in sets of triplets, or 4^3, allow 64 different codons or units of instruction. It is known that slightly more than 50 of these triplets are code words for the formation of specific amino acids. Since only 20 amino acids are commonly used in proteins, many of them obviously have redundant codons serving to produce them. Perhaps a remaining dozen codons act as punctuation marks in the total message. It is now considered that the DNA code in a more complex animal may contain anywhere from one million to five million base pairs. Thus it becomes not only possible but in fact proven that the four letters of the code of life may be used to write words containing hundreds or perhaps even thousands of letters. Each gene is considered to contain several hundred base pairs, but the sequence of pairs is as important as their number. Since our four letters can be arranged in any possible combination, this seemingly simple alphabet contains enormous potential for communicating complex messages.

The deciphering of the sequence of base pairs allows man for the first time to understand how life is coded, to recognize that mutations frequently involve substituting one base pair for another, and finally to comprehend how chromosomal duplication really occurs. Under the obsolete theory of the gene, one chromosome duplicated itself by creating an identical counterpart. In the DNA code it is known that the two spirals of the molecule of DNA somehow unwind themselves, splitting the pairs of attached bases, and then by attracting the missing bases, sugars, and phosphates from the nuclear pool of material reconstruct the missing helix and its bases. So in the code of life the process of chromosomal reduplication is a matter of complementary action. This discovery

Plate 3-2
The DNA double helix as seen through an electron microscope.

allows us to understand in relatively simple chemical terms how chromosomes do duplicate themselves, whereas in the theory of the gene it remained a matter of some mystery covered by graceful figures of speech.

Mitosis Is the Basic Form of Cell Division

In the higher bisexual animals there are two kinds of cell division. The basic type of division is called *mitosis* and occurs in all of the body cells to produce growth, maintain function, and repair damage. The other type of cell division is called *meiosis,* which occurs only in the sexual cells of physiologically adult individuals and functions to produce the *gametes,* or matured sex cells necessary for reproduction. The genetic code is of course carried in the nucleus of every cell, both body and sex cells alike. The two types of cell division are very similar in their elements. It is evident that body cell division, mitosis, is the basic process, and that the division occurring in sex cells, meiosis, has come into being through a few simple modifications of mitotic division.

The differences between the two cellular processes of division can best be understood by looking at what they produce. Mitotic division of body cells proceeds in such a way as to ensure that all of the resulting daughter cells shall be genetically identical to the mother cell from which they are derived. This, of course, is excluding rare events such as mutations and other accidents occurring in the process of division. It is designed to produce genetic identity, that is, to maintain the code constant, in all the trillions to quadrillions of cells which are found in the body of a grown person. The modifications which have produced division in the sex cells, that is, meiosis, operate so as to result in an almost infinite variety in the code of the mature sex cells which it produces. We have a striking contrast between the maintenance of the genetic code in unchanged form in body cell division and the scrambling it undergoes to produce new variations in every mature sex cell. That these totally opposite results can be produced by two types of division so similar in their essentials is remarkable.

Since mitotic division represents the basic process, we shall begin with it and briefly examine its various sequences. Let us start with a cell, not a real living one, but a very simplified model, which contains only those components which we need to understand the process of mitosis. Students of cellular biology will be horrified at the simplicity of this model and the liberties with which we construct it. Our imaginary model consists of a cell, microscopic in size, which is bounded by an outer membrane enclosing only three elements important to our discussion. In the center is the *nucleus,* which within its own containing membrane encloses the materials of heredity. Surrounding the nucleus is a complex substance called *cytoplasm,* whose ingredients do not concern us. The part of the cell outside of the nucleus, embedded in this cytoplasm, is the *centriole,* which acts to organize the poles of the cell in division. The reader should feel relieved that we are intentionally omitting mitochondria, endoplasmic reticulum, lysosomes, Golgi bodies, and even more formidable entities. In our version of the mitotic division of body cells we shall consider only the nucleus, the cytoplasm, and the centrioles, and the bounding membranes.

Mitotic division goes through a series of stages, all of which have their proper name in cellular biology. For our purposes we can eliminate all but one, that of *interphase.* This is the stage in which the cell is found prior to beginning its division and the state to which it returns after the act is completed. It is called the resting phase of the cell since during it the cell is not dividing, but in actual fact during interphase the cell is proceeding with all its normal biochem-

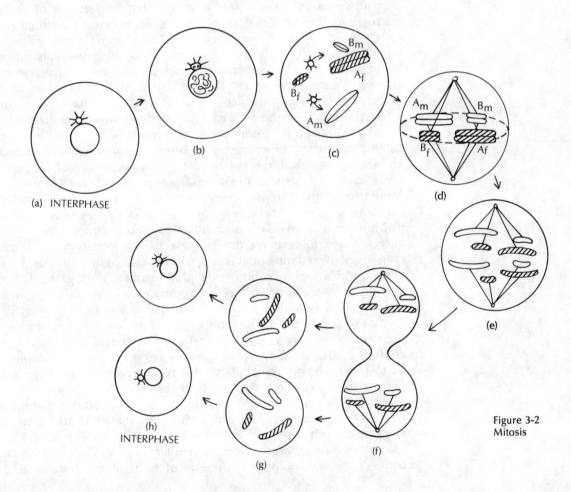

Figure 3-2
Mitosis

ical functions, serving very actively as a miniature biochemical factory. In Figure 3–2 we have diagrammed the process of mitosis in a series of eight consecutive stages. Beginning with interphase (a), we see the cell in its resting stage bounded by its membrane, and containing the nucleus in its center and nearby the centriole. Proceeding to step (b), some change is shown in that the nuclear membrane now shows visibly on its surface long twisted black lines. Shortly thereafter the nuclear membrane dissolves and there appear long wiry structures which slowly shorten and condense to become rod-like so as to be visible for the first time as shown in stage (c). It should be noted that the centriole has divided into two portions, each of which is moving toward the direction of the future poles, by which they establish the axis of the cell for its division. At the same time, there are four chromosomes shown in our model, a long pair and a short pair. Chromosomes always come in pairs in higher organisms and man is characterized by 23 pairs. Each pair of chromosomes contains different genes, that is, sections of code instructions. Furthermore, in bisexual species every individual receives one chromosome in each of the 23 pairs from his male parent, the other one from his female parent. Since all individuals differ genetically, it is to be expected that each chromosome derived from an individual's parents will carry different code instructions, or different genes. And this is true. For this reason we have shown in stage (c) the chromosomes of each pair derived from the individual's father as dark and those derived from his mother as light. They are labeled A_f and A_m to indicate the parent of origin for each of the chromosomes of a large pair. In a like fashion the small chromosome, labeled B, also has the same subscripts to indicate where they came from in the parental generation. Looking carefully the reader will note that each of the two pairs of chromosomes shows a line down its central axis. This is to indicate that the chromosomes had each already undergone the process of reduplication, so each consists of two strands, that is, the pair consists of four strands. This chromosomal reduplication took place earlier, during the latter portion of interphase. So the resting phase includes chromosomal reduplication as well as its own other vital activities.

After some period in which the chromosomes move at random in the protoplasm, they come to rest as indicated in stage (d). The centriole, after dividing in two parts, has now moved to establish a true polar axis; perpendicular to this axis in mid-cell is an imaginary equatorial plane. The individual chromosomes slowly settle to rest on this equatorial plane. It is significant that the positions that they take on this equatorial plane are random and bear no relationship of any kind to each other, either in terms of pairs or between pairs. The reader will note emerging from each polar centriole thin lines which connect the centriole to midpoints of each of the reduplicated chromosomes. These thin lines are physically real and are called *spindle threads.* Proceeding with stage (e) we see that the spindle threads have shortened and have begun to pull each section of every reduplicated chromosome toward the poles established by the centrioles. The chromosomes are pulled through the viscous protoplasm and are literally bent from the resistance it offers. Proceeding to stage (f) we see that the spindle threads continue to shorten so that the chromosomes are completely clear of the equatorial region. Under the microscope it can be seen that the equator is constricting in order to begin division and separation. In the next to the last portion of the division, stage (g), the constriction has proceeded so far as to produce two descendant, or daughter cells, each of which contains the complete genetic code found in the original, or mother cell. Both have the complete large pair, one of which is derived from the individual's father, the other from his mother, while the same is true of the small pair. In our final stage (h) we now see the nuclear membrane reconstituted and hiding the chromosomes, the centriole nestled against it, and the new cells looking very much like the original cell. There is only this difference. By the law of the conservation of matter, each of the new cells is approximately one-half the size of the mother cell. This condition does not continue long, for through the absorption of body fluids of one kind or another, the two new cells slowly grow to the size characteristic of that type of cell. Thus we finally have two body cells, identical in size, and the content of their genetic code as the result of the mitotic division of one cell. The length of time involved in mitotic division varies, but ranges from about one half hour to three hours in duration, depending on the type of cell.

The doubling process does not seem dramatic at first sight, but when it is continued through many divisions the result becomes astronomic. In the course of twenty years, man grows from one cell to an estimated ten trillion cells.

While the sequence of events in mitotic descriptions has been known at the descriptive level for years, many fundamental aspects remain unexplained. It is clear that the centriole should divide into two new and separate entities but how it does so is unknown. Nor do we know in a materialistic sense why the two new centrioles should be able to orient themselves at the opposite ends of a polar diameter, nor yet what forces drive them there. What causes the nuclear membrane to dissolve and how the chromosomes come to shorten and thicken at the appropriate time remain mysterious as processes. The emergence of spindle threads from the centrioles can be seen, but the process which produces this event is unknown. Nor do we know how the reduplicated chromosomes come to lie peacefully on the imaginary equatorial plane as determined by the poles. The force which causes the spindle threads to shorten and

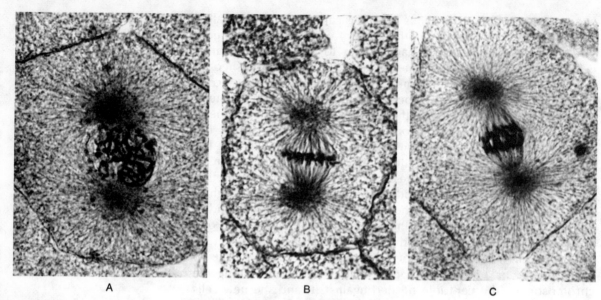

A   B   C

Plate 3-2
Six Photographs of Mitosis in
the Cells of an Embryonic
Whitefish.

more or less simultaneously the equator of the cell to constrict remains to be discovered. We are left, then, with a clear-cut description of the sequence of the stages in which mitotic division proceeds, but as yet we know almost nothing as to the how and why of its various steps. Obviously much more basic research in cellular biochemistry and molecular biology is needed to provide even approximate answers. Some of the consequences of mitotic division are given in Supplement No. 4.

On the Principle of Transformation in Evolution

One of the most useful general principles leading toward an understanding of evolutionary regularities involves the idea of *transformation* as opposed to *origination* in the development of organs and structures in new forms. Both the bones of the fossil record and the soft parts of comparative anatomy in living animals clearly demonstrate that evolution proceeds through the change in shape and function of organs and structures. It is as though changes can be made more quickly in the code of life by instructions which modify existing organs rather than in attempting to create totally new ones. The principle is of interest here, for meiosis, to which we shall turn shortly, is a transformation of the basic mitotic process of cell division.

Both the fossil record and the anatomy of living fishes show clearly that what were originally lungs in early fishes became swim bladders in later, more advanced forms. These swim bladders served primarily to regulate the specific gravity of the fish and provided internal control of buoyancy relative to the water in which they lived. Yet a further transformation occurs in some fish by which

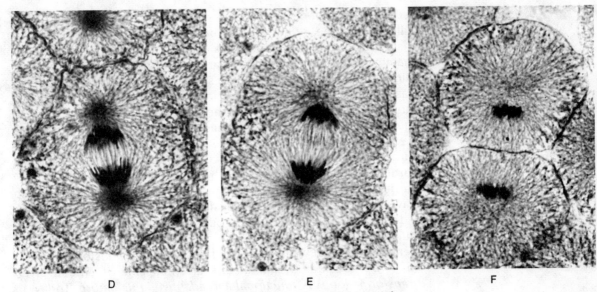

D   E   F

the swim bladder also serves as a resonating chamber to pick up vibrations in the water so that these can be transferred by a chain of bones known as the Weberian ossicles into stimuli affecting the brain of the fish.

Yet another example of this diversifying transformation involves the evolution of the five fingers of the extremities of the four limbs of the original terrestrial amphibians into a wide variety of specialized structures. All of the five-toed, four-toed, three-toed, two-toed, and one-toed terrestrial animals are the products of modification achieved primarily through elimination of superfluous digits and the strengthening of the remaining ones. Similar reductions and elongations account for the bony structure of the wings of birds, the flying reptiles of the Mesozoic era, and, of course, the mammalian bats.

One of the more complicated forms of transformation involves what happened to the seven gill arches present in primitive jawless types of fish and still retained today in that living fossil, the lamprey eel. The gills, of course, are openings through which water flows to provide oxygen for the capillary structures of the gills themselves. The openings are bounded on either side by gill arches which consisted of cartilage in the early fish. The evolutionary record makes it quite clear that the first or most anterior of these gill arches converted in the later fish, as in the shark, into the upper jaw or maxilla, and the lower jaw or mandible. During the course of reptilian evolution further transformations occurred, and in mammals such as man the remnants of this first mandibular gill arch have been transformed to provide additional structures in the form of the first two bones of the inner ear, the malleus and incus. Obviously to convert excessive mandibular structure into bones of the inner ear required long, very gradual series of transformations. The second gill arch of the primitive fish became converted into the hyomandibular cartilages in the shark and provides three small bones in mammals and ourselves. These are the third ossicle, the stapes or stirrup bone of the inner ear, the styloid process springing from the base of the skull, and the hyoid bone at the top of the throat structure. The remaining gill arches retain their original function in the shark, which in living forms has five gill arches. But in man these have contributed to the various cartilaginous structures in the throat, ranging from the thyroid cartilage at the top, to the tracheal cartilage at the posterior end. The blood vessels which originally carried blood to the gills to be reoxygenated in the primitive fish have by a series of gradual transformations and eliminations provided the basic structures of our own system. Throughout the hundreds of millions of years, and many modifications of the genetic code that were involved in this series of transformations, it must be considered that *at any given instant of time a given transformation was functional and adaptive in its nature.* Today, in the human embryo at the end of the first month, the ancestral gill arches are represented in the developing individual who at this same time has a two-chambered heart, as existed in the ancestral fish.

These examples are not evidence of undue conservatism in evolution, for they must be understood in terms of the development of the living organism. For example, all mammals, including ourselves, in their embryological stages, retain a *notochord,* a cartilaginous rod which was present in the early ancestors of the true vertebrates, those with segmented backbones. It might seem ridiculous to retain this outmoded structure in such highly organized creatures as the mammals. But if we look at development as a process, it becomes clear that the cells which form the notochord are very closely connected with organizing and bringing into being the essential structure which forms the long axis of the fetus. That is, the spinal cord, the brain, the heart, kidneys, and segments of muscle are all involved. Since the notochord provides this organizing function, its cells must be retained in the early fetal stage of mammalian development. The explanation of many so-called vestigial organs probably lies in the role they play in furthering the organization of other structures in the growth of the embryo. These ideas are important when we later turn to those transformations which have converted mitotic division into meiotic division.

Some Basic Genetic Terms

We have now reached the stage where it is necessary to understand and use some of the basic vocabulary of genetics. The carriers of the code of life are the *chromosomes,* originally named because they were color-staining bodies. They exist in the nucleus of each cell in pairs, the number of which varies with the species of organism. Man has 23 pairs, and if all 23 pairs were laid out end to end, they would represent two linear arrangements of the total genetic code. In their total length there would be a great many different sets of instructions, or genes, by the theory of the gene. Each gene ordinarily is necessary for the healthy development and functioning of the individual. At this time it is necessary to introduce the concept of the *locus,* derived from the Latin and spelled *loci* in the plural. It means place in its original sense, and it still means this in terms of position on the chromosomes. Let us assume for a moment that all of the million or so base pairs in the genetic code of man are punctuated so as to provide perhaps 50,000 messages or genes. In Figure 3–3 we have diagrammed a pair of chromosomes and we are focusing attention on Locus 1000 for the purposes of illustration. Both chromosomes of this pair at Locus 1000 carry a gene. Since one chromosome came from this individual's father and the other from his mother, they may well, and usually do, contain slightly different versions of the gene. If the messages should be identical the condition is called *homozygous,* for the genes are alike. But if Locus 1000 should contain slightly differing messages, the condition is called *heterozygous,* meaning the genes are different. It is important to realize that the instructions carried at Locus 1000 on this A pair of chromosomes relate to a given biochemical process, and even though there may be slight differences in the message in the heterozygous condition, they are both concerned with the same basic process. Thus the instruction at Locus 1000 on one chromosome might program the chemistry to produce the blood group A. The individual might have two *a* genes at this locus. In this case his *genotype,* or true genetic makeup, would be AA, which is homozygous for the gene *a.* If, on the other hand, while receiving an *a* gene from his father he also received an *o* gene from his mother, his genotype would be *ao* and his condition would be heterozygous. Since in Mendel's terms the gene *a* is dominant over the recessive gene *o,* he would show a phenotype, or visible trait in the form of the blood group A. But because he contains two slightly differing sets of instructions at this locus, he could contribute either the A instructions or the O instructions to his children, and as will be seen later, the chances are equal or 50 percent for each of these cases.

We are now prepared to grapple with the concept of the *allelomorph,* or, as it is more conveniently called, the *allele,* in the theory of the gene. The concept is really not difficult. Essentially, alleles are a series of genes which may be found at any given locus. But each variable locus has its own series of allelomorphs. At Locus 1000 which we are considering, there are three common types of genes which fit here. They are the blood group genes *a,* *o,* and *b.* No individual may possess more than two of these, for a pair of

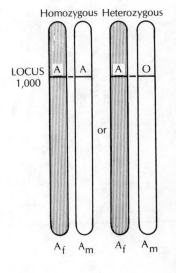

Figure 3-3
On the Nature of the Locus

chromosomes provides only two slots at a given locus into which genes may fit. But in dealing with populations a much larger number of alleles may be found. In the fruit fly, *Drosophila,* the single locus which carries genes determining eye color may have sixteen different alleles which produce faintly differing sets of instructions. Thus, eye colors in this little fly may range all the way from white to bright red. The term allele is simply a convenient one to designate all the genetic variants which may be found in a population at a given locus. There seem to be two basic kinds of loci in man as well as in other creatures. Those which contain two or more variant genes or alleles are variable loci. Their exact number is not known but an interesting method of estimating by Lewontin (1967) and Harris (1969) suggests that in man the variable loci may approximate one-third of the total number. Estimates of similar value have been made for such unlike animals as mice, fruit flies, and horseshoe crabs. But this leaves two-thirds or about 67 percent in which no deviation in the instructions has been detected. It would seem that these nonvariable loci represent those portions of the code of life in which no variation is tolerable to the development and life of the organism. It is likely that there is a locus concerned with the development of the backbone in man and other vertebrates. Any changes induced by mutational forces in these instructions may produce individuals who do not go beyond certain stages of fetal development. There is an unfortunate condition known in man as *anencephaly,* in which the infant comes to full term fetal life but with no brain. The face and the base of the skull are present, but there are no cerebral hemispheres. This is obviously a lethal condition and death follows birth very shortly. Whatever locus controls brain development obviously does not tolerate such deviations and very likely is a nonvariable locus. The nonvariable loci are concerned with the programming of instructions vital for life and characteristic of a given species, and they are more or less unchangeable. The variable loci, on the other hand, are concerned with the less vital functions; this kind of genetic variation may indeed improve fitness for individuals as well as for populations. It is also worth noting that racial differences in man and other animals are involved only in the variable loci and thus are limited to less critical sections of the code of life.

Some Modifications in the Process of Mitosis

Just as the principle of evolutionary transformation is applicable to organs and structures, so it may be applied to processes which are also coded in DNA. Mutational changes in the base pairs of the DNA code can produce shifts in protein synthesis and consequently alterations in the pattern of existing processes and so in their products. Changes in structures necessarily imply that the developmental processes which produce them have also been changed. Mitotic division arose in evolution as a reproductive process by which one cell divided into two descendant cells and so on, to produce the lineages of asexual reproduction which we know as *clones*. The production of clonal descendants by single cells differs somewhat from the production of cells in more complex creatures. Both represent a form of mitosis, but an unknown transformation has occurred in the complex forms.

We are now concerned with the important but poorly understood problem of the origin of *multicellular* life. There is no doubt that original living forms appeared as simple molecules which, over a very long period of time, became elaborated into single cell life. Today all of the highest animals and plants are multicellular, meaning their bodies are comprised of many cells. This change might have been achieved in several ways. Among lower forms of life, there exist animals which are simply large aggregations of single cells. The Portuguese man-of-war is a gaudy marine illustration of such colonial life. Its gaily colored bubble floats innocently on the ocean's surface, while below it trail poisonous tentacles. Some evolutionists use this animal, which is allied to the jellyfishes, and other similar but simpler examples as models to illustrate the origin of all multicellular forms. But an alternative hypothesis is more attractive. If the process of mitotic division had simply been altered so that two daughter cells resulting from its division adhered to each other instead of separating, then the origin of multicellular life could be rather simply explained. A number of theorists believe that this is the way in which many-celled animals actually originated. Since the development of a body with many cells opens vast new avenues of evolution in terms of size, complexity, and differentiation of structure, this was a most important event. But we are dealing here with a state of affairs which occurred millions of years ago, and neither hypothesis can be tested in direct fashion in the laboratory.

In us and most other multicellular animals, mitosis continues to be the process by which growth is achieved and cellular replacements are obtained. But bisexual reproduction required a new type of cell division to be developed. For if the sex glands were to go through ordinary mitotic division, they would produce *sperm* in the male animal and *ova* in the female, each of which would contain the full number of 46 chromosomes characteristic of the species. When a sperm penetrated the membrane of an ovum to produce a new individual, this *zygote* would be characterized by 92 chromosomes, or 46 pairs. With each new generation this system would involve a doubling of chromosome numbers so that it would quickly become astronomical, and so biologically unimaginable. To avoid the doubling of chromosomes in each generation of reproduction, it is obvious that the functional sex cells, the sperm and the ovum, must have no more than one half of the ordinary number, but that this complement should contain a complete DNA code of instructions. This is called the *haploid* condition, as opposed to the *diploid* condition, which exists both in normal body cells and in zygotes. We can therefore specify that cellular division in the sex cells must somehow be arranged so that only one chromosome in each pair is allotted to the maturing sperm and ovum. Evidently the program for mitotic division must be changed in order to achieve this end.

Evolution of Bisexual Reproduction

With the great clarity that comes with hindsight, we can say that the evolution of bisexual reproduction arose to answer the need for greater variability in natural populations. This idea rests in part upon the fact that as the life cycles of most single-celled animals are examined, it is found that their reproduction is generally through simple mitotic fission. Many instances occur, however, in which two of the little creatures will adhere to each other and by transformation, transduction, or conjugation, exchange portions of their DNA. Such a jumbling of their code of life clearly introduces variability and in a very crude way simulates the effects of bisexual reproduction. If we remember that reproduction by fission produces clones of single-celled plants and animals, among which each individual is genetically identical to the progenitor of the lineage, then the reason for the sexual phases becomes clear. Barring mutational accident, all clonal species would consist of individuals in which the genetic code is identical. Owing to the fact that mutations do occur, a given species will contain a variety of clones which differ slightly genetically. These compete with each other and survival goes to the fittest cell type under a given environmental circumstance. Thus generally single-celled forms evolve through competition between clones. These types of populations have limited variability and consequently evolve slowly in terms of their length of generation; but their rate of fission is usually so rapid that their evolutionary change in absolute time may also be rapid.

Bisexual reproduction, above the level of single cell life, produces such variability among offspring as to overwhelm the imagination, yet the evidence is clear that the variations are of about the right magnitude in each species. The meiotic division of cells in the sex glands must be arranged not only to produce functional gametes without any increase in the chromosomal numbers of the species, but also it must operate so as to maximize variations in the genetic code. The ways in which the basic mitotic process has been transformed to produce meiosis are astonishingly simple, considering the great consequences of the changes. It is interesting that trial and error processes, such as characterize most of evolution, work to produce optimum results.

The Process of Meiotic Division

The results attained by relatively simple modifications of the basic plan of division can be illustrated by the meiotic divisions going on in the testes of an adult male. It is very similar in females except for the modifications which have been developed in that sex to concentrate the yolk material into one maturing ovum. The process of meiosis without crossing over is diagrammed in Figure 3–4. We have ignored all those cell structures which are not concerned with the production of mature sex cells. In stage (a), interphase, or the resting stage, we find the cell enclosed within its own membrane, and containing within it only the nucleus and the centriole. The process of meiosis begins as the granular appearance of the surface membrane of the nucleus is replaced by a series of tangled, thread-like structures. Although they are not visible, the chromosomes have already undergone their reduplication back in the late portion of interphase. As we come to stage (c), the nuclear membrane has dissolved, the thread-like structures have shortened and thickened to become rod-like chromosomes, and now the two pairs move randomly about the protoplasm of the cell. At the same time the centriole has divided into two bodies which are already moving toward the appropriate position where they will establish the future polar diameter of the cell. So far, these stages are identical with those which go on in mitosis.

Stage (d), which is called *synapsis,* marks the first important divergence from the pattern set by mitotic division. It will be recalled that in the latter the reduplicated chromosomes settled separately and at random on the equatorial plane of the cell. Here in meiosis an altogether different pattern is followed and it has enormous consequences in terms of variability in the population. For now the pairs of chromosomes settle to the equatorial plane and the members of each like pair fuse together or *synapse.* This pairing is such that the sequence of the DNA code in each of the fused chromosomes is the same from one end to the other. These are reduplicated chromosomes so each pair is in the four-strand condition. Shading as well as letters indicate which chromosome in this pair comes from this individual's father, and which from his mother. It will be noted that as the two pairs happened to settle on the equatorial plane, the chromosome derived from the father is below in pair A and above in pair B. Since the process goes on at random, it should be stated that pair B might just as easily have aligned themselves relative to pair A with the paternally derived chromo-

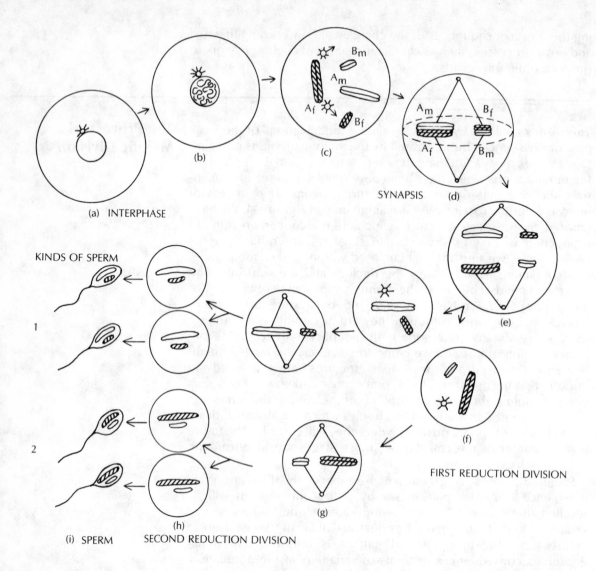

(a) INTERPHASE

(b)

(c)

SYNAPSIS   (d)

(e)

(f)

FIRST REDUCTION DIVISION

KINDS OF SPERM

1

2

(g)

(h)

(i) SPERM   SECOND REDUCTION DIVISION

Figure 3-4
Meiosis Without Crossing-Over

some on the bottom. The consequence of these two alternative types of positions will be discussed a little later.

Note that in this diagram the spindle threads which have gone from each of the poles now number only two from the upper pole and the same number from the lower. This is only one-half of the number of spindle threads found in the mitotic diagram, and the reduction results from the fact that the chromosomes here are synapsed rather than lying separately on the equatorial plane. Returning to the random way in which the pairs of chromosomes align themselves on the equatorial plane, it should be stressed that this *random assortment* holds good for as many pairs of chromosomes as the species has. We must always fix one pair as a point of reference, but the second pair will align itself with regard to the first pair at random, meaning that one position is just as likely as the other. And so the third pair will be random in aligning itself with regard to the other two. This same randomness affects all pairs of chromosomes, that is, 23 pairs in man, and it leads to producing highly variable sex cells.

Stage (e) represents the initial phases of the first reduction division in meiosis. The spindle threads have shortened and the chromosomes, having separated, are now being dragged toward their appropriate poles. Shortly, the cell will begin to constrict at the equator, and the result of the *first reduction division* is shown in Stage (f), where two new daughters, each containing one-half of the matter present in the original mother cell, are now complete entities. Each contains a polar body or centriole and one of each pair of duplicated chromosomes. It is immediately evident that these new cells contain different hereditary material than in the original cell from which they were derived. For the combination of synapsis and random assortment has so positioned the hereditary material that the lower daughter cell in the large chromosome contains only the code derived from this individual's father, where the small chromosome has genetic instructions derived only from his mother. Obviously, the opposite is true for the upper cell. It is evident that the sex cells produced further in this process will contain different genetic codes.

As the process of meiosis goes on, the centrioles divide again, establish polar positions, and attach their spindle threads to each of the reduplicated chromosomes as shown in Stage (g). Again without indicating the constricting that precedes the actual division, each of these daughter cells now goes through the *second reduction division* and in turn produces two new cells for each of the old ones. This time the spindle threads pull apart the reduplicated chromosome so in the four immature sex cells each chromosome pair is represented by a single strand. This is the haploid condition, for instead of having two pairs of chromosomes we now have two chromosomes, one from each pair. In the final stages in the male, most of the cytoplasm contained in these four cells is discarded and the nuclear material, including the chromosomes, is fashioned into the sperm head to which is attached a little lashing tail to give it *motility*, or the power of individual movement. While each cell which undergoes the process of meiosis produces four mature sperm in the male, the reader will note that these are of only two kinds. The top pair contains a large chromosome whose material is derived only from the individual's mother, while the small pair contains the individual's father's genetic code. Of course, exactly the opposite is true for the bottom pair.

The consequences of random assortment are diagrammed in Figure 3–5. The upper portion shows the same arrangement as is indicated in Figure 3–4, and the production of four sperm of which there are two arrangements of code. But the lower portion of the diagram has the B pair of chromosomes in the alternate position, with the maternally derived chromosome of the pair uppermost and

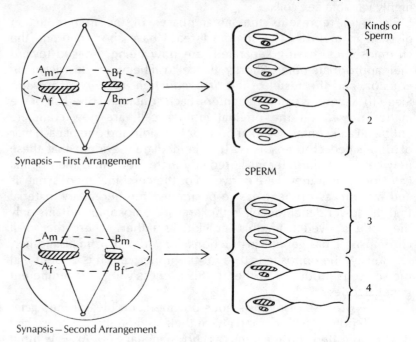

Figure 3-5
Random Assortment
Increases Gametic
Variety

the paternally derived on the bottom. When this arrangement proceeds through the two-reduction division, it produces four sperm as its total output, and these are of two kinds. But it will be noted in comparing them with those produced by the first arrangement that they are very different. The top two sperm contain only chromosomes derived from this individual's mother, while all of the coded material in the bottom pair is derived from his father. Taken all together, the two alternate positions of the second pair of chromosomes due to random assortment can provide four different kinds of sperm. In any individual male both arrangements are present many times over. So we must conclude that with two pairs of chromosomes four different kinds of gametes can be produced. It follows that with only one pair of chromosomes involved, only two kinds of sperm will be produced. The curious student may use pencil and paper to ascertain how many different kinds of sperm will be produced when three and four different pairs of chromosomes are involved in meiosis. If he has done this correctly, he will find that the numbers are 8 and 16, respectively.

Evolution proceeds so that successful animals usually have optimal characteristics, that is, they are well suited for the life they live. Therefore, it is interesting to note the differences of consequences in the human male as compared with the female. A sexually matured man produces sperm so rapidly that 300 million (300,000,000) can be produced very quickly, that is, in several days. This prodigal production of sperm apparently represents the necessary number needed to achieve reliable fertilizing of the human female. More than this would be wasted, fewer would be insufficient. The human female represents an opposite extreme. Considering that full reproductive competence is reached at about 17 years of age and may last on the average until about 47, a woman has about 30 years of effective fertility. Her ovulation is initially irregular but in a few years attains full reliability. During the last few years before the onset of menopause her ovulatory reliability again diminishes until it finally ceases. With an ovum maturing in each 28-day lunar cycle, the average woman still produces fewer than 400 matured gametes during her entire lifetime. The problem of human reproduction is further complicated by the fact that after being fertilized in the fallopian tubes, the new zygote spends some time descending into the uterus. Five days pass before it becomes implanted effectively into the intrauterine wall and thus receives dependable energy support from the mother. During this period of independence the fertilized egg must be self-sustaining. To this end it receives a concentration of almost all of the cytoplasm present in the original dividing ovarian cell. In woman, meiosis, of course, goes through two reduction divisions and so produces four end products. Three of these are nonfunctional polar bodies, containing essentially only nuclear materials, and are waste products. The first reduction division in a woman produces one oversized, immature cell, which contains virtually all of the cytoplasm derived from the original ovarian cell, and an impoverished polar body, which contains only nuclear materials. During the second reduction division, the cytoplasm is again concentrated into but one of the daughter cells, which becomes the functional ovum. The other becomes a polar body. In a like fashion the polar body resulting from the first reduction division again divides in a mechanical way to produce two further terminal polar bodies. With the retention of the matured ovum safely within the body of the woman, evolution has adjusted the process of meiosis so as to arrange for a proper conservation of the protoplasm, with its yolk material, to insure success in reproduction. While this is in marked contrast to what meiosis produces in the man, we can see it is just as well adjusted in terms of evolutionary needs.

Some Aspects of Human Reproduction

Measures of Genetic Variability in Man

We have seen that meiosis operates so as to produce two genetically different kinds of sperm from the first pair of chromosomes. Each subsequent pair of chromosomes has the same capability. The first pair serves as a reference point, and all subsequent pairs have two possible positions through the operation of random assortment. This allows us to set up a simple equation to predict how many different kinds of sperm will be produced with variable numbers of chromosomes: $N = 2^x$. In this expression of equality, N is the number of genetically different kinds of sperm produced, 2 is the base power of the equation reflecting the fact that each pair of chromosomes can produce the two different types of gametes if considered independently, while X is the exponent determined by the numbers of pairs of chromosomes occurring in the species. Ignoring crossing-over as we have done up to this point, it is simple to calculate that an organism such as the fruit fly, *Drosophila,* which has four pairs of chromosomes, would produce $N = 2^4$ or 16 different kinds of sperm. If an animal has 10 pairs of chromosomes, then 1,024 different kinds of sperm are the consequence of meiosis. Such exponential functions rapidly increase their values. In man we are dealing with 23 pairs of chromosomes so our calculation yields 8,388,608 differing versions of the genetic code in the sperm of the human male. This number is impressive but not mathematically large and we have only begun.

Let us consider what meiosis produces in the human female. A woman matures only one ovum per lunar month under ordinary circumstances (before the Pill began to induce excessive numbers of multiple ovulations), and each ovum contains only one variant of her genetic code. But the interesting point, and one easily over-looked, is that the single ovum she produces at proper periods is derived at random from the total potential of her 8,388,608 possible kinds of gametes. Since it is a woman's genetic potential, rather than the actual production of ova which is important in discussing genetic variability, it may be concluded that every woman has the same potential genetic variability in her ova as does the man in his sperm.

The Variability Available in a Single Human Mating

As impressive as this gametic variability is, it is barely a beginning. Let us estimate the potential genetic variability of a single human couple. A man's sperm unites with and fertilizes the woman's ovum. This *zygote* represents the beginning of a new individual and through subsequent mitotic division develops into an embryo. After nine months of intrauterine development it is born as a new individual. Where does the infant stand genetically in terms of the total variability available? The sperm that succeeded in fertilizing the ovum was chosen essentially at random from about 8 million different arrangements of the father's genetic code. The mature ovum that was fertilized was also chosen at random from a potential genetic store of the same magnitude. Thus, if our model man and woman were to be endowed with an infinity of time and inexhaustible energies, they could conceive an enormous number of genetically unique children. The equation expressing this number is $N = 2^{23} \times 2^{23} = 2^{46}$. We are still somewhat short of what mathematicians lovingly call "large numbers," and without calculating the last equation, it is closely equivalent to 70 trillion (70,000,000,000,000) different kinds of children. This number is much larger than our limited senses allow us to comprehend, but let us try. Imagine yourself in a helicopter high over a very large football stadium crowded with 100,000 people, all with their faces upturned toward you. One may get an overall sense of the meaning of 100,000 in these terms, but try to visualize 700 million stadiums, each filled with 100,000 people! The meaning of numbers of this size escapes us, but we may approach another direction with a figure of speech.

Today the total population of the world is nearly 3 billion human beings. Our imaginary couple, released from the confines of time and energy, could produce more children than 20,000 times the size of the world's present population and all of these children would be genetically different. Further, we might consider man capable of one billion different matings. This is a small proportion of reality, for we must consider that every man on this planet has a roughly equal chance of marrying any woman. The number of permutational matings available among two billion adults is astronomical. The magnitude of this genetic variability is full of evolutionary consequences. Evidence to be considered elsewhere suggests that even variability of this magnitude would not allow evolution to proceed indefinitely, for there are forces which constantly tend to reduce it. Nonetheless, organic evolution could go on for a large number of generations with no more variability than the species now contains.

The Consequences of Crossing-Over

Crossing-over may be defined as the exchange of homologous blocks of genes between adjacent strands of the synapsed chromosomes. Through processes which are not understood, pairs of strands may break at identical places, that is, between the same loci. In some cases the detached fragments of the strands are later reunited to the opposing strands from which they have been broken (Figure 3–6). This is a recognizable crossing-over. If the broken portions of the chromosome strands reheal in their original positions, this would not constitute crossing-over, for no exchange of sections of the genetic code is involved. Such a break and restoration of the original condition could not be detected by breeding experiments. Since the exchanged strands involve different genetic instructions, the consequences of crossing-over are detectable in altered Mendelian ratios among the offspring, and the reality of crossing-over can be determined by experimental methods.

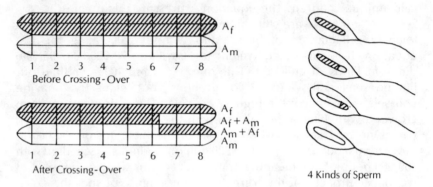

Figure 3-6
Crossing-Over and Its
Gametic
Consequences

Before Crossing-Over

After Crossing-Over

4 Kinds of Sperm

Our model, a single pair of reduplicated chromosomes, illustrates the basic process. In the top portion of the diagram they are shown synapsed but prior to crossing-over. They are labeled as in previous models to indicate whether they are derived from the father or mother of the individual involved and only eight loci are indicated, of which loci 6, 7, and 8 are of particular interest in this case. A sizable minority of the loci in man are heterozygous so that a good deal of genetic difference is to be expected in the code sequence of the pair of chromosomes.

Let us imagine that in the two central strands breaks have simultaneously occurred between loci 5 and 6. These detached strands do not move far away but in reattaching themselves they have fused to the wrong strand, as shown in the central portion of Figure 3–6. Let us inventory the types of code now present in the four strands. The top shaded strand is not involved in crossing-over and still contains the totality of the code derived from the individual's father. Fused to it is a strand, a mosaic of code, that is derived from the individual's father between loci 1 and 5, but now has genetic instructions derived from the mother in loci 6, 7, and 8. In the third strand from the top the situation is the reverse, for homologous genes have been exchanged. The fourth and bottom strand is not involved in crossing-over and maintains intact the coded instructions derived from the individual's mother. The condition of crossing-over can be detected at certain stages of cell division in meiosis by the appearance of an actual cross or an X between the strands. Such sections are known as *chiasma,* if but one occurs, or *chiasmata* for two or more. The chiasma responsible for the crossing-over in Figure 3–6 is not shown, however, an example of chiasma is shown in the synapsed pair of chromosomes in Plate 3-4. Also shown in Figure 3–6 are four different kinds of sperm derived from this single pair of chromosomes, twice the number which would be present without the act of crossing-over. They have resulted from the first reduction division of the synapsed crossed-over chromosomes diagrammed above, followed by a second reduction division to produce the haploid stage. This is no less than a doubling of variability due to the crossing-over of homologous blocks of genes.

A single crossing-over involving the two adjacent strands of the four available in the duplicated chromosome is sufficiently complicated for our purposes. But there can be double and even triple crossings-over between adjacent strands of a single pair of synapsed chromosomes. Further, three or even four strands may be involved in double or triple crossing-over, and so the variability latent in this type of scrambling of the genetic code is truly staggering.

Experimental evidence indicates that crossing-over is both frequent and regular. As a general proposition we may consider that it occurs at least once in every pair of chromosomes involved in a meiotic division. For the purposes of recalculating the increase in variability resulting from crossing-over, let us use the minimum assumption that each pair of chromosomes shows one two-strand crossing-over during the process of meiosis. Thus with 23 pairs of chromosomes, there will be 23 crossings-over, one affecting each pair. Since a single pair of chromosomes will produce four different kinds of gametes, and the same consequence results independently for every other pair, we can easily calculate the resultant variability. The equation previously used is modified by simply altering the base number from 2 to 4 to express the fact that now each pair of chromosomes will produce four different kinds of gametes by its own reduction divisions, $N = 4x$. This simple shift in the base power of the equation now gives us truly astronomical figures for the resultant variability in gametes. In man with 23 pairs the equation reads $N = 4^{23}$. Consequently about 70 trillion different kinds of sperm will be produced by each man, while each woman has a potential of the same number of different kinds of ova, even though she only matures one in each lunar cycle.

In human beings the number of different kinds of genetically unique offspring potentially reproducible by each couple, again given unlimited time and no diminution in physical energy, is $N = 4^{23} \times 4^{23} = 4^{46}$. This results in a number large enough to satisfy mathematicians since it approximately equals $5 = 10^{27}$, or an inconceivable 5 octillion. Such is the measure of genetic variability

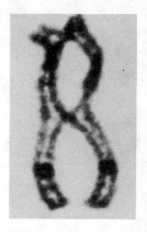

Plate 3-4
Crossing-Over in a Synapsed Chromosome Pair. Two chiasmata resulting from crossing-over are visible.

potentially available to a single couple. The variety that might come about through mate exchange at random and ad infinitum carries the calculation beyond our ability to comprehend. And all of this is still a minimal estimate, for our assumption involves only one crossing-over of each of the 23 pairs of chromosomes.

A Summary of the Differences and Consequences of the Two Processes of Cell Division

We have seen that mitotic divisions produce uniformity in the code of all their descendant cells, and that meiotic division results in bewildering variety among the gametes produced. Let us now consider how this transformation was achieved. Both processes are remarkably similar in their early stages aside from the fact that they go on in different types of cells. In both body cells and sex cells the divisions are preceded by a state called interphase, during the late stages of which the amount of DNA is duplicated invisibly, but indicating that the chromosomes have also been duplicated. In both body cells and sex cells the centriole divides to establish the polar axis for the division, and perpendicular to this in each case is an imaginary equatorial plane. Now the first difference appears. In mitosis the reduplicated chromosomes fall onto the equatorial plane separately and in random positions. In meiosis they settle down upon it and the homologous or like pairs of chromosomes fuse together or synapse. This basic difference results in the possibility of crossing-over occurring in meiosis and the necessity in the meiotic process of two reduction divisions in order to produce the haploid gametes. It also results in cutting in half the number of spindle threads that go from the poles to the reduplicated chromosomes. Crossing-over may occur in mitosis, but since blocks of homologous genes would contain the same genetic code, it would not be identifiable and it is apparently not known whether it exists or not. In any case, it has no consequence for us. But the random assortment of the chromosomes in synapsis and their crossing-over in meiosis allow the basic process to create almost infinite variability among the gametes produced. Because of the difference in the number of reduction divisions, mitosis produces but two daughter cells, where meiosis produces from each sex cell four gametes in the male and but one gamete and three polar bodies in the female. The reader may profit from making up a small table in which the various differences between mitosis and meiosis are inserted.

A Brief Note on Familial Genetics

Mendel's discovery of the particulate nature of inheritance led ultimately to the establishment of the biological science of genetics.

During the first decades of the twentieth century the new field was primarily concerned with the elaboration of its knowledge and in developing the now obsolete theory of the gene. Insofar as that theory affected man, it represented the building of the foundations for familial genetics, a field allowing prediction concerning the nature of offspring from parents of known genetic constitution. This type of inheritance is not basic to the study of evolution but it contains much of interest to the citizen and to all future parents. Stern (1973) covers these principles well.

Bibliography

CRICK, F. H. C.
1966 "The Genetic Code: III." Scientific American, 215 (April):55–62.

HARRIS, H.
1969 Enzyme and Protein Polymorphism. British Medical Bulletin, 25:5–13.

HOLLEY, ROBERT W.
1966 "The Nucleotide Sequence of a Nucleic Acid." Scientific American, 214 (February):30–39.

KORNBERG, ARTHUR.
1968 "The Synthesis of DNA." Scientific American, 219 (April):64–78.

LEVINE, R. P.
1962 Genetics. New York: Holt, Rinehart and Winston.

LEWONTIN, R. C.
1967 "An Estimate of Average Heterozygosity in Man." American Journal of Human Genetics, 19:681–685.

MOORE, JOHN A.
1963 Heredity and Development. New York: Oxford University Press.

STERN, CURT.
1973 Principles of Human Genetics. San Francisco: W. H. Freeman and Company.

SUTTON, H. ELDON.
1965 An Introduction to Human Genetics. New York: Holt, Rinehart and Winston.

WATSON, J. D.
1953a "Genetical Implications of Structure of Deoxyribonucleic Acid." Nature, 171:964–967.
1963 "Involvement of RNA in the Synthesis of Protein." Science, 140:17–26.

WATSON, J. D. and F. H. C. CRICK.
1953b "A Structure for Deoxyribose Nucleic Acid." Nature, 171:737–738.

WILKINS, M. H. F.
1963 "Molecular Configuration of Nucleic Acids." Science, 140:941–950.

YANOFSKY, CHARLES.
1967 "Gene Structure and Protein Structure." Scientific American, 216 (May):80–94.

Mitotic Division: The How and Why of Nature's Cellular Factory

The process of mitotic division serves to produce growth in the individual, to replace worn-out cells, and is involved with the process of aging as well as some causes of death such as cancer. The mitotic divisions which produce growth occur fastest in the fetus, slow down after birth, continue at a low frequency to provide replacement functions during adulthood, and apparently diminish even more in very old age. When the vertebrate egg is fertilized by a sperm, the two nuclei of the gamete unite and create the full chromosomal complement of the zygote, or new individual. This cell mitotically divides into two daughter cells, both of which undergo fission to produce a four-celled stage, and in ensuing mitotic divisions produce eight, sixteen, thirty-two and sixty-four cells. These first six mitotic divisions proceed without any protoplasmic growth, and so the final mulberry-like cluster of cells is about the same size as the original fertilized egg. But once the zygote becomes effectively implanted in the uterine wall of the mother its protoplasmic content grows, and so does the total mass of the new individual. Rates of mitotic division vary from species to species and further differ with the various types of cells in a given organism. The process of mitotic division may last anywhere from several minutes to dozens of hours depending upon the kind of animal and the type of cell in which it takes place. In humans, the full-term fetus is about 2 billion times as large as the original egg.

The total number of cells in an adult human have not been counted, but are estimated to total about 100 trillion in all or 10^{14}. This seems to be a very large number, but if we take a simple exponential equation, in which $N = 2^x$, to represent what happens in mitosis, the results are startling. The first five mitotic divisions only produce 32 cells in all, but the next five raise the number to 1024. A sequence of 20 mitotic divisions produces an individual containing more than a million cells. It has been calculated that a sequence of 46 mitotic divisions is all that is needed to produce the trillions of cells in each one of us. This does not allow for the ongoing divisions in special cells which constitute replacement, but it does measure the number of mitotic divisions needed to reach full adult growth.

This statement is generally true, but wrong in some details. We are born with essentially the full number of nerve cells, or *neurons*, that make up the central nervous system and the brain. The growth in the brain that occurs after birth is a consequence of absorption of body fluids rather than division. Since we are born with a full complement of nerve cells which cannot be replaced by further mitotic divisions, it is evident that their damage through physical accident, or by drug abuse, is final and allows for no restoration. In some parts of the body neurons project fibers several feet in length. Thus when a motor nerve in the arm is cut, regeneration is possible through the growth of the severed neuron and it may in time restore function to the hands and fingers. But this is regeneration of the severed neuron, and not a product of cell division. The neurons of the brain are affected by time and many millions are lost through degenerative processes in the aged. This damage does not produce evident loss of function until the seventh and eighth decades of life, when these irreversible damages produce the failing memory, the faltering coordinations, and a slowing down of intellectual processes that we associate with the very old. The cause of this loss is unknown.

Most other body cells continue mitotic divisions throughout life, and some, such as the cells which produce red corpuscles of our blood, divide at surprisingly high rates. These cells have a short life span, lasting but three to six months, and being vital components, they must be replaced by ongoing and constant mitosis. This capacity provides man and other mammals with an interesting kind of adaptability, which shows itself when we move to high altitudes. At 10,000 feet the density of the atmosphere is about one-half of that at sea level; consequently its oxygen content is accordingly reduced. Since the red corpuscles function to capture and carry oxygen to all parts of the body, it is obvious that if one is taken rapidly to an altitude of 10,000 feet, physical capacity will be lowered. Where it is possible to drive into high mountains by car, attempts to climb immediately frequently result in what is known as mountain sickness. But if the better part of a day is spent in acclimatizing at 10,000 feet, the hiker can then move off with a 50-pound pack with very little discomfort. During the hours involved in adjusting to the high altitude, the red marrow of the body responds to the reduction of oxygen pressure by speeding up the rate of mitosis and effectively doubles the number of red cells normally present. This type of physiological accommodation is a striking manifestation of the biological flexibility of each of us as individuals. For those who were born in a high altitude, more complex adjustments are involved.

All of us are born into the world and upon death leave it. The causes of the processes of aging are but little understood. At one time medical experimenters thought that they had demonstrated that human cells grown in flasks, that is, in vitro, possessed immortality. These experiments were in error, and today it is known that cells removed from the human body age in vitro just as they do in us. Two general hypotheses are advanced to explain aging today. The first accounts for our physical degeneration through a random pattern of neuron damage which may arise from a variety of sources, including cosmic rays. The alternative hypothesis considers that life span is coded in DNA, and thus is an inheritable attribute. Which hypothesis is more nearly correct is not yet known. But there is experimental evidence to suggest that animals which live fast, that is, whose metabolic rates are high, tend to die earlier than those who live more slowly. It is almost as though there were just so many heartbeats in the average life and individuals tend to live longer if their heart beats more slowly. This doctrine smacks of a kind of biological predestination, and until the evidence for it is better, none of us should be unduly apprehensive. Yet the idea that the length of our life may be coded into our physiological rate of activity is an interesting one.

Apparently we can age at different rates in different parts of our bodies. Man and other mammals usually manifest aging by the depigmentation or graying of hair, but some families are prone to very early graying, and indeed their members may be white-headed at thirty, without its affecting life expectancy. Whole races, particularly the Mongoloids, are very late in getting gray head hair, and yet they certainly live no longer under the same circumstances than other populations. Physical hardship probably accelerates

the signs of aging and may shorten life expectancy. Among Australian Aborigines, women of 45, who live under precolonial conditions, look like old hags. Their skins are wrinkled, their lips hang loose, their eyes are half-blinded by the disease tracoma, and their scalps bear innumerable scars that come from fights with their own sex. They have lived a hard life, and it may well be that they are biologically considerably older than their calendar years.

The fertilized egg from which we all begin ultimately differentiates into a wide variety of different cells, blood corpuscles, neurons, the chitin of the hair and fingernails, enamel, cement and dentine of the teeth, to name only a few. Since the dogma of mitosis insists that each process of division results in the exact reduplication of the entire genetic code of the individual, it has always been something of a mystery how cellular differentiation could proceed under this system of growth. Investigations of the genetic code are now sufficiently advanced to indicate that the primary or structural genes are subject to modification in their activities by other sections of the code. In fetal growth these controlling mechanisms actuate some genes in some regions of the body and suppress them in others. Growth and cellular differentiation thus may be compared to the performance of complex music of the pipe organ. This type of model reconciles the fidelity of reproduction which goes on in mitosis with the necessity of the regional differences which we know occur in growth.

The nature of regional differentiation is well illustrated by the changes that occur in the thymus gland, located where the throat meets the chest. Until recent years its function was unknown and it was thought to be left over in the process of evolution. The gland is large in the newborn infant, decreases in size in childhood, and becomes vestigal in the adult. This seemingly senseless development and subsequent degeneration was hard to explain until the function of the gland was discovered. Now it is known that the thymus gland has very important roles in the individual. Primarily, it produces antibodies which prepare the newly born child to face the biochemical hazards of the outer world; a large active thymus gland is his first line of defense in this change of scene. Since the growing individual develops a set of antibody defenses through exposure to external substances, the gland in time loses its importance. The growth and decline of the thymus represent changing genetic instruction to the differentiated cells which comprise it. If the process still remains somewhat mysterious, at least the DNA code provides a basis for explanation.

Cancer is becoming the scourge of mankind. Already the second greatest killer in the United States, its mortality rates are increasing. No direct cures have yet been found, and only a few of its many forms respond to types of medical therapy now available. The problem is especially pressing, for it represents a form of death that is prolonged, painful, and degrading to the individual. In essence the disease known as cancer may be described as the process of mitotic division gone wild. Presumably starting in one cell, it soon affects a region, and before long death may spread to the whole ravaged body. Its painful and lethal effects are due not so much to its direct assaults as to the fact that its uncontrolled pattern of growth dislocates other organs.

The initial cause of cancer remains unknown. Two opposing hypotheses dominate the field today. Researchers with a genetical bias tend to view the altered rate of cell division as a consequence of a mutation occurring in a specific body cell. This throws the regulating mechanisms out of kilter and the process of division goes awry producing a kind of overgrowth. An opposing hypothesis suggests that the disturbing influence is due to an invasion by a virus, which, in taking over the mechanism of a cell to produce its own assembly line parts, somehow disturbs nearby cells so that their rate of division is changed. This idea has received considerable support because some forms of cancer have been experimentally transmitted through viruses. Time may prove that both mutations and viruses can cause cancer.

Cancer raises another problem from the evolutionary point of view. Until the past century mankind by and large has lived in a regimen that gave an average life expectancy ranging between 25 and 35 years of age. This statistic needs a little explanation. It is a calculation by insurance actuaries including within the range of data the age at which infants die as well as adults. Thus its final numerical value reflects high infant mortality directly. *But even with this reservation people died young everywhere in the world until Western medicine prolonged life. Thus during the millions of years of human evolution men and women generally died prior to attaining old age, indeed before there was much risk of developing cancer. Consequently, the disease of cancer has not been exposed to the restraining influences of population selection.* Certainly individual genetic variability affects the risk of succumbing to cancer; indeed one medical authority has claimed that all of us have had cancer, but most of us have the biochemical capability of suppressing it. But with cancer largely occurring after the end of the reproductive period, the genetic variants involved are neither selected for nor against.

The student interested in this subject is referred to Leonard Hayflick's article, "Human Cells and Aging," in *Scientific American*, 218, (1960): 32–37.

Some Examples of Simple Mendelian Genetics

A new branch of human biology was founded in 1900 when Karl Landsteiner, in his laboratory in Vienna, noticed that the red cells of some individuals were *agglutinated,* that is, clotted, by the serum of other people. This landmark finding occurred in the same year that Mendel's work on the particulate nature of inheritance was rediscovered by DeVries, Tsermack, and Correns. The clotting of the red corpuscles, showing immune reactions in the presence of the appropriate serum, is very like the body's own defenses against invading bacteria or proteins.

Understanding the inheritance of the various blood group genes is based upon comprehension of the antigen systems that characterize the red corpuscles and the corresponding antibodies which occur in the serum of the blood. The red cells, or corpuscles, are the most numerous solid bodies in the blood, and the average adult human has about 30 billion red cells. Men have about five million per cubic millimeter of blood and women have ten percent fewer. Unlike the corpuscles in many other animals, those in humans have no nucleus. Shaped like a thick disk, concave on both sides, with a large surface for its very small volume, the cell is well designed for its primary function, which is to carry oxygen bound to hemoglobin for the general body needs. The surface of the red cells has a variety of chemical properties, for the known blood group series now numbers about 30 and each of these is characterized by several antigens that react vigorously on the surface of the red cell. *Antigens* are protein substances which cause certain cells in the animal body to form antibodies. The *antibodies* are also proteins that react specifically with the antigen molecules and so inactivate them. The nature of the reaction is so specific that each antigen can only be inactivated by a unique antibody. Most antigens represent foreign substances, so the development of appropriate antibodies is an important defense mechanism giving the invaded creature immunity against the introduced antigen. Thus, the study of the blood groups falls into the province of immunochemistry.

The clotting discovered by Landsteiner was the visible manifestation of the reaction between a specific antigen and the correct antibody. By a series of cross-matching tests, he demonstrated that men could be classified as having one of four different blood group types. These types are what are called *phenotypes,* representing the visible characteristics of the blood in the clotting situation. They have been labeled O, A, B, and AB. Subsequent investigators demonstrated that these four blood group types were inherited and the exact form of inheritance was determined through a series of three genes, any one of which could replace the others at a given position in the genetic code. These interchangeable genes, called *alleles,* generally represent a series of additional modifications of a given section of the DNA code. They tend to do the same thing in slightly different ways. In a few cases it has been demonstrated that the alleles result from a change in a single base pair in the gene. These three allelic genes, all situated at a single locus, are technically labeled L^o, L^a, L^b. But this is a clumsy notation and for teaching purposes we can represent the three alleles as $\underline{o}$, $\underline{a}$ and $\underline{b}$.

Since any one of the three allelic genes will occur in the same position on each of the chromosomes constituting a pair, it is obvious that every individual contains two allelic genes from this series. His *genotype,* in contrast to his phenotype, must thus be expressed in terms of two such genes. If the individual has two like alleles, his genotype is *homozygous,* and there are three such types in man: $\underline{oo}$, which gives the phenotype O; the genotype $\underline{aa}$, which gives the phenotype A; and the genotype $\underline{bb}$, which provides the individual with a phenotype B. In addition, there are three possible combinations of two differing kinds of alleles, and such individuals are called *heterozygous.* They consist of the genotype $\underline{ao}$, which produces a phenotypic A; genotype $\underline{bo}$, which gives the phenotype B; and the genotype $\underline{ab}$, which gives a phenotype AB, which is the rarest of the four blood group types.

We may now write the rules of dominance for the series: the genes $\underline{a}$ and $\underline{b}$ are *codominant,* that is, both manifest themselves in the phenotype when present, and both of them are *dominant* over $\underline{o}$, which is a *recessive* gene and shows only when present homozygously. This could be written $\underline{a} = \underline{b} > \underline{o}$. It is worth noting that dominance does not confer higher values on the biochemical action of a gene or make it superior or more common. In fact, in all the populations of the world the recessive gene _o_ is commonest in the population gene pool, while _a_ and _b_ are in a minority role. In general, the _a_ gene is commoner than the _b_, but there are some Asiatic populations in which this is reversed.

The student can now draw up a simple table to make it easier to translate genotypes into phenotypes. The genotype _oo_ is always the phenotype O. But genotypes _aa_ and _ao_ both give the phenotype A, and by ordinary methods of blood group testing, they are indistinguishable from each other. In a like fashion the genotypes _bb_ and _bo_ give the single phenotype B and of course the genotype _ab_ gives the fourth blood group type, AB. The reverse process, determining the genotype from the phenotype blood group, is not so simple. It is obvious that the blood group types O and AB reveal their genotypes with no ambiguity. But a phenotypic A may either be the homozygous or heterozygous genotype and the same is true for B. In actual practice, it is necessary to use mathematical methods to estimate what proportions of the blood groups A and B are homozygous and heterozygous, respectively. In the following examples, the necessary information for determination of the genotype is given. Since we shall deal with series of multiple alleles from time to time, it may be useful for the student to calculate the number of possible genotypes in which the alleles can combine in random unions at conception. This value is given by the following equation:

$$N = \frac{1}{2} \times N (N + 1)$$

Thus, with a series of three alleles the number of genotypes is:

$$N = \frac{1}{2} \times 3 \times 4 = \frac{12}{2} = 6.$$

Provided with the rules of dominance and the ability to convert from genotypes to phenotypes, it is now possible to show what kinds of offspring given parents can have. In predictions of this sort, it is important to realize that the kinds of offspring that are possible represent statistical predictions in an abstract sense. But the predictions are sound if based upon large numbers of families so that the errors of sampling will be cancelled out in the total pooled data. The larger the series of families, the closer the reality will come to the predicted values. Let us begin with a mating in which a phenotype A male is married to an O woman.

1. Phenotype	A♂		=	O♀
2. Genotype	_aa_		‖	_oo_
3. Gametes	_a_			_o_
4. Zygotes		_ao_		
5. Phenotype		All A		

We will assume that we know that in this case the male is homozygous for the gene _a_. Family predictions are best set up in a sequence of the preceding five steps. The nature of the zygotes which can be produced by this A male and O female with the various kinds of sperm fertilizing the various potential types of ova is given on line 4. In this instance we have a single type of sperm and a single ovum; all the potential zygotes have the genotype _ao_. This zygotic line, of course, assumes that fertilization occurs at random between all the types of sperm and all the potential types of ova. Since _a_ dominates _o_, which does not appear in the heterozygous phenotype, we conclude that all the children of this couple will be A phenotypically. This mating corresponds to one of the primary crosses of Mendel between pure lines of peas, and here the dominant gene has obliterated the recessive one in the first generation of offspring. There is no Mendelian ratio in this example, for 100 percent of the children are phenotypic A's.

Mendel demonstrated that genes do not disappear, but may remain invisible in the phenotype if they are recessive in nature. Let us now take an example in which two heterozygous A individuals marry. They are of the same phenotype as the offspring in the last example, but since we will not commit incest even in a model, let us presume that they are unrelated individuals.

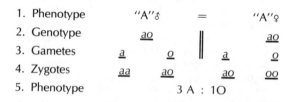

1. Phenotype	"A"♂		=		"A"♀
2. Genotype	_ao_		‖		_ao_
3. Gametes	_a_	_o_		_a_	_o_
4. Zygotes	_aa_	_ao_		_ao_	_oo_
5. Phenotype			3 A : 1O		

Each of the A's is surrounded by quotation marks to indicate that it is heterozygous. Hence the genotypes are _ao_, and meiotic division gives two kinds of sperm and two kinds of ova. In each parent, 50 percent of the gametes will be _a_ and the other 50 percent will be _o_. Random fertilization is a little more difficult now, but it can be achieved in two rather simple fashions. Let the _a_ sperm fertilize the _a_ ovum to produce an _aa_ zygote, but it has an equal chance of fertilizing the _o_ ovum and so will produce an _ao_ zygote. On the other hand, the _o_ sperm can fertilize the _a_ ovum to produce an _ao_ zygote, or equally likely, will fertilize an _o_ ovum to produce an _oo_ zygote. The way that we convert these four offspring into phenotypic children, remembering that _ao_ gives an A phenotype, we end up with a phenotypic ratio of three A's to one O. This is the simple Mendelian ratio which always occurs when both parents are heterozygous for a dominant and a recessive gene. It will be noted that in such families two-thirds of the A children are heterozygous but this is not detectable in their phenotypes. It does suggest that the heterozygous genotype _ao_ is more frequent than the homozygous genotype _aa_ and in populations this is true.

This example has not only given us the famous three-to-one ratio which both Darwin and Mendel discovered independently, but it also demonstrates the reappearance of the recessive O phenotype which had disappeared in the first example. This is not the creation of something genetically new, but merely the return to a visible phenotypic level of genes which had been present all the time. This is one reason why in real human families, children sometimes provide their parents with pleasant surprises and at other times unpleasant ones, for if both parents contain recessive genes, then some of the children may be homozygous for them and reveal the fact phenotypically. Unhappily, more recessive genes are harmful than beneficial. In a large number of families in which both parents are heterozygous, the homozygous recessive offspring in this case would be predicted to occur in 25 percent of the children. Of course, this is one of the reasons why parents with any doubts should seek professional genetic counseling before conceiving children.

Let us try another single factor example of familial inheritance to show a different kind of Mendelian ratio, a heterozygous A man and a heterozygous B woman.

♀ / ♂	_a_	_o_
_b_	_ab_	_bo_
_o_	_ao_	_oo_

Again, the phenotypic line shows quotation marks around both of the phenotypic labels to give us this information. The genotypes are easily written and each parent produces two kinds of gametes in equal numbers: _a_ and _o_ sperm for the man, and _b_ and _o_ ova for the woman. To produce the zygotes, let us introduce a simple but foolproof device known as the checkerboard. Write the male gametes across the top of the checkerboard and the female gametes vertically along the left side. The checkerboard will then contain two rows and two columns, providing four boxes within its boundaries. Each box is filled in by combining the gametes found in the two margins. The upper left-hand box becomes an _ab_, the upper right-hand box a _bo_, and so on. This shows clearly, from the number of gametes that each parent provides, how many different kinds of zygotes they can produce with random fertilization. The zygotes are then transformed to the fourth line and converted below by the rules of dominance to the phenotypes. In this case we have one AB to one A to one B to one O. Each of the four human types of blood group is present in the children in equal frequencies of 25 percent each or 1:1:1:1. This is another of the characteristic ratios found by Mendel when crossing parents who are heterozygous for two different alleles. Half of these children show characteristics not present in the phenotypes of the parents; thus, the phenotype AB is a new combination and the phenotype O is the reemergence of the homozygous recessive gene which did not show in the parental couple.

These examples of familial inheritance make the rules of heredity seem far too simple. Mendel's focus on single-factor inheritance enabled him to detect the regularities in general heredity, but inheritance becomes more complex in dealing with two factors, that is, allelic series of genes located in different portions of the genetic code. Two factors, situated on two different pairs of chromosomes, will segregate randomly at meiosis and will combine in a zygote randomly, again segregating in the phenotypes of offspring.

Our second factor is another blood group series of alleles, that of the MN. Appropriately

1. Phenotype	"A" ♂	=	"B" ♀
2. Genotype	_ao_	‖	_bo_
3. Gametes	_a_ _o_	‖	_b_ _o_
4. Zygotes	_ab_ _ao_		_bo_ _oo_
5. Phenotype	1AB: 1A: 1B: 1O		

Ratio   1 : 1 : 1 : 1

enough, these blood group antigens were also discovered in 1927 by Landsteiner and his student, Philip Levine. These blood group types are not as commonly known to the general public, for they cause no difficulty in transfusion and hence are not listed on the dogtags of GIs. Again, the antigens are carried in the surface of the red cells of the blood, and these can be clotted with the proper specific antibody serum. There is a complex series of alleles at this locus, but our interest is limited to two, the genes _m_ and _n_. These can combine into the pairs that occur in the human chromosomes which carry them in three fashions. The genotypes are _mm_, _mn_, and _nn_. Studies of familial inheritance show that _m_ and _n_ are codominant, or mutually dominant, so that each expresses itself in the phenotype when present. Thus, we can readily convert the genotypes into phenotypes and they become, respectively, M, MN, and N. The world distribution of this pair of alleles is more extreme than that of the ABO blood series. While many of the major populations of the world show about equal frequencies of the genes _m_ and _n_ in their gene pools, others show marked deviations. Among Mongoloids and American Indians, _m_ reaches very high values, sometimes exceeding 90 percent of the genes in the pool. Among the Aborigines of Australia and Melanesia, _m_ is a rare gene, usually consisting of less than 20 percent of the content of the gene pool and in a few local populations has been lost completely.

Where the parents are homozygous, then even two-factor problems of family inheritance can be simple, as in an AM man and a BN woman.

1. Phenotype   AM ♂   =   BN♀
2. Genotype   _aa_ : _mm_   ‖   _bb_ : _nn_
3. Gametes   _am_   ‖   _bn_
4. Zygotes   _abmn_
5. Phenotype   All ABMN

Our form is slightly changed, for in addition to the lines separating the marital partners down through the production of gametes, it is now useful to use a colon to separate the unlinked allelic series of genes, and this has been indicated in the diagram on the genotype level. Because of their homozygosity, the man only produces _am_ sperm and the woman only _bn_ ova. Hence, only one type of zygote, _abmn_, can be produced, and all of the children are ABMN in phenotype.

A more complex example of two-factor inheritance involves parents who are heterozygous at both loci. The male is AMN and his wife is a BMN. Since both the "A" and the "B" in this couple are heterozygous, the genotype can be directly written as _aomn_ in the man and _bomn_ in the woman. Remembering that meiosis proceeds with two pairs of chromosomes which assort at random, we can then produce the correct gametes, which are four in number for both persons.

1. Phenotype   "A"MN♂   =   "B"MN♀
2. Genotype   _ao_ : _mn_   ‖   _bo_ : _mn_
3. Gametes   _am_   _an_   ‖   _bm_   _bn_
           _om_   _on_   ‖   _om_   _on_
4. Zygotes [See checkerboard below]
5. Phenotypes 1ABM: 2ABMN: 1ABN: 1AM: 2AMN: 1AN
     1BM: 2BMN: 1BN: 1OM: 2OMN: 1ON

♀ / ♂	_am_	_an_	_om_	_on_
_bm_	_abmm_	_abmn_	_bomm_	_bomn_
_bn_	_abmn_	_abnn_	_bomn_	_bonn_
_om_	_aomm_	_aomn_	_oomm_	_oomn_
_on_	_aomn_	_aonn_	_oomn_	_oonn_

Life is becoming too complex for simple zygote production, and so let us turn to a checkerboard which in this case is a four by four table, and hence contains 16 compartments. The geometry of the checkerboard indicates that with two genetic factors the maximum number of different combinations which the offspring can show genotypically will be 16. But as we put our gametes out on the proper margins in this case, and introduce them into the boxes for which they are marginal, we find that some genotypes are reduplicating themselves. In the end, our 16 boxes when converted to phenotypes yield 12 different ones. The ratio of 1:2:1: is repeated four times. Because of the particular arrangement of gametes on the margins of the checkerboard, we have in fact four replicative cases of simple Mendelian ratios in the form of 1:2:1. But this will not always work out, for it depends upon the arrangement in the margins. This ratio can just as well be expressed as 2:2:2:2:1:1:1:1:1:1:1:1. We are now obviously dealing with patterns of reproduction too complex for anything but careful work with pencil and paper. The size of the checkerboard depends upon the number of heterozygous loci in each parent. In our example both were double heterozygotes, but if the man had been a double heterozygote and the woman a single heterozygote, then she would only have had two kinds of gametes and our checkerboard would have become a four-by-two arrangement containing only eight cells.

A three-factor situation vastly complicates the problem. We will use the Rhesus factor, a blood group locus also identified by Landsteiner in 1940, a full forty years after his original paper on the subject and while working with a student named Alexander Wiener. The RH factor, better known than the MN series because of its association with infant deaths, is characterized by many variable antigens, but we shall deal with only two, *D* and *d*. Landsteiner and Wiener had immunized a rabbit with blood from a Rhesus monkey, thus obtaining an antibody serum which clotted the blood of 85 percent of the White patients tested in New York hospitals. The negative reaction was found to be due to an antigen, *d*, for which the immune serum has never been found. The two genotypes *DD* and *Dd* are called Rhesus positive and the homozygous recessive genotype *dd* is called Rhesus negative. If this system were to be fully developed, more than 100 different genotypes are possible, so we shall review it from a simple level, as a third and unlinked allelic series of genes.

If both of the parents were heterozygous for all three of the blood group loci, each could produce eight different kinds of gametes and our checkerboard would have 64 squares in it. A three-factor example of inheritance in which less than the maximum amount of heterozygosity is present in the parents is shown with an OMN Rh positive man and an ABN Rh negative woman. It is known that the man has an Rh negative mother, so he must be heterozygous for the recessive gene *d*. Accordingly, we write his genotype

oomnDd, while his wife is *abnnddd*. In the genotype line of the diagram, the genes of the three loci have been separated by broken lines, indicating that they will assort randomly in mitotic division. The man produces four different kinds of gametes and his wife, two. Upon converting the contents of the checkerboard from genotype to phenotype, we find that there are eight kinds of children possible by this mating. It is important, with three factors, to be certain that in the production of gametes, all possible combinations are realized, since the unlinked chromosomes do assort randomly in meiosis.

1. Phenotype   O MN"Rh+"♂ = AB N Rh−♀
2. Genotype   *oo*:*mn*:*Dd*   *ab*:*nn*:*dd*
3. Gametes   *omD*   *omd*   *and*
             onD   *ond*   *bnd*

4. Zygotes [See checkerboard below]
5. Phenotype
   1AMNRh+: 1AMNRh−: 1ANRh+: 1ANRh−:
   1BMNRh+: 1BMNRh−: 1BNRh+: 1BNRhO
        Ratio 1:1:1:1:1:1:1:1

♀／♂	*omD*	*omd*	*onD*	*ond*
and	*aomnDd*	*aomndd*	*aonnDd*	*aonndd*
bnd	*bomnDd*	*bomndd*	*bonnDd*	*bonndd*

Through these examples, the student should now be aware of the full meaning of Mendelian ratios and why the presence of whole numbers in the ratios is so important. Mendel could translate the simplicity of his ratios among the offspring into an operational theory about inheritance. He realized that it was particulate in nature and hence in essence foreshadowed the theory of the gene, a fact which has not greatly changed as we translate things today in terms of the genetic code.

CHAPTER 4

HOW TO
UNDERSTAND
THE FOSSIL
RECORD OF
EVOLUTION

Since evolution consists of changes in systems in time, the measurement of time is of primary importance. The present scale of geological time has been perfected by field work by geologists and in recent decades by laboratory work in absolute and relative methods of dating directly. These advances have revolutionized geological chronology, and datings have become more exact with the passage of time. With evolutionary progress, changes occur more rapidly. Fossils record these changes and give some indication of past environments which supported life. The flow of energy through living systems is an important part of understanding the relationships of plants and animals to their environment, and it too can be studied through the fossil record.

The Evolution of Geological Knowledge

Our knowledge of the evolution of life on our planet is approaching a reasonably satisfactory completeness. For most of the major categories of animals and plants we know the point of origin in time and sometimes in space. In well-represented lineages, such as the horse, we may understand much of the detailed branching evolution. It would be a mistake to claim that the fossil record approaches completeness, but nevertheless it provides a substantial framework for our present purposes.

Fossilized examples of animals and plants have been known to many men in many places, but the understanding of their real meaning awaited the orderly evolutionary growth of geological knowledge. This required escape from rigid and erroneous interpretations of the biblical version of creation into a new world of empirical questioning and hypothesizing. The retarding influence of the Dark Ages was only beginning to lift by the middle of the eighteenth century, and the origins of modern geology came after that time. Careful observers a century earlier had correctly interpreted the meaning of fossils, and recognized that superimposed strata of rock revealed important implications about the duration of the Earth's existence. But their conclusions were too early to have much impact in their own times. By the latter half of the eighteenth century a gradually dawning enlightenment allowed man to develop an increasing interest in his physical surroundings, and to begin to read order into the available evidence.

Many men contributed to the infant science of geology, but one of the most influential was A. G. Werner, an important professor in the late eighteenth century at the Freiberg Mining Academy in Germany. As a first-rate organizer of geological data and material, he modified earlier classifications by setting up five types or series of rock, which ranged from the primitive and early to the volcanic and late in an orderly fashion. His basic concepts rested upon the belief that the all-encompassing world ocean had gradually receded to its present proportions and during the process had precipitated out all of the forms of rock now visible in the earth's crust. This emphasis of an oceanic derivation of all rocks and minerals led Werner and his disciples to become known as the "Neptunists." His scheme had considerable success, but it was incorrect, being based upon an unnatural framework of classification. Consequently, it may even have acted as a retardant of the development of stratigraphic geology, and the proper arranging of various earth levels in an accurate relative time scale.

In Scotland a contemporary of Werner's by the name of James Hutton advanced geological knowledge in a more scientific fashion. But the dominance of the Neptunist philosophy obscured Hutton's conclusions for a long time. The importance of Hutton's work lies primarily in the fact that by combining field observations with laboratory experiments he demonstrated beyond doubt that many of the igneous rocks arose through the cooling of melted rock, instead of as a marine precipitate as claimed by Werner. The real importance of Hutton's contribution, however, lay in the fact that in studying the water-laid deposits in his native Scotland, he recognized that many of the characteristics of ancient sediments recurred in the modern and unconsolidated deposits. This led him to state the principle that "the present is the key to the past," meaning that conditions which were responsible for ancient geological features still prevailed today. His principle of *uniformitarianism* did not mean that all geological activities proceeded at a uniform rate, but rather that the laws of nature uniformly applied to such processes.

Modern geology has been based primarily upon the science of *stratigraphy,* that is, the studying of earth strata, to provide a relative dating of its events. It was advanced near the end of the eighteenth century by the Englishman William Smith. As a boy he had begun by collecting the fossils from the beds near his home, and in later years from a variety of sedimentary formations further afield. He was a good observer and noted that while some fossil types ran through several successions of rock beds, others were distinctive and limited to a single bed of deposit. In time he became able to distinguish different formations by the types of fossils they contained. Smith was a successful engineering geologist and so gained experience over broad areas in the British Isles. His work culminated in his "Geological Map of England and Wales with Part of Scotland," which plotted the regional distribution of 31 major rock strata.

Charles Darwin's friend, Charles Lyell, is known as the father of modern geology. He built upon Smith's principles and conducted further stratigraphic investigation in the field. The first edition of his pioneer textbook, *The Principles of Geology,* was published in 1833, and set a pattern of geological classification which was to be followed for the next century. Its importance arose from the fact that Lyell gave to Darwin and to more modern investigators a relative time schedule reaching far back into early Earth history. In his voyages on the *Beagle,* Darwin was able to satisfy himself from his own field observation of the essential correctness of the ideas of Lyell and Smith.

The earth science of geology has progressed rapidly since the work of such pioneers as Lyell and those who preceded him. Its sequences and chronology are given in Table 4–1 with dates as they were determined in 1974. The serious student should be familiar with the details of the chart, for time is one of the major dimensions in evolution. The sequences are divided into a hierarchy in which the *eras* represent the broadest classification; this in turn is divided into *periods;* and where knowledge is refined, the last may be divided further into *epochs,* as in the Cenozoic era. The ending -zoic, which occurs in all three of the more recent eras, is derived from the Greek root meaning life, and so in ascending order the eras mean old life, middle life, and recent life. The era here given as Precambrian sometimes is designated as consisting of the Archeozoic and Proterozoic eras. But the Precambrian era covers something like three billion years, its subdivisions are but poorly known, and its fossil record is most meager. For our purposes the era name of Precambrian is adequate. The three latest eras are in turn divided into periods and the sequence, duration, and age of their beginning provide a useful scale by which to understand evolutionary process. On the right-hand portion of the chart are detailed a few of the events which are most important to students of human evolution. Much of the Paleozoic evolution of marine life is ignored, except for the appearance of the vertebrates. In like fashion the evolution of plants is only represented by those changes which ultimately affect human evolution.

The Cenozoic era is broken down into two periods, the Tertiary, which includes the first five epochs, and the Quaternary, which includes the last two. The whole of the Cenozoic covers approximately 70 million years in time. Its last epoch, the *Recent,* or as it is sometimes known, the *Holocene,* will continue to unfold into future millennia. The common portion of the names of the epochs, -cene, is derived from the Greek root meaning recent. Hence the *Paleocene* is the epoch which is the oldest portion of the recent era. Paleocene deposits were discovered later than the others, so they had to be inserted at the bottom of the sequence. *Eocene* means the dawn of recent forms, *Oligocene* means few recent forms, *Miocene* means less of recent animals, *Pliocene* refers to more recent life,

The Geological Time Chart

while *Pleistocene* refers to the most recent life. These Greek derivations are not important to remember but the sequence of the epochs is, for the evolution of the mammal and more particularly the order of the primates to which we belong lie embedded in this time scale.

A number of important changes in chronology which have occurred in recent years are incorporated in this chart. Prior to the advent of absolute methods of dating, geologists generally agreed that the Recent epoch had a duration of approximately 25,000 years. Proper dating methods show that this should be reduced to 10,000 years. On the other hand, the Pleistocene was judged to be much shorter than it was in reality. American geologists believed in a long Pleistocene, considering this epoch to probably have had a duration of about one million years. For historical reasons that are not too clear, English geologists preferred a much shorter chronology, and their estimates ranged from one-half a million to one-quarter of a million years for the whole Pleistocene. Those who believed in the long chronology have been closer to correct, but in recent years the very content of the Pleistocene has changed through international agreement among geologists. An important series of *beds*, or rock strata, known as the Villafranchian, whose type site is in Italy, has been taken from the Upper Pliocene and incorporated as the earliest part of the Pleistocene. The reason involved turns upon the ease of identifying the beginning of the Pleistocene. As originally defined it was considered to begin with the earliest ice advance of the glacial sequence found in Europe. Such ice sheets are absent from most of the rest of the world, and so it was difficult to determine whether a given set of beds belonged in the Pliocene or the Pleistocene. The problem was solved by turning to the types of animals contained in the beds to determine their identification. The Villafranchian deposit of Italy, and its time equivalents throughout the rest of the Old World, saw the appearance of the first modern genera of such important big mammals as the horses, the cattle, the camels, and the elephants. The Villafranchian beds took much longer to deposit than the whole remainder of the Pleistocene sequence so that that epoch now has a duration of about three million years, and may be even further extended back in time with future research. It still remains the geological period in which early men first became prominent in the fossil record. The definitive evolution of the horses and camels took place in the New World and these animals later moved across a land bridge connecting Alaska with northeastern Siberia. At the same time the elephants and cattle which originated in the Old World crossed in the opposite direction. Thus the appearance of the Villafranchian fauna occurred virtually simultaneously in both hemispheres.

TABLE 4-1

Scale of Geological Time

Era	Period	Epoch	Began Millions of Years Ago	Duration in Millions of Years	Some Important Events in Life of the Times
CENOZOIC	Quaternary	Recent	.01	.01	Modern genera of animals with man dominant.
		Pleistocene	3(2.5–3.0)	3	Early men and many giant mammals now extinct.
	Tertiary	Pliocene	10	7	Anthropoid radiation and culmination of mammalian specialization.
		Miocene	25	15	
		Oligocene	40	15	Expansion and modernization of mammals.
		Eocene	60	20	
		Paleocene	70(±2)	10	
MESOZOIC	Cretaceous		135	65	Dinosaurs dominant to end; both marsupial and placental mammals appear; first flowering plants appear and radiate rapidly.
	Jurassic		180	45	Dominance of dinosaurs; first mammals and birds; insects abundant, including social forms.
	Triassic		225	45	First dinosaurs and mammal-like reptiles with culmination of labrynthodont amphibians.
PALEOZOIC	Permian		270	45	Radiating primitive reptiles displace amphibians as dominant group; glaciation widespread.
	Carboniferous		350	80	Amphibians dominant in luxurious coal forests; first reptiles and trees.
	Devonian		400	50	Dominance of fishes; first amphibians.
	Silurian		440	40	Sea scorpions and primitive fish; invasion of land by plants and arthropods.
	Ordovician		500	60	First vertebrates, the jawless fish; invertebrates dominate the seas.
	Cambrian		600(±20)	100	All invertebrate phyla appear and algae diversify.
PRE-CAMBRIAN	Not well established		Back to Earth origins 4.5–4.9 billion years ago		A few soft multicellular invertebrates in latest phases. First known fossils as early as 3.3 billion years ago.

Methods of Dating   The time scale of Earth history has been established by two general methods: first, techniques which allow strata and their contained fossils to be dated relative to each other in an order proceeding from older to younger; and, second, absolute methods of dating which provide ages in terms of actual years. The methods of relative dating which we will consider include the *stratigraphic method,* the *fluorine technique,* and *amino-acid dating.* Absolute dates are determined by stable time clocks provided by the constant rate of disintegration found in a variety of radioactive substances. Three of the most widely used methods of absolute dating involve the measurement of radioactive carbon-14; the distintegration of radioactive potassium into the heavy gas, argon, as one of its end products; and fission track method. The uranium disintegration cycle is of great use for very long-term absolute dating.

Relative Dating Techniques

Stratigraphic Dating.   This was the method first used to establish the relative sequences by which beds were laid down in earth prehistory. It depends upon the identification of different strata of rock in terms of their positions relative to other beds. It involves the important geological principle of superposition, in which it is assumed that the younger beds overlie the older ones. Except where Earth movements have been severe enough to produce dramatic foldings, the technique works very well. The fossils contained in each strata are of great assistance in providing proper identification as William Smith demonstrated many decades ago. Since most geological strata originate through erosional deposition, no matter what their subsequent history may be, it is useful to measure their depth to give an approximate estimate of their age. The rate of deposition varies both with the agency producing it and other local conditions such as gradients and the like. The thickness of beds served to provide the rough relative chronology for the pioneer geologists. The maximum depth of all the geologically identified rock beds gives the remarkable total figure of more than 452,000 feet. This is a vertical depth slightly in excess of 85 miles. As we know from recent evidence from our moon and photographs of Mars, this type of erosion and deposition does not occur on all planetary bodies. It is a consequence of an atmosphere of considerable density and with a high water vapor content. Our moon has virtually no atmosphere and that of Mars is so thin that the erosional features with which we are familiar cannot occur upon it.

The order of superposition of beds relative to each other provides the basis of a time sequence; under most conditions only later strata overlie earlier ones. Geologists have used a variety of data to check their age estimates. The rate of deposition of mud, as in the Nile River delta, has been measured and projected backward into geological time to provide an estimate for the rate of formation of earlier deposits. Another approach is involved in the calculation of the rate at which the salt content of the oceans of the Earth has increased. Since a variety of assumptions is involved, these estimates yield different dates. Although each of the various methods contained its own inaccuracy including biases in the mind of the estimator, nevertheless geologists were able to provide a reasonably satisfactory scale for the timing of Earth events. But more modern methods of dating have helped to correct the earlier estimates. Pleistocene deposits have a maximum depth of 6,000 feet, and took three million years to accumulate. But the Jurassic beds reach a maximum depth of 44,000 feet and required 45 million years for their deposition. This is a rate of about 1,000 feet for each one million years of time and occurs at only one-half the rate that it did during the Pleistocene. In the ancient Cambrian era the bedding depth of 44,000 feet required 100 million years to be laid down. This is less than half the rate of deposition seen in the Jurassic. Therefore, it is clear that uniform rates of deposition cannot be assumed.

Fluorine Dating. Another type of relative dating method is that involving the fluorine test. It is not particularly useful in terms of major geological problems of time, but it has been most helpful in determining whether the remains of fossil men belonged to the beds they were found in or were placed there in later times. It depends upon the principle that the fluorine present in the ground water percolating through the bones will be deposited at a rate which directly corresponds to the length of time involved. But the method has its disadvantages, for the amount of fluorine not only differs greatly from one river system to another, but can be affected by very local features which affect the ground water systems. The method is not a new one but it has recently been useful in solving certain problems in dating human fossil remains. All three fossil finds in question come from England and are known as Piltdown man, the Galley Hill skeleton, and Swanscombe man. They illustrate, respectively, a notorious hoax, an intrusive burial, and an authentically ancient find. The Piltdown remains long troubled students of human evolution, for they consisted of the major part of the thick-boned vault of a large-brained skull, which was primarily modern in its appearance, and an accompanying primitive lower jaw. They were found in beds of the Kent Plateau by Charles Dawson in gravel variously estimated as Late Pliocene to Early Pleistocene in date. Accompanying them were some very primitive-looking stone tools heavily patinated through surface alterations, and a variety of mammal bones of appropriate age. A huge fragment of elephant leg bone had been shaped into a very primitive pointed tool. If Dawson's dates were correct, it represented the earliest human find at the time. The interpretation proved difficult, for no one really expected to find a modern-looking brain case combined in the same individual with a jaw which was positively apelike in shape. The matter was finally solved when both the skull and the mandible were subjected to a fluorine test. The human bones contained only as much as one-fifth of the fluorine found in the animal bones from the same deposit. This was conclusive proof that the human bones did not belong in the early strata. Later improved tests by Weiner (1955) demonstrated that the mandible was fresh bone, while the skull bone contained about the amount of fluorine generally found in Late Pleistocene deposits. Hence, the association between the jaw and the skullcap was demonstrated as false. Further examination showed that the jaw was indeed that of a young adult orangutan which had been stained chemically to give an appropriately ancient appearance. The separate canine tooth associated with the find had been filed down to alter its apelike appearance and also stained to give it the color of antiquity. There was no doubt that an elaborate hoax had been carefully prepared and foisted off on the great anatomists of England, who had disagreed about the details of reconstruction but all of whom had accepted its antiquity. To this day, the joker has not been positively identified.

Galley Hill man was accidentally discovered in 1888 by workmen removing gravel from the 100-foot terrace of the Thames River not far from the city of London. The finds consisted of a virtually complete skeleton, and the terrace was known to be Middle Pleistocene, dating from the second interglacial period, then thought to be as old as 250,000 to 300,000 years. The finds came to the attention of Sir Arthur Keith, perhaps Britain's most distinguished anatomist, and were hailed as evidence of completely modern kinds of men living in those ancient times. Indeed, the fossil provided Keith with much of his evidence for the early evolution of *Homo sapiens*. The fluorine content of these bones, tested much later, ranged from only 0.2 to 0.4 percent, whereas in animal bones from the same terrace the much higher values of 1.7 to 2.8 percent were realized. These findings indicate that the Galley Hill bones were buried intrusively in the top level of the 100-foot terrace.

Only 500 yards away in a gravel pit in the same terrace were found the Swanscombe remains, consisting of skull fragments which allowed the restoration of the mid and posterior portions of the cranium. Again the bones were modern in form. If they could be demonstrated to have been laid down with the terrace, then an important second interglacial human fossil would be confirmed.

Oakley and Ashley-Montagu (1949) conducted the fluorine test upon these skull bones and found that they contained 1.9 to 2.0 percent and so belonged in this second interglacial terrace. The bones were determined to be those of a female, and the large capacity, which has been variously estimated at some 1250 to 1300 cubic centimeters, provides a surprisingly modern type for such ancient deposits.

Amino Acid Dating.   Relative dating upon changes in amino acids is another new technique (Bada et al., 1974) which utilizes the organic substances remaining in fossil bones. Through time L-amino acids are slowly converted to D-amino acids, thus the ratio of D- to L-amino acids in a fossil steadily increase with time. The rate of conversion is subject to certain environmental factors such as temperature, so the D/L ratios give only relative dates until calibrated against some absolute dates, as provided by radioactive carbon. Various amino acids change their D/L ratios at different rates: aspartic acid is one of the most useful with dating range from 5,000 to 100,000 years. Its D/L ratio checked against C^{14} gave an estimated date of about 50,000 years for a skull from Del Mar in southern California. This would place the entry of people into the Americas about twice as early as previously estimated.

Absolute Dating Techniques

Absolute dating techniques are based upon the well-established fact that a number of radioactive elements disintegrate, that is, lose electrons, at a constant rate throughout time. Thus they become very accurate geological clocks. From the number of possible radioactive time clocks three elements have proved the most useful: uranium, potassium, and carbon. A good summary of these methods is provided by W. W. Bishop (1973).

Uranium Dating.   Uranium occurs in the form of a number of *isotopes,* that is, forms having nuclei with the same number of protons, but differing in neutron count. Thus, three uranium isotypes are known as U^{235}, U^{238}, and U^{239}. The radioactive isotopes of uranium lose their electrons very slowly so that one-half of their weight is lost in a period of 4,560,000,000 years. As radioactive uranium disintegrates, it passes through a series of temporary stages and finally is reduced to elemental lead, which itself occurs in the form of several isotopes. The rate of radioactive decay of uranium is so slow that it serves as the best absolute dating method for very ancient events, including the older periods in Table 4–1. It is by this method that the age of the planet Earth has been established as between 4.5 and 4.9 billion years.

Potassium-argon Dating.   This method of dating is based upon the fact that potassium is a mixture of isotopes, K^{39}, K^{40}, and K^{41}. Potassium-40 (K^{40}), which occurs as about 0.01 percent of the mix, is radioactive and loses electrons so as to produce as one of its end products the gas argon-40 (A^{40}). The half-life of potassium-40 is 1,300,000,000 years. Thus it provides a radioactive clock which ranges from very old events to those as recent as a few hundred thousand years ago. Those rocks which provide the best samples have been heated to very high temperatures, so that any A^{40} which has previously accumulated is driven off. Consequently, any A^{40} atoms which subsequently build up within the rock after that heating period provide a measure for dating it. With carefully chosen samples of igneous rock, accurate dates are obtainable. It should be understood, of course, that any radioactive method of dating involves not only slight errors within the instrument, but errors in counting the rate of remaining radioactivity. For this reason the dates are always given followed by a plus or minus estimate of the error.

Fission Track.   The fission track method of absolute dating is new and promising. It depends upon the detection of trails of intense damage in minerals and natural glass resulting from the fission of a uranium nucleus. U^{238} decays at a slow and steady rate, so that if the fraction of uranium atoms that have fissioned in a given sample can be determined, the age of the rock can be computed. A variety of chemicals are used to etch the minute tracks and improve their visibility under the microscope. Suitable materials include uranium-rich accessory minerals derived from obsidian, pumice, and other heated rocks. The method is now applicable to the past 5 million years of time, and its time depth may be increased. The method had already provided a broad check on the K-A dates for Bed I at Olduvai Gorge.

Carbon Dating.   One of the most useful radioactive time clocks, carbon-14 (C^{14}) was developed by Willard Libby, for whom it won the Nobel Prize in chemistry. It is based upon the fact that all living things contain a fair amount of elemental carbon which occurs in the form of two isotopes, C^{12} and the relatively rare C^{14}. The former is stable, the latter radioactive, with a half-life of 5,720 years. This allows absolute dates within the last 50,000 years, and by the method of enrichment an extension backwards to perhaps 70,000 years. Carbon-14 is a radioactive isotope formed high in the atmosphere through the bombardment of nitrogen atoms by cosmic rays. The circulation of our atmosphere is such that radioactive C^{14} becomes uniformly spread throughout the lower levels of the air everywhere on our planet. It is absorbed by plants along with stable C^{12} during their normal metabolic processes. Animals get both types of carbon

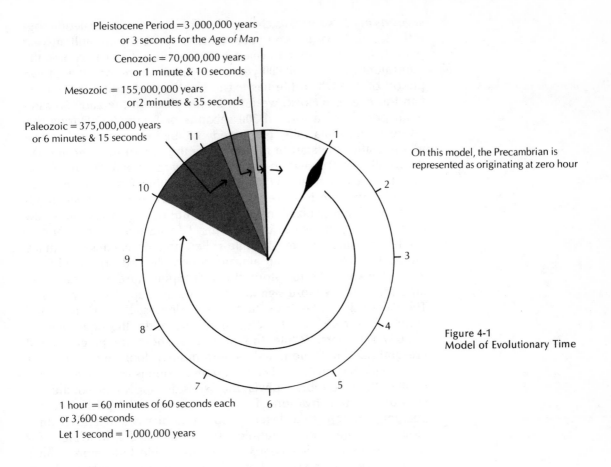

Pleistocene Period = 3,000,000 years
or 3 seconds for the *Age of Man*

Cenozoic = 70,000,000 years
or 1 minute & 10 seconds

Mesozoic = 155,000,000 years
or 2 minutes & 35 seconds

Paleozoic = 375,000,000 years
or 6 minutes & 15 seconds

On this model, the Precambrian is
represented as originating at zero hour

Figure 4-1
Model of Evolutionary Time

1 hour = 60 minutes of 60 seconds each
or 3,600 seconds
Let 1 second = 1,000,000 years

atoms from the plants they eat and the air they breathe, and both remain in equilibrium in their bodies until death. At that time the carbon-14 begins its radioactive disintegration, thus starting the dating clock. The amount of residual C^{14} in an organic sample declines with age in a logarithmic fashion. Again, samples must be chosen with care, for contamination of them can arise from a variety of sources. The best samples consist of charcoal, but with the proper techniques bones and shells also provide good estimates of age. Collagen, the organic constituent in bone, is particularly useful in dating these materials.

An Evolutionary Clock

Many natural processes show a quickening pace with the passage of time, and organic evolution is one of them. In Figure 4-1 the dial of a clock is used to illustrate the point. Let us mentally watch the minute hand proceed around the clock face for one hour. It has gone through 60 minutes, each consisting of 60 seconds, or a total of 3,600 seconds in all. Now let us change the clock to serve as a paleontological record. If we let one second stand for each one-million-year period, then the hour traveled by the minute hand becomes the equivalent of 3,600,000,000 years, or slightly more than the known period during which life has existed on Earth. Starting the minute hand at twelve o'clock, we let it sweep through the early Precambrian period for a total of 300 seconds or 5 minutes. During this interval unicellular life originated 3.3 billion years ago or five minutes after the start of the clock. By the end of the Precambrian the fossil record shows no more than a few soft-bodied multicellular invertebrates of relatively simple organization. Obviously most of evolutionary time has been consumed in achieving relatively modest-appearing organizational advances. But perhaps we should view evolutionary progress in other terms. Life started at the level of a single molecule, and most single-celled organisms consist of many molecules, arranged in an amazingly complicated organized fashion. If we knew the full story, this vast expenditure of Precambrian time might show more significant results than appear at first sight. The Paleozoic era begins with only ten minutes left in the hour and continues for six minutes and fifteen seconds, the equivalent of 375,000,000 years. While the period is short compared to the Precambrian, it is nonetheless enormously long and during its interval life has diversified considerably. It opens with the existence of all of the invertebrate phyla, then sees the appearance of the first vertebrates, the invasion of land by plants, followed by the appearance of terrestrial insects and amphibians, and ends with a radiation of the primitive reptiles. The Mesozoic era, which lasts for two minutes and 35 seconds, brings us within little more than a minute from the end of the hour. During this long 155,000,000 years, evolution passes through the age of reptilian dominance and goes on to the catastrophic end of the dinosaurs. Midpoint the first mammals and birds appear and in final phases the first flowering plants appear in the fossil record and radiate rapidly. Finally, the Cenozoic, the era of recent life, consumes the last one minute and ten seconds of the clock hour. During the Cenozoic's 70,000,000 years, life in the form of the dominant mammal passes through a brief archaic period, and then rapidly evolves, diversifying to occupy a wide variety of niches. The time of the advent of man is in doubt today, but if we accept his first appearance as no earlier than the beginning of the Pleistocene, then he has been on Earth no more than 3,000,000 years, or a mere terminal three seconds of our evolutionary hour. We are, indeed, latecomers.

The quickening pace of organic evolution is so important a principle that it is worth diagramming in another fashion, as in Figure 4–2. Against the 4,000,000,000-year horizontal time scale is plotted an evolutionary scale of achievement ranging from one to seven. The intervals in this scale are not numerically equal, would be very hard to assign values to, and must be considered as purely arbitrary. Nevertheless, it will not be far from the truth and certainly will serve to reveal the quickening tempo of the evolutionary process. If we start with unicellular life as a scale of one and shown at its earliest known date of 3,300,000,000 years ago in the fossil record, then our trend line proceeds very slowly upward to the value of two on our scale, which represents the beginning of multicellular life. This has been conservatively plotted and begins about 800,000,000 years ago, which is earlier than is shown by the fossils themselves. Small, soft multicellular animals first appear in the fossil record in Precambrian times, but they probably do not represent the beginning of this level of organization. Fish as the first vertebrates are shown with the scale value of three appearing about half a billion years ago. This marked a rather sharp upswing in our trend line, and this increases as succeedingly advanced levels of organization are achieved in the amphibians, reptiles, the earliest of mammals, and finally man. While the exact position in time and the intervals in our arbitrary scale do not allow exact identification of the form of our trend line, it clearly is an increasing curve, involving acceleration, and exponential in shape.

The quickening tempo of change is real and evident in organic

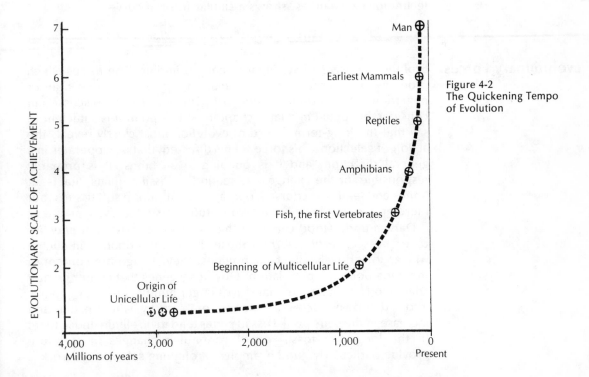

Figure 4-2
The Quickening Tempo
of Evolution

evolution but its causes are not altogether clear. Perhaps as creatures become more complex they lend themselves to a wider variety of modification and hence the rate of evolution can accelerate. If we turn to the code of life, it will be recalled that some types of viruses are constructed by DNA codes with a small finite number of instructions. The complexity of the code must increase as organisms evolve into multicellular forms, larger in size, and developing more specialized attributes. In large complicated animals like ourselves the DNA code may contain up to a million genes, or message units, as a figure of magnitude. But it is reasonable to consider that the complexity of the code of life increases in much the same fashion as the trend line shown in Figure 4–2.

Or we may view the accelerating changes in living forms in terms of a figure involving an alphabet. If the alphabet contains only two letters, A and B, these taken two at a time and disregarding the ordering produce only three combinations, AA, AB, and BB. If the alphabet is increased to three letters, the varied combinations again taken two at a time rise to six, which is a doubling of variant three-letter units. With each new letter added, the total number of paired combinations increases exponentially. This alphabetical model may well be analogous to the alphabet of DNA upon which life is based and so explain its quickening tempo. In later sections it will be shown that the evolution of human culture, as shown by its technological changes, shows a similar form of curve.

Evolutionary Forces

Evolution consists of systematic changes in time. The forces which produce these are *selection, mutational pressure, gene flow* or migration, and *random genetic drift.* These will all be discussed in considerable detail in a later chapter. At this point it is sufficient to say that the long-term record of evolution most clearly reveals the action of selection. This force seems directed, that is, it operates in a general direction, and it is ongoing at all times. It is primarily responsible for the changes we see in the fossil remains, that is, in bones and teeth. Of course it operates equally upon soft tissues, but these are not available for macroevolutionists to study.

Darwin understood very well that selection consisted of *effective differences in fertility.* When these differentials operate in such a way as to variously affect phenotypes, they change the content of the gene pool of a population. Obviously genes that confer greater adaptation so that an animal leaves larger numbers of young to live into their own successful period of maturity will prosper and increase in the gene pool. Less fit genes tend to be eliminated in time. In the long-term fossil record continuing changes in structure provide particularly good examples of ongoing selection at work.

Mutations are abrupt alterations in the genetic code, either in the form of point mutations, which involves a single base pair, or as rearrangements of long sections of the DNA molecule. Sizable mutations are apt to disorganize the individual organism and in most environments prove disabling. Mutations which produce smaller changes in the organization of DNA may under some conditions prove superior to the old configuration and thus confer added fitness. The concept of changing environments is involved here. In a stable one mutations probably are not beneficial to individuals. On the other hand, species tend to benefit from a continuing stream of mutational changes since this provides a potential variability to offset that diminished by selection. Populations will lose their potential for readjusting to changing conditions if species variability is not maintained by mutational pressure. Both in short-term and long-term evolution environments are constantly shifting so that a species which becomes impoverished in terms of its basic variation faces the danger of extinction. Therefore in the long run mutations serve as a kind of insurance for the species to meet shifts in future environments.

Gene flow involves genetic exchange between populations. In long-term evolution it serves two functions. First, it maintains a degree of genetic similarity between the populations in which the exchange is ongoing, therefore it produces species homogeneity. This does not mean that little genetic variation occurs within the species, but merely that it is not so great as to produce lowered fertility between subunits of the species. The second aspect of gene flow involves the formation of new species. This occurs where genetic exchanges are constricted by some type of isolation. This is diagrammed in Figure 4–3. There it will be seen that as an initial population straddles a mountain range, the amount of gene flow between the extremes of the population is restricted. As time progresses the throttling of gene flows increase and with further passage of time new genetic variants are added to each population. Given sufficient time the original species may evolve into two different ones, with reduced and ultimately extinguished fertility between them. This kind of process can be observed among living organisms, and so microevolutionists find the idea acceptable. Long-term evolutionists formerly argued that while such a reduction in gene flow might produce new species, it could hardly account for the higher taxons such as genera, families, or orders. All that is required to produce these higher levels of difference is greater periods of time. It is now generally conceded that the reduction of gene flow is the initial stage in taxonomic differentiation.

The fourth of the evolutionary forces goes by the cumbersome name of *random genetic drift.* In essence it involves what are called accidents of sampling as these occur in populations that are numeri-

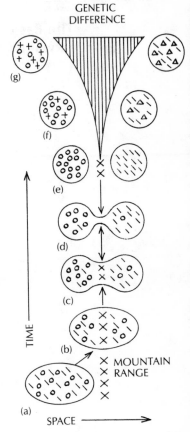

Figure 4-3
A Model of the Process
of Speciation

cally small. The effects are to produce deviations in the frequency of genes in the population pool. These are unpredictable in direction and in the structures effected. There is no doubt that drift does play a role in both short-term and long-term evolution, but its effects are difficult to see, and so its existence has frequently been denied by paleontologists. However, if one views these sampling accidents as operating in conjunction with selection and possibly starting a new trend in selective change, drift no longer appears without consequence. This combination of drift and selection operating upon chance changes in genetic makeup of populations is called *genetic coadaptation,* and in this form it certainly is important in the fossil record. It not only provides more frequent opportunities for the evolution to quicken, but also provides a greater variety of directions in which selection may operate.

The Basic Energy Pyramid

Before proceeding to examine the details of the fossil record, it is important to understand one of the limiting factors which places a rigorous ceiling upon the numbers of different kinds of animals that live within a given environment. The restraining factor involves the energy relationships between various types of animals and the total energy available to them. Solar radiation supplies all of the energy available on the surface of the Earth. Incoming sunlight in part becomes trapped within the Earth's atmosphere as if it were a greenhouse. The portion that strikes the surface of the Earth represents the energy income from which all life activity on the planet must be budgeted. The angle at which sunlight strikes the surface of the Earth is important. Sunlight strikes the Earth's surface in a generally perpendicular fashion at the equator. In the polar regions it comes in on a long slanting angle and thus much less solar energy strikes a given equivalent land area in the high latitudes. This basic geometry explains why the total mass of life is great in the equatorial regions and necessarily scant and impoverished at the poles. Variations in the amount of solar energy striking the earth result from the tilting of the planet's axis seasonally as it orbits about the sun. When the North Pole is inclined toward the sun, more energy reaches the Northern Hemisphere, and we enjoy the season called summer. When the same pole is tilted away from the sun, less energy reaches the Northern Hemisphere, its climate cools, and we have the season known as winter.

Green plants are the primary converters of the energy of sunlight into forms available to animals. Their efficiency in this conversion is very low, and in most terrestrial environments plants convert less than three percent of the solar energy into their own tissues. Efficiency may drop even further in aquatic environments. Careful studies of a lake in southern Wisconsin show that 50 percent of the incident solar energy was lost or reflected, another 25 percent was utilized in vaporizing water from the lake's surface, another remaining one quarter was primarily spent in altering the temperatures of the lake's waters. Only 0.8 percent of the solar energy received was directly utilized by the green plants in the lake.

There are some environments which receive no solar radiation and there are no green plants in them. Such areas can only be inhabited by animals, a few nongreen plants such as fungi, and some very simple single-celled forms of life such as bacteria. They depend for their life energy upon organic materials that are somehow brought in from elsewhere where solar energy has been received. Caves and their darkness have an interesting but very sparse population of animals and of course are of no great importance in evolution. The lower layers of soil represent another environment without light, but the animals living there usually receive energy sources that sift down from above. The abysmal depths of the oceans are again a special environment. In very clear waters sunlight may penetrate as deep as 2,000 feet, but generally it contributes little significant energy below 600 feet in depth, so there are few or no green plants living lower in the oceans. Its depths do contain a grotesque variety of marine animals, but they are all dependent in one form or another on energy resources which sink down from the more active upper life zones of the oceans.

Ecology is the relationship between a species of organism and its total environment, organic and inorganic. Even for simple animals and plants the relationships are always complex. In practice ecologists try to study simpler situations, such as those that exist on an Arctic tundra, a small pond, or perhaps a saltwater swamp. Even there the array of interrelationships between the various organisms are overwhelmingly complex. In a mature and complex community, such as a tropical rain forest, a temperate broadleaf forest, or a rapidly flowing river, relationships become almost unmanageable. When studying an environment, ecologists consider the *biomass*. The biomass is the living weight of the total community including all stored food. It is usually expressed in terms of weight per unit area. An even more narrowly defined concept is that of *annual productivity*. This may be expressed as total productivity, above ground productivity, or below ground productivity. When these attributes can be measured, they allow a detailed study of energy flow in the natural community.

All energy transformations are inefficient, whether they occur in the machines made by men or in organisms in natural communities. The simplified energy pyramid shown in Figure 4–4 is sometimes called the *trophic pyramid*. Not represented at all is the incoming solar energy which strikes the Earth's surface. Less than one percent

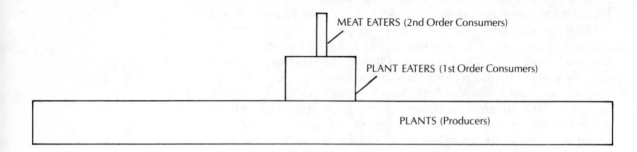

MEAT EATERS (2nd Order Consumers)

PLANT EATERS (1st Order Consumers)

PLANTS (Producers)

Figure 4-4
Simplified Energy Pyramid

of this is converted into the level of plants or producers. Obviously the plant eaters, or first order consumers, can only have a very small biomass relative to the total tied up in plants, for they must live on the expendable surplus out of the plant's annual net productivity. The plant eaters themselves support a population higher in the tropic pyramid, that of meat eaters, or second order consumers. Again, the carnivore must be relatively small in biomass compared to the herbivores which they consume. In a very general way it is considered that in a well-balanced system the biomass of one trophic level can be only a small fraction of the level upon which it subsists. While the total numbers of animals must be limited by the available energy, more specifically it shows why meat-eating animals must always be much rarer than those upon which they feed. Finally, all the energy in the three trophic levels indicated in Figure 4–4 is untimately returned to the Earth by the decomposers, little animals and plants which convert the dead bodies into simpler organic materials and return them to the soil.

In nature, energy flow is not quite as simple as the model indicates. For example, insects and/or other cold-blooded animals including the reptiles convert about 20 percent of the energy in the food eaten to tissue growth and/or muscular activity in their own bodies. Among the warm-blooded animals, the mammals and the birds, only about 3 percent of the ingested food is used in these direct ways. The rest of course, is not wasted, but it is expended in maintaining the constant temperature of the animals involved. The evolution of warm-bloodedness opened a wide variety of territorial niches for birds and mammals, since it allowed them to penetrate the moderate, cold, Arctic, and Antarctic zones.

Bibliography

BADA, J. L., R. A. SCHROEDER, and
G. F. CARTER.
    1974 New Evidence for the Antiquity
         of Man in North America De-
         duced from Aspartic Acid Race-
         mization. Science. Vol. 184, pp.
         791–793.

BISHOP, W. W.
    1973 The Tempo of Human Evolu-
         tion. Nature. Vol. 244, B. A.
         Supplement, August 17, pp.
         405–408.

OAKLEY, K. P.
    1964 Frameworks for Dating Fossil
         Man. Chicago: Aldine Publish-
         ing Co.

OAKLEY, K. P. and M. F. ASHLEY-
MONTAGUE.
    1949 The Galley Hill Skeleton. Bulle-
         tin of the British Museum of
         Natural History, No. 1. London.

WEINER, J. S.
    1955 The Piltdown Forgery. London:
         Oxford University Press.

CHAPTER 5

THE
EVOLUTION
OF THE
EARLY
VERTEBRATES

Multicellular life had perhaps been established two hundred million years before the first vertebrates appeared in the Ordovician. While their origins are not totally clear, the fish soon achieved dominance in the waters of the planet, producing a wide variety of basic types. One of these proved particularly well preadapted to developing a capability for life on land and so the amphibians arose. They developed greater fitness toward purely terrestrial living. The great radiation of the reptiles resulted, filling the Mesozoic era with a great variety of successful types. Living with them, and relatively inconspicuous, were the early mammals and birds.

How Fossils Are Formed

The fossil record is made up of a wide variety of relics of past life. Animal fossils usually consist of hard parts such as teeth and bones which have been preserved through chemical infiltration, with little or no alteration in form. Original organic materials in them are replaced by dissolved minerals in percolating ground waters. In a very general way the degree of replacement increases with antiquity. Silica, lime, and iron compounds are the common replacing minerals, but under some special circumstances bones may become agatized or even opalized.

Sometimes the disintegrating animal serves as a mold into which minerals percolate to preserve the form perfectly. Such replacement fossils may involve the leaves or branches of trees, the bones of animals, and even soft-bodied animals. In the Precambrian Ediacara beds of Australia a variety of multicellular invertebrates have left their fossilized impressions in the sandstone. In western Canada there are sandstone casts of dinosaur skins which show the arrangement of scales. In south Germany in the Jurassic, fine-grained lithographic limestone was laid down in ancient lagoons. From these beds come two skeletons of the earliest birds (*Archaeopteryx*). The circumstances of preservation were so favorable that the impressions of the flight feathers of the wing and the feathers of the tail were left perfectly duplicated in the fine-grained stone (Plate 5–1). In another instance the fossil of a flying reptile of the same age (*Rhamphorhynchus*) was recovered along with the impression of its wing membrane and the tail with a racquet-like enlargement at its end to facilitate steering in flight. Under unusual circumstances fossils may even present evidence of methods of reproduction. The American Museum of Natural History in New York has a remarkable slab which contains the complete skeleton of a female marine reptile (*Ichthyosaurus*) in whose body cavity are preserved the skeletons of seven of her unborn young. The fossil demonstrates

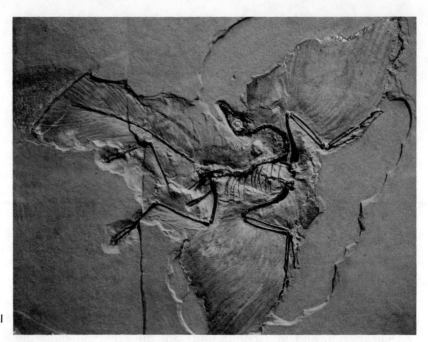

Plate 5-1
The Fossil Archaeopteryx, showing impressions of all this ancestral bird's feathers.

that these active and dominant marine reptiles gave birth to live young, not in the mammalian sense of today's porpoises, but through the retention of the eggs, a process known as ovoviviparity.

Fossil evidence turns up in a variety of other forms. Both in Siberia and Alaska preserved mammoths have been found in ground which has remained frozen since the last glacial advance. Many varieties of insects are preserved within amber, which is the hardened resin of ancient coniferous trees. Footprints of ancient animals, including dinosaurs, may harden to form a mold which is later filled with fresh mud. The casts of such tracks tell a good deal about the locomotion and customary gait of long extinct forms (Plate 5-2). The red Triassic sandstone of the Connecticut Valley is well known for the variety of dinosaur tracks which have been found in its beds. Fossil excreta, known as *coprolites*, may give a clue to the diet of ancient animals. Smooth and polished pebbles have been found inside the rib cages of dinosaurs where, as gastroliths, they aided in their digestion, just as gravel in the gizzard of a chicken helps it to crush grain. The scale of fossils varies from submicroscopic unicellular forms which must be magnified many tens of thousands of times in order to be seen, up to individual bones which may be six or more feet in length as in some of the giant terrestrial animals. Both the small and large fossils tell their story and provide evidence for the progress of the evolution of life from a point near its inception down to the present time.

Life in the Paleozoic Seas

With the beginning of the Paleozoic era, the fossil evidence of life becomes more abundant. In the Cambrian sea there existed primitive members of most of the phyla represented in later times. Animals were represented by jellyfish, sponges, various exoskeletal forms such as jointed worms, sea urchins, crustaceans, ancestral arachnids, and trilobites, highly successful Paleozoic creatures which became extinct in later times (Plate 5–3). The latter look like but are not closely related to the little sow bugs or pill bugs familiar to everyone as decomposers of decaying vegetable material. Ancestral sea urchins were also present. The Cambrian was a very long period, lasting about 100 million years, but no vertebrate fossils are known from it.

In the succeeding Ordovician the shell-bearing ancestors of the modern squids make their appearance and were among the most impressive carnivores on the scene. The first fossils of vertebrates also date from the Ordovician but they represent inconspicuous, small armored fishes living in freshwater environments. They were jawless, flat armored, bottom feeders known as the *Agnatha,* a label which denotes their lack of jaws. Judged in the light of their times, they could not be estimated to have the great future which later events unfolded for them.

Plate 5-2
Tracks of a late Devonian amphibian from Victoria, Australia. The animal probably resembled *Ichthyostega.* It walked without dragging either its belly or its tail. The feet are padded, but at least four terminal digits can be detected.

Where Did the Vertebrates Come From?

The *vertebrates* represent a new basic form of organization and so are placed in a separate phylum. They are characterized by an internal skeleton (endoskeleton) usually consisting of bone but in some forms comprised of cartilage. In a few forms, the notochords, the stiffening rod consists of unsegmented cartilage, but in the true vertebrate the backbone consists of a series of bony vertebral elements which stiffen the body against such fore and aft forces as

Plate 5-3
Trilobites were bottom-living invertebrates, usually several inches long, and were the dominant animals in the Cambrian seas.

are encountered in moving through water, but still provide the flexibility needed for swimming and other motions. The bony backbone provided fish with a stiffened body which was yet flexible enough for swimming. In the terrestrial vertebrates, where the body weight must be borne upon the limbs, the vertebras have evolved a number of processes which serve as points of attachment for various back muscles. Some of these also enclose a spinal cord and protect it from damage.

The soft-bodied ancestors of the vertebrates were probably present in the Cambrian period but do not show in the fossil record. Evidence in the form of small scales and bones documents the presence of fish in the Ordovician, but this scanty evidence is followed by a gap in the record. Only in the Late Silurian and Early Devonian do more complete remains of fossils of bony fish become common. When pages are missing from the actual record of the evolution of life forms, it is customary and necessary to turn to the comparative anatomy of living animals to see what suggestions they may provide as the basis for a proper reconstruction of the lost events. No living forms are literally ancestral to any other, but some have been conservative, evolving very slowly, and so retaining in somewhat altered form the same types of structure present in much earlier periods. A little animal known technically as *Amphioxus*, and belonging to the class of notochords, seems reasonably to represent one of the missing basic vertebrate ancestors. It is a small, translucent creature living in shallow tropical seas where it spends much of its time buried in the sand with only the front end of its body out so as to conduct its filtered feeding on small organic particles. Only a few inches long, it is rare in most regions, but near the coastal port of Amoy in China it is common enough to be sold as food in the marketplaces.

Amphioxus (Figure 5–1a) interests us because it belongs to a class of animals known as the notochords. It shows a dorsal, tubular nerve cord lying above its cartilaginous notochord. Although the latter is a continuous structural member, both sides of the body are covered with segmented muscle masses. It breathes through gills which are of the vertebrate type, although instead of having only five or six pairs of openings, it may have several dozen. A part of the evidence linking this humble notochord with the higher vertebrates is that they, including man, have a period in fetal development in which a notochord is laid down, only to be resorbed later and modified into other structures. This embryological evidence is strongly suggestive of vertebrate derivation from some early noto-chord.

Vertebrate ancestry may be traced even further back by searching diligently among other living creatures. The tunicates, or sea squirts, are improbable-appearing as vertebrate ancestors. As adults (Figure 5–1b) they are relatively formless, with two external openings. One opening pumps water into the animal's body, where a filtering apparatus strains out food particles; the other orifice discharges it.

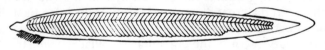

(a) *Amphioxus* (notochord)

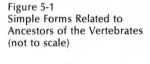

Figure 5-1
Simple Forms Related to Ancestors of the Vertebrates (not to scale)

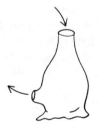

(b) Adult tunicate (sea-squirt)

(c) Larval tunicate (sea-squirt)

They are firmly attached to rock and so live a motionless life. They have no spinal cord, brain, or even much of a nervous system. But the immature or larval tunicate (Figure 5–1c) is a totally different-appearing animal. It is small, free-swimming, and organized about a bilateral plane of symmetry. Its tapered swimming tail shows segmented muscles, a notochord, and a nerve cord reminiscent of those of *Amphioxus.* This larval stage serves the primary function of allowing the individual tunicate to move around and find a proper place in the environment before settling down into its sedentary adult life. It is conceivable that such a larva, refusing to grow up, retained its tail and its active locomotion throughout life; it is further possible that the unfound and unknown ancestral vertebrates passed through such a stage of evolution and progressed through the retention of the activity of the larval form. Soft-bodied animals are so rare in the fossil record that it is unlikely that such a hypothesis will find direct confirmation in the future.

The First Vertebrates in the Fossil Record

The animals that we call fishes were the first vertebrates, and they included a wider variety of forms than we usually recognize among living types. All had backbones, lived in water from which they extracted oxygen by means of gills, and moved through the use of fins and undulating bodies rather than by limbs. Structurally these early fish covered a great diversity of forms and can be best divided into four categories. These are, in the order of their appearance in the evolutionary record: the Agnatha, the Placodermi, the Chondrichthyes, and finally the Osteichthyes.

The Agnatha, the first known vertebrates in the record, were jawless, with the mouth consisting only of a small hole or crosswise slit. It suited them only for filter feeding on the bottom of freshwater lakes and streams, for the physiological evidence derived from the kidneys of modern fish makes it clear that these earliest of vertebrates evolved in fresh water rather than in the oceans of the Earth. They were curious little animals, protected against invertebrate carnivores by scaly armor coats. Among them were the ostracoderms, or shelled-skin fish, one of which (*Hemicyclaspis*) is shown in Figure 5–2a. Their heads were low and tended to be circular in plant form. Typically they lacked paired fins, but may have had small spines at the side of the head, or a pair of flaps in the same region. In size they ranged from six to twelve inches in length. Their armored coating seems to have been evolved to help them escape from such aggressive carnivores as the freshwater eurypterids, or water scorpions. The Agnatha today are represented by two descendant forms, the hagfish and the lamprey eel. Both are rather degenerate structurally.

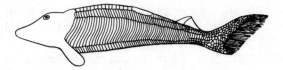

(a) *Hemicyclaspis*—a flat-bodied ostracoderm

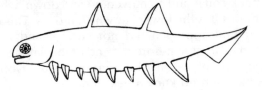

(b) *Climatus*—an acanthodian placoderm

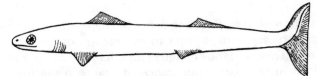

(c) *Cladoselache*—an early chondrichthye

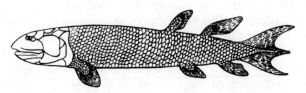

(d) *Eusthenopteron*— a lobe-finned crossopterygian

Figure 5-2
Early Forms of Fish
(not to scale)

Out of these early jawless fish evolved the placoderms, equipped with a functional jaw in a wide variety of body forms. They too evolved in fresh water and underwent considerable evolutionary change in that environment. They included many grotesque forms, quite unlike any modern fish. They were usually covered to a variable degree by armor scales. The processes of mutation and selection seem to have been producing among the placoderms a wide variety of experimental types, as reflected in overall bodily structure, and in the bizarre arrangement of their fins. These had a tendency toward paired fins, but included a variable number of extras as well. The small acanthodian placoderm (*Climatus*) was only three inches in length (Figure 5–2b). These were among the more orthodox-appearing placoderms but would be called unusual by modern standards. During the 50,000,000 years' duration of the

Devonian the little freshwater placoderms migrated to the ocean and there produced some fantastic giant forms. One of these, an *arthrodire* of the genus *Dinichthys,* reached the appalling length of 30 feet and apparently was the dominant marine animal of its day. One of the unusual structural features shown in this gigantic marine carnivore was the fact that its upper jaw, indeed its entire head, hinged upon its neck collar and so moved up and down, whereas in all modern vertebrates it is the lower jaw which moves. This unusual experimental structural approach did not survive and all of the placoderms went to extinction before the end of the Paleozoic. They represent a series of interesting but unsuccessful evolutionary experiments and have left no known descendants.

The Chondrichthyes

The evolution of jaws, enabling fishes to change from filtered diet to boldly seized carnivorous forms of feeding, produced an evolutionary spurt as a result of the change in food habits. One of these spurts is manifest in the early appearance of sharks in the fossil record. First evolving out of an unknown placoderm ancestor in fresh water, they soon descended to the ocean and became completely marine in habit. The oceanic environment at first was rather barren of vertebrate life and so the early sharks underwent no great radiating evolution. One of the best known, *Cladoselache* (Figure 5–2c), is a well-streamlined free-swimming form that foreshadowed the shape of later sharks. While an active carnivore, nonetheless *Cladoselache* likely provided much of the food of his gigantic contemporary, the arthrodire, *Dinichthys.*

Modern sharks have a stiffening structure, including their skulls, made of cartilage rather than bone. But there is reason to believe they originally did contain a bony endoskeleton which later, and which for unexplained reasons, gradually became converted into cartilage.

The structure of *Cladoselache* reveals a number of advances in organization. Steering is now achieved by two sets of paired fins. Longitudinal stability is improved by two dorsal fins. And the structural portion of the tail is upturned to provide the basis for a broad propelling tail or *caudal fin.* The origin of their jaws is one of the most interesting of the structural improvements, for it provides a classic example of the transformation of one structure into another. All primitive vertebrates had paired skeletal bars lying on either side of the throat and serving as structures to separate the gill openings. The gill bars consist of upper and lower portions. The structure of some of the early Devonian sharks, such as *Cladoselache,* indicates that their jaws are *homologous,* that is, structurally like the primitive gill bars of earlier fish and have evolved from them to serve their new function. The primitive jaws of sharks were provided with teeth which consisted of denticles, or toothlike scales, arising in the skin along the jaw margins. Such dermal teeth still characterize modern sharks and serve their bloody functions well.

The sharklike fish are worth a momentary digression from our main theme of tracing the evolution of the ascent of man. By the end of the Paleozoic era the vast majority of sharklike fish had left their ancestral freshwater homes for the ocean, there shedding their armor and gaining mobility. The first representatives of the modern sharks appeared in the Carboniferous and for several geological periods were none too common because of the scarcity of suitable food fish. Later, when the vast migration of advanced bony fish entered the seas, these became the main food supply for the sharks. From Late Mesozoic times to the present, the sharks have been a conspicuous, although apparently not very large, element in the marine world. They are aggressive animals and will strike at almost any moving object. Their olfactory senses are well developed and it has been claimed that sharks are but a machine driving a nose through water in search of food. Actually, they have a brain relatively larger than those of other fish. The basic pattern of reproduction among sharks involves large eggs, protected by a horny shell, so that fertilization must take place internally. A few sharks evolved the capability of retaining their eggs internally and giving live birth.

The mackerel sharks are most active, for they feed upon very swift fish. Sharks range in size from a few feet to a recorded 36 feet. Even though the modern white shark is capable of taking seals and men in a single gulp, there lived in the Tertiary a related species, approximately 65 feet long, with a gaping six-foot jaw opening equipped with enormous teeth.

The Evolution of Bony Fish

The fourth and most successful class of fish is the Osteichthyes, or the bony fish. The last term is something of a misnomer, for all of the earliest vertebrates were bony, but it does serve to distinguish the modern bony fish from the sharklike fish and the degenerate lampreys. In the Osteichthyes the body became completely enclosed in bony scales, while the head and shoulders were protected by stout bony plates. These plates of bone were derived from the skin embryologically, and in the course of evolutionary time they evolved into the dermal bone of the mammals and man. Their evolution provides a good example for *Williston's Law,* or as it more aptly could be called, rule, since it is based upon descriptive generalization. The rule states that as evolutionary time progresses, the number of elements or parts of a given structure tend to be reduced. If we apply Williston's rule to the dermal head plates of the early bony fish, where they numbered approximately 150 separate bones, we find that they have been reduced to 28 in man. This generalization tends to apply rather broadly to evolutionary processes and so is a kind of useful reminder of the trend involved.

The early bony fish were lungfish and possessed a pair of primitive lungs. These organs were apparently adaptive since the Devonian climate was subject to violent alternation of seasons and so freshwater fish frequently found themselves in a stagnant pool and suffering an oxygen shortage. Obviously the evolution of a membranous lung sac connected with the throat which enabled the fish to come to the surface of the water, gulp down air, and so restore its oxygen supply, would have considerable selective advantage. The three living lungfish are limited to the upper Nile basin in Africa, the swampy regions of the Gran Chaco of South America, and a single river system on the eastern coast of Australia. All are found in environments where lungs serve this useful function (Plate 5–4).

The primitive paired lungs of the early bony fish evolved through transformation into two very different types of organs. First, through enormous elaboration of their inner structure they provided proper oxygen exchange mechanisms for the early amphibians and all later land vertebrates descended from the primitive Osteichthyes. Here the arrangement was relatively unchanged but the interior compartmented cells were greatly enriched. Second, among the bony fishes the primitive paired lungs were gradually changed into a single sac which in time became the air bladder of the modern bony fish. The air bladder serves primarily as a hydrostatic organ whose function is to bring about changes in the specific gravity of the body of the fish.

Plate 5-4
This Australian Lungfish, a living fossil, retains some of the structural features of its ancient ancestors from the Middle Devonian.

This is very different but highly adaptive in the way of life of the modern fish. Interestingly enough, a few fish, such as the so-called walking catfish of Southeast Asia, combine both functions in their swim bladder. These animals are able to slither around for considerable distances on dry land, and even do some of their feeding there. The swim bladder of this catfish has its anterior portion subdivided so as to function as adequate but rather inefficient lungs for these terrestrial excursions.

From Devonian times on, the bony fish were divided into two groups, the *ray-finned fish,* which have evolved into the modern dominant types, and the *fleshy-finned fish,* which lost their importance, except for the fact that they gave rise to the amphibians and all subsequent terrestrial vertebrates. The ray-finned fish are among the most successful of vertebrates, and were we interested in the total story of evolution, their radiation would have an important part in it. The early lobe-finned fish form a natural group in which the paired fins have common fleshy bases instead of spreading out in flexible, horny rays. Their lungs, central nervous systems, and circulatory systems were reminiscent of those found in living amphibians such as frogs and the early amphibians from the fossil beds. Among such fish the *crossopterygians* generally had a swept-up type of tail common in many early fish. In the structure of teeth as well as in their skeleton they compare closely with the most primitive of the amphibians which evolved from them. In spite of avoiding the trap of a specialized diet, these early freshwater fish passed on to extinction in Mesozoic times. The modern living lungfish are, of course, survivors from this general group. An ancient form is shown in Figure 5–2d.

Among the crossopterygians was an evolutionary side branch, the *coelacanths,* that migrated from fresh water into the oceans, where they evolved new habits and internal structures, including the conversion of primitive lungs into swim bladders. Coelacanths occur in the fossil record from the Late Paleozoic through the whole of the Mesozoic. Since none occurred in the last 70,000,000 years of the fossil record, there was every reason to believe they had become extinct. In one of the great surprises of this century, in 1939 a commercial fisherman, deep trawling with a net off the coast of South Africa, brought up a five-foot fish, deep-bodied and covered with large bluish scales. Recognizing its uniqueness, he brought it ashore, and had it mounted by a taxidermist. Fortunately, a zoologist saw the specimen, recognized it for what it was, a true living coelacanth, which he called *Latimeria* (Figure 5–3a). Delayed by the outbreak of the Second World War, Professor J. L. B. Smith finally was able to institute a hunt, and he was rewarded by the discovery that living fossils occurred in some numbers in the deep waters of the north end of the great island of Madagascar on the east coast of

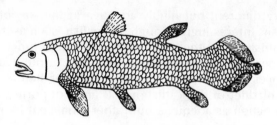

(a) *Latimeria*—a living coelacanth

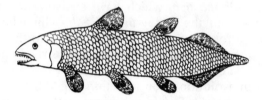

Figure 5-3
Types of Coelacanths
(not to scale)

(b) *Diplurus*—a 200,000,000 year old coelacanth

Africa. Since then more than 30 specimens have been recovered, and they are being intensively studied for the knowledge that their soft parts and biochemistry can give us of Mesozoic fish of their kind. *Diplurus,* a coelacanth that lived in the Triassic period some 200 million years ago, is shown in Figure 5–3b. The two fish are very similar in form in terms of basic structure. The living fish shows the fleshy stems of all the fins more fully developed than in the ancient one, and the fleshy portion of its tail has expanded for more efficient propulsion. But all in all, the Triassic fish looks a little more like a modern one than does *Latimeria.*

Toward Life on Land

Earth's age has been estimated at 4.5 billion years and before one billion years had passed, the continents and oceans of its surfaces had been formed. In their initial stages the surfaces of the continents must have been almost as stark as a moon landscape. The upthrust of rock would have produced jagged contours to be smoothed out as the forces of the weather blotted them by erosion and filled hollows with the deposition of sediments. But land without a covering of plants is not only formidable but uninhabitable.

Because plants are the converters of solar energy into more utilizable forms, they must have preceded the emergence of animals onto the land surfaces of the Earth. The place of their emergence is uncertain but undoubted terrestrial plant fossils are found in Upper Silurian rocks. These earliest land plants had a very simple form but possessed a rudimentary system for conducting water from their roots, through stems, to their aerial structures. Some stages of the transition are missing from the fossil record, but it seems probable that some of the green algae living in the brackish waters of estuaries, or possibly further upstream in fresh water, gradually adapted themselves to a fully terrestrial life. By Carboniferous times, about 100 million years later, the swampy surfaces of the Earth were densely covered with luxurious forests of primitive seedless plants, including ferns, fern trees, and horsetails—a primitive segmented, reed-like plant. A few of these ancient forms are still preserved in the flora of today. Trees evolved and became abundant during the Devonian and so contribute largely to the coal forests of the following period. Once plants adapted themselves to the more rigorous conditions of life on land, they went through their own ramifying paths of evolution and so set the stage for the emergence of animals.

Coming ashore shortly after the first wave of plants was established were the *arthropods,* which include spiders, centipedes, and the broad category of insects. They were following the well-tried evolutionary principle of going where there was free available food and so they primarily must have consisted of plant-eaters. But among their numbers there may have been a few carnivorous forms, for the energy source of terrestrial insects themselves could not long go unexploited. Some of the impressive insects of this early day included cockroaches of relatively gigantic size and dragonflies with a wingspan of two feet. Living in the freshwater streams and pools of that time were lobe-finned fish, themselves carnivores, who subsisted by eating smaller fish and no doubt the aquatic larvae of various insects. Equipped with lungs, they could survive in stagnating pools by coming to the surface to gulp air. The exact incentives which prodded these fish into evolving into land-based amphibians will never be reconstructed with certainty. The search for a new food supply, as provided by the terrestrial insects, has been suggested. On the other hand, A. S. Romer (1960) has hypothesized that the variable climate of the Devonian would cause the recurrent drying up of the pools in the intermittent rivers of the day. Fish with normal rayed fins would either die on the dry bottom of the pool, or perhaps would bury themselves in the mud as do living lungfish under similar circumstances. But those fish with better developed fins—that is, structurally strengthened by a complex of bones—would have the option of clambering out onto land and searching for other pools up or down the dry riverbeds. On such forced excursions our crossopterygian ancestor could have eaten the stranded fish along the shores of the drying pool, and the insects of the nearby swamps. In any case, the amphibians did evolve out of such fish, and for a brief period of evolutionary time were the dominant animals on land.

The Earliest Amphibians

Figure 5-4
The Oldest Known Amphibian, *Ichthyostega*, from the Late Devonian of Greenland

The earliest amphibians in the fossil record occur in the Late Devonian of east Greenland, which was then a warm and moist land mass. They are known as *Ichthyostega* (Figure 5–4) and are almost completely intermediate between the ancestral crossopterygian fish and later and more evolved amphibians. The head of *Ichthyostega* is somewhat flattened, as in its fish ancestor, but the eyes are directed more upward than laterally. The fleshy fins of the fish had been converted into modest but terrestrially adequate limbs, each bearing five digits. Its tapered, laterally flattened tail carries an external finlike structure revealing its ancestors. The teeth of the ancestral lobe-finned fish, *Ichthyostega*, and later labrynthodont amphibian descendants are characterized in cross section by complex infolding of enamel in the basic dentine of the teeth. This early amphibian reached a length of three feet and judging from living forms would have had a moist and scaleless skin. Very likely it spent much of its time in the water, emerging occasionally onto the moist surfaces of the surrounding land. Its ecological niche must have been relatively narrow and yet the animal provided a good beginning for the development of later terrestrial vertebrates.

During the following Carboniferous period the amphibians were not only dominant terrestrial animals, but they radiated to produce a wide variety of forms. Characteristically, they were all carnivorous, ranging from small and almost completely aquatic forms to relative giants which superficially resembled fat, squat, short-tailed crocodiles. A moderate-sized form, *Cacops* (Figure 5–5), lived in the Early Permian, and must have competed with the early reptiles. It was equipped with sharp labrynthodont teeth, limbs sufficiently well muscled to indicate an almost completely terrestrial mode of life, and a series of heavy scales down its dorsal ridge as a kind of armored defense. *Cacops* undoubtedly was reproductively still bound to return to moist places to lay its eggs.

Figure 5-5
Cacops, a well-developed early Permian amphibian, suited to a terrestrial life

The Sources of the Failure of the Amphibians

In our modern world the amphibians are represented by only a limited number of rather insignificant survivors, including the salamanders, frogs, and toads. The latter two forms, known as the *anurans,* show a surprising range of successful adaptation extending from Alpine regions in the temperate zone into inhospitable deserts in the tropics. The amphibians have fallen to a state of insignificance among modern terrestrial vertebrates for two reasons. Modern amphibians are characterized by smooth glandular skin through which they literally do a small amount of breathing, but which at the same time allows them to lose water unless they stay in moist environments. But, more importantly, amphibians have remained conservative in their basic reproductive processes and so are restricted to live near areas with free water.

Reptiles Are Well-Adapted Land Animals

The first primitive reptiles which appear in the fossil beds of the Carboniferous do not differ greatly in appearance from the more advanced forms of amphibians such as *Seymouria* (Figure 5–6). Originally considered to have been an early stem reptile, or *cotylosaur,* this animal is in fact an advanced terrestrial form of amphibian. There is fossil evidence for a lateral line in the body of *Seymouria,* and some larval forms with external gills seem to have been discovered. But in bony structure there is no real dividing this very advanced amphibian and the most primitive of reptiles. The real change in grade or structure between the two involved a new form of reproduction, the amniotic egg. Its essentials are diagrammed in Figure 5–7, where an embryonic crocodile is shown cradled in his secure, if complex, surroundings. The egg-laying reptiles operate at a much lower reproductive risk ratio than most of the amphibians, for their fewer young have been endowed with greater advantages prior to birth. The developing embryo is contained with the *amnion,* an envelope which is liquid-filled, and so the young reptile still develops in a watery environment, although a very protected one. A second sac, the *yolk,* is connected to the digestive tract of the fetus and provides it with a large reserve supply of food to serve through its protracted period of development. The *allantois* is liquid-filled and into it the waste products of the embryo's body are deposited. Inside of the flexible but stiffening shell is yet another membrane, the *chorion,* which is adherent to the inside of the shell and with it acts as a lung, allowing the passage of oxygen from the outer world into the embryo and the diffusion outward of the carbon dioxide

Figure 5-6
Seymouria, an advanced amphibian that was structurally a reptile

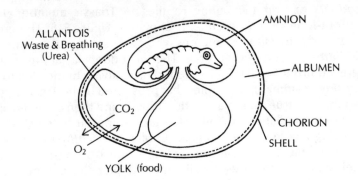

ALLANTOIS
Waste & Breathing
(Urea)

AMNION

ALBUMEN

CO_2

O_2

CHORION

SHELL

YOLK (food)

Figure 5-7
Amniotic Egg of the Crocodile

created by its growth processes. This complex egg freed the reptiles from any dependency upon water for their reproductive activities, and so they came to inherit the terrestrial world of the Mesozoic era. It is of more than incidental interest that the embryo in placental animals, including ourselves, shows a very similar series of arrangements to insure its safety and growth. But in the mammals the large yolk is no longer needed, for the chorion has been modified in the mammals into the placenta, which, implanted in the mother's uterine wall, provides both for the fetus's energy needs in terms of food and oxygen and as an outlet by which to scavenge its waste products.

The Great Reptilian Radiation

There is a general evolutionary principle which states that where energy sources are unexploited, some animal will evolve to utilize them. When progressive crossopterygians evolved into terrestrial amphibians, the latter entered a new kind of living space unoccupied by any other vertebrates. They rapidly diversified structurally and adapted to new ways of life within the overall terrestrial way. Such branching diversification is known as an *adaptive radiation*. Such events have occurred many times in the course of evolution and are recorded in the fossil record. Examples are given in Supplements No. 7 and No. 8.

The great reptilian radiation of the Mesozoic is one of the more spectacular examples of this type of event. The first reptiles evolved out of progressive amphibians in the Carboniferous but did not establish clear-cut dominance as terrestrial vertebrates until the Permian. Then the stem reptiles, or *cotylosaurs,* prospered and became the founders of the first reptilian radiation. Evolving out of the stem reptile were no less than five greatly diversified groups of descendants which had become well differentiated by the beginnings of Mesozoic time. The main line of evolution led through the *thecodonts* or primitive ruling reptiles. A Triassic member, little three-foot-long *Ornithosuchus,* is shown in Figure 5–8. The group consisted of active little carnivores who evolved bipedal locomotion to aid in the pursuit of their prey. The thecodonts lay at the stem of the great dinosaur radiation. A second diversifying branch led to the *therapsids,* mammal-like reptiles that prospered in the Triassic, gave rise to the first true mammals in the Jurassic and then declined. This branch of the radiation obviously has great importance to us. A third diversifying line led to the fishlike reptiles, the *ichthyosaurs,* while yet another led to the *plesiosaurs,* a group of long-necked and unlikely marine carnivores. Finally, the fifth branch from the cotylosaurs led to the turtles. It is one of the wonders of organic evolution that such generalized animals as the rude reptiles could evolve in so many directions and produce such totally different end products.

Figure 5-8
Ornithosuchus, a primitive thecodont or "ruling reptile," a Triassic form ancestral to the later dinosaurs

Radiation of Dinosaurs

The thecodonts, or ruling reptiles, started off rather unimpressively in the Triassic, but rapidly gave rise to two great orders of dinosaurs, flying reptiles, the ancestral birds, lizards, snakes, and rhyncho-cephalians. Dinosaurs, or "thunder lizards" actually consist of two quite separate orders of reptiles. The first of these consists of the *saurischians,* which have a reptile-like pelvis, while the other great order is that of the *ornithischians.* The essential differences in structure are shown in Figure 5–9. The saurischian pelvis is charac-terized by one of the three bones which make up that structure, the pubis, projecting alone and at right angles to the backbone. This is the reptilian form of pelvis. The saurischian went on to become the typical reptile of today. In the ornithischian the shape and position of the pubis are quite different, for it lies more or less parallel to the backbone with its posterior portion in close juxtaposition to the ischium. Both of these dinosaurian orders evolved members of giant size. This is so common an evolutionary tendency that it is known as *Cope's Law.* Not all evolutionary lineages become enlarged in size, but many of them do, so that *Cope's Law* stands as a reasonable descriptive generalization. Later we shall find that it applies also to the evolving mammals.

The ornithischian reptiles were all herbivorous in diet and in-cluded a number of bizarre forms which the student will recognize in Figure 5–10. *Stegosaurus* was a 20-foot reptile whose back was covered with two rows of bony plates for defense, while four great spikes jutted from the big tail, which no doubt gave a lethal lash. *Stegosaurus* had a remarkably small head for his body size and a brain of commensurate size. Related forms included the armored dinosaurs, of which *Ankylosaurus* was a representative form. Back, head, and tail were covered with bony plates, while great horns projected more or less randomly from the perimeter of both sides. The tail again ended in a knob which could give a clublike blow in defense. These animals were not able to retract their heads and legs into a shell, but no doubt they could crouch on the ground and be difficult to deal with. The horned dinosaurs generally lacked body armor but had defensive neck frills of bone and a variety of horns on their heads. *Triceratops* is a well-known form, but *Styracosaurus* is shown for the sake of variety. This, too, was a formidable reptile with a great projecting nose horn. Finally, there is a whole series of large bipedal dinosaurs in which the front teeth were lost, to be replaced by a ducklike bill (*Parasaurolophus,* Figure 5–10). These duckbilled dinosaurs were apparently amphibious, for casts in the fossil record show that some of them had webbed feet and flattened tails which would have facilitated swimming. *Parasaurolophus* had a bizarre closed-end hollow bone tube projecting backward several

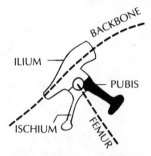

(a) Saurischian (reptile-like)

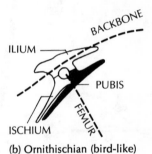

(b) Ornithischian (bird-like)

Figure 5-9
Two Types of Dinosaur Pelves Shown Schematically

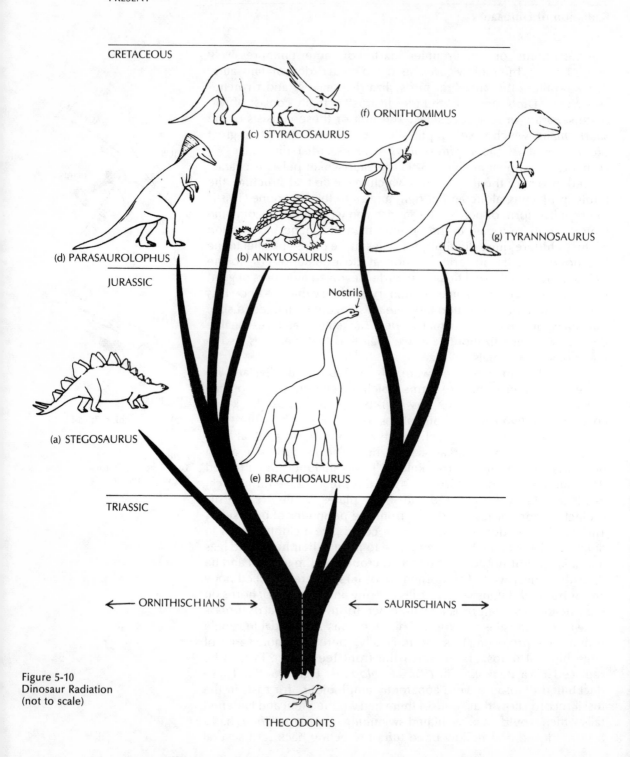

PRESENT

CRETACEOUS

(c) STYRACOSAURUS

(f) ORNITHOMIMUS

(d) PARASAUROLOPHUS

(b) ANKYLOSAURUS

(g) TYRANNOSAURUS

JURASSIC

Nostrils

(a) STEGOSAURUS

(e) BRACHIOSAURUS

TRIASSIC

← ORNITHISCHIANS →   ← SAURISCHIANS →

Figure 5-10
Dinosaur Radiation
(not to scale)

THECODONTS

feet, the function of which is not known. The duckbilled dinosaurs were sizable animals ranging from 20 to 40 feet in length.

The idea that evolution is irreversible is widely held but incorrect. Both orders of dinosaurs demonstrate that this is not true. For each has evolved from the little bipedal thecodonts which ran swiftly on their hind feet. Most of the herbivorous dinosaurs, and especially those which grew to great size, reverted to the quadrupedal form of locomotion to support their great body weight. The sequence, then, goes from primitive quadrupedal reptile-like *Seymouria* to the bipedal thecodonts, and then in many cases back to quadrupedal descendants. Clearly, this involves a reversal in the type of locomotion. The erroneous idea that evolution is irreversible is known as *Dollo's Principle.* G. G. Simpson (1949) has correctly pointed out the error in this statement and substituted for it his own accurate formulation, which is that *evolution is irrevocable.* Essentially, this involves the idea that when a series of complex evolutionary changes has occurred, and been incorporated in the DNA code of life, it is unlikely that in any future series of changes they will be completely undone and the original code completely restored. While a single mutation may be and usually is reversible, a complex set of instructions in the code cannot be undone, although they may be modified in the future. The doctrine of irrevocability does not prevent simple reversal in overall structural features such as the return to quadrupedalism seen above. There are numerous examples in the fossil record, one of which involves tooth form in the mammalian whales. The ancestral reptiles from which they descended had undifferentiated teeth of sharp and pointed conical form. In one of the early Eocene whales, *Zeuglodon,* the teeth had become laterally flattened and carried crests of three or more cusps. Since this Early Cenozoic whale is considered to be the ancestor of our present-day whales, it must follow that the simple conical teeth of the living porpoises and their relatives represent an evolutionary reversion to an earlier type. Thus the form of these teeth has gone through a reversal but this does not nullify the doctrine of irrevocability.

The other branch of the dinosaur radiation involved the saurischian order of reptiles. Among its members are included the giant quadrupedal and semiaquatic form of which *Brontosaurus* is probably best known. This animal had a bulk of only 30 tons more or less for its 70 feet of length and is relatively dwarfed by *Brachiosaurus*, an 80-ton animal that held its head 30 feet or more above the ground. *Brachiosaurus* presents several points of interest. First of all, this largest of all land animals evolved early, that is, in the Jurassic period, thus again illustrating Cope's Law. Second, remembering that it is descended from bipedal ancestors, we note that its front legs are longer than its hind members, an unusual condition among the dinosaurs and an even further reversal than merely returning to a quadrupedal gait. Its nostrils were on top of its head, suggesting an amphibious life.

The saurischians included bipedal members, as well as the ponderous quadrupedal types. Some of these were no larger than a chicken, but *Ornithomimus* was a swift, ostrich-sized Cretaceous bipedal carnivore. Having lost its teeth, it evolved a ducklike bill which no doubt allowed it to feed on a variety of foods and quite possibly on the eggs of other dinosaurs. The most terrifying dinosaur, *Tyrannosaurus,* was the most formidable of all the land carnivores. This great bipedal saurischian reached a length of 50 feet, stood 20 feet high, and had a four-foot skull containing a fearful battery of sharp but laterally flattened conical teeth. Its forelimbs had been reduced so in size that they could neither bring food to the mouth nor support the animal in a quadrupedal position on the ground. It is not certain whether they remained really functional. Obviously this great carnivore had a secure place in the order of the Cretaceous world, and it must have been preying upon the giant herbivores of the day.

Evolving from an unidentified ancestor in ruling reptile stock were the *pterosaurs,* or winged reptiles, which were common by Jurassic times. They showed two basic types, long-tailed and short-tailed. The former were generally moderate in size, with *Rhamphorhynchus,* which has been mentioned earlier, being about two feet in length, with perhaps a three-foot wingspan. The short-tailed types of pterosaurs were more advanced and ranged in size from the dimensions of sparrows to great forms with a wingspread of 27 feet.

Plate 5-5
A Small, Racket-Tailed Flying Reptile, *Rhamphorhynchus,* of Jurassic age. The wing surfaces and steering tail are shown in this restoration of the fossil.

The smaller ones no doubt could flap their wings, which would allow them to dive into the water for their fish and then take off as modern seabirds do. But the great ones, such as the *Pteranodon*, must have primarily been limited to gliding flight in the manner of modern albatrosses. In all cases their legs were ill-suited for walking but could have functioned well as clutching organs, so they may have rested much like bats. Nothing is known about their reproductive methods, but it seems likely that they must have retained the eggs to give birth to living young. Membranous wings possess a vulnerability in that when torn, through either accidents or fighting, they would not be expected to heal. But this has not proved to be a disadvantage in the mammalian bats. Presumably, the pterosaurs filled their aerial niche well until competition from the evolving and more advanced birds drove them to extinction.

Marine Radiation

Once a given living space is well filled, it is difficult for other animals to move into it unless their stage of development gives them some special advantage. The oceanic environment has been well filled by the bony fish, which became dominant in the Devonian, and have undergone a series of their own radiations in later time. Hence it is of interest that the Mesozoic reptiles could evolve successful marine forms after having been structured for a terrestrial life. The return of so many reptiles to the seas indicates that the reptiles are superior to fish in some aspects of their structure and organization; otherwise they could not survive in competition with the fish by generally feeding upon them. Some of these marine lineages evolved directly out of the stem reptiles in Late Paleozoic times. One of these, the turtle, produced giant marine forms which today are largely herbivorous in diet. More spectacular descendants from the cotylosaurs include the sharklike ichthyosaurs, and the bizarre plesiosaurs. These and other marine forms of reptiles are shown in Figure 5–11. The ichthyosaur (Figure 5–12a) evolved from unknown ancestors into rapid-swimming fish-eating forms, ranging in size from eight to seventeen feet overall. Their external forms provide a good illustration of evolutionary *convergence*, developing the similar form from different origins. As shown in Figures 5–12b and 5–12c, they remarkably resemble both the swordfish among fish and the porpoises among mammals. In all three cases the requirements of rapid locomotion through the water have produced beautifully streamlined forms. While the swordfish was naturally endowed with paired fins from its early fish ancestors, the ichthyosaur evolved its paired fins from limbs originally adapted to locomotion on land. Porpoises have evolved steering fins out of their anterior pair of limbs but suppress the hind pair. All three animals, fish, reptile, and

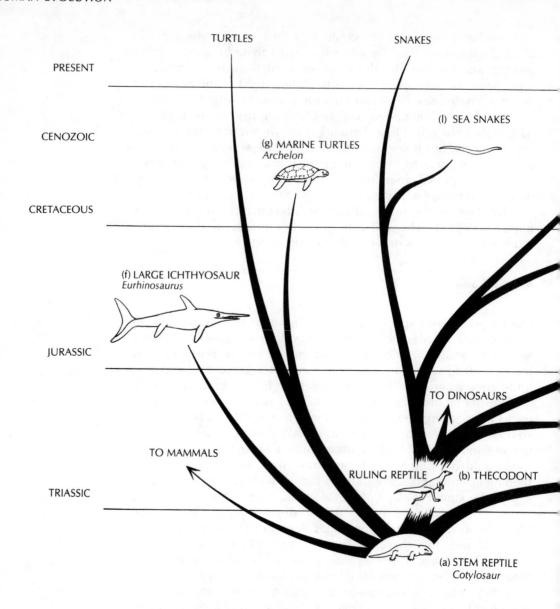

mammal, have evolved dorsal fins, which are required by their type of swimming motion to provide longitudinal stability. All three have effective caudal members for propulsion but differ in the details of their structure. In the swordfish the tail is a symmetrical, rayed structure. In ichthyosaurs the vertebral column is bent abruptly downward to provide a comparable structure. This angular displacement of the terminal backbone was for many years considered to result from some accident in fossilization until finally an ichthyosaur fossil was found which preserved the outlines of its skin, including

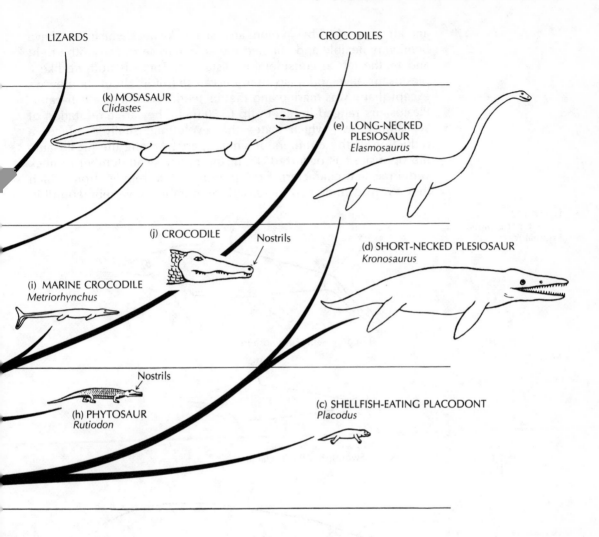

LIZARDS

CROCODILES

(k) MOSASAUR
Clidastes

(e) LONG-NECKED
PLESIOSAUR
Elasmosaurus

(j) CROCODILE

Nostrils

(d) SHORT-NECKED PLESIOSAUR
Kronosaurus

(i) MARINE CROCODILE
Metriorhynchus

Nostrils

(c) SHELLFISH-EATING PLACODONT
Placodus

(h) PHYTOSAUR
Rutiodon

Figure 5-11
Marine Radiation of Reptiles
(not to scale)

its dorsal fin, and clearly revealed that the downbending was normal and functional (Plate 5-6). This type of tail was evolved twice among marine reptiles, for a seagoing crocodile, one of the *telosaurs*, has evolved a fishlike tail by the same means. The porpoises and whales are propelled by a different type of tail, in which the flukes are horizontally placed and contain no supporting bony structure.

The *plesiosaurs* were of such curious proportions that they are difficult to describe. They usually possessed a flattened but unarmored body from which protruded great rowing flippers, both fore and aft, topped off by an elongated snakelike neck which must have been very flexible and enabled the animal to seize fish to the right and to the left at considerable distances. There is nothing like a plesiosaur alive today so that we can tell little about its lifeways except that it was marine and that its teeth reveal it as a fish-eater. Plesiosaurs ranged in length up to 50 feet. They are a refutation of Williston's Law, which states that evolution usually produces a reduction in the numbers of similar parts. Among the plesiosaurs the opposite has occurred. From the rather small number of neck vertebrae that characterize the primitive stem reptiles from which they evolved, they had increased the number of vertebrae until in

Figure 5-12
Convergent Evolution of
External Form

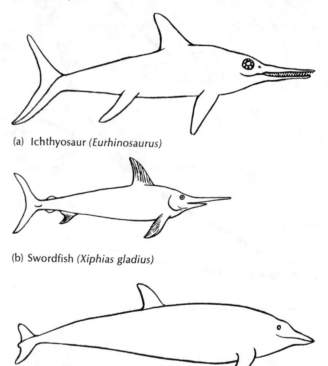

(a) Ichthyosaur *(Eurhinosaurus)*

(b) Swordfish *(Xiphias gladius)*

(c) True's Beaked Whale *(Mesoplodon miras)*

one form there are no less than 76 cervical vertebrae. By the same token they lengthened their all-important paddling flippers by increasing the number of finger bones involved. There are usually five finger bones in a generalized terrestrial animal but the flippers in some plesiosaurs increased to twelve. Classed as a plesiosaur but quite unlike them in form is *Kronosaurus,* a short-necked plesiosaur known from Mesozoic beds in Australia. As indicated in Figure 5–11, this great marine reptile looked something like a toothed whale, and, probably, like a modern sperm whale, fed on giant squid. Its great skull is twelve feet long and it propelled itself with enormous flippers. An unrelated marine reptile, *Placodus,* was rather generalized in appearance but specialized in diet. This eight-foot marine reptile evolved crushing teeth and apparently lived on a diet of shellfish. Aside from the fact that it must have operated in warmer waters than the walrus, it filled a surprisingly similar ecological niche.

A further series of marine lineages evolved as radiations from the thecodonts, or ruling reptiles. The *mosasaurs* are nothing more than gigantic marine lizards. They reached a length of 40 feet and apparently swam by undulations of their body while steering with two paired fins derived from their terrestrial limbs. The mosasaurs must have been effective in locomotion to have earned their living feeding on active and large bony fishes. An instance of *parallel evolution* is provided by the crocodile-like reptiles, the *phytosaurs,* which evolved in the Mesozoic. The early crocodile-like phytosaurs looked very much like modern narrow-snouted crocodiles, such as the gavials, except for the fact that their nostrils were placed in a little mound just in front of their eyes. Crocodile nostrils are in the forward tip of the snout and air is channelled back into the throat above a long palatal bone which separates the breathing passages from the mouth. The crocodiles are perfectly capable of continued breathing with their mouths open under water as long as their nostrils protrude above it. Mesozoic crocodiles included some gigantic forms with skulls six feet in length and bodies approximately 30 feet long. The phytosaurs seemed to have lived much like modern crocodiles in preying upon fish, mammals, and water birds which they could obtain along the banks of rivers. Finally, among the other descendants of the ruling reptiles, some of the snakes have evolved into fully marine types. There are fifteen genera of sea snakes living in tropical waters and eating small fish, which they paralyze with powerful venom. All in all, the marine radiation of the

Plate 5-6
Ichthyosaur fossil with impression of skin. The sharply angled tail vertebrae are normal.

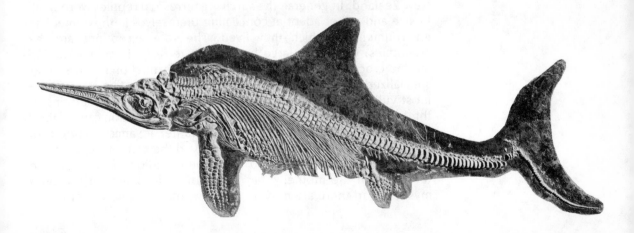

reptiles was more diversified than that achieved by the later mammals.

The Extinction of the Mesozoic Reptiles

The reptiles were the dominant vertebrates on land, in the air, and in the sea during the entire Mesozoic. The cause of their mass elimination at the end of the Cretaceous is unknown. All of the spectacular forms of marine reptiles became extinct even though nothing very dramatic happened to the bony fish living in the same seas. The ichthyosaurs had become extinct earlier, by the end of the Jurassic for reasons that are equally obscure. All of the terrestrial dinosaurs and the flying pterosaurs failed to survive the transition into the Cenozoic. The fossil evidence makes it quite clear that these mass extinctions were not the result of competition from the evolving mammals, for although the latter represented a superior grade of organization, they seemed to have had nothing to do with the passing on of the great reptiles. The reptilian brain was considerably improved over that of the amphibians, their reproductive methods were both varied and in many ways advanced, and while they had not attained a constant temperature, nonetheless many of them must have been approaching the four-chambered heart, as is found among modern crocodilians, and so their energy level must have been high. Under the warm conditions of the Mesozoic it has even been said they may have operated somewhat like warm-blooded animals. Some change in climate did occur at the end of the Cretaceous but hardly enough to explain the drastic elimination of the great reptiles. Perhaps some clue can be found by examining those types of reptiles that survive down to the present time. They include the turtles, the snakes and lizards, the crocodilians, and a single member of the rhynchocephalians, little *tuatara*, a primitive reptile which survives on some of the offshore islands of New Zealand. In general, the surviving terrestrial reptiles were small in size and rather adept at concealing themselves in the immediate environment in which they lived. The sole exceptions are the crocodiles, which are large but inhabit terrestrial waters. The big marine crocodiles, the teleosaurs, went to extinction like the other specialized Mesozoic reptiles. Since the reproductive stage is the most vulnerable stage in the lives of all creatures, it is tempting to think that somehow the reptilian mode of reproduction caused their elimination. The reptiles had achieved rather advanced methods of reproduction which, at the worst, involved the amniotic egg, and at best included live birth based upon the retention of these eggs in the body of the mother. The causes of their extinction remain a mystery, and there are no very good hypotheses to account for it.

Bibliography

COLBERT, EDWIN H.
1961 Dinosaurs: Their Discovery and Their World. London: Hutchinson.

ROMER, A. S.
1960 The Vertebrate Story. Chicago: University of Chicago Press.
1968 The Procession of Life. Cleveland: World Publishing Co.

SIMPSON, G. G.
1944 Tempo and Mode in Evolution. New York: Columbia University Press.
1949 The Meaning of Evolution. New Haven: Yale University Press (revised edition, 1967).
1953 The Major Features of Evolution. New York: Columbia University Press.

The Effect of Continental Drift Upon General Evolution

In 1912, Alfred Wegener, a German scientist, published the hypothesis of continental drift. According to his ideas, in the Carboniferous there had been a single world continent consisting of all the major land masses. Later in the Jurassic, it began to drift apart, and in later times the continents gradually took their present day positions. Wegener's ideas had some firm grounding in the physical nature of the Earth's crust, both oceanic and continental. They further explained why certain animals and plants in regions now far apart in Africa, South America, and

Australia were so remarkably similar. Today we know that Wegener was essentially correct, for it is now possible in terms of theories of plate tectonics, measurements of true expansion in the ocean floor due to extrusion of basalts, and other new types of evidence to conclude that the German scientist was very nearly right in his ideas. He was merely a half century ahead of time.

Of concern to us is the form of the original land masses, their connections, and when these were sundered. Figure 5–13 shows the ancient world

Figure 15-3.
Ancient Continent of Pangaea
Before Sundered by Plate
Tectonics

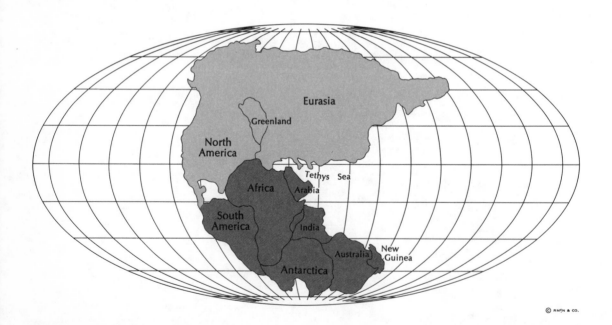

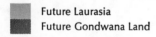
Future Laurasia
Future Gondwana Land

continent of Pangaea as it has been reconstructed by fitting together the major continental land masses. It will be noted that beginning on the lower right corner New Guinea and Australia are connected broadly with Antarctica, which in turn was in contact with a separate India, Madagascar, Africa, and South America. Proceeding upwards, this major land block comes in contact with North America, with Greenland connecting it with Europe and the conjoined major portion of Asia. Note that the larger Indonesian islands, such as Borneo, are the eastern terminus of this ancient continent. The upper and lower arcs of Pangaea are separated by the great extent of the Tethys Sea, which may have reached as far west as the Mediterranean Sea, which is a modern relic of it. This general arrangement of land remained unchanged throughout the Permian and the whole of the Triassic. Thus, to the early part of the Mesozoic era, or from about one hundred eighty million years ago, Pangaea remained unrifted, and so many of the animals of those times were widely distributed on it. For example, a mammal-like reptile *Lystrosaurus* is found in the early Triassic of South America, Antarctica, India, and even Indochina. A labrynthodont amphibian and two mammal-like reptiles which occur elsewhere in Pangaea have been discovered at Coalsack Bluff, a bleak, bare mountain ridge rising above the Antarctic ice mass. Their presence there speaks of broad land connection in the ancient world continent and a much warmer climate throughout the planet.

With the beginning of the Jurassic, Pangaea began its breakup with the southern continents of South America, Africa, Antarctica, Australia, and the peninsulas of Arabia and India constituting what has been called Gondwana Land. To the north, Eurasia and North America together became known as Laurasia. Some contacts remained, but they were constricted so that the reptilian faunas of the Jurassic and Cretaceous became more localized in those two great continental blocks than had the ones prior to the breakup.

By the end of the Cretaceous period, all of the continents as we now know them seem to have attained their own separateness. The North American and South American plates had drifted to the west, and the South Atlantic became a relatively broad ocean. Antarctica and Australasia had withdrawn from their close proximity to Africa. Madagascar drifted east from Africa. Arabia and India began to drift toward their ultimate positions as separate land masses. Eurasia remained intact. This is the time at which the mammals began to assert themselves, and their distribution during their great radiation in the early Cenozoic reflects the consequences of these continental breakups.

Further information on continental drift can be found in Edwin H. Colbert, *Wandering Lands and Animals* (New York: E. P. Dutton and Co., Inc., 1973); and A. Hallam, "Continental Drift and the Fossil Record," *Scientific American,* Vol. 227, No. 5 (1972), pp. 56–66.

Ecological Niches and Animal Diversity

There are about one million species of animals on our planet today. This variety is the result of a long-term pressure to evolve animals which are suited to live in a particular ecological niche, that is, with a special life style. The methods which animals use to earn their living vary greatly in terms of their size, their preferred foods, their rates and times of activity, and, of course, their structural specializations. The groups of animals that exist together form an ecological community tied together by many interdependent relationships. Such *ecosystems* tend to increase in complexity over long evolutionary periods of time. An ecosystem exists in a kind of gently oscillating balance in which the numbers of all species vary a little in time but tend to be restored in numbers as corrective forces set in. This dynamic state of ecosystem balance exists everywhere except where man has upset the system in recent years.

The number of ecological niches has never been counted, but each of the one million species of animals must be considered to have a unique one. The tendency of organic evolution to produce a great variety of animal life can be illustrated by a striking figure of speech which is attributed to Thomas Huxley, the nineteenth-century defender of Charles Darwin. In essence Huxley said to visualize a barrel as the total environment of the planet, and to fill it to the brim with apples. The barrel is now full, and the apples represent the primary broad ecological niches which are filled first as evolution proceeds. Even though the barrel is filled in one sense, much space remains unutilized between the apples. These available and unused spaces allow narrower ecological niches to be exploited between the apples in our figure, and this can be done by sifting small pebbles into the crannies. Such specialized niches are more numerous than the primary ones, but only become filled after the occupation of the latter. As overall adaptation proceeds further, new varieties of life evolve into yet more specialized niches so that sand can be put into the spaces left empty by the apples and gravel. The small size of the grains of sand indicates a very narrow life-way, involving great specialization, but it opens up a host of new opportunities that adaptable animals seize. Even with careful packing the barrel will yet contain some unfilled spaces between the grains of sand, indicating that the total environment is not yet completely utilized. This figure of speech is intended to show that ecological niches vary in scale from very broad primary ones to fantastically narrow specialized ones. But our figure needs one further change, for with the passage of evolutionary time we must in our imaginations let the total available environment expand considerably. Thus our barrel should have flexible sides in order to increase its interior content. Originally the barrel represented only the waters of the earth. As plants, insects, and amphibians adapted to life on land, the total size of the barrel enlarged greatly. Finally, the development of the reptiles, and subsequently the mammals and birds, required a further enlargement of the total available environment.

Animal ecologists simplify the analysis of the total ecosystem by thinking in terms of models which involve important food chains. A *food chain,* explored briefly in Chapter 4 (Figure 4-4), forms a convenient kind of illustration. The lowest link in the chain is represented by green plants, which are the primary producers of energy transformed from sunlight. The second link lies above and represents a plant-eating animal and so a first order consumer. The third link in energy terms is that of the primary carnivores which eat the herbivores. Subsequent links may be added to the chain in favorable environments, but their total number is limited. The most complex predator chain so far investigated contains only five animal links. More frequently the number of links in nature numbers three or four. One grassland food chain begins with grass which is eaten by grasshoppers, which themselves are consumed by grasshopper mice, which in turn may fall victim to the striped skunks. At the top is the great horned owl, which includes skunks on its normal menu. Since the amount of energy which passes from one link, or *trophic level,* to the next is of the order of only 10 percent of the food eaten, it follows inevitably that as one proceeds up the links of a predator chain, the animals at each level tend to become larger in size and certainly rarer in numbers.

Other examples will help to illustrate the nature of the predator chain. Most insects are plant eaters and they occur in such great numbers and so wide a variety of species that they represent a considerable energy resource. Evolution has produced many types of insect eaters, but among the mammals the bats evolved early in the Cenozoic in virtually their modern form and have subsisted ever since as nocturnal carnivores which earn a very good living by locating their flying prey through sonar, and then catching it through their superb flight abilities. Bats do not constitute a very great biomass, and so one would expect them to be the final link in a short predator chain and so be immune from being eaten themselves. But there is a general principle in evolution that *where free energy is available some animal will evolve to eat it.* At least two different kinds of bat-eaters have become adapted to extend this predator chain. One of these is the so-called leopard bat of central Australia, which has become a specialist in eating other bats. It is enlarged in size, with a wing span of about 12 inches. The leopard bat is naturally few in number, and, for reasons that are less clear, limited in range. On the other hand, an even less predictable bat-eater has evolved among the hawks. The bat hawk (*Machaerhamphus alcinus*) is a pointed winged hawk with a body length of 16–19 inches and pointed falcon-like wings. It has a far-flung but interrupted range, living in Africa below the Sahara, the offshore island of Madagascar, and in Indonesia. This curiously specialized bird of prey earns its living by flying only at dawn and at dusk near bat roosts. As the bats leave or come to their quarters, the hawk catches them in its talons with a falcon-like swoop, puts the victims in its oversized mouth, and swallows them whole while continuing its flight. The wide geographic distribution of this curious hawk indicates that it has found a successful way of earning a living, but it is obviously very specialized. It is significant that it is the single species in its genus.

The top link in any predatory chain is always interesting regarding the question of how its numbers are controlled so that it remains in a state of balance with the rest of its community. The African lion provides an interesting example. On the great Serengeti Plains of East Africa live the greatest game herds in the world. Consisting primarily of gnus and zebras, and supplemented with other varieties of antelope, the game herds total more than 1,000,000 sizable grass-eating animals. Living upon them at the top of the chain are 1,000 resident lions which have staked out permanent territories on the plains, and an additional two hundred lions which wander seasonal-

ly, following the herds as they move in and out of the plains to other grassland reaches. George Schaller, who has spent several years studying these lions, estimates that on the average each adult consumes 15 pounds of meat daily. Since the lions eat only the best of the meat and leave the rest of the carcass for such scavengers as the hyenas, a 500-pound zebra would provide one day of food for a pride of 10 lions. These great cats must therefore kill about 365 equivalent sized prey each year to feed the pride. In a rough way it may be estimated that each lion by itself consumes between 30 and 40 major game animals annually. For the victims to be able to afford this sustained loss they obviously must number about 1,000 for each lion that is supported. Predators must always be rare compared to the numbers of the animals they eat.

The way in which the spaces in the barrel are consistently filled is illustrated by the data summarized by Bernard Rensch (1960) on the variety of birds in four different regional ecosystems. The tabulation excludes all birds of prey, but includes all others, whether they are found high in the forest canopy, in its middle or lower levels, or on its ground surfaces. The regions involved (1) a central European forest, (2) a tropical rain forest from northeastern Brazil, (3) a tropical rain forest from East Africa, and, finally, (4) a similar rain forest in Indonesia. The total numbers of species ranged from 45 to 106, and both extremes occurred in rain forested areas. But the distribution of birds by size showed surprising uniformity. The species of birds were classed in three categories as small, medium, and large. It would be anticipated that the small birds would comprise the largest proportion of the total avifauna, and they do. But the surprising thing is that their representation is regular in these four totally different areas, ranging only between 55 and 60 percent of the total list of birds. Medium-sized birds, in energy terms, should be less frequent and this is true. Again their contribution to the total bird fauna varies surprisingly little, falling between 24 and 30 percent. Finally the large birds are a natural minority in all four regions, but again the range is narrow, extending only between 13 and 18 percent of the total. In rough terms there are about one-half as many species of medium-sized birds in these four environments as there are small ones, while the same relationship holds between the rare large birds and the medium-sized ones. Obviously size is a matter of some importance in the distribution of the utilized ecological niches.

When evolution has produced a successful predator fitting into a broad ecological niche, it shows a tendency to diversify by producing new species which are merely larger or smaller models of the primary type. E. G. Hutchinson (1959) demonstrates that a surprisingly constant ratio in size exists between such related forms. Examining a number of animals and birds which showed a series of related forms living in the same area, he found that the larger forms tended to be double the mass of the next smaller ones. There is no real reason for this relationship to represent a small whole number, but the empirical facts indicate that it is true. Presumably this separation by size is of just the proper magnitude to insure that the food requirements of the two related forms do not seriously overlap. Expressed in terms of linear dimensions, the larger animal is about 126 percent longer than the next smaller one, a value which cubed gives twice the volume. In Hutchinson's data the average linear dimension was actually 128 percent greater.

The short winged hawks, or accipiters, provide a series confirming *Hutchinson's Rule*. Living in the deciduous and coniferous forests of North America are three accipiters so alike in appearance, including plumage markings, that size is the only distinguishing feature. These long tailed, fast flying forest hawks outfly and outmaneuver the birds upon which they prey, for their tails provide acrobatic steering. Among these hawks, as in all others, the females are considerably larger than the males, so that the range goes from the smallest males to the largest females, respectively. The largest member of this series of pure predators is the goshawk, a truly regal hunter much beloved by falconers, which takes game birds up to the size of large grouse and mammals up to the largest of the hares. Below it stands Cooper's hawk, in which the largest females just equal the size of the smallest male goshawks. These birds live on medium sized prey such as flickers, quail, and the smaller mammals. Finally, the miniature sharp-shinned hawk is only slightly bigger than the robin, with the smallest males being only 10 inches in length, and the largest females 14 inches overall. They assail the smaller birds and rodents with unbelievable ferocity. When we test Hutchinson's Rule we find that the total length of the goshawk, averaging both sexes, is 135 percent greater than that of Cooper's hawk, and the latter are 142 percent greater than the sharp-shinned hawks.

Together these size ratios average 138.5 percent, which indicates that each species in this series is a little more than twice the size of the closely related form lying below it. These values exceed Hutchinson's Rule in magnitude, but the consistency in their size differences is in accord with his findings. The fact that the male and female accipiters vary so much in size explains why the size gap must be a little greater than is usual between the related species. Similar examples of Hutchinson's Rule can be found among the dog and cat families where series of forms occupy the same space. Like most evolutionary rules, this one has its exceptions, but it does serve to illustrate that size differences are important in the filling of niches of the same type.

The diversity of animals is so great that it is tempting to think that in a given environment all the niches have been filled by evolutionary radiations. But the introduction of alien animals into a new environment demonstrates that this is not true. One and a half centuries ago the European rabbit was introduced into Australia, and their numbers exploded until all the available environment was saturated. At one time the exploitation and controlling of rabbits was the seventh biggest industry in Australia. But our interest is ecological, and the great success of this introduced animal clearly indicated that the native Australian marsupials had not evolved animals to fill this particular niche.

In our own country the English sparrow, which is actually a type of house finch, was introduced about a century ago. Everywhere in the environment associated with human dwellings, this sparrow has become a dominant, if noisy and messy, bird. The western linnet, also a type of house finch, today maintains itself only in areas that are too rural for the urban English bird to take over. In the brief time that the English sparrow has been in North America it has shown evolutionary changes in both size and color. This includes the Hawaiian population, which has only been there for about fifty generations. It is probable that there exist animals which could fill new niches in most environments outside of their home domain. Huxley's barrel is very full, but still contains empty spaces.

The student who wishes to pursue this point of view will be interested in E. G. Hutchinson, "Homage to Santa Rosalia or Why Are There So Many Kinds of Animals?," *American Naturalist,* 93 (1959), pp. 145–159.

An Explosive Adaptive Radiation as Illustrated by the Hawaiian Honey Creepers

One of the most important recurrent events in general evolution is that of adaptive radiation. It occurs when a group of animals reaches an environment which offers new and previously unexploited food opportunities. If the unfilled ecological niche is broad, its new occupants may diversify in a large number of directions so that both behavior and structure become highly varied in their descendants. During the first phases of an adaptive radiation, differences between the various lines occupying the broad niche are slight but with the passage of time increase in magnitude. A great many varieties of new types of animals arise, but later the less efficient forms are eliminated, and so in the end fewer but more highly differentiated species and genera result.

Oceanic islands provide some of the best examples of adaptive radiations. The finches that reached the Galapagos Islands hundreds of miles off the coast of Ecuador found themselves with little competition and so evolved adaptively to fill the most important niches for small birds in that archipelago. But the adaptive radiation of the honey creepers, the *drepanids,* into 35 species is even more spectacular. The Hawaiian Islands lie more than two thousand miles in all directions from any possible source of immigrants. While they have been successfully reached repeatedly via air and water by seeds and spores of plants, only a small number of birds and no terrestrial mammal succeeded in doing so unaided by man. Birds are sometimes carried far out to sea by a storm path, but small animals are only ferried over great distances of water on logs or floating masses of vegetation. In both cases the chance of a pair of animals of the same species arriving simultaneously at the same point in the island is incredibly small. For this reason zoologists concerned with such problems usually invoke a pregnant female as the original immigrant. Once arrived, she may serve as the founding mother of a whole new lineage of animals if the food resources are adequate in amount and type.

The Hawaiian Islands are geologically young, with Kauai the oldest at 5.6 million years of age, and Hawaii the youngest at 700,000 years. Since the original immigrant honey creeper could not survive until after the islands became well forested, its initial radiation may have started much later in time. The ancestral honey creeper was probably one of the nectar-sipping birds that occur in tropical South America. One of these was carried early enough to the island so that its

Figure 5-14
Beak Adaptation in Honey Creepers

(Amakihi)
Lopox virens virens

(Kona)
Psittirostra Kona

Flower of
Clerniontia Grandiflora

(Akialoa)
Hemignathuus obscurus obscurus

(Akiapolaau)
Hemignathuus wilsoni

descendants radiated explosively to fill almost all of the niches available to birds of this size. They diversified into no less than 35 recognizable forms, indicating a high rate of evolutionary change. Diversification involved bill form more than anything else, for their size range is not great and their colors are rather monotonously red in one subfamily and green in the other. Four honey creepers, including three of the most extreme forms, are shown in Figure 5–14. First of all, the Amakihi is a generalized bird living on nectar and insects and apparently retaining the ancestral form of bill. A group of finch-like drepanids evolved with heavy bills for seed cracking and fruit eating. The Kona shows the extreme form of this development. The Akialoa developed a long curved bill with upper and lower portions of equal length. It is known to particularly favor the flowers of the various lobelias which themselves have strongly recurved flower tubes. It is primarily a nectar feeder. Finally, there is the Akiapolaau, a specialized half-bill who fills the niche of the woodpecker. The lower bill is short but very strong and used to gouge out bark or soft wood in seeking beetle grubs. The recurved upper bill is used to tap lightly on the tree and so locate likely spots for excavation.

The great diversity developed by the honey creepers in part results from their early arrival on the island. The living birds indicate that there were an original 16 colonial invasions and most of these must have happened later in time. Of the niches available for small perching birds, the descendants of the honey creepers occupied almost two-thirds. Obviously the adage that possession is nine-tenths of the law works ecologically as well.

Adaptive radiations occur at all points in evolution and result in spectacular successes and dramatic replacements. The primates have gone through their own series of radiations and so, of course, have primitive men. The fossil record indicates that living men today are nowhere near as varied as were Middle Pleistocene hominids, and so it may be judged that competition has eliminated the less viable forms of men. This is the general pattern to be followed in adaptive radiation and man is no exception to it.

CHAPTER 6

THE
EVOLUTION
OF THE
MAMMALS

The radiations of the reptiles seemingly gave the Mesozoic a fully occupied environment. With their extinction, the inconspicuous, arboreal, nocturnal, and furry little mammals did not appear to be potential replacements. But the special attributes of the mammals allowed them to radiate repeatedly, to fill all the vacated reptilian niches, and to create many new ones of their own.

One of the many surprises in the course of organic evolution is that some reptiles began to evolve in a mammal-like direction early in their existence. The first evidences of such a shift are found surprisingly early, in the Late Carboniferous. Shortly after the first reptiles appeared, some of them, known as *pelycosaurs,* showed dental differentiation to the extent that the canines became enlarged, the front teeth began to take on incisor-like form, and the cheek teeth consisting of premolars and molars began to show evidences of grinding function. They were slimmer in body build than the usual stem reptile, but their limbs still sprawled out at the side, and they seemed to be semiaquatic and to feed on fish. Some changes in the skull also foreshadowed the mammalian condition. They were among the dominant carnivores of the times.

Continuing into the Permian, there was a radiation of mammal-like reptiles of more advanced structure. These *therapsids* occur in many forms and in great abundance in the red Karroo beds of the later Permian and Triassic of South Africa. They were still close to the main line of reptilian evolution, but the dominance of the ruling reptiles had not yet been established. In a number of bony features they are good illustrations of Williston's Law. Those bones of the skull ultimately lost in the mammals were becoming reduced in the therapsids. In the lower jaw, or mandible, the dentary bone, which was the only remaining one at the mammalian level of organization, is enlarged in the therapsids and the other posterior bones reduced. The stem reptiles show five digits for each limb but they contain more elements than do the mammals. The primitive count of toe bones in the pelycosaur foot going from the big toe to the small one is 2–3–4–5–3. In the primitive mammal, and man as well, the count is 2–3–3–3–3. This modern formula was attained among the therapsids. Further, the old single condyle, by which the skull had been articulated with the backbone in the earlier reptile, was replaced by a double condyle in the mammal-like reptiles, as it is in all of their descendants. Teeth show further differentiation in a mammalian direction. A marked shift occurred in the limbs, as is shown in Figure 6–1a, in the therapsid *Lycaenops.* In this progressive carnivore both the knees and elbows had been shifted inward so as to begin to approach the modern mammalian condition in which they are

The Pathway Toward the Mammals

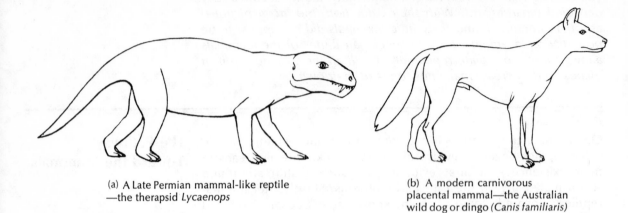

(a) A Late Permian mammal-like reptile
—the therapsid *Lycaenops*

(b) A modern carnivorous
placental mammal—the Australian
wild dog or dingo (*Canis familiaris*)

Figure 6-1
Evolution of the Mammals

directly under the body. This advanced mammal-like reptile in life might have appeared to be something akin to a scaly, thick-tailed, earless dog or possibly a sparsely hair-covered advanced reptile. There is no evidence as to whether the purely physiological features of the mammal were yet incorporated in these mammal-like reptiles, but it is not unreasonable to hypothesize that the changes involved in reproduction and temperature control were evolving at about the same pace as the visible structural traits. The evidence from the Karroo beds is quite clear that mammalian features were being evolved in a whole series of therapsid lines, and that with the passage of time each lineage was coming closer to the mammalian structural status.

Today we recognize the mammals as the dominant terrestrial vertebrates. But the mammal-like reptiles that gave rise to them did not really profit from their own structural advances. While the therapsids flourished during the late stages of the Permian, by the Triassic they were on the decline and disappeared from the record in the early part of that period. Apparently, they could not compete on even terms with even the primitive ruling reptiles or thecodonts which were to evolve into the dinosaurs. The latter seem only to have had the advantage of bipedal locomotion at this time, and they do not show the essential changes in structure which led in the direction of the mammals.

The first true mammals appear after some gap in time but are a part of the general Lower Mesozoic record. They occurred in late Jurassic and Cretaceous deposits, and their total remnants could be held in two large cupped hands. They are represented by fragments of jaw, a few skulls, and no complete skeletons. The poverty of the record is perhaps matched by their faunal insignificance. These earliest of true mammals were very small in size, ranging from the size of a kitten downwards to that of a mouse. During the 100 million years in which they must have existed during the Mesozoic, they seemed nowhere to have been able to compete with any of the reptiles and gave no evidence during this long time of any consequences of their superior physiological structural organization. It is generally considered that they survived during the age of dinosaurs only as a consequence of their small size, their likely nocturnal habits, and a tendency to hide away in the brush or the trees. The reptiles completely dominated the more important and visible ways of surviving, and the mammals made no progress toward displacing them. In a sense the evidence strongly suggests that the possession of an ecological niche combined with large size makes it difficult to be displaced from possession. If our evidence ended with the final years of the Cretaceous, we would be forced to conclude that the reptiles were the superior vertebrates, and the almost invisible mammals an unsuccessful experimental path in evolution.

The Emergence of the Mammals

The subordinated mammals of the Mesozoic underwent their own small radiation, for they are represented by no less than six different *orders,* or groups of families in a taxonomic sense. Since these distinctions are based primarily upon dental differences, we will not go into detail here. Three of these mammalian orders became extinct in the Jurassic, while the *multituberculata* disappeared in the Late Eocene. Present-day orders of mammals go back to other Cretaceous ancestors.

Living mammals can be divided into three broad types, based upon differences in their method of reproduction. The living *monotremes* are today reduced to the duckbilled platypus and the spiny anteater, or echidna, and are limited to Australasia. Reproductively they are anomalous, for they are the only egg-laying mammals, although the young after hatching are nursed by their mother. It is generally agreed that they represent a separate line of mammalian evolution, and many authorities consider them to have evolved from a different stock of mammal-like reptile than have the others. Indeed, G. G. Simpson says that the living monotremes may be viewed merely as reptiles masquerading as mammals.

The second great group of mammals is known as the *marsupials,* or pouched mammals. They characteristically give birth to young which we placental mammals would regard as a barely formed embryo. But the little marsupial finds its way by blind instinct into the pouch of its mother, there fastens itself firmly to a teat, and continues immobilized in this position until it has grown to an adequate size to occasionally emerge from the pouch and ultimately to a stage where it is independent of its mother's care. These marsupials were visible in the Cretaceous and may have been the most numerous types of mammals in the Late Mesozoic. In the later age of mammals, or Cenozoic, they persist but under rather special circumstances. In isolated South America they evolved into the dominant carnivores, including forms that mimicked wolves and even the saber-toothed cats. But their dominance ended there with the advent of more modern mammals which entered South America over the reestablished land bridge in Pliocene times. In South America, as in North America today, marsupials survive only as modest-sized and largely arboreal forms. On the other hand, by random luck they were the first mammals to reach Australasia and there they became the dominant terrestrial animals. The radiation of marsupials in Australia is an interesting manifestation of evolution in its own right. An original immigrant form, not unlike modern opossums, which really are very little changed from Cretaceous times, upon reaching Australia filled all of the niches available to this type of organization. The grazing and browsing hoofed animals of other parts of the world are there represented by a wide variety of kangaroos, which range upwards from rabbit-sized animals to the great gray kangaroo which stands six feet high and weighs more than 200 pounds. Now extinct forms of marsupials, the *diprotodons*, were rhinoceros-sized marsupials which survived until the advent of man upon the Australian continent. Another terrestrial marsupial evolved into the equivalent of the placental mole, while the wombat evolved as a large burrowing marsupial rather like the placental woodchuck. Among the marsupial carnivores there were forms that mimicked the wolf, small cats, and a small wolverine as they occur among placental forms. Arboreal marsupials evolved into a great variety of animals, ranging from mouse-sized ones upward through the equivalents of our flying squirrel, and a wide variety of phalangers, which do not find exact equivalents in the placental world. Even the kangaroos got into the arboreal act, for in the rain forests of north Australia and New Guinea two species of kangaroos clamber effectively from branch to branch and are seemingly well adapted to this mode of life. But despite the success of their ramifying evolution, these Australasian marsupials achieved their success only because placental mammals did not reach that area in early days. When the placental dog, in this case the dingo, shown in Figure 6–1b, reached Australia as a companion of the incoming Aborigines, it quickly drove to extinction two of the marsupial carnivores, the wolflike Tasmanian tiger and the wolverinelike Tasmanian devil. Both forms survived in Tasmania, which the dingo did not reach because it could not cross the intervening water gap. In more modern times the introduction of cats which have grown wild, or feral, as well as the English red fox, has driven the native cats to the verge of extinction. The evidence is very clear that the more highly evolved placentals can outcompete the marsupials and cause their extinction.

The Success of the Placental Mammals

The Cretaceous period saw the evolution of insectivores, which did not differ too much in size or in appearance from such living forms as *Tupaia* shown in Plate 6–1. These small, furry-tailed, long-snouted little mammals are close to the founder stock from which all other placental mammals evolved. Presumably their great success originally lay in their advanced placental method of reproduction.

No less than 16 orders of placentals have survived into modern times. The 12 most important of these are indicated in Figure 6–2. Recognizing that a taxonomic *order* represents a group of families which have evolved to fill a rather specific type of *ecological niche,* it is worth briefly summarizing the lifeways of the successful placental mammals. Of course, the order of the primates is first in our mind,

Plate 6-1
Tupaia, showing furry tail, clawed extremities, and elongated muzzle.

for we belong to it. The *primates* are primarily arboreal animals, of which only the larger types have returned to life on the ground. The bats are highly specialized flying mammals of small to moderate size, which had attained their modern form by Eocene times and also presumably their specialized sonar senses. It is interesting that among the largest tropical bats, the original insectivorous diet has become replaced by a fruit-eating diet, again illustrating the principle that large animals almost of necessity must be herbivorous. The rodents are one of the most successful mammalian orders, and their way of life shows in their enlarged, constantly growing central incisors, which make efficient chisels for the opening of such hardshelled foods as nuts. Their niche has been considerably broadened by the evolution of the flowering plants, the angiosperms, which came into being in the Early Mesozoic, providing seeds suitable for rodent food. Both in terms of the numbers of species and the number of individuals, the rodents are the most successful of the mammals. The *lagomorphs,* or hares and rabbits, were long classed with the rodents because of their chisel-shaped incisors, but they are now placed in a separate order and have a long independent evolutionary record of their own. Their resemblance to rodents seems primarily a matter of convergence.

The order of the *carnivores* is both broad and structurally interesting. Generalized primitive carnivores known as *creodonts* emerged in the Paleocene and continued as the dominant flesh-eaters for some 30 million years. In time some of their members became more specialized and now the living order includes all of the dogs, cats, bears, weasels, civets, hyenas and raccoons. The marine carnivores, or *pinnipeds,* including the various seals and walruses, split off early and represent one of the successful mammalian returns to the waters of the seas. While modern bears weighing three-quarters of a ton may seem large carnivores, an Eocene creodont with the appalling name of *Archesuchus andrewiisi* attained the formidable length of 12 feet. In addition to its name, which commemorates the paleontological explorer Roy Chapman Andrews, the giant creodont must have been as unpleasant on a face-to-face basis as any mammalian carnivore in the fossil record.

The whales, or *cetaceans,* represent another return to exploit marine resources. The original whales were toothed, and no doubt fish-eaters. This structure and habit persist among the smaller whales and the sizable squid-eating sperm whale today, but the other great ones have evolved a whalebone or baleen sieve by which to strain out the minute organisms that comprise the plankton. Feeding upon this rich and basic resource, the baleen whales have grown to great size, including the 100-foot blue whale, which is the largest animal to ever have lived on earth.

The elephants, or *proboscidians,* are animals with long prehensile

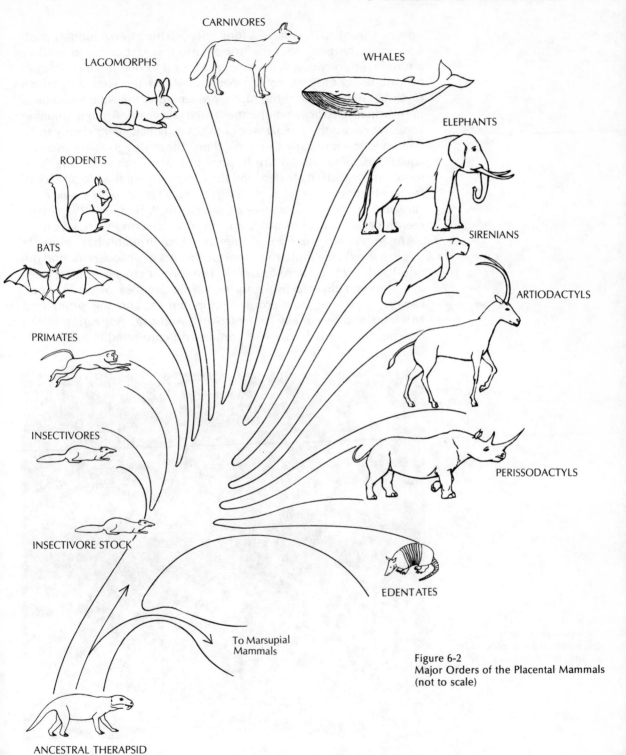

Figure 6-2
Major Orders of the Placental Mammals
(not to scale)

noses or trunks. They have a long and illustrious evolutionary past, with forms ranging from the tropics into the Arctic. But today they are reduced to two species, one in Africa and one in India. Even in historic times their range had contracted, for the elephants which Hannibal used to cross the Alps were tamed in the Atlas Mountains of northwest Africa, where they then survived. Unlike them in appearance are the *sirenians,* or sea cows, which seem clearly to be related to them in Early Cenozoic time. These sluggish, herbivorous aquatic mammals today are tropical in their range, with one form found in the Gulf of Mexico, and the other ranging the Indian Ocean from Africa to Australia. A third giant siren, Steller's sea cow, survived in the Arctic waters of the North Pacific until the seventeenth century. Easy to hunt, it went to extinction in a few years.

The great variety of hoofed animals, which probably have multiple origins, are divided into the even-toed forms or *artiodactyls,* and the odd-toed forms or *perissodactyls.* The former cloven-hoofed types include such diverse living forms as the families of cattle, deer, camels, pigs, the ponderous hippopotamus, and the pronghorn antelope and mountain goat of North America. Appearing in the Eocene, they early became important and continued in a dominant

Plate 6-2
Eohippus, the Eocene ancestor of the horse. There are four toes in front and three on the rear limbs.

position until the Pliocene, when they expanded greatly in variety. They are perhaps numerically the most successful of the larger types of mammals. The odd-toed hoofed animals, or perissodactyls, include one-toed horses and three-toed rhinoceroses and tapirs. For a brief period in the Eocene they were the dominant ungulates but shortly thereafter began to contract, a trend which has continued down into modern times. The even-toed animals have better sprung feet and many were also *ruminants*—cud-chewers—and this digestive difference may have been important in their ultimate survival. Both groups of hoofed animals are characterized by some similar trends in their evolution. Originally small-sized mammals, they have progressively increased in size, thus following Cope's Rule. The reader will recall that Eocene horses were the size of a fox-terrier and evolved into the modern wild horses standing between three and four feet high at the shoulder. The even-toed animals show a similar trend but culminate in even greater forms, such as the gaur, a wild ox from India, and the American buffalo, each of which reaches a ton in weight.

With increasing size comes the need for better and longer-lived teeth, for much more food is required to keep these large bodies fueled. When the grasslands of the world expanded with the changing climate of the Miocene, many types of both forms of ungulates changed from a browsing habit to a grazing one. Twigs and leaves eaten by browsers are relatively clean and so cause no unusual abrasion of the teeth. Grass, on the other hand, is always dust covered and further contains a great deal of silica, which is one of the hardest of the elements. Hence a grazing diet requires dental modification and many lines of both perissodactyls and artiodactyls responded to this need by slowly evolving high-crowned, or *hypsodont,* teeth. Simpson, who is the world's expert on the evolution of the horse, among other creatures, has calculated that assuming an average rate of mutation, the observed evolution of the hypsodont tooth in the horse must have involved a minimum of 300 different mutations progressively increasing tooth crown size. Crown height in horse upper molars evolved from eight to forty millimeters; when this is divided into 300 steps, the average magnitude of change in each mutational step would be 0.1 millimeters, or about $^4/_{1000}$ of an inch, which is the thickness of many grades of sheet paper. (See G. G. Simpson, 1953, for the complete discussion of this and a host of other interesting evolutionary problems.) Mutational changes of this size are to all effects invisible. The example provides evidence of the nature of continuous variation in evolution and, perhaps even more important, demonstrates that structurally important, indeed vital, mutational changes tend to be very small in their individual effects. The magnitude of overall change is shown in Figure 6–3.

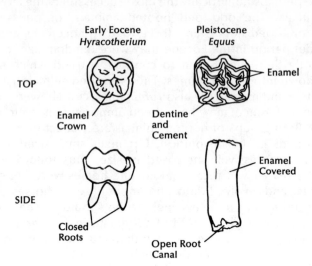

Figure 6-3
Low and High Crowned Horse Molars
(not to scale)

Dental Differentiation

The changes in tooth form in macroevolution are worthy of a brief study for they are the basic data of general evolution. Many marine animals of simple organization live by the elementary process of siphoning water through their body apertures and straining out the minute food particles carried in the moving stream. But others have gone on to develop more elaborate feeding habits. Molluscs have developed a rasping toothed tongue called the *radula.* Among the arthropods, both marine and terrestrial, there is a great variety of mandibles both for capturing and cutting up the food. The cephalopods, including both the squid and the octopus, have strong, horny beaks. But it is among the vertebrates that the greatest diversity occurs. The earliest fish were jawless, but the conversion of the anterior gill arch into a functional pair of jaws marked a great advance, for it not only enabled their owners to capture active food but to hold it and in some cases to reduce it to chunks suitable for swallowing. More recent bony fish have teeth consisting of a series of sharp spikes which are designed to grasp food and hold it until it can be swallowed. With such a dentition, victims must be swallowed whole, and so their food sources are generally limited to a size considerably smaller than themselves. Greedy individuals often overestimate their capabilities. I recently saw washed up on the shores of Lake Michigan a seven-pound silver salmon that arrived there dead because of a ten-inch forage fish stuck in its throat.

In special cases, evolution produces special solutions. Food is so scarce in the depths of the ocean that some of the fish living there have developed jaws so extensive, and stomachs so elastic, that they are able to feed upon fish considerably larger than themselves. Some fish, including the rays, subsist on a diet of shellfish and so have evolved teeth in the form of great grinding blocks in order to crush them. The parrotfish have developed enormously powerful jaws and projecting teeth to enable them to break down the limey structure of coral reefs and so feed upon the animals that build them. The sharks, an ancient and remarkably successful group, have evolved dermal teeth, from the bony ossicles which are embedded in all of their skin. Such teeth not only form a never-ending series of replacements but are so effective in their cutting capability that sharks can prey upon animals too large to be swallowed whole. Their tactic is simple: they grab a mouthful and, with rotary motions of their tail, twist their body so as to cut the chunk of meat out. No wonder men consider sharks very unpleasant fish.

Among the early terrestrial animals, the amphibians retained a rather simple type of dentition in which the teeth showed very little regional differentiation. But one important group among them, the *labrynthodonts,* are named from the fact that their teeth became complicated through infoldings of the enamel in the dentine. The early reptiles who succeeded them were enabled by improved means of reproduction to expand into many more ecological niches, and so to earn their living in a wider variety of ways. Many reptiles retained the basic dentition of simple cone-shaped, undifferentiated teeth. Such a dentition is called *isodont,* because the teeth are so alike. Modern crocodiles fall into this group, even though there is a little size differentiation in the anterior portions of their jaws. But while such teeth serve well for carnivores, the reptiles early evolved a variety of plant-eaters, and so dental evolution necessarily occurred. By the Late Mesozoic such herbivores as the duckbill dinosaur had developed dental systems featuring several hundred teeth so arranged in serial position in the jaw that worn teeth could be constantly replaced by others erupting below them. This system provided unfailing grinding surface for large-bodied animals which require substantial food intakes. Reptiles generally are able to replace worn or lost teeth.

The most significant reptilian line consisted of the so-called mammal-like reptiles which did evolve later into the true mammals. In them we see the beginning of the regional differentiation of the teeth which was ultimately to provide the mammals with the basis for their great dental diversity. Some of the early therapsid reptiles were beginning to show the enlargement of the isodont teeth in the region where the future canine teeth were to be developed. The cheek teeth became multicusped to provide grinding surfaces. The regional differentiation that culminated in the mammals was well under way.

With the emergence of true mammals in mid-Mesozoic times, dental differentiation reached new levels. In the placental mammals the generalized or primitive dentition consisted of 44 teeth arranged in a regular order. If we split the upper and lower jaws into halves, then the dentition can be represented by four quadrants. Beginning at the front center of one of these, the generalized mammalian pattern consisted of three teeth called *incisors,* which had characteristically taken on the sharp-edged form of chisels. Immediately behind them in the arcade there universally remained a single enlarged single-cusped *canine* tooth. While reptiles might have more than one canine tooth in each quadrant, the mammals never do. Behind these were arranged four generally small teeth called *premolars,* usually characterized by two cusps, of which the outer one is usually higher than the inner. Today's dentist calls them bicuspids. In the rear of the dentition are the big cheek teeth, the *molars,* which in the original generalized mammals numbered three on each side of each arcade. They have multiple cusps and originally served to grind and break down the animal's food so as to prepare it for mastication, for one of the great advances found in the mammal is the dental capability of breaking food down to such size that it can be easily swallowed and more easily digested.

The rational aspects of progressive dental evolution are nicely demonstrated by the relation between tooth function, form, and position in the mouth. Figure 6-4a gives the dentition of one of these generalized mammals. It will be noted that the anterior chisel-shaped incisor closes the most rapidly and so it is appropriate that the incisors in the anterior position should serve as chisels. The next time the reader bites a chunk out of an apple, let him look at the chisel-like marks left behind in the fruit. The next tooth behind, the elongated canine rising well above the level of the other teeth, serves as a grasping, piercing, or tearing tooth. These functions are admirably served by its single sharp cone and its relatively forward position. In addition to its uses in food seizure and preparation, it is the primary fighting tooth of most mammals. It is significant that in man the canine tooth has been reduced to the general level of the others, although it still retains its primitive form with a single cusp, now much more blunt. It has lost its old functions, but it can be shown that genetically it is a very stable tooth, and that in evolution it has been very conservative. Next in order, the premolars serve an intermediate function in that they both grind and shear the food. Their position and rate of closure are well suited for this and in many mammals they are not unduly specialized. The final teeth of the cheek area, the molars, have forms that relate closely to the kind of food eaten. In the generalized mammal they were not specialized, with the cusps remaining more or less low in relief, and indicating an omnivorous diet. But with specialization these cheek teeth reveal much about the animal's life-style. Among the herbivores, all classes of animals that turned to grazing developed *hypsodont* or high-crowned teeth with complexly infolded enamel. Among flesh eaters of all types, whether they are the dogs and cats among the placental mammals, or their counterparts independently evolved among the marsupials, the cheek teeth became reduced in number while one or more of them became greatly enlarged and so shaped as to operate efficiently in the shearing of flesh. Positioned as they are in the rear of the mouth, the molars are capable of exerting great

Figure 6-4
Dentition of Generalized Mammal (a) and Man (b) (not to scale)

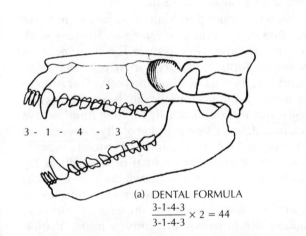

3 - 1 - 4 - 3

(a) DENTAL FORMULA
$$\frac{3\text{-}1\text{-}4\text{-}3}{3\text{-}1\text{-}4\text{-}3} \times 2 = 44$$

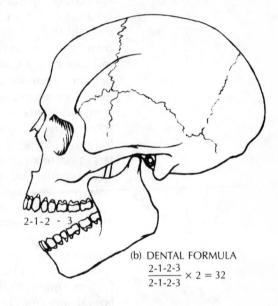

2-1-2 - 3

(b) DENTAL FORMULA
$$\frac{2\text{-}1\text{-}2\text{-}3}{2\text{-}1\text{-}2\text{-}3} \times 2 = 32$$

pressure, for they travel slowly and through but a short distance as the jaws close. The proof of this simple mechanical principle will quickly be brought home if the reader will take a hard-shelled nut and crack it with his teeth. No one ever places a hazelnut in the incisor, canine or premolar region of the mouth. It is always instinctively placed between the molar teeth, for it is there that sufficient pressure can be generated to open it.

So great is the variety of teeth among ancient and living mammals that they serve better than any of the other parts of the bony skeleton to reveal relationship through lines of descent. There has been a general evolutionary tendency among the mammals for the basic number of 44 teeth to be somewhat reduced, though some of the toothed whales provide the exception in that their teeth have shown a numerical increase. But among terrestrial mammals reduction may affect the number of incisors and in some herbivores the upper incisors have been replaced by a pad to assist in plucking the grasses. The canine can be vastly elongated, as it is among animals where it is much used for fighting or essential for food procurement. It is rather interesting that enormous elongation of the canine occurred not only in saber-toothed cats, but in an analogous but unrelated saber-toothed marsupial in South America, and among some ponderous early Cenozoic herbivores known as uintatheres. The function of the tooth was so important to these animals that in both of the latter forms great bony flanges grew down from the lower jaw to provide protection against their breakage. Premolars are frequently reduced in number but have not become markedly specialized in form in most mammals. But the molars show the greatest variety of forms. It is these teeth which seem to reflect most directly the relationships between diet and dental structure. Our own low-crowned molars, as well as those of the primates generally, indicate they are adapted to a variety of foods.

In Figure 6–4b, a stylized human dentition is shown to indicate how it compares with that of the primitive mammal. Incisors in all four quadrants of the mouth have been reduced in number from three to two. Man's canine tooth is reduced in size so that it no longer protrudes functionally above the adjoining teeth. Our premolars, as in most of the higher primates, number no more than two in each region. And finally, our molars, normally three in number, are now in many human populations being numerically reduced. At some future time two may be the normal number of molars in man. The 2–1–2–3 dental formula of man gives a total of 32 teeth as his normal complement. But as the pressure of selection is lifted to maintain functional teeth, no doubt the numbers will continue to reduce. In addition to the loss of the third molar as a common feature, many individuals are born without one or more of their lateral incisors. This tooth is genetically unstable and perhaps in time also will be lost to our species. Certainly a diet in which frozen milk shakes are the hardest component is not calculated to maintain dental integrity. Hunting and gathering peoples, both past and present, show dental signs of great stress, much abrasion and little reduction in numbers. These people obviously use their teeth so vigorously that a loss of function would be detrimental to the individual. Among Australian Aborigines the teeth not only served as vises in a number of situations, but were also used in the retouch flaking of flint tools.

Poikilotherms and Homeotherms

One of the surprising features of the evolution of the mammal was the rate at which some lines attained a giant size. They emerged from the Mesozoic as rat-sized animals and quickly filled by mammalian radiation the vacuum left by the extinct dinosaurs. A now extinct group of animals known as the uintatheres had reached the size of an elephant by the end of the Eocene. Small-brained and heavy-bodied, they did not last long but their role in nature was replaced by a series of other ponderous mammals. By the Early Oligocene, one of these, known as *Brontotherium,* or the thunder animal, reached a height of eight feet at the shoulder, a length of 15 feet, and had a ponderous body. The evolutionary life of these rapidly evolving perissodactyls came to an end by mid-Oligocene time, with their total history running its course in 25 million or fewer years. In the Oligocene landscape there existed the largest terrestrial mammal to ever evolve. This was a giant hornless rhinoceros known as *Baluchitherium,* which reached a height of 17 feet at the shoulder and an overall length of 27 feet. This odd-toed browser did not survive long and left no descendants. It is a remarkable testimony to the power of mutation and selection that squirrel-sized Mesozoic mammals could in a relatively short time reach such gigantic proportions. Giantism characterized many of the mid-Mesozoic reptiles, giving them the advantage of much bulk and a relatively small surface area from which to lose heat by radiation. It has been hypothesized that in the mild climate of their times they virtually operated as warm-blooded animals.

This brings us to the interesting question of what distinguishes the so-called cold-blooded animals, properly called *poikilotherms,* from the warm-blooded ones, or *homeotherms.* The cold-blooded animals include all of the invertebrates, the fish, amphibians, and reptiles. Their body temperatures vary, approximately following that of their surroundings. When active, their metabolic rates approach those of the mammals but they lack the capability of producing and maintaining a constant heat level. In general, the cellular efficiency of poikilotherms is determined by the external temperature of their environment plus the internal heat added as a result of muscular activity. This leaves cold-blooded animals with periods of vulnerability. When their body temperatures fall, it handicaps their ability to move, their sense of responsiveness, and in a sense their personalities. A cold rattlesnake is something like a man with a hangover hiding his head under the blankets. He neither senses the outer world nor wishes to be a part of it. In this condition a rattlesnake is easy to avoid or to kill. But put the same reptile in the open on a sunny day when the air temperature is 90° F. and you will see a totally different animal. He becomes active, responds to all sensory stimuli, and is a malevolent and aggressive being. Such a serpent is best avoided unless you carry a long-handled shovel. For the rattlesnake in the heat of midday is functioning as he should, since his body chemistry is going on at its most efficient level and it will kill you if it can.

The warm-blooded animals, the mammals and the birds, have the great advantage of having evolved the regulatory basis for maintaining a constant body temperature. In fact, barring illness their temperature is so constant that the slightest deviation is taken as a manifest evidence of an upset system. Our own physicians are very quick to note that the rise of our temperature from 98.6° F. by even a single degree indicates something is systemically wrong with us. Homeotherms possess the capacity to respond to stimuli with immediate bodily activity without a warming period. The only exceptions consist of such animals as bats and hummingbirds, who go into a torpid state during a part of their daily cycle, and those animals that may hibernate for some months during the winter. In general, even a sleeping man is capable of immediate direct action as any reader will know who has been frightened out of his sleep by some truly startling stimulus, such as a fire or an armed intruder. With a jolt of adrenalin the body is mobilized for immediate action.

One of the interesting aspects of general evolution is that animals embodying expensive processes are among the most successful. For example, a warm-blooded animal in the Arctic has an overall metabolic rate measurably higher than the same species living in an equable climate. Obviously a sizable increase in food intake is involved, but the fact that the Arctic does contain terrestrial mammals and birds demonstrates that the expenditure is valid. It is a part of the evolutionary process of trying to fill all possible ecological niches. These warm-blooded animals were able to invade the colder land surfaces which had turned back their predecessors, the reptiles. Temperature regulation was one of the great systematic changes by which progressive evolution opened up new frontiers for animals to inhabit.

At a more immediate level, it has been calculated that for a man living in a mild climate such as southern California, some 75 to 80 percent of the food eaten is expended in energy terms in maintaining a constant temperature. This is a high price but success has validated it. One of the consequences of the evolution of temperature regulation is that selective processes have pushed the normal temperatures of both animals and birds to a point just below the lethal range. Remembering that ordinary proteins begin to addle at 130° F., it is obvious that living materials cannot approach this temperature as a normal range. In fact, the tolerable temperature is considerably lower, for in man prolonged fever ranging from as low as 106° to 110° may leave permanent brain damage. The generalized relationship between cellular efficiency and actual temperature is plotted in Figure 6–5. Cellular efficiency is expressed as arbitrary units. The trend of the relationship between the two variables is quite clear. *Van't Hoffs Rule* states that for every rise of 10° C., the rate of chemical reaction is doubled. The Rule holds for the

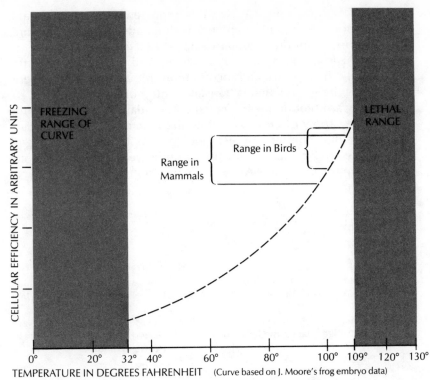

FREEZING RANGE OF CURVE

LETHAL RANGE

Range in Birds

Range in Mammals

CELLULAR EFFICIENCY IN ARBITRARY UNITS

0° 20° 32° 40° 60° 80° 100° 109° 120° 130°

TEMPERATURE IN DEGREES FAHRENHEIT   (Curve based on J. Moore's frog embryo data)

Figure 6-5
Range of Tolerable Body
Temperatures

production of carbohydrates in plants and the rate of cell division in eggs. While this sounds like a linear or straight line relationship between rate of reaction and temperature, it actually is a geometric one. The point is that biochemical activity measured by the rate of cell division is geometrically proportionate to temperature. It can be seen in Figure 6–5 that among the mammals normal temperatures cluster around 100° F. Man is a large animal and has a normal temperature of 98.6° F. Very large animals like elephants have a temperature a degree or so lower. Very small mammals such as mice and shrews normally have a temperature a few degrees over 100° F. These temperatures range inversely with respect to body size and with good reason. Large animals contain a relatively smaller skin surface compared to their body mass than do small animals. Correspondingly, they lose less heat in a relative sense than do the little ones. Animals like mice with their high losses of heat through radiation must maintain higher internal temperatures for normal physiological functioning. The same general principle holds true for birds; however, none of these temperatures normally much exceeds 110° F., and one may say this represents the normal safe ceiling of body temperature above which an advanced animal dare not go without suffering cellular damage.

In conclusion, the mammals show marked advances over the reptiles in limb structure and locomotion, in the circulatory system which fuels the energy that is translated into bodily activity, and, of course, in warmbloodedness. Dental differentiation makes possible a very wide variety of lifeways among the mammals. Other specializations are described in Supplement No. 9. With these advantages it is not surprising that the mammals underwent a series of rapid radiations in the Cenozoic. An example of adaptive radiation is given in Supplement No. 10.

Bibliography

KARTEN, BJORN.
  1968 Pleistocene Mammals of Europe. Chicago: Aldine.
MAYR, ERNST.
  1966 Animal Species and Evolution. Cambridge, Mass.: The Belknap Press of Harvard University Press.

SIMPSON, GEORGE G.
  1953 The Major Features of Evolution. New York: Columbia University Press.
  1967 The Meaning of Evolution. New Haven: Yale University Press (revised edition).

The Progressive Evolution of Some Functional Systems

Progress implies advancement to a higher state, and this is readily seen in the improvement of bodily functions and structures as we go from early, lower organisms through the higher animals including man. Such improvement involves *specialization* in the sense that there is departure from the ancestral condition, but the degree of specialization need not be and usually is not extreme. Genera that become overspecialized are prone to go to extinction and hence are not in the main line of evolution. On the other hand, considering organisms as a whole, primitive or *generalized* features are not always eliminated, for they often serve broadly adaptive functions and may in turn lead to new stages of structural progress. The primate hand is a good example of this, for if the generalized ancestral condition of five digits had not been retained, it might not have developed a functional opposable thumb, and in time become able to make and use tools. In this sense we may say that in the earlier primates the five-digited hand represents a clear case of structural *preadaptation*. There is no implication that this generalized type of hand was part of a design toward the future evolution of men, for in any given stage a structure must be essentially adaptive to be retained. Five-fingered hands were optimal for most of the primates and this is why the feature continued in their evolution. It is interesting that among the most acrobatic forms, such as the gibbon and the spider monkey, the size of the thumb has been reduced while the other fingers elongate to provide a better hand for grasping branches in their rapid locomotion. These are instances of moderate specialization within the primate order.

We owe the term *anagenesis* to the great German evolutionist Bernhard Rensch. A technical designation for progressive evolution, anagenesis usually involves a consideration of straight line or *phyletic* evolution proceeding through time and marked in general by an increase in complexity. It also involves a tendency toward structural centralization in the interior of the body, a more efficient division of labor among the organs, and a series of other changes which improve the individual functioning and adaptation of the animal. Rensch also gives us the term *cladogenesis* to describe the type of evolution characterized by the branching of its lines of descent. Here the emphasis is upon the differentiation of different genera and species as they adapt to different ecological niches. Such diversified evolution may proceed rather slowly in time but very frequently it occurs in almost explosive terms, in which case it is labeled an *adaptive radiation*.

Animals eat, a process which involves the capture of exterior energy and its assimilation into the body. In some forms of single-cell animals the tiny creature simply engulfs the food particle by wrapping its body about it. Among simple multicellular animals such as the flatworm and jellyfish, there is a body cavity forming a sac into which the food can be moved for digestion and the waste products later excreted. In more complex animals a straight line digestive system was evolved with a new opening, the mouth, added to the gut which eliminates waste products through the primitive opening, or *anus*. The gut itself becomes biochemically specialized, with enzymes secreted in the foreportion to aid digestion, while the terminal portions are devoted to absorption of the broken-down food. Among the vertebrates, such primitive living forms as the lamprey have an intestinal tract so poorly developed that its internal surface for its entire length is enlarged only by one simple lengthwise fold. In the higher types of fish and amphibians, there is a marked increase both in the length of the intestine and in its differentiation, thus resulting in the enlargement of the area of its inner surface. The reptiles show a series of fine folds. The mammals have a vastly increased absorbing area owing to the evolution of efficient *villi*, tiny fingerlike projections about a millimeter in diameter which lie on the small intestine in enormous numbers and consequently give its interior a velvetlike appearance. The area of absorption is further increased for each villus contains thousands of smaller projections upon it. The progressive evolution of the digestive tract has provided the higher forms with the means of retaining larger amounts of energy from the food eaten.

All animals are capable of motion to some degree and at some stage of their life. Even among single-celled free-living animals there is some ability to position themselves in their chosen environment, and it is achieved by a surprising number of simple devices. Some project their bodies forward while pulling in their behinds in order to achieve new positions. Others have bands of thread-like projections of protoplasm, called *cilia*, which beat in rhythmic fashion to produce tiny displacements of their bodies in a chosen direction. But while the powers of locomotion in single-celled animals may provide them with simple option of movement, in general they are carried along by the moving mediums in which they live, whether these be fresh or marine water or the bloodstreams of men.

The evolution of multicellular animals allowed them to grow to greater size and to evolve more complex structures, including those which aided movement. In some marine animals only the larval forms are free living, and in maturity they settle down on the ocean bottom to lead an essentially stationary life. Examples are the coral-building animals, the sea ferns and sponges, and the sea anemones. Some multicellular animals have developed spectacular rates of locomotion in water, on land, and in the air. Generally, the power of locomotion has evolved to efficiently suit the needs of the organism's ecological niche. Those who live by the sea may have seen a simple coelenterate, the jellyfish, coming up to the lights on piers at night, with their mantles or "petticoats" gently swishing rhythmically to drive them slowly upward toward the surface of the ocean. When this motion ceases they slowly descend. Jellyfish possess the limited capability of moving only in a vertical direction in their marine environment, and consequently are totally at the mercies of currents. Sudden storms may drive great numbers of them onto the beaches, where they are stranded. They have no effective control over their lateral movements.

As rates of motion increased, particularly among the vertebrates, the shape of the body changed so as to present the least resistance to motion. This is particularly true among water-living animals where the density of the medium is great and hence resistance to motion high. But it is also found in high-speed flying animals. Among terrestrial animals the requirements of supporting structure seem to dominate shape considerations.

While most of the free-living marine invertebrates move at rather slow rates, there has evolved among the carnivorous cephalopods—the squids and the octopi—a form of jet propulsion. The squids, ranging in length from a few inches to more than 50 feet, are largely wide-ranging carnivores and must move rapidly to feed upon a wide variety of fish. Consequently, they have evolved into a streamlined form, with steering planes to give maneuverability. Their good lensed eyes have evolved independently from those among the vertebrates and provide a nice instance of parallel evolution. The squid has neither an external skeleton nor a true backbone, so its high speed in water required another solution to prevent its body from losing its optimum shape through water resistance. Its gelatinous flesh maintains its form due to the *pen*, a much altered vestigial remnant of the shell which once protected the ancestral cephalopods. This pen—lightweight, rigid, and hollow-celled—stiffens the squid's body so that its streamlined form is retained. The pen is nothing less than the familiar cuttlefish bone which you may have seen in your grandmother's canary cage, where it was placed to provide the calcium for the bird's health.

The earliest fish were rather slow bottom feeders. But they met their basic problems so success-

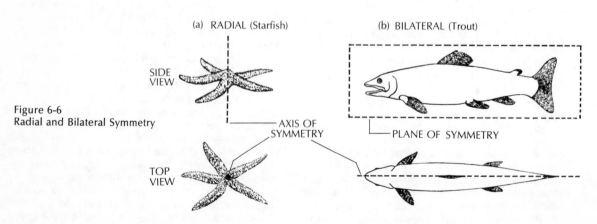

(a) RADIAL (Starfish)

(b) BILATERAL (Trout)

SIDE VIEW

TOP VIEW

AXIS OF SYMMETRY

PLANE OF SYMMETRY

Figure 6-6
Radial and Bilateral Symmetry

fully that their descendants evolved into forms which swim at speeds up to 30, 40, and even 50 miles an hour. Even the sluggish earliest fish had to evolve proper instruments to guide their motions through their three-dimensional environment. Lensed, image-forming eyes were a part of their original instrumentation. Swimming originally involved sending a wave-like motion down the length of the body through the alternating contraction of lateral muscle segments, and its efficiency was improved by the development of a posterior or caudal fin by which they sculled their bodies more rapidly through the water. And, of course, they needed steering devices. Early fish possessed two paired fins placed well apart at both the front and rear portions of their bodies (Figure 6–6). These were essentially not swimming fins but steering surfaces which operated much like the diving planes on modern submarines. They could be used independently to make rapid turns and to direct vertical motion.

These paired fins are of particular significance, for they were the forerunners of the paired limbs of all subsequent terrestrial animals. Although their original functions involved steering the fish's body, they were fortunately preadapted to evolve into the fore- and hindlimbs of the first land vertebrates. Indeed, the first amphibians are little modified fish in their general appearance, and the basic structure of the amphibian limbs is of great importance (see Figure 6–7). The fore-limbs differ from the hindlimbs in the mode of their attachment to the body, but the two pairs are otherwise similar in structure. In the fore-limbs of amphibians and other generalized vertebrates, a flat bone called the *scapula* anchors them to the thorax of the body. The scapula is embedded between layers of muscle so that it is securely attached to the rib cage and yet capable of flexible motion in wide variety. The hind pair of limbs is attached through a ball-and-socket joint to a complex bone, the *pelvis,* which is firmly knit to the backbone through a series of ligaments and articulations. This provides a relatively rigid point of attachment of the hind legs to the body, an important point since the propulsive power of most terrestrial animals is primarily developed by the hind legs.

In both pairs of limbs a single bone articulates with the scapula or pelvis, respectively. In the forelimbs it is known as the *humerus,* and in the hindlimbs as the *femur.* These bones are generally circular in cross section and hollow in construction so as to provide a maximum of strength and rigidity for a minimum of structural weight. They also conveniently serve an added function in housing red marrow, which through mitotic division provides the red corpuscles of our blood. The humerus articulates to the scapula as the femur does to the pelvis through a shallow ball-and-socket joint which gives the bone a wide range of possible motion.

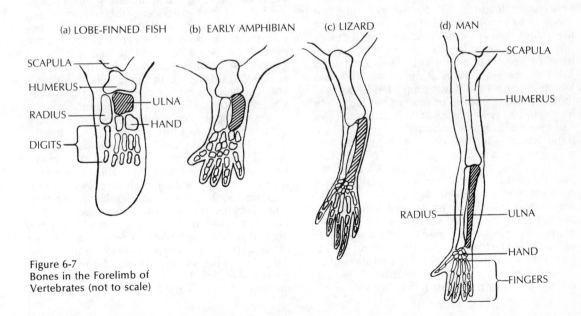

Figure 6-7
Bones in the Forelimb of
Vertebrates (not to scale)

In each pair of limbs the third segment consists of paired bones. In the forelimb these are the *radius* and *ulna* and in the rear limb the *tibia* and *fibula*. This basic pattern is present in all ancestors of all terrestrial animals including the birds, although considerably altered in all flying vertebrates. The interesting question arises as to why this section of both limbs should be characterized by paired bones, since in a structural sense they use materials less efficiently in that adequate strength is provided at the cost of somewhat greater weight. The answer lies in the nature of early amphibian locomotion. In these early *tetrapods*, the four limbs were extended in bent positions from the sides of the body, which moved with a fishlike motion as the animal progressed forward. The limb bones in such amphibians are small and so weakly muscled as to barely allow the animal's belly to be lifted off the ground. In the wiggling motion characteristic of these generalized amphibians the feet need firm contact with the ground. The paired bones in both sets of limbs have the capacity of rotating, which is necessary for this kind of locomotion.

Look at your own hand, and note its five fingers, the primeval or basic number of digits also common to the ancestral amphibians. This primitive number indicates that in this feature man is very generalized in contrast to such features as the brain in which he is highly specialized. Man, like so many other animals, possesses a kind of mosaic or patchwork of specialized and generalized features. Of course, the question arises as to why the number of digits should be five rather than any other number. Was it design produced by selective pressure or was it some form of accident? In the ancestral lobe-finned fish the steering portion of the fin was stiffened by longitudinal rays. It was supported by a fleshy lobe containing a series of small bones judged to be homologous with those of the hand and upper and lower arm. It was anchored to the body by a bone which ultimately developed into the amphibian scapula and pelvis. The fleshy and bony nature of the support base of the fin indicates that they must have been used under some circumstances to provide support for the body out on land. With hindsight it is easy to judge that five stiffening members were the most efficient. But perhaps we should examine informally some of the design factors. Clearly, a membrane will be most efficiently stiffened with the least amount of structural support if a single member is passed through its center, dividing it into two symmetrical unsupported portions. Each of these in turn can be further stiffened by one additional member on each side. It is evident that with the central supporting structure the total number of members will be odd, whether they remain at three, proceed to five, or even on to seven. In our ancestral lobe-finned fish, five seems to have been the efficient number.

Except for the specialized jumping amphibians, such as the successful toads and frogs, all modern and most ancient ones seem to have moved slowly. Compared to the relatively free locomotion found among fish, this might seem to be a retrogressive condition. But such a judgment would not take into proper account the difference in the media in which the two animals move. In water a fish's body is buoyed by the weight of the water displaced, which nearly enough produces a state of essential weightlessness. On land an amphibian is only supported by the weight of the air displaced by its body, an amount that is negligible. Hence the full gravitational force of the amphibian's weight must be supported by its limbs, and so it moves under vastly more difficult conditions. As the ancestral lobe-finned fish slowly evolved into the early amphibian, it is only to be expected that we would find their paired limbs small, poorly muscled and seemingly inadequate. Their position jutting out from the sides of the body would confer some mobility but only at a very low rate. Many of these early amphibians spent much of their time in the fresh water of their environment, and we may consider that in most cases they proceeded slowly and awkwardly on land. Their form of life was a compromise and even today with all our technological skill, aircraft, tanks, and automobiles designed to operate as amphibians are relatively inefficient compared to those designed to operate in only one medium.

Whereas the amphibians were, and remain today, compromises through their partial adaptation to both aquatic and terrestrial environments, their descendants, the reptiles, became primarily land-based in their orientation. This involved the strengthening of limbs and changes in their proportions. Such generalized reptiles as the little fence gecko serve as satisfactory models to illustrate the shifts that occurred. Freed from the original tyranny of having to return to water or moist places for reproduction through the evolution of an improved egg, the reptiles moved into terrestrial niches. This required the conversion of the water-losing skin of the amphibian into the dry one of the reptile, which served as a barrier against such losses. Among the reptiles both the fore- and hindlimbs became elongated and more powerfully muscled. As a result, they could move across ground with their bellies held well above it and at considerable speeds. Today's little geckos scamper away in fright at a very respectable rate. Larger lizards such as the monitors show surprising bursts of speed for unexpectedly long periods of time. I have seen a six-foot desert goanna in western Australia take alarm and head off across the sandhill country at a rate that must have approached 25 miles per hour. Accurate figures for these big lizards do not seem to be available, but they are clearly capable of rapid movement and stand closer in this respect to the mammals than they do to the amphibians. Even though the limbs remain jutting out from the sides of the body, they are so powerfully muscled as to support the body easily and provide the animal with high mobility.

During the Mesozoic period a number of evolving reptilian lines became bipedal in their locomotion, in a fashion analogous to that of modern running terrestrial birds. This shift in posture required a balancing tail carried free of the ground and allowed the forelimbs, freed of locomotor responsibilities, to become much reduced in size. Many of the ruling dinosaurs, including both carnivores and herbivores, adopted this style of progression. The modern birds trace their ancestry to one of the more lightly built bipedal dinosaurs. Even today a number of smaller lizards when alarmed take off at maximum speed, running swiftly on their enlarged hind legs. Interestingly, in two entirely different regions of the world, Central America on the one hand and Southeast Asia on the other, there have evolved bipedal lizards who earn their livings along the edges of rivers. In both cases they are capable of running so rapidly bipedally that they can rush off the bank of the narrow river and continue their way across the surface of the water without sinking and reach the opposite shore. In Central America these water-running lizards are irreverently called "Jesus Christ" lizards.

Even in early Mesozoic times, a number of lines among the reptiles began to take on mammal-like attributes, so they are appropriately called the mammal-like reptiles. Their teeth began to undergo differentiation in terms of new functional requirements. Their bodies showed changes involving locomotion. Limbs which formerly jutted out from the sides of the body, somewhat like pairs of broken oars, moved toward positions more nearly under the body. This shift was not completed in the mammal-like reptiles, but they were clearly foreshadowing a true mammalian posture. Limbs placed directly beneath the body hold it higher off the ground, and their movements are more efficiently transmitted into body progress since there is no lost component of circular motion as with side-positioned legs. It is interesting that for all these progressive changes the mammal-like reptiles did not dominate the Mesozoic, even though their descendants, the mammals, did in later times. The Mesozoic remains firmly in the hands of the various types of dinosaurs which were themselves descendants of earlier small bipedal reptiles.

Modern mammals characteristically have their limbs completely positioned beneath their bodies and so gain efficiency in movement. Horses

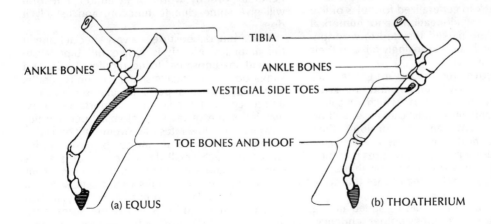

TIBIA

ANKLE BONES

ANKLE BONES

VESTIGIAL SIDE TOES

TOE BONES AND HOOF

(a) EQUUS

(b) THOATHERIUM

Figure 6-8
Parallel Evolution
(not to sclae)

and dogs, both of which move fast enough to race, illustrate this posture well. In addition to the new position of the legs, there were other accompanying changes. The lower sections of the legs become elongated relative to the upper portions. This distal elongation, which primarily affects the bones of the foot, characterizes all rapidly moving terrestrial animals, and can be taken as a measure of the velocity of which they are capable. This tendency is exemplified in Figure 6–8, where two animals only distantly related and coming from very different early Cenozoic ancestors, end up with very similar-looking feet. The reduction of the toes from five to one in the horse, *Equus*, is well known. Less appreciated perhaps is the fact that buried deep in the sides of his foot are tapered vestigial bones representing the remnants of the second and fourth toes, respectively. A very much smaller animal, one of the strange forms evolving in isolation in South America, was little *Thoathurium*. It is illustrated here to show the same tendency for the bones of the foot to become greatly elongated, for the side toes to be eliminated, and for the animal to move rapidly on the tip of the remaining central toe. The little South American pseudo-horse carried the reduction of the side toes even further, and all that remain are two tiny sharpened splints of bone as shown in Figure 6–8. In addition to showing the characteristic elongation of the lower extremities, this pair of animals nicely illustrate evolution proceeding along almost identical paths. *Parallel evolution* involves changes over time among related animals which began at similar structural levels but which developed comparable, however different, characteristics in the end. *Equus* and *Thoathurium* had very different ancestors, but presumably both go back to generalized formulas of five digits. A series of elongations and numerical reductions as time passed resulted in very specialized feet which are surprisingly alike in their anatomical details.

A steady selective pressure toward greater running capabilities is easy to understand. In a given population of herbivores, those which are a little faster, even if only in a statistical sense, will be most likely to survive and to contribute their qualities to the next generation. By the same consideration, among those carnivores that tried to run the herbivores down and eat them, the faster carnivores ate better than those that were slower.

With the evolution of increasing speed among terrestrial animals there goes another tendency,

that of the muscle masses to lie higher upon the limb. The principle of inertia is involved here, and it is easy to see that heavy muscles carried on the further ends of the extremities would require much more energy to move them through space at the same velocity. Without worrying about the actual mechanics, the student can visualize which of the following options he would prefer if running for his life! Would a ten-pound weight strapped to each ankle impede motion more than the same pair of weights belted to the hips? It is obvious that the weight carried high on reciprocating limbs is carried more easily and impedes progress less. Evolution has generally favored the principle that rapidly running animals develop muscle masses lying high on their legs. Again the horse provides a useful and well-known example. On the foreleg the primary muscles lie high on the humerus and almost seem to be an integral part of the animal's chest. The lower portion of the leg and the vastly elongated feet consist primarily of bones and tendons and look it. The hind leg of the horse illustrates the point even more strikingly. Its large muscle masses, frequently called hams, lie primarily about the upper bone, the femur. These muscles are much bulkier than those in the foreleg and show directly what proportion of the horse's total propulsion comes from his hind legs. The tibial and foot region again are bones and tendons. This progressive evolution of the muscle masses to higher positions on the limb requires the lengthening of tendons to operate the more distant bony masses to which they attach. This is in contradistinction to slower moving animals in which the muscles are distributed so as to operate the bone immediately below them. It is to be expected in the generally orderly world of evolution that form will give some clue to function, and usually it does.

Man's bipedal posture is a very special thing in the animal world. It is unlike the bipedalism present in dinosaurs, living reptiles, birds, and kangaroos in that man's spinal column is parallel to his legs instead of perpendicular to them. This difference is of fundamental importance in explaining some of man's anatomical peculiarities as well as his slow rates of movement. For a large animal he moves poorly and can be easily outdistanced by such small domestic creatures as his own dogs and cats. Since the evolution of his bipedal posture represents very drastic series of changes, if it did not improve his mobility, it must have been under other selective pressures. And, of course, it does provide the all-important freeing of man's hands from the functions of locomotion to serve as tool-holding and tool-using organs. This anatomical and subsequent cultural revolution totally changed man's way of life and his future evolutionary potential.

The relationship between form and function in movement is manifest in many portions of the animal world. The tail of a fish tells much about his speed of swimming. If we limit our discussion to marine fish, the inshore dwellers, those of the kelp beds, are relatively sluggish fish whose tails are short and broad like those of fresh water fish. These proportions are correct, for their way of life does not require very rapid locomotion. But if we turn to the most active of the deep-sea fish, members of the tuna and swordfish families, they invariably show an inordinately high and narrow tail which usually has the form of a symmetrical and elongated crescent, as shown in Figure 6–9. Such tails have what aeronautical engineers call a high aspect ratio. This term is taken from the design of airplane wings and it is simply the ratio of the length of the wing to its average breadth. At subsonic speed planes designed for high performance flight or the breaking of long-distance records are invariably characterized by high aspect ratio wings, for these promote the greatest efficiency. What is good in the air apparently is also good in the water. Rapid swimming pelagic fish have high aspect ratio tails for equivalent reasons.

The same general principles apply to the wings of birds. All soaring birds tend to have high aspect ratio wings. It is true of the soaring hawks, it is even more extreme in the ever-gliding buzzards, and becomes most marked among oceanic birds like the albatrosses which glide on interminably, riding the updrafts of air from the waves. Among the last, aspect ratios reach and even exceed a ratio of ten to one. Even among undistinguished little birds, these differences of form are significant. Within closely related species, the nonmigratory bird will have shorter and more rounded wing tips, while those species that migrate, and hence need to have more efficient flight, evolve longer and more pointed wings, that is, wings with a higher aspect ratio. And finally the most skillful fliers of all the aerial carnivores, the falcons in all their forms and sizes, are characterized by long wings which taper sharply to a point and so again have a high aspect ratio. Some of these forms are shown in Figure 6–10. Structures generally evolve to suit a particular way of life, and if their meaning can be read, the variations in nature become orderly.

Man's brain represents the evolutionary culmination of the central nervous system, but many lowly animals have nervous systems of one sort or another which are essential for their success in life. Since the fossil record of the invertebrates

(a) Kelp Bass

(b) Sail Fish

Figure 6-9
Tails Reveal Speed
of Swimming (not to scale)

Figure 6-10
Wings of Birds
(not to scale)

(a) Song Bird

(b) Falcon

(c) Albatross

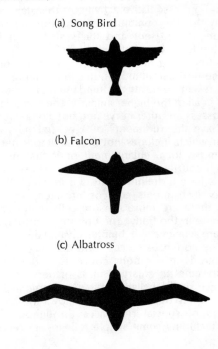

has many missing pages, it is convenient to compare the anatomy of living members of the various invertebrate phyla to study their nervous systems.

The simplest single-celled forms of life have their own mobilities and responsiveness. Protozoans, when prodded by the point of a needle, alter their direction of movement to a new course. Although we understand little about it, they obviously can conduct impulses of some sort through their entire minute bodies. Proceeding to simple multicellular animals, the sponges provide a curious model. These animals consist of an aggregation of cells that operate quite independently of each other, but they do function as a community since one cell can transmit stimuli to another. This conduction of impulses across intervening cells may stand as an analog of a primitive nerve cell.

A little higher up the scale of animal life stands the phylum of the coelenterates, which do show a nerve network. In the jellyfish, corals and their relatives, the nerve net consists of nerve cells, or neurons, all very much alike and spread rather thinly but evenly throughout the outer layers of the animal's body. The main response to external stimulation is a local contraction of the net. Such an event is signaled throughout the whole nerve net. The nerve system operates by conduction, which is slow, but it does permit simple reflexes, although no associative activity. At this stage we can say that the minimum essentials of nervous reaction are present and the system is quite adequate for animals such as the jellyfish.

The free-living flatworms, the planarians, provide the next useful model. In a very simple way they present some of the fundamental features found in all of the higher animals. The planarians still possess an outer nerve net, but along with it they have the rudiments of a central nervous system which includes not only a simple nerve cord but a local enlargement of it that might through courtesy be called a brain. These little animals have a definite front end in which their sensory instruments are concentrated, and in which there are more sensory cells than occur elsewhere in the body. Their photoreceptors, or eyes, are sensitive to light intensity and direction, but have no lenses so can form no images. The nerve cords in the body converge in the front end, forming an enlarged mass of nerve tissue which we shall label the brain. As simple as the planarian nervous system is, it allows the coordination of special responses throughout the body, including some simple reflexes. In recent years experiments have shown that these little flatworms have some learning capacity in terms of responses to changes in light.

The nervous system of the flatworm provides the basis for enormous elaboration and differentiation in the evolution of the higher animals. In them, the concentration of neurons in a central nervous core and in adjacent masses called ganglia sets the pattern. Ganglia are connected by a series of branching nerves to the peripheral nervous system. The system becomes differentiated into sensory nerves which carry the messages of stimulation to the central system, and motor nerves which carry back the responses to be made. With this differentiation within the system, there is an accompanying increase in associative neurons as well as an increasing complexity of nerve roots and connections. The front end of the nerve cord becomes increasingly enlarged and more complicated as the sense organs in the front end of the animal undergo evolutionary elaboration. The number of associative neurons increases, tracts are developed, and in time what we may appropriately call the brain provides the central coordination for complex responses.

Man is inclined to believe that lesser animals have no brains and hence no sense or feelings. Many an otherwise sporting fisherman allows his trout to die by suffocation in his creel, believing that they feel nothing in dying. Hence it is perhaps worth pointing out that well below the vertebrate level of organization a functionally adequate nervous system often occurs. Among the arthropods, including the insects, the central nervous system is functionally well developed and culminates in a brain in the animal's head. In the more advanced molluscs, such as the squid, the central nervous system is well developed, and a brain allows the animal fairly complex learning patterns and some simple decision making. While animals with backbone have much more highly evolved brains, it is perhaps worth remembering that even such lowly creatures as ants are endowed with brains adequate for their needs.

But we are primarily interested in the evolution of the vertebrate brain, for much of the vertebrates' phyletic success results from being endowed with a superior central nervous system. The primitive vertebrate brain shows three irregular swellings in a hollow nerve cord at its anterior end. These are, respectively, the forebrain, the midbrain, and the hindbrain. Even though the human brain is much changed in their proportions, these three basic regions can still be distinguished in it. In Figure 6–11 these regions

Figure 6-11
Evolution of the Brain in the
Vertebrates (not to scale)

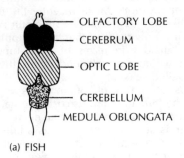

(a) FISH

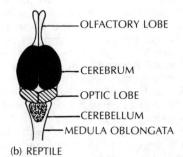

(b) REPTILE

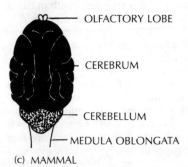

(c) MAMMAL

are diagrammed in the fish, the reptile, and the generalized mammal. The forebrain consists primarily of the olfactory lobes and the *cerebrum,* or cerebral hemispheres. The midbrain is primarily identified with the *optic lobes.* The hindbrain contains the *cerebellum* and the *medulla oblongata.* The forebrain receives the sensory stimuli of the nose, the midbrain is connected with the eyes and receives the visual stimuli, while the hindbrain has six to eight pairs of nerves that connect with other scattered receptors and the muscles of the head. All three regions of the brain are connected directly or through chains of neurons with the medulla oblongata, which grades imperceptibly down into the spinal cord.

The three fundamental portions of the brain show vast changes in relative proportions from fish to mammals. The sense of smell is important in all three classes, but in fish the olfactory lobe is relatively larger than in the reptiles or the mammals. This is primarily the consequence of the great expansion of the cerebrum, which is small in the fish, relatively larger in the reptile, and overwhelmingly dominant even in the simplest of mammals. In this progressive evolution the cerebrum overgrows the olfactory lobe and the midbrain. The latter in the fish, represented by the optic lobe in Figure 6–11a, is the preponderant region of the brain. These animals clearly depend primarily upon their sight for their living. In the reptile, the optic lobe is relatively much smaller compared to the expanded cerebral hemispheres. In the mammal, the midsection of the brain, that is, the optic lobe, is completely buried beneath the expanding cerebral hemisphere, and even functions have changed. For among the mammals the visual stimuli are now received in the rear portions of the cerebral hemisphere, that is, in the occipital lobes. As viewed from above, the optic lobes have disappeared entirely in the mammal, although the midregion is still present in the base of the brain. The hindbrain, represented by the cerebellum and medulla oblongata, acts as a kind of message center with impulses from both directions between most of the body and the brain passing through it. Today we speak of this region as the lower center of the brain, but the importance of its functions should not be underrated.

In lower animals, behavior is largely genetically determined, but it can often be modified by learning experiences. The same statement applies to the lower vertebrates, and it may be reasonably claimed that fish are guided through instinctive behavior, and that learning processes are limited, as shown by the small size of their cerebrum. Reptiles are considerably more advanced in this respect, although it is fair to claim that they behave primarily in an instinctive fashion. But with the arrival of the mammal we find that the cerebrum has expanded enormously in response to learned behavior becoming dominant over instinctive behavior. At the same time the *neopallium,* or *neocortex,* which made its first feeble appearance in the advanced reptiles, now spreads over the cerebral hemispheres. The neocortical cells are involved in association in the coordination of all kinds of impulses from various other receptors and brain cells, though it is not involved with olfactory functions. While the neocortex first appears in the reptile, its real expansion begins with the mammals. It is of some interest that brain casts of the earliest mammals and birds of the Mesozoic indicate that their brains, if we judge them by modern standards, are primarily reptilian in organization rather than approximating those of their modern descendants. Casts of the interior of the skull of these fossils cannot give us fine structures, or indicate quality in improvements, but they do reflect the relative size of the various portions of the brain. Their organization suggests that the evolution of the structural organization of bones and muscles proceeded faster than did the evolution of the brain. By the time we come to so simple a modern mammal as the shrew, the brain has diverged widely from the reptilian type. While the olfactory bulbs remain relatively large, they no longer terminate in long stalks. The cerebrum has become greatly enlarged but is not yet completely covered by the neocortex. The midbrain is already reduced in size and squeezed between the forebrain and the hindbrain.

In the more highly evolved mammals such as dogs, horses, and man, the olfactory bulbs undergo further relative reduction, although in dogs and horses the sense of smell is still of considerable importance. But the dramatic change in the higher mammal involves the expansion of the cerebral part of the forebrain, so that it overrides both the olfactory lobes and the midbrain. Now the cerebral hemispheres are completely covered by the neocortex. Since the neopallium is primarily involved with learned behavior, its area has been increased in the progressive mammal by infolding or the production of convolutions on the surfaces of the hemispheres. Without changing the contained volume of the cerebrum, its surface area and hence that of the neopallium are markedly increased. Primitive mammals lack these convolutions, but they increase in complexity in the advanced mammals, culminating in man, and, interestingly enough, the whales. In fact, it is claimed that the convolutions show greater complexity among these aquatic mammals than they do in human beings. Yet there is no doubt that in terms of body size the brains of man are larger and more complex than in other animals.

In spite of our vast technological progress it remains true that we do not totally understand nerve action, let alone how the brain works. A variety of models has been proposed to explain the functioning of the human brain. At one time it was said to be analogous to a complex telephone switchboard. We now know that this was an oversimplification. More recently the brain has been likened to a complex miniaturized computer. It is now realized that the computer is also a false analogy in terms of brain functioning. Indeed, we do not really understand the brain in any detailed sense.

Let us return to some of the things we do know about the human brain. Its cortex contains about 10 billion neurons, or elemental nerve cells. This number is so enormous that as adults we can afford to lose hundreds of thousands of neurons in later years of life. Only in very old age are intellectual capacities affected by this loss. Very little is known about the operation of the cortical neurons except in a few special areas. In the midregions of the cerebral hemispheres lies the central fissure. Here our knowledge is reasonably secure since it is based on operations on man and experiments on monkeys. The motor area of the brain lies in front of the central fissure and the sensory area just behind it. Mapping of the motor area bounding the anterior rim of the central fissure indicates that starting with the top it enervates the toes of the body, and proceeding downward it is concerned with the trunk, hands, and mouth, and finally ends with mastication. It is interesting in terms of man's cultural specialization, which involves both tool-using and talking, that the motor regions given over to the hands and to the face, particularly the tongue, are disproportionately very large. On the posterior side of the central fissure the sensory regions start at the top with the genitalia, followed by the toes, the rest of the body, and ending up with tongue, pharynx, and inner abdomen. Again, significantly, the hands and face are represented by disproportionately large regions. This is a specifically human pattern and would not be found in other mammals, even if the mapping were possible.

The human neocortex is enormously complicated, but we know very little about it apart from our knowledge of the limited motor and sensory areas. It used to be claimed that the neocortex of the frontal lobe was the center of the higher associations in the brain. But more recent studies indicate that most of the neocortex of the brain is really involved in the function of association. When the frontal lobe is removed or injured, other regions take over its functions. The temporal lobe houses the auditory area, and

Broca's speech area is centered in the left frontal lobe. There is no known reason for this one-sided position for the speech center. In fact, the two hemispheres, and their slight asymmetry, pose special problems. It is not clear why the right side of the body should be represented on the left side of the brain and vice versa. We do know that larger bodies require somewhat larger brains, but the relation is mathematically rather complex and not a simple linear one. It serves to explain why human populations of low body weight have smaller brains than those with large bodies. The relationship between body size and brain size allows us to comprehend why in all human groups the brains of men are larger than those of women, for their bodies are, too. But in a given population and a single sex, larger brains show no convincing correlation with intellectual capacities. Some very bright people have been endowed with relatively small brains. Since we do not even understand the functioning of a single neuron, it is unlikely that in the immediate future we shall comprehend how ten billion of them act in concert. Indeed, we do not even know how memory is stored in the brain. But current research suggests that qualitative differences in brain function are more apt to turn upon biochemical differences in the brain's operation rather than either gross size or differences in anatomical detail. Until the time comes when scientists have enough knowledge to comprehend the real manner in which the brain operates, we shall do well to accept it as a very specialized end product in mammalian evolution.

SUPPLEMENT NO. 10

What the Evolution of the Horse Really Shows: A Complex Adaptive Radiation

George Gaylord Simpson (1953) angrily declared that several generations of students have been misinformed about the real meaning of the evolution of the horse. Since it is the best known evolving line among all the vertebrates, the record of the horse is of particular importance. The picture is usually presented as one of relatively straight line evolution, beginning with little Eohippus, a multi-toed small forest browsing animal, and passes through the development of three-toed grazing horses and in later times ends with the modern horse, *Equus*, which is single-toed and large in size. It is usually stated that the side toes underwent a gradual reduction in number until the one-toed stage was reached. Much of this story is incorrect, and the real evolution of the horse provides some points of general interest.

Instead of being simple, the ascent of the horse was very complex, more so than the simplified diagram given in Figure 6-12 indicates. There were no less than three major radiations involved in the course of its evolution. Horses shifted from one adaptive zone, that of animals browsing on forest leaves and shrubs, to plains grass eaters or grazers. And while marked changes occur in various structural features of the horses, nowhere do they happen at a single uniform rate. Finally, the record gives the usual history of disaster, for of the 16 distinct genera which can be identified in the fossil record, 15 of them went to extinction leaving no descendants, while one, *Equus*, survived into modern times. Wild horses and asses are surviving with difficulty, and only the zebra can be found in large numbers today.

The evolution of the horse starts with an Early Eocene genus, *Hyracotherium*, popularly but incorrectly known as Eohippus. Standing as high as a medium-sized dog, this little animal really looked as much like a potential ancestor for future tapirs or rhinoceroses as for the horse. The axis of its foot was through the third digit, but it had already lost one side toe on the forelimbs and two on the rear. It walked on its toes in a *digitigrade* manner but the foot was padded as in our dogs. In the course of many millions of years of time this ancestral horse evolved into *Miohippus*, by this time the size of a large dog. The digital pad was now reduced to three toes on both sets of limbs, and the animal continued to be a forest browser.

Late in the Oligocene or Early Miocene two separate radiations arose from such a horse. One involved a variety of three-toed browsing horses which became moderate-sized by horse standards. Of the four genera involved, two became extinct at the end of the Miocene and the other two barely lasted into the Lower Pliocene. There was no further room for browsing horses.

The other radiation, much more successful, involved leaving of the woodland niche of the browsing animals and moving into the now expanding grasslands of the Miocene world. No less than six separate lines of three-toed grazers radiated in the Middle Miocene, with two of these dying out in the Early Pliocene, two in the Late Pliocene, and one lasting into the Pleistocene. The sixth branch of this radiation produced the one-toed grazing horses. These underwent their own radiation in the basal Pliocene, which produced no less than five separate generic lines. Three of these appeared in South America when immigrant horses from North America were able to reach that continent. All but the living genus *Equus* went to extinction. The general moral of the evolutionary record of the horse is that most of the divergent genera in a radiation proceed to extinction. *Extinction is a normal part of the evolutionary fabric, and we should expect it ultimately to operate in human evolution.*

The details of structural change within the evolving lines of horses themselves provide certain lessons. It is frequently stated that horses showed a gradual and steady increase in size throughout their evolution. As Simpson points out, there was no increase in size during the long Eocene, then a rather rapid increase in the following period and much diversification thereafter. In the Miocene and Pliocene no less than two genera of horses became larger than the present one, and four were smaller. The latter obviously represent a reversal of the general trend in size. The evolution of the foot mechanism proceeded by rapid and abrupt changes rather than gradual ones. The transition from the form of foot shown by miniature Eohippus to larger consistently three-toed *Miohippus* was so abrupt that it even left no record in the fossil deposits. The latter as a browsing forest animal still had a padded foot. When some of the evolv-

ing lines of horses moved into the new adaptive niche of plains grazers, their foot structure changed very rapidly to a three-toed sprung foot in which the pad disappeared and the two side toes became essentially functionless. Finally, in the Pliocene the line leading to the modern one-toed grazer went through a rapid loss of the two side toes on each foot. This evolution was not gradual and constant. It proceeded by rapid jumps.

Teeth must evolve to suit the needs of the animal which uses them. In the Eocene the premolars of the little horse Eohippus had already taken on their characteristic molar-like form. Very little change in this regard occurred in the rest of horse evolution. The height of the crown remained low in the cheek teeth as long as horses browsed on the relatively soft diets provided in a forest environment. The only increase in crown height was that to match the increase in body size, for bigger eaters need longer wearing teeth. But among the Miocene horses that became grassland grazers there was a very rapid abrupt increase in crown size, the teeth becoming markedly *hypsodont,* with roots remaining open until old age. Such an adaptive change was required by the new diet, for grass not only contains the glass-like substances silica, but is covered with dust and dirt. Grazing involves an abrasive diet. At the same time the crowns of these hypsodont teeth became complicated with many folds of enamel separated by softer cement so as to wear down in such a way as to always provide a rough grinding surface. The elaboration of crown complexity was not regular in rate, and some of the extinct grazing horses had more elaborate ones than surviving *Equus.*

Simpson emphasizes that the evolution of the horse shows no trend that continues through its history, nor any trends that affect all the various lineages at one time. Any detectable trend continued in the same direction and at the same rate for periods no longer than 15 million years out of the 65 or so million years for which we have the record. Men have been too quick to see regularities where they did not exist. Since the record of fossil man is so scanty compared to the fossil horse, many of the seeming regularities will no doubt become much more complicated as further finds become available.

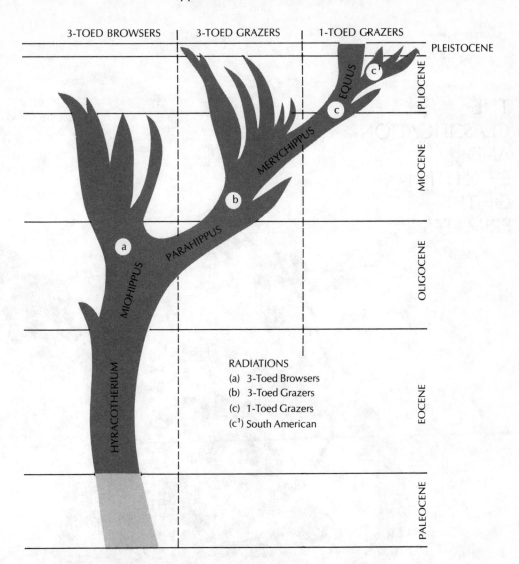

3-TOED BROWSERS | 3-TOED GRAZERS | 1-TOED GRAZERS

PLEISTOCENE

PLIOCENE

MIOCENE

OLIGOCENE

EOCENE

PALEOCENE

HYRACOTHERIUM

MIOHIPPUS

PARAHIPPUS

MERYCHIPPUS

EQUUS

(a)

(b)

(c)

(c¹)

RADIATIONS
(a) 3-Toed Browsers
(b) 3-Toed Grazers
(c) 1-Toed Grazers
(c¹) South American

Figure 6-12
Radiation of the Horses

CHAPTER 7

THE CLASSIFICATION AND EVOLUTION OF THE PRIMATES

The primates originated from generalized mammals late in the Mesozoic and finally evolved into man, the dominant animal on this planet. During the last seventy million years of the Cenozoic, the primates have prospered as they evolved from lower to higher grades. After their rich radiations in the early epochs of the Cenozoic, the lower primates became replaced by higher forms in both hemispheres. By the Miocene the Old World monkeys had reached the grade of anthropoids, or manlike apes. Seemingly within that same period the great dryopithecine radiation in Africa produced the first hominid, or man, in the form of Ramapithecus.

The Variety of Living Primates

When the great eighteenth-century classifier, Linnaeus, came to monkeys, apes, and men, he recognized them as being much alike and placed them all in an order which he called *Primates,* after the Latin term "primus" or the first. This reflected egocentric thinking in his day, but we now recognize that man has reached a new plateau in evolution and it is appropriate that the order which contains us should be given priority.

Taxonomically the primates are distinguished from all other placental mammals by a combination of rather curious diagnostic traits. Their extremities generally end in flat nails instead of claws. Their shoulders are positioned by a strut-bone known as the clavicle, which is adaptive especially in broad-chested forms for it preserves a wide range of arm motions. A complete ring of bone encircles their eyes; in many mammals this protective device is incomplete. Their teeth come in four series: incisors, canines, premolars, and molars. The penis is pendulous, while a scrotum contains the testes. There are always two breasts high on the chest. Man and the other anthropoids which have been tested have lost the ability to synthesize vitamin C, that is, ascorbic acid. For primates constantly eating fresh vegetation this is no handicap, and so they do not die of scurvy.

It is the combination of these features which define the primate order. For example, the clavicle is a conservative bone retained in such animals as amphibians and reptiles, but lost in most rapid terrestrial mammals. The hairless, clawless fingers and toes of the primate not only are useful in grasping, but also for feeling. They are highly sensitive to touch, and this is reflected in areas of the brain. Bats and bears share the pendulous penis with primates, and a wide variety of mammals carry their testes externally in a scrotum. Elephants and seacows which give birth to a single offspring also have two breasts, but they are positioned differently. Except for the very lowest grades in the order, primates are easier to tell by sight, than to define anatomically.

Before examining the fossil record of the primates, it is helpful to look among the living members of the order which provide models suitable for comparison. There is a general agreement that the living shrew, of which *Tupaia* (Plate 6–1) is representative, presents a good Mesozoic model for the ancestor of the primates as well as other living mammals. This small, arboreal, and nocturnal animal is insectivorous and has changed but little from the forms living some 80 million years ago. Its fundamental position in the lineage of mammals makes classification difficult. Most authorities place it among insectivores, while a few still include it as a very generalized member of the primates. But its exact taxonomic position is unimportant, for it certainly is a reasonable living representative of the placental stem of mammals.

Four general levels of general organization, or *grades,* are recognized among the primates. The simplest and the earliest evolved is that of the *prosimians,* which include such living animals as the lemurs, the lorises, the galagos, the pottos, and finally the several species of tarsiers. The range of the prosimians is limited to the Old World, and almost entirely to the tropics. They inhabit essentially the same ecological niche occupied by the ancestral primates, the canopies of second growth in forests. A great majority eat fruit, a few eat leaves, more eat insects along with occasional lizards and birds' eggs. Most are quadrupedal in their gait, but a few are vertical climbers and are capable of relatively great jumps. The lesser galago (*Galago senegalensis*) can leap more than twenty feet with ease and operates as a birdlike nocturnal predator (Plate 7–1). Unlike most prosimians it has shifted its habitat to savanna and bushlands in Africa extending well into the temperate zone in South Africa.

Most of the lemurlike prosimians live on Madagascar, the world's fourth largest island which lies several hundred miles east of the central African coastline. They reached this island early, and there they radiated into essentially empty ecological space. The lorises are found primarily in East Asia, and their place is taken in the tropics of Africa by the pottos, which vaguely resemble large-eyed teddy bears. In addition Africa has a number of galagos, or bush babies, named from their wailing like an infant in the night. The Asiatic lorises include the grotesque slender lorises, with their stilt-like limbs, and the slow lorises, named for their rate of movement (Plate 7–2). The tarsiers are limited to three species all living in Southeast Asia. Their anatomical specializations include a great lengthening of their hind legs and feet, and these together with their padded digits allow them to leap through forests of their environment (Plate 7–3). They, too, are primarily nocturnal and insectivorous. The generally long muzzles and furry faces of prosimians give them more of a foxlike look than the other primates (Plate 7–4). Their furry coat has resulted in the evolution of trocumbent lower incisors which act as a

Plate 7–1
This Bush Baby, or lesser galago, is a highly active, nocturnal prosimian with widespread range in Africa.

Plate 7–2
A Slender Loris, which moves slowly through the nights of Southeast Asia on stilt-like limbs

Plate 7–3
Tarsier, with eyes for nocturnal life and padded digits to improve its grip

Plate 7–4
Ring-Tailed Lemur Mother with
Infant in Jockey Position

Plate 7–5
This Cotton-Top Marmoset lacks
the prehensile tail of the more
advanced ceboids.

grooming comb in these animals. Seeing these animals in a zoo fails
to give the human observer any sense of kinship.

The monkeys of Central and South America illustrate a higher
grade of organization than the prosimians, and through both their
activities and their short-muzzled bare faces we recognize them as
our collateral relatives. They migrated from the Old World by
crossing some type of filter-bridge in the Oligocene and thereafter
underwent a successful radiation in the tropics of this region. They
are known as *platyrrhines,* since the nasal cartilage between their
nostrils is wide. The larger and more numerous forms are known as
ceboids, after their principal genus, *Cebus.* There are a variety of
smaller forms known as *marmosets,* which have furry tails and give
birth to twins (Plate 7–5). Most of the platyrrhines move as arboreal
quadrupeds, but the spider monkey is primarily an arm-swinging
animal (Plate 7–6). In diet, they all are fruit-eating, and a majority

Plate 7–6
This Spider Monkey, with its elongated limbs, shows why the species was given its name. Note the use of the prehensile tail as an extra hand.

Plate 7–7
A Red-Howler Monkey, in the act of producing one of his social calls.

round their diet out with insects. The howlers are also leaf-eaters (Plate 7–7).

The primary radiation of the platyrrhines in South America has resulted in a wide range of sizes. The largest are the howlers going up to 22 pounds in adult males, while the smallest of the marmosets can curl up in the palm of one's hand. The bulk are intermediate in size, as typified by the capuchins, which are the typical organ-grinder's monkey. There is nothing remotely approaching an ape-like level of organization among them, so it is quite clear that neither man nor the great apes could have evolved from this grade of monkeys in the confines of the New World.

The higher primates of the Old World contained two structural grades: the *cercopiths*, or monkeys, and the *hominoids* or manlike forms which include both the great apes and man. The cercopith grade contains a wide variety of monkeys which occupy two primary

Plate 7–8
Gibbon Brachiating

ecological niches. The arboreal forms include such forms as the colobines, the guenons, and the mangabeys of Africa, and the various langurs of southern Asia. The terrestrial monkeys of both Africa and Asia may be divided into two major groups on the basis of size. The smaller of the ground-living monkeys are known as macaques; all the larger ones are baboons. Both are aggressive. Baboons range throughout most of Africa south of the Sahara, while the far-flung macaques are primarily concentrated in Asia, where they range from the tropics out into the continental islands and as far north as North China and Japan. Most of the arboreal monkeys are fruit- and leaf-eating animals, while the terrestrial monkeys add a variety of foods found on the ground to this basic diet. Some local groups of baboons even eat the newly born young of small antelope, and others have developed the practice of hunting large rabbitlike animals, the African hare.

While the primates today are as an order tropical to subtropical in distribution, it is interesting that one species of macaque still dwells in the temperate climate of North Japan, with its snowy winters. In northwestern Africa, the macaque known as the Barbary ape is limited to the Atlas Mountains, with a related group surviving in southern Spain. The reader may recall the British legend that the

Plate 7–9
A Pair of Adult Orangutans, showing the great sex difference in
size characterizing these pongids
and demonstrating fist-walking.
Note the facial flaps on the male
which may restrict his vision but
presumably serve to intimidate
his rivals when spread laterally.

naval base of Gibraltar will remain under their flag as long as any
monkeys live on its rocky heights. It is rumored that the Gibraltar
monkey troops were secretly reinforced from time to time with
macaques imported from North Africa. The various ground-living
cercopiths are interesting in their own right, but they have developed their own unique ecological niches over a 30 million year
period and so do not serve as models for early men.

The *hominoid* grade includes all four living apes, or *pongids,* and
all of man, *hominids,* old and new. The hominoids are generally
large in size, have no tail, and manifest the highest development of
intelligence yet evolved on the planet Earth. The gibbon of Southeast Asia is the smallest of the apes, with body weight ranging from
12 to 25 pounds for both sexes in several species. The gibbons have
developed a specialized form of locomotion, known as *brachiation,* in which they swing from arm to arm through the canopy. They
are the most adept aerial artists of the primate world and have been
seen swinging over twenty-foot gaps between branches (Plate 7–8).
They are monogamous, with a diet limited to fruit, and so they have
defended territories large enough to allow tropical fruits to ripen in
series throughout the year. The orangutan, the other pongid living
in Asia, lives in a very restricted area in the islands of Borneo and

Sumatra. Its numbers have been reduced to an estimated total of 10,000, for men still shoot the females to take the babies for export to zoos of the world (Plate 7–9).

Tropical Africa is inhabited by two great apes which differ much in appearance but are closely related in an evolutionary sense. The gorilla occurs in two forms, one which inhabits the lowlands, and the other the mountains. Big males reach a weight of 400 pounds, and so are the largest of living primates. Socially they live in bands headed by a mature, or silverback, male. Despite their fierce appearance, these great pongids are in fact gentle, placid, and good-natured. Chimpanzees are smaller in size and livelier in temperament than gorillas. They live wherever their large cousins do, although they are more numerous and range more widely. Chimps show some regional variation in size and color pattern but probably should be grouped together as a single species. They are so familiar as trained animals, both on the screen and in the zoos, that their behavioral capabilities need no special comment here. In watching these extraordinary primates no one can doubt that they are closely related to us. This impression is confirmed by the fact that students of anatomy, biochemistry, and molecular biology all agree that the two African great apes of all the primates are the most closely related to man. The chimpanzee seems to show a small edge in this regard over the larger gorilla.

The Pattern of Primate Evolution as Seen in the Fossil Record

Compared to other orders, G. G. Simpson (1949) has rated the primate record as neither very good nor yet very bad. The first recognizable primate fossils appear in the Paleocene as a part of the great mammalian radiation filling in the lifeways left vacant with the extinction of the ruling reptiles.

The picture of primate evolution comes into clearer focus as time passes. The Paleocene deposits of both North America and Europe contain a substantial number of prosimian species, and their numbers expand rapidly in the following Eocene epoch. The fossils indicate that a primary radiation was underway in the early Cenozoic. The ecological niche of prosimians is the forest canopies of the tropics, and presumably it always has been. Their presence in the now temperate regions of North America and Europe is considered to indicate that both regions enjoyed a much warmer climate than they do today. For more than 30 million years the prosimians were the preponderant primates, and during this interval they radiated to form at least nine families, of which six continued on to the end of the Eocene. Some would have looked very like the lemurs today, notably the North American form, *Notharctus,* while others evolved in divergent directions. One form evolved specialized incisors like that of the rodents. Quite independently and later these specialized chisel teeth evolved yet a second time in the prosimians, for they are present today in the deviant aye-aye in Madagascar.

During their Eocene period of abundance, prosimians apparently swarmed all over the available tropical land masses. But following this epoch their remains virtually disappeared from the fossil record, and it is uncertain whether this apparent scarcity corresponds to their real decline or whether sampling was faulty. Probabilities favor the theory of scarcity, for the higher grades represented by ceboids in the New World and cercopiths in the Old World were evolving as animals better suited to occupy the same living space. The living forms indicate that the prosimians have remained generalized on the whole, failing to progress in the development of higher types of brains and not shifting to new adaptive niches.

The great island of Madagascar provides us with an example of a second prosimian radiation, an event which accounts for the importance of living lemurs in its limited fauna. Today no less than 10 genera comprising 21 species exist there. There is no indication of how the ancestral lemur arrived on Madagascar, but it must have involved one of the rare accidents by which islands receive occasional immigrant stocks. In all probability a pregnant female lemur found herself by accident upon a log somewhere in the great Zambezi River of the mainland and drifted across the several hundred miles of Mozambique Channel to make landfall on Madagascar. The chances of such an event occurring are so slight that Simpson has called it the *sweepstakes lottery.* Obviously a great many wet and unhappy lemurs must have gone to sea before a pregnant one ended up safely on the island. Once safely ashore the new immigrant found a refuge which contained few enemies and no competitors. There, in what was probably Oligocene times, the immigrant stock underwent a new radiation, which is largely responsible for the wide variation found among living prosimians. Most of the products of this radiation remain in an arboreal niche and included both day living and night feeding animals. They ranged in size from that of a mouse up to that of a gibbon. A few lines became largely terrestrial in habit. Because there were few large herbivores on the island, the radiation produced a giant lemur, *Megaladapis,* which grew to the size of a Saint Bernard dog. It became extinct only in fairly recent times when the swampy rain forests in which it lived were invaded and destroyed by competing men who arrived from Indonesia about two thousand years ago.

The Madagascar lemurs ranged from primarily insectivorous, through the omnivorous, to the exclusively herbivorous. One of the specialists evolved into *Daubentonia,* or the aye-aye. It is about the size of a cat, has long naked ears, large eyes such as are found in nocturnal animals, and a long bushy tail. Its incisor teeth have been reduced to the central ones in both jaws and these have greatly enlarged and elongated into much extended roots which provide continuous growth during life. They are very like the gnawing incisors of rodents. Even more specialized is the middle finger of the hand, which is much lengthened and terminates in a long claw. The aye-aye lives upon the grubs of boring insects and listens for their activity as they eat their way through wood. Once the grubs are located, the aye-aye either uses its elongated third digit or its gnawing teeth to obtain its food. This is a very specialized ecological niche, and the animal's structure has evolved to accommodate it. The more generalized prosimians provide better insight into the basic form of Eocene ancestors.

South America became cut off from North America by the sinking of the connecting land bridge early in the Cenozoic. When and how it received the founder lemur is unknown, for the fossil record is faulty. Nonetheless, the variety of ceboid monkeys living in South America today indicates a substantial radiation which began by early Miocene times. The ceboid monkeys as compared to their prosimian ancestors show a definite enlargement of the brain to respectable proportions in terms of their body weight. Some of the ceboid monkeys have performed so well in laboratory experiments as to deserve a high behavioral rating. One animal became as skilled in tool use as a test chimpanzee. The enlargement of the ceboid brain involved a remodeling of the skull with the reduction of the long prosimian snout and the turning of the orbits from the primitive more lateral position to one directed to the front as they are in the hominoids. The fact that most of the larger South American monkeys developed long and functionally useful grasping tails is a unique event in primate evolution. In effect it grants these animals a fifth hand, for they not only use it in locomotion but can actually hang suspended by it. These South American monkeys all live in the canopies of tropical rain forests, and in most of their domain the land surface lying below is wet and forbidding. In the same general environment there evolved a porcupine, an anteater, and opossums also equipped with prehensile tails. That four different families of mammals should evolve grasping tails in the same general region suggests but does not prove that powerful selective forces were at work in their evolution.

Old World Primate Radiation

This is the area in which men evolved, so it is a critical one for examination. In the basal beds of the Cenozoic are contained a variety of prosimian forms, and it is from these that all of its higher

Figure 7-1
Present Distribution of Nonhuman Primates

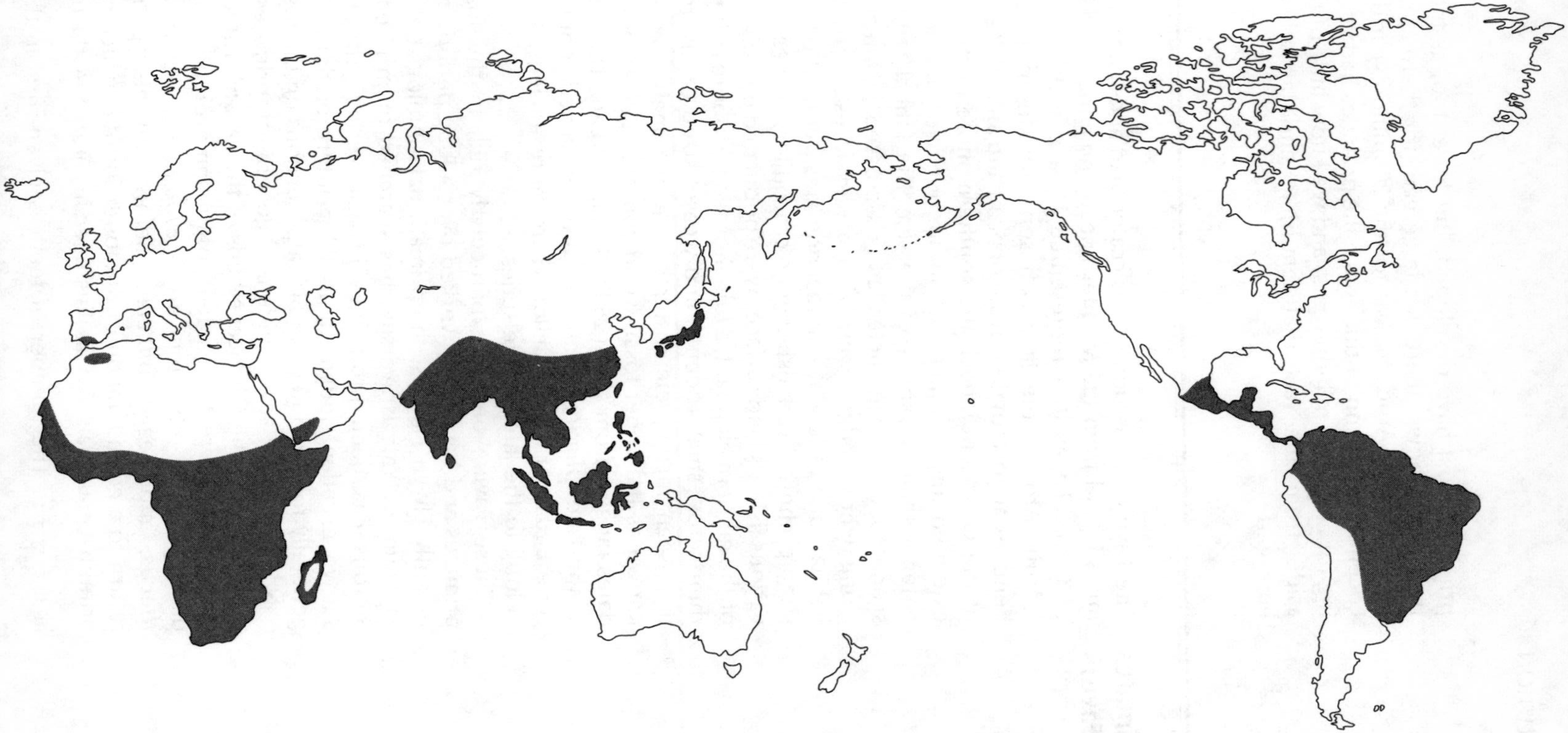

primates must have developed. But here, too, as the fossil record frequently shows, there is a large gap, for there are no significant finds in either Africa or Asia for the 20 million year duration of the Eocene. By the end of this epoch the primates had been differentiating for almost 30 million years, and we know little of their evolution aside from the fact that lemurlike and tarsierlike forms had emerged.

Fossil Primates from the Fayum

The Fayum is a region of desiccated badlands lying some hundred miles inland from the Mediterranean and about 60 miles southwest of Cairo. Today it is represented by a brackish lake called Lake Qârun. Around the lake is a series of desert escarpments and benches almost totally devoid of plant and animal life. But at the end of the Oligocene epoch the southern shore of the Mediterranean extended this far inland, while interior rivers flowed northward to enter a shallow estuary of the sea. The faunal remains allow the scene to be reconstructed as crisscrossed by meandering rivers along whose edges were dense rain forests providing an ideal habitat for a variety of arboreal primates. Along the rivers and in the swamp there lived fish-eating crocodiles, turtles, dugongs, and various fish. Its land surfaces were populated by a pig-sized ancestor of the modern elephant, a four-horned hervibore as big as a modern rhinoceros, small rodents, and weasel-sized carnivores. No doubt this is an incomplete roster of the mammals present. The beds known as the upper fossil-wood zone have given potassium-argon dates ranging from about 25 to 27 million years. This places the beds in the Late Oligocene, and the area contained some lower Oligocene beds perhaps 6 million years more ancient. They may even belong to the late Eocene series.

These Fayum beds are particularly rich in the fossils of early primates and have been worked over a long period of years. More recently Elwyn Simons, the primate paleontologist at the Peabody Museum of Yale University, has searched them intensively. Since Simons has been productive in discovering critical fossils, and shown excellent judgment in their interpretation, his views are generally followed here and in later sections on the fossil primates.

No less than seven different genera of primates have been removed from the Oligocene beds of the Fayum. Three of these are of marginal interest to students of human evolution and will not be discussed. The other four are *Aeolopithecus, Apidium, Propliopithecus,* and *Aegyptopithecus.* Before considering these, let us return to a consideration of the tooth formula which characterized primitive mammals. As was illustrated in Figure 6–4, the dental formula is $\frac{3\text{-}1\text{-}4\text{-}3}{3\text{-}1\text{-}4\text{-}3}$. The number of teeth is given, first in the upper jaw and then in the lower jaw or mandible. The formula indicates in this case that each jaw is characterized on each side by three incisors, one canine, four premolars, and three molar teeth. Thus, these Mesozoic mammals had a total of 44 teeth.

The evolution of the mammals shows two general dental trends. The first of these involves increasing differentiation of tooth form regionally so that the different clusters can serve different functions. The type of dental differentiation varies in different lines of descent. A second trend involves the general reduction in the total numbers of teeth present. In most of the New World ceboids the general dental formula is $\frac{2\text{-}1\text{-}3\text{-}3}{2\text{-}1\text{-}3\text{-}3}$. This gives a total of 36 teeth, the reduction resulting from the loss of one incisor and one premolar in each side of both jaws. Further dental reduction has occurred in most of the Old World monkeys and higher primates, including man. Their formula is $\frac{2\text{-}1\text{-}2\text{-}3}{2\text{-}1\text{-}2\text{-}3}$, for a total of 32 teeth. It is worth noting that continued reduction characterizes some human populations, for there is a marked tendency to suffer the loss of the third molars, or wisdom teeth as we call them. This tendency is present in a rather large minority of individuals in both modern White and Mongoloid populations. Since these two groups will contribute heavily to the human world of the future, it no doubt means a continuing trend toward the reduction of human molars to a series of two, and thus a total of but 28 teeth in each head.

Aeolopithecus, known from a single mandible found in the upper fossil-wood zone, seems to foreshadow the specializations of the gibbons which certainly descended from it. It had a short face and dental characteristics indicating a fruit-eating diet.

Apidium is one of the interesting primates occurring in the Fayum deposits. It is dentally well known, for its remains are represented by 50 lower jaws and four upper ones. These show a formula of $\frac{2\text{-}1\text{-}3\text{-}3}{2\text{-}1\text{-}3\text{-}3}$, similar to that in the New World monkeys, but not necessarily an indication of any relationship. *Apidium* was about the size of a squirrel and had a short face. The cusp patterns of its molar teeth have some unique features indicating that Apidium represents a side branch of primate evolution. Further, the same traits are found in *Oreopithecus,* a hominoid of the Pliocene that seems to have evolved parallel and separate to our own ancestral line. Thus *Apidium* represents the apparent Oligocene ancestor of a separate evolutionary experiment in the production of a manlike grade of structure.

Propliopithecus is one of the most controversial of the Fayum primates. It has been considered to be ancestral to the modern gibbons. Others relate it to the other hominoids. Some speculate that it went to extinction as an evolutionary side branch. Simons has been ambivalent in his opinions, and now seems to consider it ancestral to the somewhat later *Aegyptopithecus*. It is known only from two mandibular fragments and a few molar teeth.

The largest of the Fayum primates was recently discovered by Simons himself (1967) and has been given the name *Aegyptopithecus zeuxis*. Its general importance lies in the fact that it lived a little later than *Propliopithecus* of the Middle Oligocene and before the very early Miocene forms of *Dryopithecus,* the genus ancestral to the chimp and gorilla. *Aegyptopithecus* was no bigger than a modern gibbon and, judging from the few foot, limb, and tail bones found in the deposit, lived an arboreal life and looked much like a monkey. Its skull is relatively complete and is shown in Figure 7–2. The teeth of this primate are of particular interest, for in shape they resemble living pongids although of course much smaller in size. Its canines are large, and the front premolars and the three lower molars increase markedly in size from front to back. Its mandible is more apelike than that of the earlier *Propliopithecus* since it is much deeper in the region below the long-rooted canine teeth. In general its skull is similar to that of a monkey except for containing the teeth of an ape. The snout is still long compared to the pongids, but the orbits of the eyes have shifted toward a forward axis. Going along with this is some expansion of the forebrain, and a reduction of the olfactory lobes. These changes suggest increasing dependence upon visual senses as opposed to nasal ones. The totality of the fossil find suggests that *Aegyptopithecus* lived an arboreal life but that some unknown aspects of its niche were already allowing for the selective directions which resulted in the evolution of the later hominoids. Thus, *Aegyptopithecus,* despite its general monkeylike attributes, may represent the earliest anthropoid in the fossil record and so man's own earliest direct ancestor.

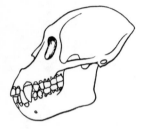

Figure 7–2
Aegyptopithecus from the Oligocene of the Fayum

The Hominoid Radiation of the Miocene

Each of the four grades within the primates seems to have undergone a basic radiation of its own in the Oligocene or Miocene. The coincidence of timing does not indicate that the stimuli for these radiations were the same. The prosimian radiation was due to the fortuitous arrival of a lemur on Madagascar to serve as founder's stock. The ceboid radiation in South America presumably involved another case of sweepstake immigration and subsequent radiating to fill the variety of arboreal niches available. But in the Old World both the cercopiths and the hominoids underwent ramifying evolution in the Miocene, and it is that period in which the grasslands expanded throughout the world. The general pattern of primate evolution is shown in Figure 7–3. Many lines are shown going to extinction even though the record is incomplete.

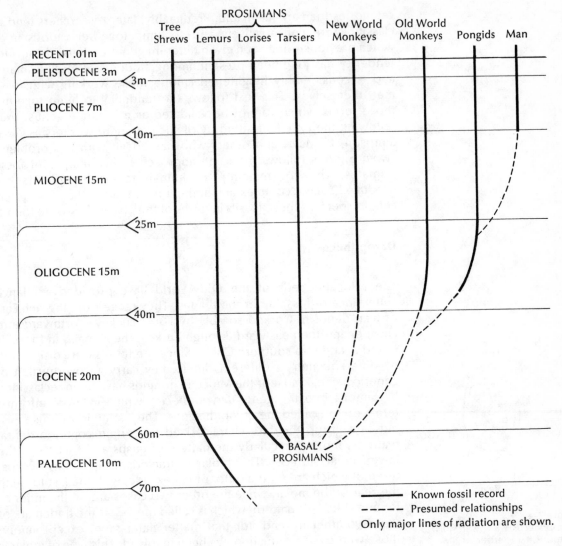

Figure 7-9
Generalized Scheme of
Hominoid Radiations

Considerations in Fossil Classification

In discussion of the fossils of this and later periods we must note the very human failing of inflating the importance of one's own discoveries. Almost every important fossil has been given a new generic name by its discoverer. This elevates the importance of the find, and inflates the ego of the fortunate scientist. This tendency in classification is known as splitting, for it tends to create more named categories than are justifiable. Predictably, later researchers tend to classify more conservatively and so lump together various finds which previously had been given different generic labels. The former tendency has long been present among students of fossil primates and is considerably heightened among those working with fossil men themselves. This text follows the tendency to lump together those forms which cannot be validated as a separate genus, thus following the excellent authority of such eminent workers as G. G. Simpson, E. Mayr, and others who have dealt with this problem. Elwyn Simons follows the same approach in his analysis of fossil primates. The "overnaming" of human fossils by splitters has produced such a complex and ambiguous situation that it is preferable to speak of these fossils in terms of the places they were found.

Dryopithecus

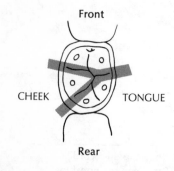

Figure 7–4
Dryopithecus Y-5 Lower Molar Pattern Characteristic of the Hominoids

The Miocene beds of the Old World have provided a relative abundance of fossil anthropoid apes. They range from the rich East African deposits worked so ably by Louis Leakey, northward into Europe, and then eastward through Turkey, the Himalayan foothills of India, and into southern China. Since the fossils were discovered by different men at different times, they carry an overburden of generic names. Most of the Miocene pongids have been reclassified by Simons into the genus *Dryopithecus*, which in more informal terms can be called the dryopithecines. They come in various sizes, but as a group they are characterized by certain common dental features, more particularly the pattern of cusps and fissures of the lower molar surfaces. This molar characteristically has five cusps, three of which are on the outer or cheek side, with the last forming the rear of the molar, while the other two cusps are on the inner or tongue side. This pattern, which is called the Y-5, is the trademark of the dryopithecine and, for that matter, later men, except among those where cusp reduction has been involved. This type of molar is shown in Figure 7–4, where the Y is indicated by overshading. This tooth first appeared in *Aegyptopithecus* in the Upper Oligocene in the Fayum and is the primary reason for considering that diminutive and monkeylike primate one of our direct ancestors.

Simons condensed more than twenty different genera, which ranged in size from animals as small as the living gibbon to some exceeding the bulk of the modern gorilla, into the single genus of the *Dryopithecus*. A medium sized form, formerly called "Proconsul" by Leakey, is now classed in the genus *Dryopithecus*. Except for *Aegyptopithecus* this is one of the most complete fossil anthropoids, for much of the skull and face are preserved. As shown in Figure 7–5, "Proconsul," to use the term colloquially to identify the specific find, is apelike, but not excessively so. Authorities generally agree that it is less extreme than the chimpanzee and much less specialized than the gorilla. It cannot be called manlike, but it is more so than any of the living great apes, a condition that renders its forerunner as a possible direct ancestor of both man and the living pongids.

The dental characteristics of the Miocene dryopithecines include a sizable canine, shearing lower premolars in which the outer cusps are much higher than the inner ones, and molars which are longer than they are wide. The molar teeth become increasingly large in proceeding from the first to the posterior third tooth. The dryopithecines have incisor teeth that are both smaller and more vertical in positioning than those of living great apes. The post-cranial skeleton is represented by only a few limb bone fragments of such a character as to indicate that the more specialized forms of locomotion, such as brachiating, had not yet been evolved. The same may be said for Miocene forms of gibbons, such as *Limnopithecus,* so that it would appear that this form of locomotion was a late and certainly post-Miocene specialization. It results in the elongation of the arm and some reduction in the size of the thumb.

In East Africa Leakey found that his dryopithecines from the lower Miocene beds were accompanied by fossils of monkeys and galagos, forms that clearly indicate that they were all forest dwellers. But rather than imagining the region at that time to consist of an unbroken rain forest, it is more likely that such forests occurred as strips along the margins of the river with intervening stretches of grasslands and scattered dry woodlands. No doubt with more time enough fossil evidence will accumulate to allow the determination of the regional ecology in such detail as to permit accurate reconstruction of the dryopithecine niche. At this time there is no reason to doubt that the various dryopithecines, both small and large, were largely herbivorous, but added a little omnivorous variety with such protein items as they might obtain.

The diet of some of the more northerly dryopithecines is more open to question. Those pongids occurred not only in East Africa, and, of course, presumably much more extensively on that continent, but have been found also in both Miocene and Pliocene deposits of southern Europe, in the Siwalik Hills of North India, and eastward into the southwestern provinces of China. Some of these areas must have been characterized by a different flora and hence different dietary possibilities than those which occurred in East Africa. Indeed, the generic title *Dryopithecus* means "oakwoods ape." In Europe these large manlike apes must be visualized as living in forests with a wide variety of trees, including chestnuts, live-oaks, and some conifers. Presumably pongids of the size of even a medium dryopithecine would have spent a good deal of time on the

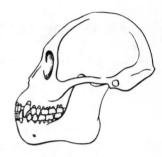

Figure 7–5
"Proconsul" from the Miocene of East Africa

ground, even if only in going from one tree to another. The question of what types of foods pongids would eat in such nontropical regions remains unsolved. Certain fruits and seeds would vary with season. Even if acorns were an important resource, as they are for many animals, the competition for them would make them available only for limited times. Certainly they must have been dependent upon ground foods seasonally. The "snow monkeys," macaques living in the north of the main island of Japan, subsist through severe winters on the inner bark of shrubs and trees, so life in Europe with a southern temperate climate could not have been too difficult (Plate 7–10).

Gigantopithecus

The Old World beds of Miocene and Pliocene times contain representatives of three other genera in addition to the dryopithecines. These are *Gigantopithecus, Oreopithecus,* and *Ramapithecus.* The first two are interesting as representing side branches, while the third leads directly to later men. *Gigantopithecus* was originally named from a few molar teeth which had found their way to coastal Chinese "drugstores" which used ground up fossil bones both as love potions and as medicine to cure various ills. Bought by the young Dutch geologist G. H. R. von Koenigswald, they finally found their way into the hands of the great anatomist Franz Weidenreich. They were enormous teeth, much larger than those of

Plate 7–10
The Japanese Macaque survives rigorous winters. Here a baby jockey-rides his agile mother while leaping across a stream. Seasonally these monkeys develop a thick winter fur.

a gorilla, and several times larger than those of any known man. In a provocative book titled *Apes, Giants, and Man* Weidenreich (1946) postulated that they had belonged to men eight feet tall, weighing more than 400 pounds, who represented an early giant phase of human evolution. Simons has summarized the Chinese finds, and described a new one from India. He reconstructs the complete animal as a gigantic herbivorous quadrupedal ape, as heavy as 600 pounds and standing nine feet tall if upright. It falls outside of the size range of both dryopithecines and early men of all varieties. Therefore, *Gigantopithecus* has correctly been given a separate generic stage, but its real evolutionary significance must await more detailed finds. Its dentition shows specializations that parallel those of the robust australopithecine. Authorities tend to consider it a specialized pongid line of evolution that went to extinction in the Middle Pleistocene of the Far East, but it adapted to life in the grasslands and survived more than five million years. It seems to have been an evolutionary success up to the time when advanced men appeared on the local scene.

Oreopithecus: The Swamp Ape Who Almost Became a Man

Oreopithecus provides more information about the complexity of the Miocene primate radiations. The first specimens were found in brown coal deposits of north Italy as early as 1870. Subsequent discoveries have found their way into museums so that the animal is known from fairly extensive materials. Until recently it was viewed with ambivalence, for it was variously considered a cercopith, a great ape, and even man. A Swiss paleoprimatologist by the name of Johannes Hürzeler became interested in the problem of *Oreopithecus,* and he determined to search further for more complete remains. With funds provided by the Wenner-Gren Foundation in the United States he was able to reopen a lignite coal mine in Tuscany in 1964 which had produced many of the earlier fossils. With a small labor force the water was pumped out of the drowned mine. After an extended period of work, and just as his funds became exhausted so that his hunt would have to cease a few days later, Hürzeler looked up at the ceiling of the mine and found there in place an almost complete skeleton of *Oreopithecus.* In this dramatic fashion a new aspect of the problem of hominoid evolution was brought into view.

The new skeleton demonstrated that *Oreopithecus* was a primate of moderate size, standing perhaps four feet high and weighing as much as 80 or 90 pounds. This is close to the size of a female chimpanzee. The evolutionary meaning of the new find was difficult to determine. As shown in Figure 7–6, its face is short and flat

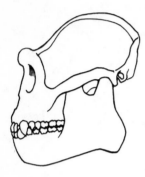

Figure 7–6
Oreopithecus from the Miocene and most recently reconstructed. It shows some hominid features and may have been a separate experiment in evolving a manlike primate.

without the elongated snout one associates with the pongids. Its canine teeth are moderate in size and smaller than those of a chimpanzee. But it is the cheek teeth or molars which cause the greatest puzzlement. Both the upper and lower sets of *Oreopithecus* molars, monkeylike in cusp pattern, strongly resemble the corresponding teeth of *Apidium,* which lived some twenty to twenty-five million years earlier. The puzzle is further compounded by the fact that the pelvis of *Oreopithecus* in certain anatomical ways resembles that of man. The upper blade, or *ilium,* is relatively broad, and its anterior inferior iliac process, which is important in maintaining upright posture, shows appreciable enlargement. The sacral canal tapers sharply, suggesting the animal had no tail, or at most a short one. Some of the animal's foot bones deviate from a pongid condition in a direction which is found in man with his bipedal gait. With these data there is general agreement among evolutionists that *Oreopithecus* was a separate evolutionary experiment which reached very close to, and perhaps attained, the hominid grade of structure. If its ancestry does go back to *Apidium,* then there would have been two evolving lineages beginning in the Oligocene which both in time evolved bipedalism and thus the potential of entering the human cultural niche. That *Oreopithecus* failed to fully attain this grade may very well result from competition from our successful ancestors. *Oreopithecus* is usually put in a separate family within the hominoids.

One further mystery remains about *Oreopithecus.* The fossil remains have consistently occurred in 12-million-year-old beds of brown coal which were laid down in swamps in northern Italy. Therefore he has been called a swamp dweller. Even though the arms of *Oreopithecus* show slight elongation and so suggest some capability of brachiation as a means of locomotion, a swamp is an odd place to find a near-man. The remains of so many individuals have been found in the solidified swamp deposits that it seems obvious that they frequently went to their death in that environment. Most swamps in wooded areas have trees approaching their watery edges but seldom extended out into them. It would not seem likely that so many of these animals would have fallen to their death while traveling arboreally around the fringes of swamps. The more intriguing possibility is that swamps may have provided the food resources for *Oreopithecus.* No other primates are known to find their food in swamps, but this is not true of all primitive men. Many hunters and gatherers find that the roots, stems, and seeds of great tropical water lilies provide a variety of staple foods. Other people use the starchy root structure of reeds and rushes as a dependable base for subsistence. Even though these dietary suggestions may be wide of the mark, we should remember that the diet of early hominoids must have been much more diverse than it is among their living pongid descendants. Whatever attracted considerable numbers of *Oreopithecus* to these often fatal swampy places remains a challenging question.

The Ramapithecines

The genus *Ramapithecus* is the most important of all Miocene finds in contributing to the solution of the early evolution of man. Ironically, the necessary evidence has been unnoted in museum drawers for many years. The specimens were originally obtained in the Siwalik Hills and described by G. E. Lewis (1937). He was something of a splitter and created many new generic terms for animals now recognized as dryopithecines. Among the genera

Figure 7–7
Distribution of Dryopithecines and Ramapithecines

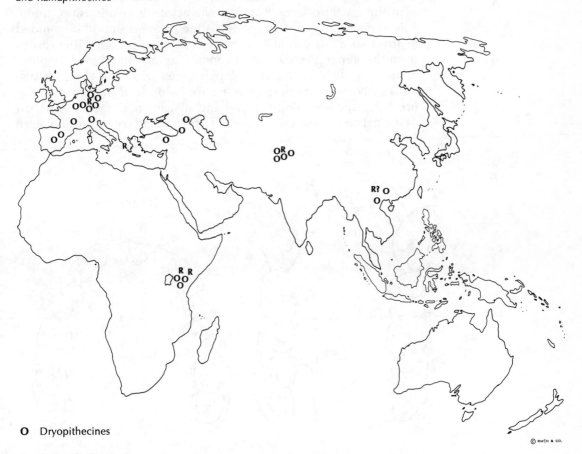

O   Dryopithecines

R   Ramapithecines

created by Lewis was *Ramapithecus,* which he noticed in some ways approached a truly hominid condition. But his observations were lost from view for many years, and it remained for Simons and Pilbeam (1965), in restudying these specimens, to reaffirm this identification and to make it stick. The palatal shape and dental relationships of *Ramapithecus* are similar to those found in man, and very distinct from those which occur among the pongids. The chimpanzee is a good example of the pongid and its palate (Figure 7–8a) is characterized by a generally elongated rectangular shape in which the front is squared off by the great projecting canines. A considerable gap between the canines and the incisors permits the interlocking lower canine tips to fit into this space. The portion of the palate occupied by the cheek teeth is in some forms constricted toward the rear. If we divide the palate into two portions, an anterior one, including the premolars, canines, and incisors, and contrast it with the posterior one, which includes the three molar teeth, then in the chimpanzee and most pongids the anterior region is larger than the posterior one. This is a characteristic relationship. In the palatal restoration of *Ramapithecus* the palate is smoothly rounded in front so as to produce an overall parabolic shape. This results from the general reduction of canine size and characterizes other early men such as the australopithecines (Figure 7–8b) and Australian Aborigine-modern man (Figure 7–8d). In *Ramapithecus* (Figure 7–8c), the australopithecines and all living men the anterior half of the palate is reduced compared to the pongid condition and more

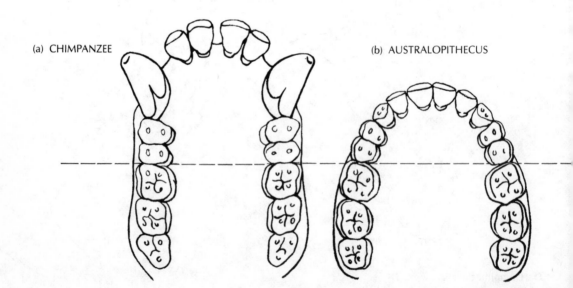

(a) CHIMPANZEE   (b) AUSTRALOPITHECUS nearly equivalent in size to the posterior portion. This results from the great reduction in canine teeth, which no longer interlock and require a gap, or *diastema,* in the dental series for their fit.

In East Africa, Louis Leakey found important fossil jaws in both Miocene and Lower Pliocene deposits. They, too, show a reduced canine tooth, smaller incisors and, in general, hominid form. Named *Kenyapithecus* by their discoverer, they have recently been reclassified by Simons as *Ramapithecus,* differing from the original only at the level of species. These new finds carry the probable direct ancestors of modern man back 20 million years. This extended ancestry contrasts sharply with the views of earlier workers, who sometimes considered that man, as opposed to pongids, could be traced back perhaps only a million or two years. The new evidence is surprising, for it indicates that a progressive anthropoid entered the unique human niche in mid-Cenozoic times, and then instead of evolving rapidly as might have been anticipated, apparently remained rather unchanged over long periods of time. Judging from the remains of the australopithecines, who are the best known and most numerous of the early men, dentition did not change much from that of the ramapithecines. The latter, despite the scanty fossil remains, show a geographical range very similar to the dryopithecine, for they are known to have occurred in both East Africa and India and southern Europe.

While the hominid palate is easy to distinguish from that of the pongid, the explanations of its evolution are not completely satisfac-

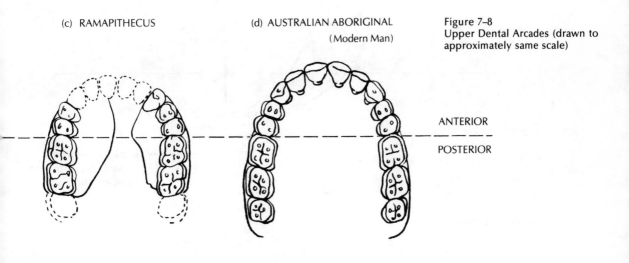

(c) RAMAPITHECUS

(d) AUSTRALIAN ABORIGINAL

(Modern Man)

Figure 7–8
Upper Dental Arcades (drawn to approximately same scale)

ANTERIOR

POSTERIOR

INDIA   AFRICA

tory. With the ancestral form beginning its entry into the human niche, it is postulated that the increase in tool using and bipedalism allowed the selective pressure which maintained the structural integrity of the big pongid canines to be released. But this alone would not explain the evolution of the structurally sound, but very much reduced, hominid canine. Therefore, in addition to this negative selection, it is necessary to consider that some type of positive selection was also operative. Primitive men universally rotated their lower jaws upon their upper to assist in grinding the hard food particles that formed a portion of their diet. Since this could not be easily accomplished with large interlocking and projecting canines, it has been reasoned that a dietary shift would introduce positive selection for the reduction of the formerly projecting canines. This kind of push-pull change in the direction of

Figure 7-3
Time Relationships among Major Categories of Living Primates

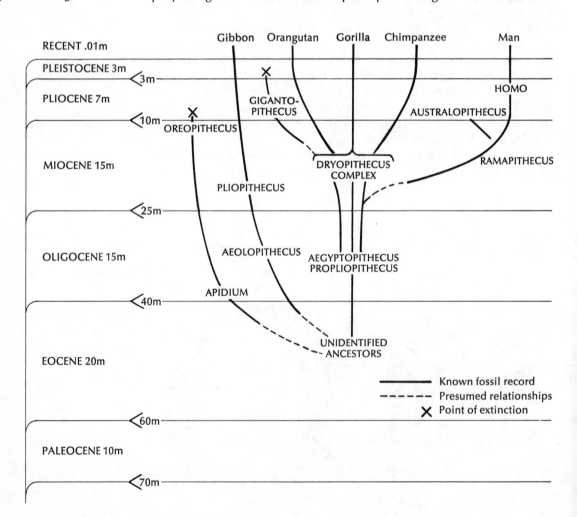

selection may *explain* the evolution of the smaller parabolic human palate, but the actual processes involved may have been more complex.

The fossil record is clear in showing that the evolution of the hominid line began in an isolated Africa, and continued there during the rest of the Cenozoic. Until Eurasia was joined to Africa by land bridges at Gibraltar and Saudi Arabia in the latter half of the Miocene, its primate fauna was limited to prosimians. After that date, terrestrial animals passed back and forth, and the advanced African forms of primates reached Europe and Asia. This is the basis for the first appearance of dryopithecines in the fossil beds of the northerly land masses. Africa seems to have been the Garden of Eden.

Bibliography

LEWIS, G. E.
1937 Taxonomic Syllabus of Siwalik Fossil Anthropoids. American Journal of Science. Vol. 34, pp. 139–147.

PILBEAM, DAVID R.
1968 The Earliest Hominids. Nature. Vol. 219, pp. 1335–1338.
1972a Adaptive Response of Hominids to Their Environment as Ascertained by Fossil Evidence. Social Biology. Vol. 19, pp. 115–127.
1972b The Ascent of Man. New York: Macmillan.

SARICH, VINCENT.
1971 A Molecular Approach to the Question of Human Origins. In P. J. Dolhinow and Vincent Sarich, eds., Background for Man. Boston: Little, Brown, pp. 60–81.

SIMONS, ELWYN B.
1964 The Earliest Relatives of Man. Scientific American. Vol. 211, pp. 50–64.

SIMONS, ELWYN B. and DAVID R. PILBEAM.
1965 Preliminary Revision of the Dryopithecinae (Pongidae, Anthropondea). Folia Primatologica. Vol. 3, pp. 81–152.

SIMPSON, GEORGE G.
1945 The Principles of Classification of the Mammals. Bulletin of the Museum of Natural History. Vol. 85. New York.
1949 The Meaning of Evolution. New Haven: Yale University Press.

UZZELL T. and DAVID R. PILBEAM.
1971 Phyletic Divergence Dates of Hominoid Primates: A Comparison of Fossil and Molecular Data. Evolution. Vol. 25, No. 4, pp. 615ff.

WEIDENREICH, FRANZ.
1946 Apes, Giants, and Man. Chicago: University of Chicago Press.

Molecular Time Clocks in Evolution

The idea of using molecular biology to measure basic mutational changes in evolution is so attractive that it is tempting to accept all of its findings without close scrutiny. Its basic premises are well described by Vincent Sarich (1971) and will be summarized below. It starts with the idea that the DNA *codons*, or sequential triplets, contain a great deal of redundancy. And this is true of the 25 amino acids considered here: eight, or almost one-third of them, are coded by four different codons. Thus valine will be produced by any of the following codons: CAA, CAG, CAT, and CAC. Fourteen amino acids are coded by either three or two codons respectively. In all cases of redundancy the two initial bases are constant, and the variability lies in the third one in the codon. More than one-half of the amino acids can be produced by either of two codons. This leaves only three in which a single codon carries the key, and one in which three codons operate. This evidence is used to indicate the so-called "degeneracy" of the genetic code. It has been calculated that about 20 percent of the point mutations affecting the code would cause redundancy, and so have no affect on resultant protein structure. This leads to the conclusion that many mutations which become incorporated in the phylogeny, or genetic history, of a species are neutral and not subject to selection, either positive or negative.

Using these ideas derived from detailed molecular analysis, Sarich has operated with serum albumin and devised a method which estimates in an approximate fashion the magnitude of molecular differences between different species. The serological reactions which the method involves are translated into *immunological distance*, or as it is called ID. Certainly when the method is applied to such groups of animals as the great and lesser panda and they are compared with the black bear and raccoon, the resultant IDs are informative. There is little doubt that the greater panda is related to the bears, and the lesser one to the raccoons.

But Sarich has worked closely with the living primates, constructed phylogenies from his IDs, and drawn rather far-reaching conclusions from his evidence. Thus the ID of seven between man and chimpanzee was originally translated to mean that both were derived from a common ancestor about 4 million years ago. Under recent

pressure Sarich has extended this estimate of branching a little further back in time. He feels so strongly that the following quotation is in order: "To put it as bluntly as possible, I now feel that the body of molecular evidence on the *Homo-Pan* relationship is sufficiently extensive so that one no longer has the option of considering a fossil specimen older than about 8 million years as a hominid *no matter what it looks like.*" In effect, the fossil record is dismissed in one sweeping gesture. As might be expected, most paleontologists find this position unacceptable, for there is a consensus that *Ramapithecus* who certainly goes back 14 million years ago and has evolved some distance from any pongid is on the direct line of human ascent. Accepting Sarich's values would also mean that none of the dryopithecines have anything to do with the ancestry of later hominoids, such as the gorilla or chimp, or of course with man. All similarities would be due to parallel evolution.

Among those who disagree with the interpretation offered by Sarich are Uzzell and Pilbeam (1971). It has been pointed out that the ID is based upon *phenetic* similarities and differences of proteins and ignores the evolutionary pathways taken by each lineage. Thus these estimates omit parallel mutations, back mutations, mutations to synonymous codons, and consequently underestimate the total number of evolutionary events, in this case base changes. These authors undertook a phyletic analysis based upon five proteins. Their evidence failed to substantiate the idea of a constant rate of base exchange and the idea of neutral mutations. Other investigators have produced data which suggest that protein evolution has slowed down in the hominoids. One factor is that the hominoids have much longer generational spans than do lesser animals, and this may to a considerable degree account for the short-time estimates obtained by Sarich. It is very probable that at the present time the assumptions upon which the idea of molecular clocking are based simply have not been fully enough investigated to yield correct time estimates for divergences of related forms from common ancestors. Accordingly all phylogenies based upon the technique will be in one way or another erroneous at present.

Aside from the above comments, a much more

serious objection to these kinds of molecular exercises is to be found in the fact that they result in what is called non-Darwinian evolution. For in the essence of the neutral mutation concept is buried in the unstated premise that mutations in general are not subject to selection, positive or negative; they are simply neutral and so can accumulate in random and nondirected fashions in the phylogeny of species. This stands in direct opposition to neo-Darwinian evolution, in which it has always been conceived that selection acts as a directing force introducing the observable changes shown in the fossil record. The issue is of sufficient importance so that a good many evolutionists and geneticists are amassing data to press the point. To this time they have primarily found that non-Darwinian evolution is untenable and that selection does act as predicted in the neo-Darwinian postulates. So the reader faces something of a paradox. Can it be that by resorting to the very molecular level of biology improper time estimates can be produced? Or does the molecular approach really offer a new and convincing set of reasons to dismiss selection and with it neo-Darwinian evolution? It looks very much as though time will bring a compromise, perhaps something in the following form. Certainly some mutational substitutions of basis may be neutral and not subject to selection in some environments, although it is difficult to see that this would be true in all environments. If the molecular evolutionists could merely restate their position in that direction it would eliminate the direct confrontation with the rest of scientific evolutionists that now exists. Certainly most of those who believe in neo-Darwinian evolution would be happy to concede that some mutations may act in the manner suggested by Sarich and others. This admission would undercut the basis for any molecular evolutionary clock, for as soon as selection is allowed to operate in DNA changes, uniformities go out the window. Neo-Darwinian evolution will no doubt survive this conflict, for the idea that evolution should proceed on the basis of chance events, or by random walk processes is not convincing.

CHAPTER 8

SOME
SELECTED
EXAMPLES
OF PRIMATE
BEHAVIOR

The monkeys and apes have intrigued men from the earliest times. In them we recognize a kind of gross caricature of ourselves, and we are fascinated by their antics in the zoo. But, in truth, all of the members of the order of primates are more or less closely related to us, and among their fossil members lie our direct ancestors. Since it is now certain that men evolved in the tropical Old World and probably in Africa, where the earliest human fossils have been found, it is appropriate to look at the kinds of adaptations found in some of the African monkeys and apes to see if they can throw any light upon the way of life followed by early men. Thanks to the great interest in primate behavior in recent years, good accounts are now available for many forms. The primates worth examining from the point of interest of our own early adaptations include several types of ground-living monkeys, and, of course, the African great apes, the gorilla and the chimpanzee.

The Social Life of the Primates

The primates are intensely social animals with few exceptions. The nonsocial primates are usually nocturnal and are generally found among the prosimians. It has sometimes been said that sex is the glue which holds primate societies together, and perhaps human societies as well. The studies of Allison Jolly (1966) on lemur behavior in Madagascar show that even among prosimians, sex is not the social essential. Instead there are a number of factors other than sex which make social living adaptive for lemurs. Anatomically they go back to the Eocene phase of primate evolution and so may some aspects of their behavior.

The ring-tailed lemur (*Lemur catta*), with its banded tail waving gently as the animals glide along on the ground, is the more visible of the two lemurs. They live in permanent troops which consist on the average of a half-dozen adults of each sex, together with their young. This prosimian has a very restricted breeding season of no more than two weeks out of the entire year so that sex can be ruled out as the bond holding their society together during the full year. The cohesive bonds in the troop consist of grooming, social play, and the attraction that infants hold for adult females and subadults. Grooming has an important social function among primates generally, and the lemurs are no exception. The thick fur of the ring-tailed lemur requires special grooming devices, and so, along with many other prosimians, it has evolved a "grooming comb" which consists of the procumbent lower incisors and canines. This comb is raked through the fur of the groomed animal with upward sweeps. It is not certain how much material advantage lies in such grooming, but it certainly has great social significance. Grooming has evolved to take an important place in troop activities.

The young of many herd animals show some degree of playfulness, but the social primates are particularly playful and spend much time in this type of activity. Playing is a social activity and acts as a kind of preparation for various types of behavior, including sexual, in adulthood. In experiments, primates deprived of the opportunity to play with their peer group grow up to be warped in their attitudes toward other animals of their kind, and are usually incompetent in the sexual act.

The ring-tailed lemurs show some dominance ordering, for there is a linear dominance chain among the males of the troop. Interestingly all the males are subordinate to the females. If diurnal prosimians go back to the base of the Cenozoic in time, Jolly's data suggests that primate society may have evolved early in the evolution of the order, without diffused sexual opportunities to hold those societies together.

Primate behavior has been intensively studied in the wilds for little more than a decade. In recent years a marked reorganization of interpretation has occurred in this area. When the first few primate studies of natural populations were published it was quite naturally assumed that the animals observed were representative of their species, and indeed that their behavior would be the same for the whole species. Today enough different local groups within species have been studied so that it is clear that behavior is not uniform in a species. Since the primates are highly adaptive, it might almost have been predicted early that their behavior would mold itself to suit the local scene in a highly suitable fashion. It is in fact now widely recognized that social structure which expresses group behavior differs widely in the same species in different habitats. Ecology can override modal structure so that in extreme environments the same species will behave very differently in response to local requirements. This point is of particular importance in conceptualizing about the societies of the first humans, for they must have been even more flexible in behavior than the other primates living today. There probably never was a single pattern of human behavior and social structure in man's emerging days. Instead one might expect that men would behave advantageously, that is adaptively, in differing local scenes.

Group Structure. All human hunters and foragers at the generalized level of economy have the biological family as the basic social unit. It consists of a man and his wife or wives and their offspring. Numerous societies allow a man to take more than one wife, but even so most men in them are monogamous all their days, for multiple wives are a luxury only attained by a few individuals. The universality of this human pattern is some indication that the earliest of men, entering their bipedal niche, must have lived in a similar type of relationship. Therefore the relations between the sexes among the adults of terrestrial primates is of interest in an evolutionary sense.

Among the higher terrestrial primates the *uni-male group* is a basic form and quite possibly an early one. According to Eisenberg (1972) it has been altered twice toward more specialized forms. In the first variant the young males growing up in the group are allowed to remain into their adulthood, as a result of increased tolerance by the dominant male. This type of primate group is called the *age-graded-male group* and seems to be a rather simple extension of the fundamental uni-male group. Finally from the age-graded-male group it is possible to evolve the specialized *multi-male group* which is characterized by an oligarchy of adult males who are roughly equivalent in age. The dominance of these few senior males affects the order between them and the rest of the group, even though ranking within the top males may not be pronounced. The multi-male troop is characteristic of the savanna baboons who were studied earliest by the primate ethnologists, and so their composition has naturally influenced anthropological thinking about the social organization of very early human groups. This accident in the history of primate study may have obscured important realities which bear upon the problem of human social origins.

The Strategy of Flight

The patas monkey (*Erythrocebus patas*) is a terrestrial colobine that has evolved an unusual type of adaptation for successful living on the ground (Figure 8–1). This shaggy, red, black, and white, long-limbed monkey lives at the desert's edges in regions more arid than those which support the larger and more aggressive baboons. In the thorn scrub they eat fruits and pods, the pulp of the tamarind tree, a variety of berries, and occasionally take the eggs of lizards and birds. Males weigh up to thirty pounds, and these slender monkeys have been timed running more than thirty-five miles an hour.

In social structure they retain the uni-male group of the arboreal colobines in East Africa, while they seem to verge on age-graded-male groups in West Africa. So here we have the same primate bridging the two forms of social organization. Ecology does not seem to play a terribly important role in structuring patas society, since it differs little from their arboreal relatives. Home ranges are of considerable size, but even so this impoverished country is able to support about ten patases per square mile. While the male patas is supreme in his group, his females form a hierarchical chain and tend to initiate group travel for foraging. They are considerably smaller than adult males.

Figure 8–1
Patas Monkey Showing Speed
Ratio of Limbs

Little is known about the actual risk of predation among these monkeys, but their behavior provides several clues. During the day the troop male remains posted as a guard in one of the stunted trees, and if danger is sighted, he leaps to the ground and takes off at high speed. He acts to decoy the predator away from the rest of the group. In addition to his great speed, he is one of the gaudiest of monkeys for his black, red, and white coat is highly visible. To cap this scheme he sports a pair of incandescent blue testicles which must be a spectacular visual clue to the predators. At the same time the females and young freeze on the ground and remain silent. At night all the members of the troop disperse, each to a separate tree, apparently to minimize the risk of being taken by a prowling feline. Presumably cheetahs may threaten them during the day and leopards at night. As befits their survival problems, their visual acuity is great. One White hunter in East Africa carried a tame male patas on the hood of his land rover, for the animal could spot lions, leopards, and other predators much earlier than any of the men in the car. These monkeys survive by stealth and inconspicuousness, and they are almost unique among the social primates in being very quiet animals. Their special adaptation has resulted in a special way of life in difficult country, but it offers no model for the evolution of very early men.

The Successful Baboons

Baboons are the biggest of the monkeys, aggressive by nature, and live in as large social groups as their country will allow. They range throughout Africa south of the Sahara, from the tropical rain forests of West Africa, down through the savanna and veldt country to the very tip of Cape of Good Hope. They are of importance since they have been put forward as suitable models for the earliest of ground-living hominids. The rationale behind this claim seems to be that as ground-living monkeys the baboons have successfully coped with their predators, and early man in facing the same problem may have solved it in the same manner. Thus predation is the key to this view, despite the fact that all baboons differ from all known men in diet, in locomotion, and usually in social organization.

Olive Baboon. Most investigators have observed the baboons of East Africa, of which the olive baboon (*Papio anubis*) is a typical form. Like most of the savanna baboons it lives in dry, but not desert, grassland country which is relieved by scattered trees which become more numerous along the water courses. There they are used as sleeping trees, and their frequency seems to primarily determine the number and range of these animals. Baboons are primarily vegetarian in diet, eating almost every variety of plant food that they can digest. The range includes a great number of seeds, berries, fruits, and even flowers, with perhaps particular attention paid to the fleshy stems and underground roots of the savanna grasses. The latter must contain a great deal of cellulose, a material man finds hard to digest. Additionally, baboons eat insects, scorpions, lizards, young birds, and birds' eggs when these are available. They have been observed killing newborn antelopes and obviously relish this meat; however, these big monkeys primarily search for vegetable foods.

The social organization of the olive baboons is striking, for it consists of tightly knit, multi-male groups which range from relatively few individuals up to as many as 185. The general order of group size is between 30 and 50 animals. Social dominance plays an important role in organizing these small societies, with the adult males arranged in a hierarchical order of dominance and providing group defense when and if needed. A close bond exists between mothers and their offspring, and these tend toward the center of the group where they are surrounded by dominant males as they move off into their feeding areas. Juveniles of both sexes seem to be expendable and silently form the fringes of the troop.

This multi-male type of social organization is largely rationalized on the grounds that it is a defense against terrestrial predation in open country. This is certainly true to some extent, although earlier accounts of baboon life in East Africa certainly overstressed predation as a formative factor. More recent accounts indicate that the hierarchical structure is not as rigid in some areas as first reported. In fact, when threatened in open forest they run, each saving its own skin. Even more interesting, though related baboons occur in the very dry country of southwest Africa, they have abandoned the multi-male group structure and forage in small uni-male bands so as to maximize the efficiency of the food search. So again, the generalization that social structure may be overridden by ecological factors is true of these ground-living monkeys.

In the Nairobi Park in Kenya, East Africa, baboon densities range from two to seven animals per square mile. Numerically this is a much higher density than is attained by human hunters in a comparable environment. The secret of these high baboon densities lies in the fact that their diet is primarily vegetarian, whereas human hunters are located higher in the energy pyramid due to the expensive meat which they eat when available. It is not profitable to view the olive baboons as providing models for very early men, for it is agreed that the latter were partially carnivorous and so their densities must have been much lower than baboons, and lower even than present-day hunters. East Africa, in spite of its vast game herds, did not support more than one hunter for every ten or so square miles, so that large social groups would not be expected among early men there.

Hamadryas Baboon (Plate 8–1).   In less favored country baboon groups become much smaller and their foraging range is necessarily extended. This reaches a maximum in upland Ethiopia, where the hamadryas baboon (*Papio hamadryas*) lives in an impoverished desert area. There it has evolved a very different type of social unit, the uni-male group. This uni-male group has apparently resulted from the evolution of coded behavior which minimizes agression which is expressed only when one's female is threatened. The male arouses fear among the females by nipping them in the neck, and the females are quickly trained to stay at heel behind their head of the family. This behavior on the part of females is not necessarily

Plate 8–1
The hamadryas baboon of highland Ethiopia is characterized socially by uni-male groups.

coded, for *anubis* females released into hamadryas groups learned to behave as they should in a rather short period of time. Some families originate through a subadult hamadryas male becoming the protector and herder of a juvenile female. Of course, when she grows up the arrangement provides a proper biological family basis. In some groups the senior male even tolerates the presence of a younger male (his son?), showing the transition to the age-graded-male group.

Night protection in this arid region is found in sleeping areas on cliff faces, where as many as 130 baboons may gather for the night. In one unusual locality 700 were observed together. But during their waking hours the hamadryas baboons of necessity behave differently from their East African cousins. The scarcity of food requires the baboons to scatter in their foraging. During the daylight, groups consist of a single adult male accompanied by one to four females with their offspring. This is numerically similar to the polygamous family of human hunters in modern times. This type of social organization is not structured by any dominance hierarchy, nor is it based upon aggression, either threatened or practiced. The family units are stable, and this is the most manlike society among the ground-living monkeys. It is interesting that it seems to be an adaptive response to an ecological situation which limits baboons to low daytime densities.

Gelada Baboon.   The gelada baboon (*Theropithecus gelada*) lives in the higher mountains of Ethiopia, above 7,000 feet. As its genus name indicates, it is quite different from the other baboons and so has a separate evolutionary history. For example, the females are unique in having bare chest skin which is red and develops a necklace of white tubercles when she is in heat, or *estrus* (Plate 8–2). This advertises her sexual availability from the front, as does the sexual skin in the rear. These animals have also evolved the uni-male group, presumably to increase the efficiency of their food search for small ground objects in a difficult environment. Crook (1966) suggests that the males serve primarily to impregnate the females and to separate them when fights occur. Like the hamadryas baboons, the gelada males show a high degree of tolerance for each other. The presence of uni-male groups in these two distantly related baboons seems due to ecological override and convergent evolution.

The Fearsome Gorilla

This giant anthropoid, the subject of so many horror tales, turns out to be an amiable, quiet, rather stoical ape whose dominance hierarchies are quietly arranged and whose infrequent sexual activities are characterized by no signs of jealousy. Gorillas live in the

Plate 8-2
This pair of gelada baboons shows great sexual differences in size. The bare chest of the female is decorated with the display "necklace."

humid forests of tropical Africa, both in lowland West Africa and in the mountains of the eastern Congo border. The latter animals, studied by George Schaller (1963), are the better known. These two widely separated populations indicate that gorilla numbers have been declining in recent millennia. This gigantic anthropoid, whose adult males exceed 400 pounds in weight and whose females weigh about half that size, is a primarily ground-living and herbivorous animal. It eats more than a hundred species of plants, with shoots, stems, leaves, flowers, and piths consumed according to taste. Fruits comprise only a very minor part of the diet.

The social units of the gorilla consist of an old dominant silverback male, a number of females and their young of various ages, and one or more subordinate adult males whose presence is tolerated. Troop size ranges between 7 and 17 animals, and the home range is 10 to 15 square miles. For the mountain gorillas the inhabited region shows a density of about one animal per square mile. Group composition tends to remain stable over periods of years except that lone males frequently join the troop and later leave it. Dominance tends to follow size in a linear fashion among males, with, of course, all males dominant over females. The latter do not seem to show any stable dominance hierarchy.

Predation is probably not an important factor in gorilla life, but one silverback male was found to have been killed by a leopard. It had lost the fingers of one hand, and there is some question how much this affected the encounter. More serious problems for the gorilla are various parasites which infest them. They include those of the blood and intestines as well as various viruses. Their diseases range from malaria, yaws, arthritis, pneumonia, heart disease, and, for whatever comfort it may be to heavy human drinkers, cirrhosis of the liver.

Both anatomy and behavior indicate that the gorillas formerly lived more extensively in the trees than they do now. The length of their arms in proportion to their body is one such sign. The fact that they build nests every night, usually upon the ground, is a trans-

Plate 8–3
Magnificent male gorilla, showing locomotor posture normal to the species

ferred arboreal habit. As infants the gorillas move rather easily through the trees in quadrupedal fashion, but the older animals are very careful of their movements off the ground. Locomotion is quadrupedal, with the anterior weight of the body carried on clenched knuckles. With their food resources almost totally restricted to ground plants, these apes are almost completely terrestrial in their way of life. It seems likely a trend toward giantism forced them out of the trees. The fact that the gorilla is tied by its food habits to the lush vegetation of tropical rain forests and mountain forests disqualifies it as a model that would illuminate how the earliest of men lived in the African savannas. Although the gorilla is said to be as intelligent in test situations as the chimpanzee, it is to the latter animal that we must turn for our protohuman model.

Wild Chimpanzees Provide a Model for Dryopithecine Behavior

Bones tell much of the anatomical story of evolution, but where possible it should be fleshed with reconstructions of the animal's behavior. Skeletons may allow conclusions about diet and locomotion, and hence the general way of an animal's life, but they cannot provide interpretive details of its social behavior. Therefore, we turn to the most similar living animal to assist in reconstructing the total behavior of extinct ones. Of all the living primates the chimpanzee behaves in the most human fashion. There is general agreement that the chimpanzee has been an evolutionary conservative in a broad sense, though showing some specialization away from the dryopithecine level in the lengthening of the arms to make brachiation easier. We take the chimpanzee as a living behavioral model for the dryopithecines of the Miocene, for there is no chance of finding a better one. Let us see how those ancient pongids may have behaved.

The most extensive information has been provided by Baroness Jane van Lawick-Goodall (1967, 1968). Better known to more people as Jane Goodall, this pretty young English protégé of Louis Leakey has devoted most of her time since 1960 to living with and studying the wild chimpanzees which reside in the Gombe Stream Reserve on Lake Tanganyika, in the country of Tanzania, East Africa. This is a region with strips of dense rain forests along the valleys of the streams, whose slopes then grade into open and deciduous woodlands on the higher ridges. At these higher elevations the trees are scattered through grasslands, which made the task of observing the wild chimpanzees much easier. This type of country is much like that postulated for the evolution of the earliest of men.

Goodall has estimated that between 60 and 80 of these manlike apes resided more or less permanently in the 30 square miles under her observation, a respectable density of 2.6 animals per square mile. In the Budongo Forests to the north where chimpanzees live in climax rain forests, their density reaches about 10 per square mile. Primitive hunting and gathering humans seldom, in the best of their environments, reach a density as great as one person per square mile. Energy relationships again provide the explanation, as noted in Figure 4–4.

These high chimpanzee densities are supported almost entirely by food obtained from the trees. Goodall's chimpanzees ate 37 different kinds of fruit, which may have accounted for 90 percent of the bulk of their diet, some 21 kinds of leaves or leaf buds, six different blossoms, four kinds of seeds, the stems of three different species of plants, and two kinds of bark. In addition, they had as hors d'oeuvres a number of insects and occasionally some red meat. They are not fearful of predation themselves, for while leopards are common throughout their range, they seem not to attack chimpanzees. Although leopards customarily kill both baboons and other monkeys, even very young chimpanzees wander off by themselves, and Jane Goodall has never found any evidence of their being attacked by the big cats. It may well be that both pongids and men have evolved passive defenses against the big carnivores, quite possibly including the nature of their own odors.

Social Organization. Chimpanzee society is both open and poorly defined. In most animal societies the composition of the group reflects the kind of social dominance which characterizes the species. Among chimpanzee dominance rules are not as rigid in form or as frequently reinforced as among some other primates as

Plate 8–4
Adult male chimpanzee, showing position of hands and feet in knuckle-walking

the olive baboons. The rules are simple in that all grown males dominate all females and young. Within adult males the order of dominance is fairly clear but may be shifted with circumstances. The grown females dominate all the young of both sexes. The consequence of this set of social rules is that the only permanent observable group is formed by a mother and her infants. It seems true that the bond between mother and child continues as the latter grows up, and so it might be better phrased that chimpanzees live in families headed by the mother alone. This of course is not quite the equivalent of a matriarchal extended family in human terms but that is a tendency among these primates. In late adolescence the young leave their mother's side to go their way, but apparently maintain a real bond of intimacy as long as the mother lives. Aside from this basic family group, the chimps are seen in small, changing clusters which may consist of a few males maintaining friendly relations, or a cluster of mothers with dependent young, or in other varying combinations. Overall boundaries of chimpanzee society may be defined by geographical barriers such as rivers and lakes, or wide stretches of open grassland, or other factors which restrict their mobility.

Sexual Behavior.   As pleasant animals as the chimpanzees are, their sexual behavior is hardly an ideal model for man, for they are completely promiscuous. Females in heat, that is, in *estrus,* have the sexual skins on their buttocks mightily swollen and illuminated by inflammatory colors. A female in such a condition will apparently take on all male comers. Jane Goodall observed in one instance that within a very short period of time a female in heat copulated seven times with six males, one of which was an adolescent. During this performance there were no indications of rivalry or sexual jealousy among the males, who generally are very tolerant of each other under all circumstances. In chimpanzee society neither the mother nor the interested human observer could possibly determine the real father of an infant. It is worth mentioning that although homosexual behavior is commonly observed among captive primates, including the chimpanzee, it is seldom seen among these wild pongids. Heterosexual relations are the rule among adults.

Births are usually spaced from a minimum of three to a maximum of five years apart. This is achieved by prolonged lactation, suppressed ovulation, and hence deferred estrus. For the first six months after birth the chimpanzee babies are completely dependent upon their mother for food, protection, and transportation. Between that age and the end of the first year and a half the infant still nurses, but also begins to eat some solid food. The dawning of social independence is demonstrated by the infants' playing some small distance from the mother, but never out of her sight. Suckling may continue up to three years of age, with the baby riding in jockey fashion and sleeping with the mother at night. Chimpanzees at this age are very active and may spend up to 90 percent of the day out of physical contact with their mothers' bodies. These animals reach sexual maturity about nine to ten years of age and apparently survive more than 30 years of age in the wild. Chimpanzees observed under captive conditions indicate that they might live to be 40, 50, or even 60 years old.

Goodall made the interesting observation that chimpanzee mothers differed considerably in their competence in handling their own infants. Some are extremely attentive and tend the needs of the infant very carefully. Others provide for their infants' basic wants, but in a heedless and insensitive fashion. Such differences may reflect the experience of old hands as opposed to new and unconditioned mothers, or it may be that the intensity of mother love, as mediated by the hormone prolactin, is variable among the chimpanzee just as it seems to be among human mothers. Early infant mortality is fairly high among the chimps, and during her field observations Goodall was able to note the case of two mothers who had lost their infants. In one case the baby animal clearly had suffered a fall, for one arm showed a compound fracture which contributed to its death. Arboreal animals are always subject to the hazard of injury or death by falling from the trees, and observers have witnessed chimpanzees falling from heights of 30 and 70 feet, respectively, and in these two cases apparently suffering no injuries. Another male fell over 30 feet, broke his neck, and that was that. Both chimpanzees and men are more adaptable than anatomists would like us to think, and one older chimpanzee with a broken wrist was seen climbing in the trees and apparently able to support himself on an adequate diet. Of course the moral is that a successful animal always has a built-in margin of safety in his lifeway.

Manlike Behavior.   Given these basic data, let us now go on to some of the more specifically manlike behavior which has been ascertained by Jane Goodall through her years of careful observation. The anatomists claim that the pelvis and legs of the chimpanzee are ill-suited for bipedal locomotion, yet the animals habitually stand, and walk, or run on their hind legs under certain circumstances. When out in the open, in tall grass, they commonly stand bipedally in order to look over the grass tops. Goodall has observed chimps doing this for as long as 15 minutes at a time. Chimpanzees frequently go from tree to tree bipedally. It is also a common mode of locomotion when carrying something. The chimpanzees run bipedally, either through excitement or fear, and Goodall has seen this done for distances of thirty yards. While it would be too much to say that bipedal locomotion plays an important part in the average day of the chimpanzee, nonetheless it is an acceptable variant form of locomotion, and may be preferred under some circumstances. There is no reason to believe that Miocene dryopithecines would be any less adept in bipedal movement.

It has long been known from systematic experiments with captive chimpanzees that they can use tools effectively; further, they can construct them out of various components; and, finally, they use tools in situations in a way that involves foresight. These tool-using capabilities have now been confirmed and extended by Goodall's observations on the wild chimpanzees. Nest building can probably be classed as tool-making, and each chimpanzee builds one every night with no more than an expenditure of one to five minutes of its time. Nest building is learned behavior, as is shown by the fact that chimpanzees born in captivity cannot do it. This is further confirmed by observations in the field which show that two-year-old chimp infants watch their mothers carefully as they build the nests, and soon they are building miniature ones of their own as a game. A learning process is clearly involved.

But chimpanzee tool-making is more complicated than nest building. On occasion chimps have been observed chewing up a handful of leaves which are then used as a sponge to dip water out of an otherwise inaccessible crotch high in a tree. More impressive tools are the "fishing rods" with which the chimpanzees obtain both termites and ants. This is only a seasonal occupation, and it was not observed by Goodall until she had been in the field several years. These activities begin with the onset of the rainy season in East Africa, when the chimpanzees begin to show interest in the local termite nests. This is the time of year when the insects are growing their wings prior to their nuptial swarming from their home nest, as they disperse to establish new ones. The termites have already prepared escape routes from the nests. These are closed over by only a thin layer of clay, so that the chimps are skilled in finding such getaway hatches, opening them and inserting their fishing poles down into the nest. These instruments consist of rather carefully selected pieces of grass, twig, or vine and usually measure somewhere around twelve inches in length. But Goodall has found, after collecting and examining several hundred of these fishing poles, that individuals show preferences both in the materials they use and the style of the tool made from these materials. This suggests that an ideal type of tool exists in the mind of each animal.

The technique of fishing for termites involves the moistening of the fishing pole by passing it through the chimpanzee's lips. It is then inserted into the interior of the nest, left there for a short time, and withdrawn. As it comes out, the rod is found to bring with it a dozen or so termites who have their jaws embedded in the pole. The chimp takes off these termites with his lips, the fishing pole is reinserted in the hole and this process goes on until the chimp's appetite is sated. The process has been observed to continue for four hours in the case of an individual animal. Goodall claims that East African termites are rather tasteless.

Goodall's chimps constructed and used tools in a variety of other ways. At certain seasons the animals also fish for weaver ants, which she says have the interesting flavor of lemon curry. For ant fishing the chimpanzees construct fishing poles 30 to 40 inches in length and bring out clusters of the little insects.

Some individual chimps have been observed under circumstances which clearly show foresight. One old male was observed by Goodall to carry his fishing rod for a good half mile from the place where he obtained his material to the termite nests where he fished. In West Africa chimps have been observed to use stones to crack open hard-shelled palm nuts. Thus in a simple but convincing fashion wild chimpanzees demonstrate the ability to make tools and to use them for specific purposes. This kind of behavior is an impressive beginning on the path toward more complex tool use, and we must grant the same capabilities to the radiating Miocene dryopithecines living some 20 million years ago.

Since chimpanzees have always been described as fruit eaters, and certainly as vegetarians in the broadest sense, it came as a considerable surprise that Jane Goodall could report a number of cases in which her chimpanzees not only ate red meat, but hunted the animals to provide it. Their victims included the young of baboons, bush pigs, antelope, and even adult red colobus monkeys. They gave every indication meat is a much prized delicacy, and they ate it by sucking and pulling it through the incisor teeth. Chimps occasionally show a spirit of altruism by giving other kinds of food to their friends, but they are very possessive with meat and indulge in no sharing. The young bush pigs and antelope are probably found on the ground shortly after birth and more or less by accident. But Goodall saw the killing of a young baboon, an act that involved a good deal of excitement and violence. An adult male chimpanzee, named Rudolf for ease of identification, grabbed a young baboon by the hind leg and beat its brains out on the ground. Three of Rudolf's cronies danced up and down, screaming in excitement. But in this case the killer climbed the tree with his victim and sat there for the next four hours, eating it by himself until he grew sleepy from his food.

More revealing is Goodall's observation concerning the hunting of a red colobus monkey. The victim-to-be was sitting in a tree when he was observed by an adolescent male chimpanzee who climbed a neighboring tree and remained there to hold the monkey's attention. A second nearly grown male chimp then climbed the tree in which the colobus was sitting, quickly ran along the branch, and seized the poor monkey by its neck. This remarkable instance seems to demonstrate a hunting plan which was carried out successfully with no oral communication. Speech is not required for successful team hunting.

It has been rumored for years that West African chimpanzees seize human infants and carry them off. This report receives horrifying confirmation from the fact that chimpanzees are actually cannibals. No less than two cases are reported of dominant males seizing infants from their chimpanzee mothers and eating them. In one instance from Goodall's reserve, five males attacked two females, the older of which was a stranger. After a big fight in the bush, "Humphrey" emerged holding a struggling baby chimp by the legs. He killed it by beating its head against a branch. He ate on the carcass for an hour and a half, and then the body was passed around to six other chimps for the next six-hour period. The legs, one hand, and the genitals were eaten. This picky feeding suggests that chimp babies are not judged to be quite as good eating as other forms of meat. An even more revolting case is reported for the chimps in the Budongo Forest where a dominant male seized a newborn chimp and ate it alive. This casual type of predation presumably goes back to the dryopithecines of the Miocene period. Even though it preadapted man's ancestors to become hunters, we have received a terrible heritage from them.

The Gombe Reserve chimpanzees represent one local group. Observers of other groups of chimpanzees find somewhat different behavior. For example, chimpanzees living on the edge of the savanna, and often traveling considerable distances into it, show rather more formalized social structure and more fixity in the nature of their groups. There again we have suggestions that ecology is a more important determinant of social structure, and this certainly would have affected our very early ancestors. In another instance a stuffed leopard with mechanical "innards" which turned its head and twitched its tail was placed by the path the chimps used in their daily rounds. Upon sighting the phony cat, the chimps became enormously excited and aggressive. They not only rushed at the leopard, but beat it with sufficiently heavy sticks to have been able to break its back. Of course this particular leopard was easy game, but it is significant that chimps will behave so aggressively toward this effective carnivore. The evidence suggests that the very early men may have been able to cope with some of their predators in a similar fashion.

Evidence of Communication.   Chimpanzees have no speech in our sense, but they do communicate through a wide variety of sounds which serve as emotional signals. These provide a kind of primitive system of communication among themselves. They further use gestures to indicate mood. Chimpanzees have been seen kissing, touching hands, and clapping each other on the back in perfectly human fashion. More involved are the "carnivals" which are reported to involve large numbers of excited chimpanzees whooping it up for extended periods of time. Perhaps the best authenticated mass social behavior is contained in Goodall's description of two "rain dances" which she had the good fortune to observe. One morning as rain clouds appeared, she watched a group of chimpanzees feeding and playing on a ridge lying across a narrow ravine. After threatening all morning, the rain finally came down in torrents. Coming down from their feeding trees, the chimps grouped themselves into two lots, the adult males who were the performers, and the females and young who as spectators climbed a tree at the top of the ridge. The performance began when one male, hair on end, stood erect at the top of the slope, then plunged down the hill, breaking off a large branch on his way. Two more males set off in the same path, hooting excitedly. One after another the male performers charged down the slope, leaping into a tree to tear off a bough, and then grasping tree trunks at the bottom of the ravine to break their onward rush. After catching their breath, each climbed back up the hill to repeat the performance over again. The shrieks of the apes, often lost in the crash of thunder, gave the dance an eerie accompaniment. The dance continued for about half an hour, and then the spectator groups climbed down from their tree perch, and the participating male dancers joined them and all disappeared over the top of the ridge. This whole spectacle is suggestive enough of the spirit and actions of some dances among simple people to allow us to label this protodancing. As a result of Goodall's observations of complex chimpanzee behavior, many anthropologists would now classify them as protocultural animals rather than as precultural ones.

And What About the Behavior of the Ramapithecines?

There are no living models by which we may reconstruct the behavior of the Miocene ramapithecines. But it is possible to use the bracketing technique by which reasonable upper and lower limits are fixed to estimate something of the range of behavior. The dental evidence strongly indicates that the ramapithecines were much more manlike than their contemporaries, the dryopithecines. As scanty as their remains are, evidence for the ramapithecines has been found in East Africa, Europe, China, and North India. In all of these places they coexisted with dryopithecines. Since it has long been established from studies in nature that two species with exactly similar food demands cannot coexist for long in the same space, a principle called *Gause's Law,* it would appear that these two types must have lived quite differently so as to have avoided serious competition for the same food. Since the coexistence of the ramapithecines and the dryopithecines persisted over millions of years, it follows that their food requirements must have become well differentiated in earlier times.

Ramapithecine behavior can be estimated as lying somewhere well above that of the living chimpanzees, but considerably below the behavior which characterizes such early men as the australopithecines. This bracketing allows some of its major features to be sketched in. Certainly these emergent men ate more meat and hunted more effectively than do modern chimpanzees. Their use of tools would have been more skilled, and they should have possessed a greater variety than Goodall observed among her friends. No doubt their hunting still remained largely fortuitous, depending upon some elements of chance. Tools made of such organic materials as wood or bone are not likely to be preserved over the long time periods involved. But the use of intentionally formed stone tools is being steadily pushed back in time, and they now occur in the basal Pleistocene deposits of East Rudolph as a well developed industrial complex, and at a date of about 2.6 million years ago. When investigators diligently search the Pliocene and Miocene deposits for human tools, it is possible they may be there for recognition. Indeed, Louis Leakey reported some evidence suggesting human tool use from the important site of Fort Ternan in East Africa, where ramapithecine jaws have been recovered from lower Pliocene layers. An angular piece of basalt, which is alien to the site and that must have been carried in, was found in association with the long bones of animals which seemed to have been cracked for their marrow. Indeed, one long bone shows a depressed fracture clearly preserved. And an animal skull has had its top battered off as though to obtain the brains it contained. One transported stone and a few shattered bones do not stand as positive proof that tool-using man was present there and then, but the evidence is suggestive. Certainly the investigation of the deposits of this age promise most exciting returns in the decades ahead.

Bibliography

ALTMANN, S. A. and J. ALTMANN.
1970 Baboon Ecology. Chicago: University of Chicago Press.

BYGOTT, J. B.
1972 Cannibalism Among Wild Chimpanzees. Nature. Vol. 238, No. 5364, pp. 410–411.

CROOK, J. H.
1966 Gelada Baboon Herd Structure and Movement: A Comparative Report. Symposium of the Zoological Society. London. Vol. 18, pp. 237–238.

EISENBERG, J. F., N. A. LUCKENHIRN, and R. RUDRAN.
1972 The Relation Between Ecology and Social Structure in Primates. Science. Vol. 176, No. 4037, pp. 863–874.

HALL, K. R. L.
1965 Behavior and Ecology of the Wild Patas Monkey, *Erythrocebus patas,* in Uganda. Journal of Zoology. Vol. 148, pp. 15–87.

JOLLY, ALLISON.
1966 Lemur Behavior. Chicago: University of Chicago Press.
1972 The Evolution of Primate Behavior. New York: Macmillan.

KUMMER, HANS.
1968 Social Organization of Hamadryas Baboons, A Field Study. Biblioteca Primatologica. No. 6. S. Karger, Basel.
1971 Primate Societies. Chicago: Aldine-Atherton.

SCHALLER, GEORGE B.
1963 The Mountain Gorilla: Ecology and Behavior. Chicago: University of Chicago Press.

VAN LAWICK-GOODALL, JANE.
1967 My Friends the Wild Chimpanzees. Washington D.C.: The National Geographic Society.
1968 The Behavior of Free-Living Chimpanzees in the Gombe Stream Reserve. Animal Behavior Monographs 1, 3: 161–311.

CHAPTER 9

CLUES
FOR THE
INTERPRETATION
OF FOSSIL
MAN

Human fossils are fragmentary, rare, and unevenly distributed in time and space. Determining their age and sex can be difficult, and determining their relationships by descent is an uncertain art at best. A lack of accurate dating of the fossils imposes great difficulties upon their interpretation. Physical barriers to terrestrial movements tend to cluster ancient people into regional groups. Entrance into the bipedal, tool-using niche left its mark in terms of structural changes in both the human body and head.

How It Began

Neanderthal man, the first fossil human to make an impact upon the scientific world, was discovered the way many important fossil men have been found—by accident. The Dussel River, which flows into the Rhine near the city of Dusseldorf, passes through a gorge just below the village of Neander. The valley cliffs of Devonian limestone were heavily quarried for industrial purposes in the mid-nineteenth century. In the summer of 1856, workmen clearing out the soil from a small cave high on the face of the cliff in order to prepare for further blasting stumbled upon the human skeleton. Fortunately, the owner of the quarry sensed that the bones might be of some importance and rescued them from the rubbish heap. The find consisted of a damaged skullcap and fifteen pieces from the rest of the skeleton. Their condition indicated that prior to being disturbed the find had consisted of a complete burial in reasonably well-preserved condition. They ultimately found their way into the hands of Professor D. Schlaaffhausen of Bonn, who made a preliminary report upon them early in 1857 at the meeting of the Lower Rhine Medical and Natural History Society. Schlaaffhausen recognized that the skullcap showed an extraordinary form which had not been seen before among even the most barbarous of living races. He believed it to predate both the Celts and the Germans and thought it derived from one of the wild races of northwestern Europe which had been conquered by the more civilized Germanic tribes.

His report resulted in considerable controversy. The great German pathologist, Rudolf Virchow, looked upon the skullcap as pathological and was of the opinion that its strange form perhaps was the result of a vitamin deficiency disease known as rickets. Others commented that the sutures, or lines where the bones joined, may have closed prematurely and so distorted its normal pattern of growth so as to produce its strange form. Some were of the opinion that the skull was that of an idiot, or possibly of a Mongolian Cossack who had died there during Napoleon's retreat from Moscow in 1814. There was a general reluctance in the world of science of that day to believe that this first fossil man could have represented a normal individual from a population of very much earlier times. Gradually these extreme opinions were forsaken, and Neanderthal man finally received recognition as being a normal member of an ancient population which had lived in Europe at the beginning of the fourth glacial period of the Pleistocene. He was the first representative of the classic Neanderthal group. An equally impressive and a more complete skull discovered in 1848 in the Forbes quarry at Gibraltar in the southern tip of Spain received no recognition whatsoever at that time; however, some years later it was brought to London and finally recognized by the scientific world as a classic Neanderthal female.

The brief period represented by the 1850s saw remarkable progress in the evolution of scientific thought. The Neanderthal skullcap had been discovered in 1856 and presented to science in the following year. Barely twelve months later Charles Darwin published his revolutionary *On the Origin of Species*. In 1859 a delegation from the Royal Society of Great Britain, including Charles Lyell, the geologist, visited the excavations of the Boucher de Perthes at Abbeville in the gravel terraces of the Somme River in northwestern France. For more than three decades the amateur French archeologist had proclaimed that the flints he found imbedded in the riverbed were manmade and obviously of great antiquity. The British visitors were convinced, and the crudely flaked fist axes were attributed to human beings destroyed in the biblical flood. Nonetheless they stood as evidence for the existence of man earlier on the earth. That same year Gregor Mendel published his research findings which remain today the foundation for the science of genetics. That so much progress should be attained in the span of five years was to some degree coincidental, but the stage had been set earlier by the relaxation of certain church dogmas, which had allowed the stirring of evolutionary ideas.

Some of the Difficulties in Interpreting the Fossil Record

In the little more than one century which has elapsed since discovery of the skullcap at Neander, over five hundred other human fossils have been recovered from various parts of the Old World, ranging through the three million years of the Pleistocene epoch. The number sounds impressive and sufficient upon which to reconstruct the details of human evolution. Unfortunately, this is not the case at this time. As important as such remains are, their interpretation is hampered by their fragmentary nature, the gaps in the record produced by the fact that they are widely scattered in both time and space, the difficulties in determining both sex and age of individuals from unknown populations, the uncertainty as to how much of a bone's form is due to inheritance and how much is due to the functional stresses imposed upon it during the individual's life, and the difficulty of setting up rules for establishing relationships within the tight family of man. The defining of a correct time scale is the most important of the unresolved problems.

Fragmentary Data.   The fragmentary nature of human remains provides a first stumbling block. The practice of burying the dead is a recent custom which goes back only a little more than the last two percent of Pleistocene time. Prior to this, human bones became fossils much like those of the bones of any other animal, that is, by the accident of local circumstances. When men died, their corpses decayed, and usually their skeletons were consumed and scattered by scavengers, or perhaps simply disintegrated due to the action of sunlight. Occasionally a few bones would find their way into deposits which preserved them long enough for the mineral replacement to make them into fossils. As with the rest of the world of the vertebrates, this means that our knowledge is to a disproportionate degree based upon teeth, a few of the denser bones such as the mandible and the cap of the skull, with an occasional limb bone to help balance the record. The other bones of the body are less likely to be preserved and so are missing from the record. Indeed, human behavior mars the record, for in all too many cases the individuals whose bones we find have been the victims of headhunting and cannibalism and so subject to unnatural mutilation.

Distribution Problems.   The distribution of human fossils is very uneven in both time and space. Dental fragments of perhaps two dozen individuals are known from the Pliocene. Owing to changes in the chronology in the Pleistocene which arose in 1971, that period will be divided into simply Early, Middle, and Late segments. In the Early Pleistocene, a period covering about two million years, the record improves, and fragments of over a hundred individuals are known, although they are almost entirely limited to Africa south of the Sahara. In the Middle Pleistocene, fossil remains again become scarce, although they are fairly widely scattered, occurring in Africa, Europe, China, and Java. Coming into the Upper Pleistocene, which may have lasted about one hundred thousand years, the record continues scanty until the final glacial period in Europe. Then the appearance of intentional burial and better preservation add greatly to our store of fossils. During the Upper Pleistocene, the finds again exceed one hundred individuals, primarily concentrated in Europe, but with a few scattered in the rest of the Old World. In a spatial sense, the record is so erratic that there is not a single fossil human fragment known for the whole of the Pleistocene in the subcontinent of India. We cannot change the unevenness of the fossil record, but this fact must be considered when reconstructing human evolution.

Difficulties in Determining Age and Sex of Fossil Remains

Human anatomy is one of the oldest of the medical sciences and certainly in many ways is completely respectable. But it is generally practiced as a historical and descriptive science and thus is not equipped to deal with processes such as evolution. Yet it contributes to the study of fossil man in several ways. Among these are its technique for estimating the age and sex of individual skeletons. Competent anatomists who are familiar with the population range of variation can estimate age and sex with impressive accuracy. Determinations of age rest primarily upon changes in bone growth and tooth eruption during the first twenty years of the individual's lifetime. The eruption of the milk teeth, and then their replacement with permanent teeth, provide a complex sequence of events which will usually serve during that period to fix the age of the individual within a year of the chronological age. Tooth formation, which can be ascertained by X-ray examination, refines dental aging even further. In reptiles which grow throughout their lifetime the long bones grow at their ends, that is, on their weight-bearing faces. Mammals, on the other hand, have an improved but more complicated method of growth. Growth continues until *maturity* and then stops. During this period of growth the long bone and some of the other bones of the body achieve their growth at the ends of the shaft of the bone close to the end caps, which are called *epiphyses* (Figure 9–1). Since the epiphyses fuse to the central portion of the bone at different ages throughout the body, they provide a very accurate division of the growth period in the individual's lifetime. Once the period of growth has ceased, then the changes that occur within the bony skeleton are largely degenerative in nature. Some cellular breakdown occurs in certain parts of the pelvis, and both the external and internal sutures of the skull and face close or grow together at varying rates. But most authorities would agree that once the age of twenty is passed, then the best of estimating can put a person only in a given decade of adult life.

In the materials provided by modern dissecting rooms, and these are usually the bodies of Whites, aging can be done within the limits outlined above. But in a fossil population where rates of growth may differ, as well as dental eruption timing and the closure of bony sutures, the estimates of aging are necessarily less accurate. It is suspected that thick-boned skulls may show suture closure at an earlier age than do our own. Thus, there may be a small but systematic bias to judge ancient individuals as being a little older than they really are. Until the fossil series becomes much more numerous, and their skeletons more complete, this is a disadvantage that must be endured.

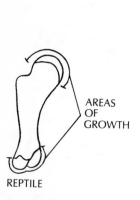

AREAS
OF
GROWTH

REPTILE

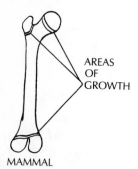

AREAS
OF
GROWTH

MAMMAL

Figure 9-1
Patterns of Bone Growth
(Femur—not to scale)

Bone variations. To the uninitiated the sexing of skeletal material always sounds a little magical. Even the sexing of some of our contemporaries has become a little risky since the advent of unisexual dress and hairdos. But bones do give clues to the sex of the individual, and these are particularly concentrated in the skull and in the pelvis. Both the vault and face of the skull are generally more robustly built in males than in females. Absolute size tends to be larger, bones thicker, and ridges for the insertion of muscular attachments higher in relief. An anatomist who is familiar with a given population can usually sex skulls with an accuracy ranging from 80 to 85 percent.

The human pelvis carries its own special clues as to sex. These arise from the fact that in the male the pelvis is a bony structure serving no more than the mechanical requirements involved in posture and locomotion. But in the human female its form is compromised by the necessity for providing passage to the young mammals which are to be born through it. This primary requirement indirectly produces changes in the other proportions of the pelvic bones. Consequently, the skilled observer can correctly identify sex pelves in something like 80 percent of a series. Where both skull and pelvis are available on an individual skeleton, the accuracy of sexing rises to something like 90 percent. But in any population some individuals having masculine appearing skulls are revealed to be females by the contours of their pelves. In life they would probably be recognized as masculinized women. By the same token, an occasionally delicate skull is revealed by its pelvis to have belonged to a feminized male. Sexing by skulls alone obviously allows errors of this type to creep into the estimates. The sexing of immature individuals is very much less secure than for adults. The secondary characteristics which distinguish males from females are not fully developed until growth is completed. Even the pelvis remains undifferentiated until the onset of puberty, when in females it begins changed rates of growth that finally produce the adult configuration. Children and infants show so few sexual distinctions that sexing should not be attempted.

The principles of sexing are well established where we know the sexual variation within a given population. But the fossils recovered are very seldom sufficiently numerous to display adequately the full range of sexual variation that characterized the living population. Since pelves are very rare fossils, most of the judging process must be based upon skulls and frequently nothing better than skullcaps or vaults. The magnitude of the sexual difference varies in living populations and it must be presumed to have done so in ancient ones. Among small and lightly built peoples, the differences between sexes is apt to be reduced, and identifying them correctly is more difficult. There are no Pleistocene populations for which the real ranges of sexual variation are sufficiently known to make the task easier. Among the Late Pleistocene burials of Europe, sexing can be done reasonably well primarily because these populations were like modern Europeans.

The fossil samples of the human population in Java are better known than in most areas of comparable size. They include eight incomplete skullcaps and a number of mandibular fragments from the Early and Middle Pleistocene deposits and eleven skulls from the Late Pleistocene. Weidenreich sexed six of the more complete skullcaps of the latter series on the basis of size and bone thickness. He judged three to be male, two female, and one indeterminate. These eleven skulls are an interesting sample, for in all but one the bases have been removed so that the headhunters could eat the brains of their victims. In modern times this was largely done for ritual reasons, rather than out of pure appetite. But certainly the fact that modern headhunters kill both men and women, and for that matter children when available, provides no information as to how the game was played several hundred thousand years ago in Java. It is conceivable that the series is totally comprised of a single sex, and all eleven individuals may have been male.

One of the effective ways of testing prehistoric data is to measure it against known materials on living people. I have measured a large number of hybrid peoples in Australia and included a sizable series of full-blooded Aborigines as a check sample of the maternal ancestors. Included were 578 adult aboriginal males and 189 full-blooded females. Maximum head length is the best measurement to represent head size among a people whose head shape is uniform. These measurements from both sexes of Aborigines have been broken into two millimeter group intervals and are plotted separate-

Figure 9-2
Range of Overlap in Head Length

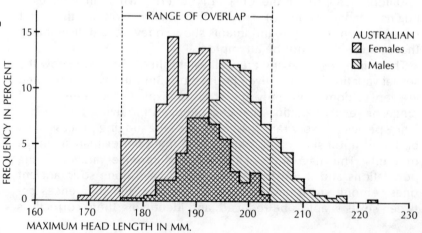

ly as percentages for each sex in Figure 9–2. The full series of males ranges from 176 to 223 millimeters in head length, while the females go from 168 to 203 millimeters. As Figure 9–2 shows, there is an overlap of 28 millimeters, which includes the majority of both sexes. The proportion of males who lie above the female range amounts to only 13.8 percent, while the number of females who fall below the lowest male value is only 5.3 percent of that total series of that sex. Expressed in another fashion, the region of overlap in maximum head length amounts to more than 90 percent of the total series, allowing less than 10 percent to be positively identified as to sex by this dimension because they lie outside of the range of overlap on the high and low end of the data. We have chosen measurements from the living, since then the sexes are known with certainty. The difference in the thickness of flesh in the living as compared to the measurements on the dry skull amounts to about ten millimeters in males and perhaps nine millimeters in females, a difference which is not significant. The point is that a series of crania cannot be sexed correctly if the judgment is based upon size alone.

On the Relationship Between Form and Function

Evolution usually produces adapted and well-suited end products. Viewing bones as structures, it is not surprising that the shapes they take reflect the uses to which they are put. This relationship between form and function is both long-term and short-term; it has evolutionary significance as well as being an individual response to the stresses endured during a single lifetime. The refinement of bone form achieved during long-term evolution is primarily brought about in genetic terms through various forces of selection acting upon the available variability. The short-term accommodation of growing forms to imposed stress primarily involves the capacity of bone to grow in such a way as to buffer stresses placed upon it. This response is nongenetic in nature and reflects the plasticity inherent in complex living organisms. Functional anatomists in recent decades have designed experiments on living animals to learn to what degree bone form is genetically determined and to what degree it can be altered by stressing it during its growth period. Such experiments are difficult to design and at the moment about all that can be concluded is that the primary form of individual bones is a matter of genetic determination, but the overall shape can be modified to some small degree by the imposition or release of the stresses normal to it. Insofar as the shape of bone is genetically determined, it may have a bearing on tracing relationships in evolution. But even where external shape appears to be similar, caution must be observed, for the variation of form in most structures is determined by many genes and so, as will be seen later, the same shapes may be determined by many different combinations of genes. To the degree that the external form of bone is modifiable by the activities and life-style of an individual, likenesses and differences can be misleading.

Bone grows in such a manner as to buffer increasing stresses. Anthropologists did not recognize this fact for many years. Correspondingly, they confused inheritable changes, such as might occur between differing races, with alterations that were clearly induced by the kind of life led. Thus squatting facets, or flattened places on the long bones constituting the knee joint, were thought to be characteristic of primitive living men. In fact, they turned out to be the result of hunting men squatting around campfires all of their life and producing the flattened planes through continuous pressure. The bending of the shaft of the human femur so that it bowed in an anterior-posterior direction was again thought to be a primitive characteristic. Likewise, the flattening of the human tibia in a lateral plane was viewed in the same fashion. Today it is known that such bone alterations as these, and a good many others, are a consequence of people's living in mountainous terrain and using their muscles in a different fashion than do people walking in a flat country. Some of the difference in the robustness of the facial bones of ancient man, as compared to his present-day descendants, has primarily been produced by a rougher diet. Ancient hunters used the teeth vigorously to masticate tough forms of food, and so the bones of the face which give support to the dental apparatus responded by growing in thickness. In modern Europeans, the face has a somewhat collapsed look by comparison, and this can clearly be traced to many millennia of soft diets which have occurred since the invention of agriculture. Similar changes are shown in the faces and skulls of zoo-born animals as compared to their wild parents. There is a reduction of facial structure and the bones supporting the processes of mastication, and a kind of doming of the vault of the skull in both wolves and the great apes. No genetic change is involved here, but pampered animals in the zoo use their teeth much less vigorously than if they were living in the wilds.

Genetically determined variations in bone structure are of more interest to us in attempting to unravel the course of human evolution. But even though we consider that much of the change that we see is due to genetic change, as though by adaptive designing, there still remain pitfalls affecting our interpretations. As an example, let us consider the ridges of bone over man's eyes. These brow ridges, as they are called, are much reduced in modern man compared to the enormously enlarged ones found in most early men and in some but not all of the great apes. First, they display sexual differences in that as a rule they are considerably larger in males than in females. Since the two sexes differ but little in their genetic makeup, this outward difference must manifest the way in which the hormonal differences in the sexes affect bony structure. Second, the question of dental stresses again arises, for it is generally agreed that this brow ridge transmits to the vault of the skull chewing stresses from the cheek and nasal bones. With masticating stresses concentrated at these points it seems mechanically reasonable that the brow ridges should be considered in part a response to chewing stresses. But the situation is not this simple, for the angle of the forehead apparently affects the sizes of the brow ridges enormously, in that most slanting foreheads seem to require buttressing to transmit the chewing stresses, while vertical foreheads can do pretty well without them. In the great apes both the gorilla and the chimp have low foreheads and come equipped with large brow ridges. The orangutan, with a relatively steep forehead, has virtually no brow ridges. But even this may not be the total answer, for the relation of the dental plane to the skull and the positioning of the face with regard to the skull base may all be factors affecting the relationship between genetically determined form and function. The differences can be observed, but as yet they have not been explained. And on top of all this the actual chewing stresses generated in each individual—this will vary to some degree with diet—may contribute to the form the brow ridges take as a structure.

To make progress in the interpretation of evolution it is necessary to assume that changes in form are primarily due to changes in genes, in spite of the reservations listed above. The way in which function molds form can be seen easily if we examine the long bones of the limbs of various animals. Quite aside from the change in proportions that is related to speed of movement, there are structural relationships involved in the efficient use of bone. For a natural organic substance bone has very high mechanical properties, its strength being equal to that of low carbon steel in terms of its relationship to weight. The leg bones of most mammals are hollow, for this is the way in which to construct a lighter bone for the same strength requirement. The fact that the shaft has also come to store the red marrow which produces our red corpuscles is merely an added dividend. But as we pass on to some of the giant terrestrial animals, more bone is needed to support their enormous weights, and so the shafts tend to become nearer and nearer to solid bony columns. Birds, on the other hand, must rigorously evolve to achieve the greatest strength for the least weight. Consequently, the bones of their wings have enlarged hollows to attain this result. One of the most striking evidences of the way in which selection operating upon mutational variation can produce an optimum structure is found in the metacarpal or primary finger bones in the wing of the graceful, gliding *buteos,* or soaring hawks. These are

(a) Finger Bone of *Buteo* or
Soaring Hawk

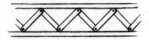

(b) Warren Truss in Airplane
Wing Spar

Figure 9-3
Internal Bracing for Strength

birds that soar on the thermal updrafts with wings held motionless. This type of gliding flight requires the lightest possible weight for a given wing area. As efficient as simple tubular bones are for general purposes, in this bone there has evolved a system of internal bracing shown in Figure 9–3 which provides yet greater strength. The inner diagonal struts connecting the outer load bearing bony beams are arranged at proper intervals and in a three-dimensional plane so as to yield the lightest bone possible for a given strength. The remarkable part about this bony design is that human aeronautical engineers came to exactly the same design, except that they were limited to running their stiffening diagonals in a two-dimensional plane. This structure is known as a Warren truss. Thus, it can be seen that maximizing principles work in both the directed changes occurring in organic evolution as they do in man's own advancing technology.

On Measures of Relationship

One of the vital questions in the study of fossil man involves the determining of true lines of descent. Starting with living populations it is necessary to trace their ancestry backward stage by stage into ancient times. In the Middle Pleistocene, where regional populations varied greatly from one another, it is necessary to try to sort out those that contributed more to modern population compared to those that contributed little or nothing. Back through the Cenozoic, the crucial question is which populations come from common ancestors. In a general evolution the major pathways of descent are reasonably well-known even though some of the pages of the record may be missing.

Homologous structure.   Animals have been primarily classified on the basis of deep-seated *homologous* structural similarities. *Homology* is a state in which the structures of different organisms show similarities due to the fact that they have been inherited from common ancestors. Among animals which have been separated only brief periods of time and have lived similar types of lives, the homologies are close and more or less self-evident. But among animals whose common ancestry lies far back in time and whose ways of earning a living are very different, the homologous structure may be masked by diverging evolutionary paths. The similarities evaluated should involve basic structures, meaning those which are least apt to become similar through converging evolution. A series of structures (Figure 9–4) illustrates divergent degrees of homology, as well as a lack of homology. The pectoral fins of modern ray-finned fish are positioned in the same place as the flippers of whales and serve similar functions in that they are steering devices aiding the animals to change their direction. Superficially they look much alike, but as we examine their deep structure, it becomes evident that their supporting and stiffening methods are very different in nature, in number, and in arrangement. The rays of modern bony fish have a series of small bones at their base which do not correspond directly to any bones found in the shoulder of the whale. Furthermore, there springs from these a series of stiffening rays which are not made of bone. In the whale the flipper rotates in the ball-and-socket joint of the shoulder bone or scapula through the head of its primary bone, the humerus. This in turn articulates with two bones, the ulna and radius, which in turn are connected to a series of small hand bones from which spring the bones of five fingers. The whole arrangement of the bones in the whale's flipper is closely homologous to that in our own arm, and the similarities are self-evident. Turning to the wing of the bat, a mammal living a very different type of life since it depends upon its nocturnal flights to provide it with its insect food, we find a bony system very like that of the whale. The major bones of its forearm correspond. These terminate in five digits that not only support the skin membrane which is its flight surface but also provide a hook by which the bat may hang. If we further examine the bones in the wings of birds and in the extinct flying reptiles, we would find the major bones strictly homologous and only minor variations shown in the arrangement of the finger bones. In modern birds these are reduced in number to three, with the second digit being dominant. In the flying reptiles of the Mesozoic the fingers are four in number, with the first three serving as hooks from which to hang, while the fourth is enormously elongated as the sole outer support of the wing surface. These forelimbs are clearly homologous even though the common ancestors of all go back more than one hundred million years in time.

Analogous structure.   The wings of the modern dragonfly (Figure 9–4d) provide flight surfaces as do the wings of bats, birds, and the extinct pterosaurs, so that they are similar in function. But if we examine the structural detail of the wings of dragonflies, and other insects, fundamental differences are evident. The wings are moved by muscle lying in the thorax with none extending out into the wing itself. The surface of the wing is chitonous in structure, material similar to our own fingernails and hair. Quite clearly these insect wings have evolved independently from those of the vertebrates and have nothing in common with them. Hence they are called *analogous,* for while they have functional similarities, they have none in structure. Likewise, though the pectoral fins of the modern ray-finned fish and the whale are similar in function, they are analogous in structure.

The story of evolution has been successfully pieced together from fragmentary evidence within rocks by using fundamental homologies to group those animals which have descended from common

Figure 9-4
Homologous and Analogous Structure (not to scale)

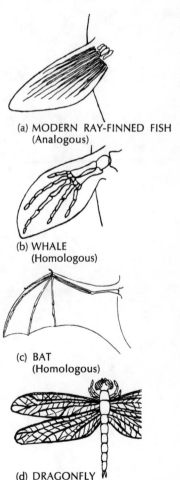

(a) MODERN RAY-FINNED FISH
(Analogous)

(b) WHALE
(Homologous)

(c) BAT
(Homologous)

(d) DRAGONFLY
(Analogous)

ancestors. In general, the process has been successful in that the major lines of descent for the higher taxons are now well established. But there have been a variety of surprises in the past and perhaps a few will occur in the future. For example, animals called porcupines today live both in the New World and in Africa. Both belong in the order of rodents and so have an ultimate common ancestor. But their presence in the two hemispheres has always posed a problem as to the paths by which they reached their present distribution. Porcupines do not occur in the North American record of fossils until the land bridge had been reestablished between North and South America in the Late Pliocene. Early students of the problem concluded that there must have been some sort of land bridge connecting Africa and South America in order to account for the distribution of the porcupines. Surveys of the ocean floor of the South Atlantic provided no evidence for this belief, and there remained the very awkward fact that most of the other animals living in the two continents were totally different. It has been determined by serological tests of blood group properties that the two porcupines are only very distantly related, while each shows a close degree of relationship with the rest of the rodents living in its own continent. The New World porcupine gives serological reactions very like that of the guinea pig, the agouti, and other South American rodents. It is now clear that the porcupines have evolved their similarities as the result of convergent evolution. Their conspicuous quills, which make them appear so similar, have also been independently evolved in two other animals, the European hedgehog, an insectivore, and the spiny anteater of Australia, a monotreme. So it is now evident that hair in four different instances has been modified to produce quills in animals that are only very distantly related to each other. Traps of this nature may for a time confuse judgments about true relationships.

Man is an animal in which the unraveling of relationships presents unusual difficulties. In general evolution the frameworks of classification are built upon middle-level taxonomic orders such as the families, subfamilies, and genera. The category of species is of course used for identification in paleontology, but usually the fossil record is so meager that population studies of species are difficult. Very seldom has long-term evolution been concerned with subspecies classification. But man, for the whole of the Pleistocene, is essentially represented by a single central species with possibly one side branch which may have attained the level of specific difference. Thus our mainline fossils are all included within the concept of a single species.

Judiciously tested structural homologues have been used to define lines of descent and common ancestors. The method fails altogether with man, for neither any living populations of men nor any Pleistocene ones show differences in structural homologues.

Using this fundamental criteria, all men living and dead belong together and are placed within a single species with a few exceptions in the Lower Pleistocene. With the monotonous likeness of fundamental structure, it becomes necessary to turn to other kinds of criteria for the ordering of human variation.

One of the most serious problems presented by human populations is the fact that they contain a very great amount of variability, both in measurements and in characteristics of form. For example, the data presented in Figure 9–2 give Australian aboriginal males ranging in maximum head length from 176 to 223 millimeters. The largest is more than 26 percent greater than the smallest. Admittedly, these data represent Aborigines from all over the continent, coming from areas in which stature varies considerably, and, of course, collected from many different breeding populations. But even where our focus is restricted to a single breeding population, men remain surprisingly variable. In northwest Australia the Ngangamarda tribe, which does constitute a single breeding population, provided the following information. One hundred and eight adult males showed a maximum head length ranging between 182 and 208 millimeters. This 26 millimeter range is more than 14 percent of the smallest in the series. Obviously differences in dimensions alone will not provide reliable guides for determining relationships among human fossils.

Similarities in details of form have sometimes been considered as providing guidelines for determining a relationship. Some years ago Carleton Coon (1962) asserted that the living races of the world could be distinguished by the number of cusps on their molar teeth and claimed that Australians had four cusps on each of their upper molars and a five-four-five cusp pattern on their lower molars. To the degree that this author meant to indicate that the Australian Aborigine had deviated less from the ancestral dryopithecine condition in terms of dental reduction than have other living races, he was correct. But in a review of his book (Birdsell, 1963) I analyzed molar cusp patterns as they actually occurred among Australians. A random sample of 100 males, five from each of 20 different tribal groups, showed that only 18 percent of that series had the pattern of five-four-five for the lower molars, while only 6 percent had all three upper molars unreduced, that is, with the full four-cusp pattern. When the two criteria were taken together, only 2 percent of the Australian series would have been classified correctly as being Australians by their molar cusp patterns. It is quite clear that each of the molar teeth involved may vary independently in its molar cusp pattern, and when taken together in combinations of six, the 100 Australians showed no less than 52 different patterns. Such variables must be granted a wide range within a regional population, and an even wider range within continental populations, or in time depth.

The problem of judging relationships is further complicated by the story revealed by blood group genes. This will be detailed in later chapters, but here it is sufficient to indicate that the common serological genes vary as much in frequency between adjacent breeding populations in Australia as they have been recorded to differ between such major racial populations as Europeans and East Asiatics, or the latter and the peoples of India. These findings emphasize the fact that human populations are highly *polymorphic* (many forms within a population) even within single breeding populations, and that man is highly *polytypic* (many types among populations) in that very considerable differences occur between different breeding populations. In the interpretation of the human fossil record, ample allowance must be given for variation between individuals, sexes, and populations.

Some Bases for Judging Relationships Among Fossil Men

Two characteristics of the fossil record assist in making judgments about relationships among ancient people. First, during the three million years of the Pleistocene, man evolved very rapidly, so that substantial differences are apparent in this time span. For example, the size of the human brain virtually tripled during this time span and so men living during the Early, Middle, and Late portions of the Pleistocene vary by amounts great enough to override differences between sex as well as individual variations within populations. Second, a modest radiation of man occurred within the Pleistocene and so a greater variety of human populations lived then than do now.

Since 1930, evolutionary theory has been changed by the introduction of genetical concepts. One of the most important developments in current evolutionary thought rests upon the fact that *the units of evolution are populations,* and not individuals. Even where the rare fossil remains are scattered, the processes of evolution must be conceived of as going on in populations. This raises the point of how well a single fossil skull can represent the population from which it has been drawn by chance. When populations differ in their overall characteristics, it is probable that the first specimen will reflect this distinctness. In all cases where subsequent discoveries have been made, the first skull found has proved to be representative of the population. For example, the skullcap of Neanderthal man is very close both in size and in its various proportions to the skulls of subsequently discovered male Neanderthals. To be sure, some small variation exists, but the first was a good representative of this population. Again, the first skullcap found in the Trinil Beds of Java established a new type of human grade, that of the pithecanthropines. Subsequent human fossils discovered from the same Javanese beds have been enough like the first one so that their differences are no larger than those between the different sexes of a single population. And this in spite of the fact that these fossils may range over a time span of possibly half a million years. These examples, which fly in the face of statistical theory, are valid, for the populations which they represent are very different from many others. Indeed, in some cases they represent different grades of human evolution.

In judging relationships among fossil man it is necessary to avoid the trap that similarity in form is always a measure of relationship. The possible combinations in the variable dimensions of the human skull and in human teeth are numerically great. Consequently, similar proportions may appear by chance among populations which are known not to be closely related. The skulls of Australian Aborigines at the mouth of the River Murray are characteristically long, narrow, low, and big browed. These same characteristics in a more exaggerated form occur in classic Neanderthal man, and some early anatomists were led to claim that the two populations were related, that is, that the Australians represented living descendants of classic Neanderthal man. Today the evidence suffices to reject this idea. Living African Negroes are commonly characterized by long, narrow, high cranial vaults. Some anthropologists have asserted that this combination of proportions always characterizes Negroid skulls and so have labeled skulls of this shape as belonging to that racial group wherever they were found. A series of burials at the tip of Baja, California, were called Negroid on the basis of their skull proportions by an eminent anthropologist even though there was no way in which Negroids could have gotten there. Another equally distinguished worker claimed to see Negroid characteristics among skulls discovered in northern Siberia, another totally unlikely place to find them. The fallacy in such cases rests upon the investigators' refusal to recognize how variable crania can be in any region of the world.

In recent years a number of statistically powerful methods of analysis have been developed to test for similarities or differences between specimens. They all suffer from one or more incorrect biological assumptions and so do not solve many of the problems in human evolution to which they are applied. A brief discussion of some of them is given in Supplement No. 12 at the end of this chapter.

Forty years ago the problems of human evolution seemed about to be solved, for the gene, the basic unit of inheritance, was brought into use to differentiate human populations. The analysis of genes answered few questions and generated many more. Today, molecular biology again beckons to evolutionists. It gives some answers to questions of relationships among living animals originating far back in time. But even the genetic code is not immune to evolutionary processes, so it does not provide a phylogenetic clock, as some proponents hold. In any case, there is no reason yet to believe it can illuminate time changes in a single species, such as *Homo sapiens*.

The best bases for judging the relationships among fossil men probably lie in combined considerations of similarities in form and correspondences in geography and in time. With these criteria, and fully allowing for variability between sexes within populations, it is probably justifiable to claim a close degree of relationship between fossil crania which are *much alike in form and size, are found in the same general region,* and are *dated within a reasonable time span.* But skulls of very similar form, coming from different continents, and from beds of very different ages are not necessarily related, so judgments about relationships must be made with due caution.

Chronology Is a Major Problem

Evolutionary studies. must be firmly based in a sound time perspective. And yet today, in 1975, students of human evolution find their biggest problems in conflicts in dating, and their inability to cross-date from one broad region to another. The Pleistocene is the age of man as well as the last great glacial stage. Climatically, as well as geologically, it is one of the most complex epochs in the time record. Some of its complexity results from our knowing it the best. The exact boundary between the preceding Pliocene and the succeeding Pleistocene offers a problem at this time. It is usually defined as the first appearance in abundance of certain northern foraminifera in the marine deposits of El Castello, Calabria, in southern Italy. This time boundary has an absolute dating of just under 2.0 million years. If man was a marine mammal, this would represent a suitable time for the beginning of the Pleistocene. But the dating of terrestrial deposits follows a different pattern. As explained earlier, the Villafranchian period begins with the appearance of modern genera of horses, cattle, elephants, and camels. Their advent is dated as approximately 3 million years ago. Since human remains occur essentially only in terrestrial deposits, it is convenient to use the beginning of the Villafranchian as the beginning of the Pleistocene for our purposes. This practice will be followed in this book.

The Subdivision of Glacial Time

Glacial events provided the first relative chronology against which human evolution could be tested, and these events still represent an important relative sequence of time for northwestern Europe. To some degree its influence extends further, since glacial advances were accompanied by worldwide or *eustatic* lowering of sea level, due to the water locked up in the ice mass. Beyond its limited area of application, the glacial chronology offers further problems in that while the relative sequences are well known, these have been poorly dated in absolute terms. Consequently, there is a difference in cross-dating of some glacial episodes with events and finds in other parts of the world.

The exact numbers of glacial advances which occurred in northwestern Europe during the Pleistocene are not known. With the terminal Pliocene there began a general planetary phase of cooling characterized by oscillations in temperature. Since the major European glaciers have been given Germanic names, it is fortunate that they occur in alphabetical order. Following the Donau comes the Günz, followed in turn by the Mindel, Riss, and Würm. The warmer periods between the glacial advances are usually labeled from the glacial periods which bound them. No hypotheses are completely satisfactory in explaining why our planet should have been plagued from time to time with periods of glaciation. Today the most generally accepted hypothesis involves a series of long-term and regular changes in the Earth relative to its orbit about the sun. The variables include the tilt of our planet's axis, which varies over a range of three degrees in long cycles of time, as well as other subtle changes such as the eccentricity of the Earth's orbit itself. Finally, there is a kind of "wobble" in the axis of the Earth which goes on in a predictable fashion in a cycle of 26,000 years' duration. The combination of these three kinds of events makes possible the mathematical calculation of how the solar energy arriving on the Earth's surface will vary in time in a series of complex cycles. Unfortunately, this very sophisticated set of mathematical calculations allows the beginning of the Günz to be 300,000 years ago in the so-called short chronology or as long as 600,000 years ago based upon long-term interpretations. Absolute dates in general indicate that the "long" chronology is the more realistic, and there is a tendency today to push the beginning of the Günz glacier back a million or more years in time. It is assumed to have begun 1,150,000 years ago, but there is no assurance that this is more than an estimation of proper magnitude.

In Figure 9–5 glacial changes, time, and the more important finds of fossil men have been arranged in a tentative ordering. On the left-hand side the sequence of events of glaciation is indicated in a generalized way to indicate colder and warmer than present conditions. The scale of temperature variation between the extremes would be of an annual mean magnitude of 10 to 15 degrees F. A long-term chronology indicates the continental Pleistocene began 3 million years ago with the advent of the Villafranchian fauna. The

The Sequence of Pleistocene Events

FIGURE 9-5
Fossil Man in the Pleistocene

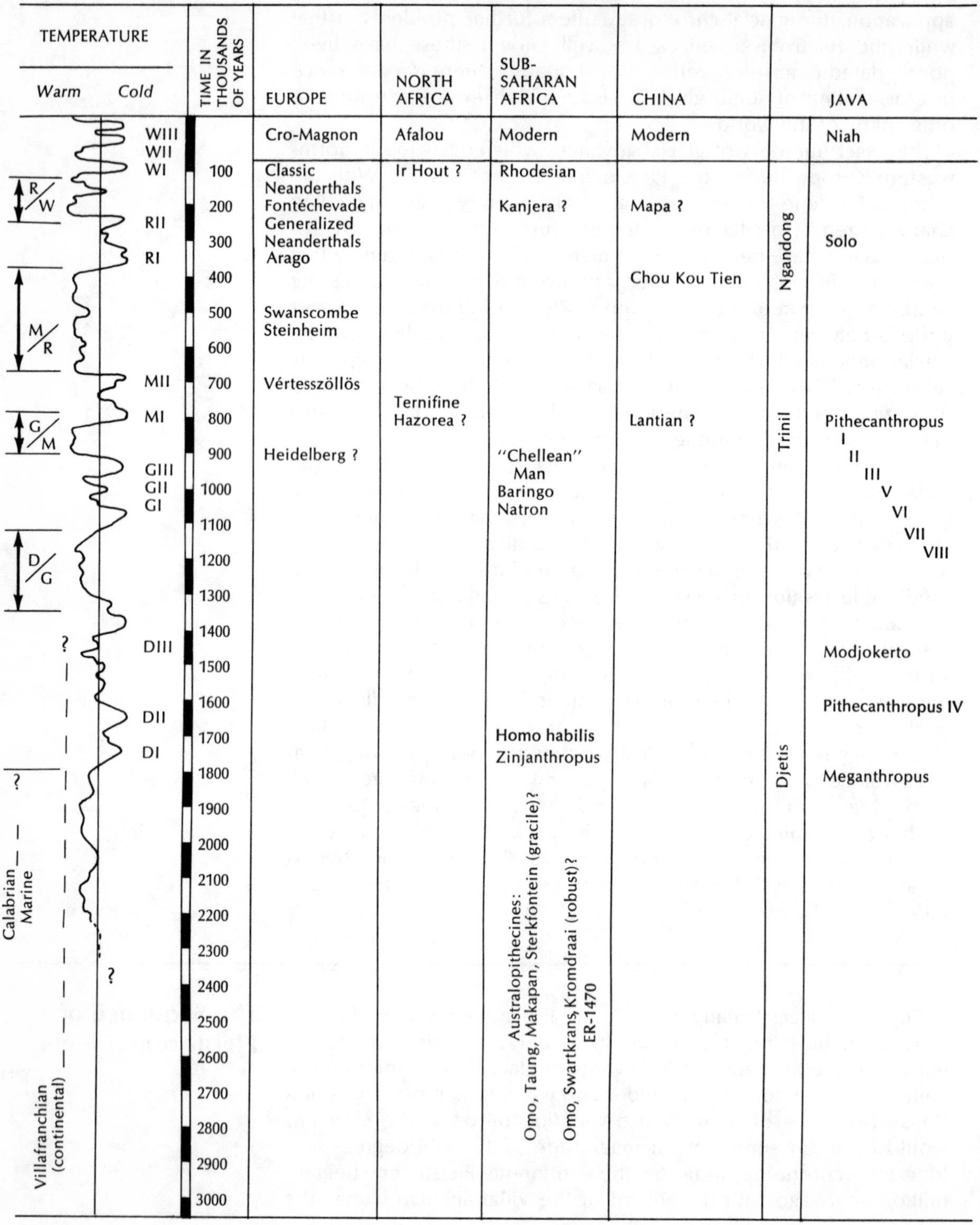

TEMPERATURE (Warm / Cold)	Glacial stage	TIME IN THOUSANDS OF YEARS	EUROPE	NORTH AFRICA	SUB-SAHARAN AFRICA	CHINA	JAVA
	WIII / WII		Cro-Magnon	Afalou	Modern	Modern	Niah
R/W	WI	100	Classic Neanderthals	Ir Hout ?	Rhodesian		
		200	Fontéchevade Generalized		Kanjera ?	Mapa ?	
	RII	300	Neanderthals Arago				Solo
	RI	400				Chou Kou Tien	Ngandong
M/R		500	Swanscombe Steinheim				
		600					
	MII	700	Vértesszöllös				
G/M	MI	800		Ternifine Hazorea ?		Lantian ?	Pithecanthropus I
		900	Heidelberg ?		"Chellean" Man		II · III
	GIII GII GI	1000			Baringo		V
		1100			Natron		VI · VII
D/G		1200					VIII
		1300					
?	DIII	1400					
		1500					Modjokerto
	DII	1600					Pithecanthropus IV
	DI	1700			Homo habilis		
?		1800			Zinjanthropus		Meganthropus · Djetis
		1900					
		2000			Australopithecines: Omo, Taung, Makapan, Sterkfontein (gracile) ?		
Calabrian Marine		2100					
		2200					
		2300			Omo, Swartkrans, Kromdraai (robust) ? ER-1470		
?		2400					
		2500					
		2600					
Villafranchian (continental)		2700					
		2800					
		2900					
		3000					

(? = uncertainty in date)

earliest events in the cooling phase of the Earth are not clearly indicated, but after some oscillation of a milder type, the first advance of the Donau glaciers began about 1,800,000 years ago. It is a complex event with three numbered advances. The following warmer period is called the Donau-Günz Interglacial and represented a period rather uniformly warmer than the present. The next glacial advance, known as the Günz, contains three peaks of cold, representing major glacial advances. They were separated from each other by short but warmer than normal periods of time. After the recession of this glacial sheet, Europe enjoyed a short interglacial period, the Günz-Mindel. This ended in a double advance of the Mindel glacier. Then followed the longest single episode of the glacial cycles, the Mindel-Riss Interglacial, or, as it is often called, the Great Interglacial. In the obsolete short chronology it was estimated to have lasted about 150,000 years, but in the long chronology we are using, it extended over the much greater span of about 320,000 years. This is a period of importance in human evolution, for it contained the greatest variety of populations yet documented for man.

This long period of favorable climate was followed by the double advance of the Riss glacier, which in turn was succeeded by the Riss-Würm Interglacial. About 100,000 years ago the last glacial advance, the complex Würm, began and reached increasing degrees of coldness in three succeeding intervals, with the greatest drop of temperature in the last. The variation in these Würmian events provides important dating criteria in Europe, and is the period there, as well as in the rest of the world, in which modern man came to be predominant. With the passing of the last ice advance, the Earth entered into an interglacial period which we are enjoying today. There are data to indicate that our climate is still warming up, that more glacial ice will melt into sea water, and so the level of the oceans will continue to rise for some millennia ahead. Perhaps seashore real estate is not as good an investment as some people think.

Cross-Dating in the Pleistocene

Even allowing for the very real difficulties of cross-dating from glacial events in Europe to fossil finds in nonglaciated regions, Figure 9–5 indicates great variations in the known time depth of human evolution in the different parts of the Old World. For Europe, North Africa, and the included Near East, human fossil remains do not extend back to more than one-third of the Pleistocene period. The same is true in China on the basis of our present poorly dated evidence. But Africa south of the Sahara reveals a series of human documents extending back through the Villafranchian into the

Upper Pliocene. In fact, there the record of the earlier Pleistocene is more complete than in later sections.

Unfortunately, due to technical reasons, cross-dating has not been possible between the australopithecine types of East Africa and those of South Africa. Methods of relative dating have yielded some good regional sequences in the Pleistocene, but again do not allow the projection of accurate cross-dating. The determination of absolute dating is still in its infancy, and there are not enough of these absolute dates to provide a firm chronology. As they become available, changes will be needed in Figure 9–5. Perhaps the most critical unknown involves how the glacial chronology in Europe really relates to the Pleistocene deposits of sub-Saharan Africa, China and Java. At the present time, this cannot be determined.

In spite of these problems it is convenient to break the Pleistocene period into three broad subdivisions. The earliest and longest of these is known as the Lower Pleistocene, beginning with the appearance of the Villafranchian continental fauna, and extending upward in time to an indeterminate boundary. In Europe, the Lower Pleistocene conventionally lasts until the end of the Günz-Mindel Interglacial period. Its boundary is not known at the present time for either East Africa or Java. On the latter island, the Djetis beds are now to be included in the Villafranchian, and so Lower Pleistocene. In a general way the Lower Pleistocene has a duration of the magnitude of 2 million years or more. The succeeding Middle Pleistocene begins in Europe with the Mindel glaciation and lasts through the Riss ice advance. Its exact boundaries in Africa are not known at present, but in Java the Trinil deposits probably belong to it. On our long-term chronology, this subdivision of the Pleistocene lasted approximately three-quarters to one million years. The Upper Pleistocene in Europe consists of the last interglacial and the final glacial advances. According to the long chronology, it began about 200,000 years ago and lasted until about 10,000 years ago. The boundary between the Middle and Upper Pleistocene is not well defined in the rest of the Old World.

Barriers to Man's Movements

Barriers to human movement help to judge actual relationships between fossil men. If we assume that human mobility suffices to allow man to move throughout regions except where barriers intervene, we will not be far from the mark. We can now plot out barriers which would tend to restrict human migratory capabilities. In Figure 9–6 the major barriers are indicated. Four major types of barriers impede human movements: water gaps, mountain ranges, continental glaciers, and belts of desert.

Figure 9-6
Areas of Glaciation and Emergent Land in Last Ice Advance

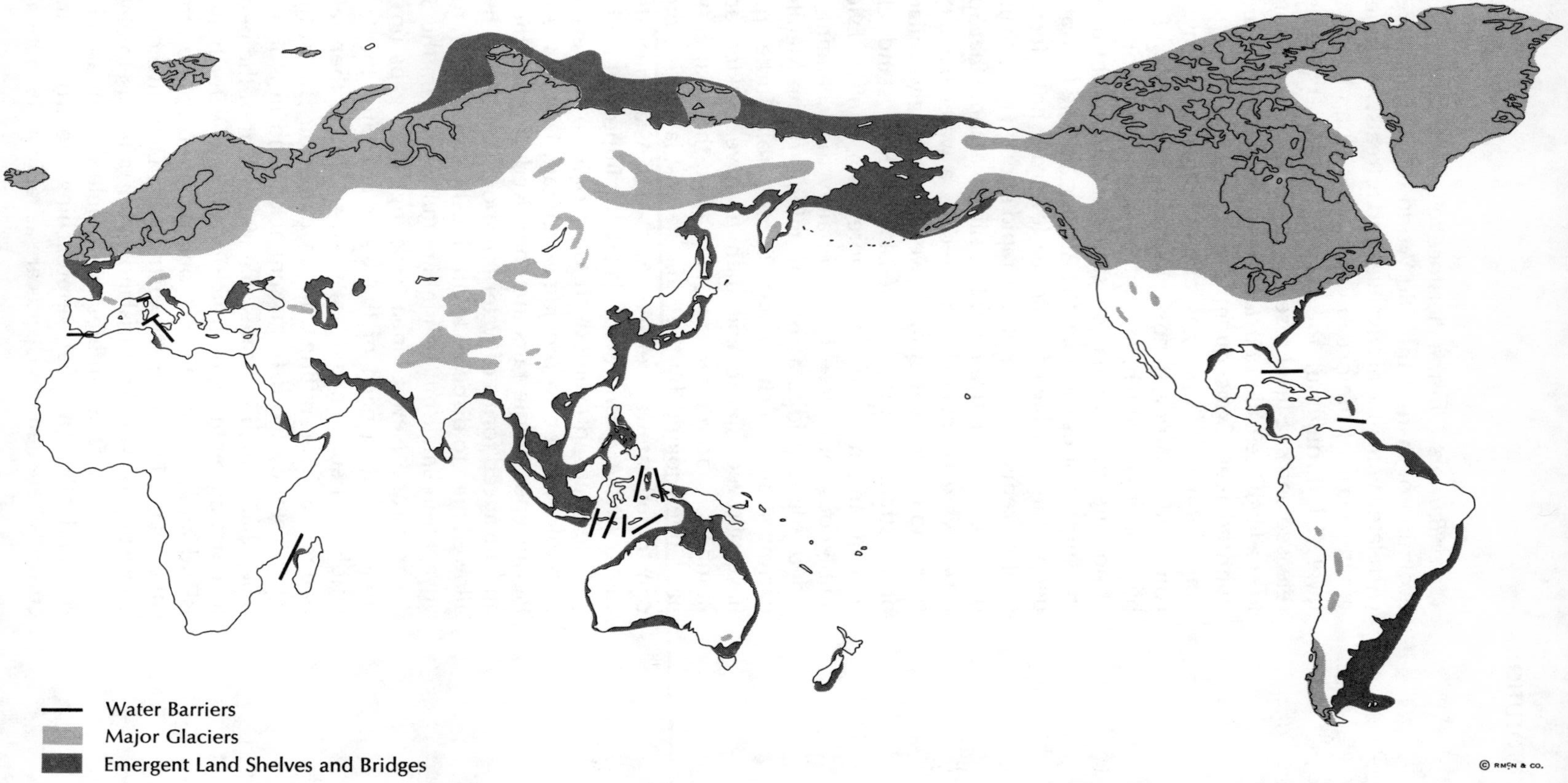
Water Barriers
Major Glaciers
Emergent Land Shelves and Bridges
© RMcN & CO.

Water Barriers. In the Pleistocene we are concerned only with the evolution of man in the Old World. Oscillations in sea level were correlated with glacial advances and retreats. As glaciers advanced, a greater mass of the ocean water was tied up in ice, and so the sea level fell throughout the world. As glaciers melted, water was released and sea level rose everywhere. These uniform movements in sea level are called *eustatic* and were of the magnitude of several hundred feet. Consequently, eustatic changes in sea level sometimes formed land bridges, and at other times drowned them. The continent of Africa stands separated from eastern Europe and Asia by a water gap which, in the Pleistocene, was constant. Apparently at no time were the narrow straits at Gibraltar bridged by land, for the surging tides between the Atlantic and Mediterranean have cut a deep channel which falling sea level did not affect. Furthermore, tidal currents in the straits are rapid enough to pose insurmountable obstacles to primitive men. It is safe to assume that Spain and North Africa were constantly separated by a water barrier, even though men and the tools they made were surprisingly similar on both sides of the straits. To the east, the island of Sicily stands close to the toe of Italy and not too far offshore from Tunisia. Evidence here is ambiguous, but there is no certainty that changing sea level provided a land bridge across the middle of the Mediterranean. The narrow straits which separate Europe from Turkey probably would have allowed Pleistocene men to have trickled across in small numbers in both directions. Africa and the Near East seem long to have been joined by the Sinai Peninsula and so they provided a constricted gateway between Africa and Asia. The water gaps further east on the Arabian Peninsula are not important for our discussion.

Water gaps prevented crossing only to the extent that man's technology failed to provide watercraft. During the entirety of the Pleistocene, simple rafts are conceived as being the most efficient and advanced forms of water transportation. Travel by raft no doubt allowed man to disperse to previously empty lands, for we know he voyaged in small but continuing numbers from the Sunda Shelf to Australia and New Guinea, crossing water gaps up to 25 miles in breadth. But at each of the water crossings the land ahead was visible and so men did not need to launch themselves upon the open seas. But rafting can be dangerous, as witnessed by the protohistoric record from Bentinck Island in North Australia (Tindale, 1962) which indicates that 50 percent of the people who tried to cross an eight-mile stretch of shallow water between the main island and a small offshore one perished in the attempt. On the other hand, in northwest Australia the Djaui live on an archipelago consisting of very small islands, and rafting is totally necessary to their lifeway. These interesting people exist in a region where tidal rise and fall is 40 feet, approaching the world's maximum, and currents between the islands are as rapid as 15 miles an hour. These

Aborigines, with no accurate way of timing tidal changes and using rafts that can be propelled only one or two miles per hour, push off into these tidal currents to travel to their destination on the next island as we would step aboard an escalator. They demonstrate that with proper knowledge and skill rafting can serve men well to cross water gaps, but no doubt it has always been moderately dangerous.

Land Bridges.   Just as water barriers inhibit human movement, so do land bridges facilitate them. At the height of glacial advances Ireland and England formed a continuous part of the European mainland, so that to the degree that Arctic weather permitted, men could have moved back and forth. Low sea levels in the Pleistocene undoubtedly connected the island of Ceylon to the mainland of India. More importantly, all of the greater islands of Indonesia became part of the Asiatic mainland. The islands of Borneo, Java, and Sumatra became part of what is known as the Sunda Shelf. Likewise, Australia and New Guinea were broadly joined by a now sunken region known as the Sahul Shelf, while Tasmania was firmly joined to the mainland. Men who lived nearby could have walked across these land bridges. In northeast Asia, the Bering Straits today effectively separate the Old World from the New. But in times of glacial advances, lowering sea level revealed a land bridge 500 to 600 miles in width across which animals and men could go in both directions from Siberia to Alaska. But the land bridge connecting Asia with the New World could not have been traversed until men were equipped to live under Arctic conditions, and archeological evidence suggests that this was achieved in the Upper Pleistocene.

Mountain Barriers.   Mountain ranges may serve as areas of refuge, or as truly formidable barriers to foot travel. The mountain ranges in Africa are rather minor in character and so did not impede human movements, but in Asia, the Himalayas, whose peaks approach 30,000 feet in altitude and whose plateaus often rise about 14,000 feet, are in a spectacular chain of mountain barriers extending to the northeast. These include such great ranges as the Tien Shan and Altai mountains. The land levels off only toward the Arctic circle. Except for a narrow arid coastal belt in western India all of the land lying to the east of these mountains was largely isolated from land lying to the west. Men could trudge across the Khyber Pass or the Dsungarian Gate and so pass from one side of the barrier to the other, but these mountains must nevertheless have imposed partial barriers to human movements. Archeological evidence indicates that the barrier was not perfect, but it is still clear that these ranges divided Eurasia into two major regions and so discouraged their human inhabitants from mixing.

Westward into Europe is a chain of lesser mountains that acted as partial barriers. The effects of the Caucasus Mountains as broken country is still felt today in the hodgepodge of languages and customs that characterize that region. It is not likely that the Ural Mountains barred movement from east to west across the great steppes which occurred on either side. Much of central Europe is moderately mountainous, but barriers are not formidable. But the Alps of Switzerland and the Pyrenees which mark the boundary between Spain and France no doubt did act as isolating factors. Not only are their passes high, but their peaks were repeatedly glaciated. But even in the case of these greater ranges men could always filter along the coastal plain and so circumvent them. It is likely that their uplifted masses slowed human movement but did not prohibit it completely.

Glacial Barriers.   Further kinds of barriers to human travel can be produced by climatic extremes. Figure 9–6 shows the maximum extent of the last or fourth glacier in northern Europe. Obviously, man cannot live under an ice mass a mile or more in thickness. But neither can he live very close to its southerly margin, for the regions adjacent to glaciers become outwash sheets of gravel with little or no vegetation and formidable weather. These periglacial belts may have been a hundred miles or more in breadth and man could not penetrate them until he had perfected Arctic types of clothing and housing.

Aridity Barriers.   Extreme aridity represents another type of barrier to human travel. Such belts have undoubtedly influenced human movement, and consequently evolution, but since their characteristics change in time, their real impact is hard to determine. The great Sahara Desert, which extends across North Africa from the Atlantic to the Indian Ocean, is today such a barrier. Its penetration requires such domestic animals as camels or modern, specially designed motor equipment. But the region varied somewhat during the Pleistocene and certainly during the earlier phases of that epoch was not altogether formidable. Scanty data seem to indicate that during the early Pleistocene the Sahara may have received enough rainfall to have presented extensive belts of grasslands upon which great game herds grazed. But with the passage of time the region dried out, and so since Pleistocene times it certainly did act to impede human travel. Thus Africa, for perhaps the last ten thousand years, has been effectively separated into a region below the Sahara Desert, and a coastal region climatically like and connected with the rest of the Mediterranean area. The river Nile always provided a narrow passage from north to south, as perhaps did the Atlantic coastal plain. But the measure of the isolating effect of the Sahara Desert is shown by the fact that one of the great major units of man, the Negroid population, evolved below the Sahara, while another, the Caucasoid, evolved to its north. The kinds of genetic differentiation represented by these major groups stand as evidence to the isolating effect of the arid belt across North Africa. Deserts occur in other parts of the Old World, but in general they are not so strategically placed and could be passed around on one side or the other. Consequently, their effect on human evolution has not been so great. In fact water, mountain and climatic barriers have helped man to differentiate into regionally different populations, but at the same time human mobility has always resulted in enough genetic exchange to maintain man as a single evolving species.

The Human Niche and Bipedal Locomotion

All animals have their own particular way of making a living and man is no exception. The human niche may be considered from two points of view, the structural and the behavioral. The human niche has required considerable rearrangement of man's skeletal structure in order to accommodate to a new mode of locomotion, that of walking bipedally on his hind limbs, or, as we call them, legs. This form of mobility has freed the forelimbs, including the all-important hands, to become tool holders as well as tool makers and to manipulate objects within the environment. The freeing of the hands opened the entire new world of culture to man, but we shall be concerned with that later. Here let us focus on the structural changes that are involved in walking upright.

There are very few parts of the human body that have escaped change as a consequence of shifting to bipedal locomotion. Some of the major changes have been schematically diagrammed in Figure 9–7 to illustrate the differences between a generalized quadrupedal animal, such as the dog, and advanced anthropoids, such as the chimpanzee, and man himself. Figure 9–7a indicates some of the changes that have occurred in the vertebral column. In a four-footed animal the backbone acts as a slightly arched bridge supported on the fore and hind limbs and sustaining the weight of the various internal organs. Such an animal, if of small to moderate size, runs by flexing its backbone sharply as its hind legs reach for the next new foothold, then straightening out the spine with a sharp muscular effort while the forelimbs balance the body in between strides. It is the hind legs which provide the major portion of the propulsive force. A surprising amount of energy in this type of running is provided by the flexing and straightening of the backbone. The spines on the vertebra of the chest (or thoracic region) and the small of the back (or lumbar region) converge toward a single central vertebra, the spine which is perpendicular to the backbone. This is the so-called *anticlinal* vertebra and it represents

QUADRUPEDAL ANIMAL   CHIMPANZEE   MAN

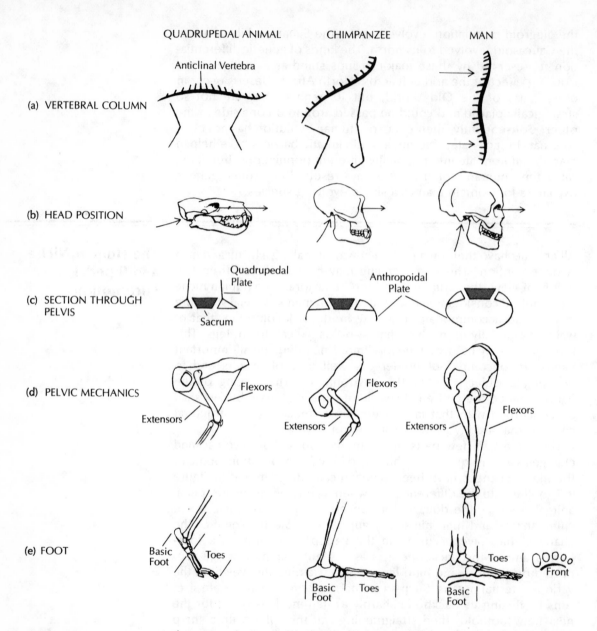

(a) VERTEBRAL COLUMN

Anticlinal Vertebra

(b) HEAD POSITION

(c) SECTION THROUGH PELVIS

Quadrupedal Plate

Anthropoidal Plate

Sacrum

(d) PELVIC MECHANICS

Flexors

Extensors

Flexors

Extensors

Flexors

Extensors

(e) FOOT

Basic Foot   Toes

Basic Foot   Toes

Toes

Basic Foot

Front

Figure 9-7
Structural Changes Due to
Bipedalism (purely
schematic—not to scale)

the point about which the backbone flexes in running. The spines of
the vertebra point toward it since the muscular forces act in such a
fashion. This general pattern characterizes most active mammals,
but it is not found in larger ones such as horses and elephants that
do not flex the back in running.

The spine of the chimpanzee provides us with an altered situation. This animal climbs trees, hangs from branches by its arms, and
progresses on the ground by a variety of ways. Its common gait is

reminiscent of shuffling, with the knuckled hand of the extended forearm and the bent hind leg resulting in a spine carried at about a 30 degree angle. Other forms of locomotion include bipedalism, although this is rare, and swinging along on the arms as though they were crutches. None involves back flexing. Consequently, the muscles of the back are used differently, the vertebral spines are in a general way perpendicular to the backbone and there is no anticlinal vertebra. The spines of cervical vertebrae in the neck are bent upward to carry the head; the rest of the vertebra form a general convex arch as in quadrupeds.

In man, upright posture has resulted in a number of changes. The vertebral spines point downward in the neck region, slant less in the thoracic region, and are perpendicular to the vertebra in the lumbar region. The basic convex arch of the quadrupedal animal is still preserved in the chest region, but there are now two sections which are concave in anterior or forward fashion. The neck vertebrae are more sharply bent than in the chimpanzee and a new concavity occurs in the lumbar region, or the small of the back. These two ultimately become fixed early in life as the baby first learns to hold its head up and then later to sit up and finally to walk.

The basic position of the vertebral column greatly influences the organization of the skull and the direction of its sensory equipment. In the quadruped the spinal cord, encased in the protective processes of the vertebra, enters the skull from the rear and shows only a slight upward component in direction. With the head thus held horizontally the eyes, nose, and mouth are also placed in a horizontal position and ahead of the brain. In the great apes the foramen magnum, or the hole in the skull through which the spinal cord joins the brain, has moved forward moderately beneath the skull so that the head is considerably better balanced on the spine. Even so the large jaws of the ape overbalance the head, and massive neck muscles are required to maintain its position properly. In man the brain is expanded, the jaws have become reduced, and with the spinal cord now entering the skull from directly beneath, the head is finely balanced upon the spine (Figure 9–7b). In consequence the neck muscles can be reduced in mass and the head has greater mobility in lateral movements. Since man hunts by sight, this is an important structural improvement.

The forelegs of quadrupedal animals tend to specialization, the degree of which depends upon the type of locomotion. In all such animals the chest is flat in a lateral fashion so that it is much deeper than it is broad. This is equally true of the quadrupedal monkeys, whether they live in trees or on the ground. In the great ape, chests have become flattened in the opposite direction, that is, from front to rear, as a result of tree climbing, although not necessarily from arm-swinging. The shoulder blades, or scapulas, are now embedded in muscles on the back of the thorax, and the arms are at the sides of the body and capable of a great degree of movement in all three planes. The human condition varies relatively little from that of the great ape, involving slight changes in proportions of bones and muscles. The great mobility of the human arm has been directly inherited from pongid ancestors. The fossil evidence suggests that the latter primates were arboreal climbers that had not structurally specialized as a consequence of brachiation.

Most of the primates retain a generalized pattern in both fingers and toes. The evolution of the hand has been fully discussed by John Napier (1962). He points out that our hands function primarily in two different ways. A so-called power grip, as in grasping a hammer, allows a tool to be used with rapid powerful motions involving arm and wrist when needed. Contrasted to this is the precision grip in which objects are picked up between the opposable thumb and one or more fingers, such as in holding a pencil. These two grips, and variations of them, have enabled man to become a master tool user. Many of the primates can use their hands in generally similar fashion but perhaps with a little less freedom of motion. Their capabilities are often underestimated simply because their anatomy is a little different. Baboons have been seen to pick the sting of a scorpion from one hand by the fingers of the other used in a delicate precision grip. If quadrupedal baboons can do it, most of our other near relatives no doubt have similar capabilities.

The pelvis, or innominate bone, is a fusion of three separate bones. The *ilium* is the top or anterior bone, the *ischium* the posterior and lower bone, which is joined to the *pubis,* which is the inferior and anterior bone. All three fuse together to provide the socket for the hip bone. In primitive mammals the ilium is a bone triangular in cross-section which parallels the backbone on each side. As quadrupeds evolved greater mobility, there grew from the top of the ilium a plate known as the quadrupedal plate, diagrammed in Figure 9–7c. This type of enlargement characterizes most active four-footed mammals. But in the primates a different pattern of enlargement occurs, with the ilium growing downward into what is called an anthropoidal plate. In man this extention of the ilium is carried to an extreme.

The structure and proportions of the pelvis are closely related to the type of locomotion. The pelvis of the generalized quadrupedal animal lies generally parallel to the spine, and the thigh bone is perpendicular to it (Figure 9–7d). The latter is moved by two major sets of muscles, the flexors, which bring it toward the body, and the extensors, which move it to the rear. The major flexors attach to the forward end of the ilium and the lower part of the femur, while the extensors are inserted into the posterior region of the ischium and again into the lower part of the femur. In these positions the major muscles are well arranged to exert their forces at efficient angles. Similar relations between pelvis, femur, and muscle are preserved in bipedal birds such as chickens or the extinct bipedal dinosaurs as well as in bipedal jumpers such as kangaroos. In all these instances the femur is essentially perpendicular to the pelvis and so the angles at which the muscles pull are optimum.

In the chimpanzee similar relations between femur and pelvis continue, as is seen in its shuffling gait. But when this animal assumes bipedal posture, even though the leg may remain considerably bent at the knee, the femur moves into a position in which the major muscles act at a disadvantage. On structural grounds alone one could predict that chimpanzees can neither walk gracefully nor run rapidly in bipedal positions. And they can't.

The bipedal posture of man is unique in the animal world, for it brings his thigh bone into position effectively parallel to the spinal column. To compensate for this displacement, and to allow both flexor and extensor muscles to regain a little efficiency in their position, the ilium in man has expanded to the fore and rear until it presents an almost wheel-like appearance from the side. This broadening of the iliac plate has allowed the flexor to attach at a somewhat more efficient angle than that presented in the bipedal chimpanzee, while the extensors—because they are inserted higher on the thigh bone—also operate at a reasonably efficient angle. None of the other primates have buttocks to match those in man. This posterior enlargement includes a fat pad, to be sure, but it primarily is comprised of extensor muscles which have become enlarged to provide man with his striding gait. We are apt to think of our own buttocks as comfortable cushions upon which to sit, but in reality they provide us with our distinctively human stride.

The lower leg and foot have been extensively modified through the evolution of bipedal posture in man. In the anthropoid the paired bones of the lower limb, the tibia and fibula, are not of equal size but more nearly so than in man. Man's tibia has been strengthened to bear more weight, while the fibula has become less robust as the need for rotating the leg has been reduced. But more dramatic changes have occurred in the foot (Figure 9–7e). In generalized quadrupedal animals the basic foot is long and carried off the ground, while the animal walks on relatively short toes. The apes and man walk on their entire feet and so-called plantigrades. This is a rather rare condition but is found as well in bears and raccoons. The bones of the foot in the chimp and man are alike in number but greatly different in proportions. In the chimp, the heel and the basic bones of the foot are short, while the toes remain elongated. In man, the heel bone has become both strengthened and elongated, the basic bones of the foot enlarged and more tightly articulated, and the toe bones shortened but strengthened. In the human stride we walk off the ball of the foot and the great toe and so naturally this first digit has become strengthened. The axis of the human foot lies between the first toe and the second, so that it is numerically divided into two very unequal portions. As the great toe has become strengthened, the outer digits have become reduced. Our little toes seem degenerate in their puny size and their usually repulsive toenail. In our culture this condition is commonly blamed upon shoes worn too small, so that they pinch and constrict this portion of the foot. No doubt there is some truth in the fact that some shoes are too tight, but it has been shown by X-rays taken on the feet of Australian Aborigines who have never worn shoes that their little toes are just as degenerate in appearance as are our own. This digit has been undergoing reduction and structural degradation for a long period of time and as a direct consequence of the way we walk.

Upright posture has eliminated the normal shock resistant features of the quadrupedal skeleton. An animal like a cat can safely be dropped 20, 30, or perhaps 40 feet and land upon hard ground with no great harm. Each joint of its four limbs serves as a shock arrestor as the body hits the ground. In man the conditions are quite different. As a rule, we walk with a fully extended leg, meaning that it is barely bent at the knee. On uneven ground in the dark unanticipated depressions and rises in the ground produce very irritating senses of shock. But we still do have a series of built-in shock absorbers which we unconsciously use to smooth our progress. In the foot itself there is a longitudinal arch extending from the heel bone to the ball of the foot. Less visible is a transverse arch running from the great toe to the fifth through the ball of the foot. Together they provide a small but useful three-dimensional cushioning in the foot structure itself. This effect is further enhanced by a fibrocartilaginous pad built into the arch of the foot. In consequence the foot itself absorbs a certain amount of shock. Also, we can utilize the cushioning of a bent knee, and people who habitually walk over rough ground favor this kind of walk. Finally, the curves of the vertebral column and the cartilage cushion between each of its separate members also give small cushioning effects. As a result, if we know where we are placing our feet, we manage reasonably well.

Many earlier anthropologists have stressed the imperfections of man's upright posture on the grounds that he has only recently attained it. Today the fossil record indicates that we have been progressing toward our unique bipedalism for perhaps five or ten million years, and attained most of the structural alterations required two to three million years ago. And yet as individuals we frequently suffer from incomplete adaptation in our upright position. Today many people suffer from "lower back" pains. These may involve anything from ruptured disks, as frequently occur in the lumbar region, to a slipped articulation between the sacrum and one pelvic bone. It is difficult to determine how much of this is due to poor evolutionary engineering, or how much of it results because many modern men and women do not keep their bodies in good physical tone, and so both muscles and tendons do not maintain structural integrity. Fallen arches are a continuing complaint that almost certainly results from poorly designed shoes, excess weight, or again possibly a lack of exercise. Finally, ruptures, which allow some of the content of the visceral cavity to emerge through several different kinds of apertures, are claimed to demonstrate the incomplete adaptation of man's erect posture. Hernias of this kind do occur, and there is some indication that failures of muscle masses to completely close off apertures—like that provided by the inguinal canal—may be the fault. But the human body, when well-conditioned, is a surprisingly well-adjusted organism.

It occasionally has been suggested that man evolved into a bipedal animal in order to move more rapidly. Nothing could be further from the truth. The tamest of house cats can run one hundred yards at 25 miles an hour. Only a trained athlete under ideal track conditions can equal this. At the distance of a mile, man's world record barely averages 15 miles an hour. Some ground monkeys run quadrupedally at 35 miles an hour, many grazing animals can run 40 miles an hour, and the cheetah has been clocked at 70 miles an hour. Man could not escape from any likely predator by running away, and he walks bipedally for another purpose, that of using his hands as tool holders.

In the altered human pelvis the crest of the ilium serves as a point of attachment for a number of muscles which are involved in maintaining the trunk upright on the pedestal provided by the legs and pelvis. These include deep muscles of the back and transverse muscles of the sides. That stabilizing the upper portion of our body requires considerable muscular effort can be demonstrated by the reader's placing his hand across the small of his back and gently bending to one side or another. What were once muscles of locomotion in quadrupeds have been transformed into trunk supporting muscles in ourselves. One of the invisible shifts brought about by the evolution of upright posture involves the support of the internal organs or viscera. In the generalized quadrupeds these "innards" are suspended from the arched backbone by membranes known as mesenteries. The arrangement is simple, with the mesenteries vertically supporting the intestines and the like against the pull of gravity. With the assumption of upright posture the mesenteries retain approximately their same position but now are supporting the visceral organs at an oblique angle and so less efficiently. The abdominal contents in man remain in place due to transverse muscles on the side and the paired rectus abdominis muscles which stretch from pubis to sternum at the junction of the ribs. As long as these muscles retain their tone, men and women are flat stomached. But with gross eating and no exercise they relax and allow man's gut to protrude as a paunch. That this is a consequence of muscle relaxation and gravitational pull is demonstrated by the fact that old anthropoids, when seated upright, also have bulging paunches. Again this seems less to be an evolutionary defect than abdication by the individual.

The Relationship Between Brain Size and Face Size

Human fossil skulls differ strikingly in appearance. Most of the differences are related to the size of the brain of the individual compared to face size, which in itself is largely dependent upon tooth size. The rapid increase in the size of the brain of man during the Pleistocene is one of the most dramatic aspects of the whole of organic evolution. Early in the Pleistocene, animals which had clearly entered the human niche averaged cranial capacities of around 500 cubic centimeters, with some crania, such as ER 1470, considerably larger. Three million years later, by the end of the Pleistocene, brain size had approximately tripled to average around 1400 to 1500 cubic centimeters. Going on at the same time, but in a reverse direction, was a more gradual reduction in tooth size. The earliest of men had palates shaped very much like our own but equipped with considerably bigger teeth. By Middle Pleistocene times some dental reduction had occurred, and with the advent of the Recent epoch there was a further diminution. With a change in diet, teeth have become yet smaller but this is another story.

Cranial capacities fall into three general grades. Capacities of 500 cubic centimeters represent the australopithecine grade of human evolution (Figure 9–8). The next clearly recognizable grade is that of the pithecanthropines, in whom cranial capacity averages 1000 cubic centimeters. Finally, *Homo sapiens* in the Late Pleistocene has a capacity of 1500 cubic centimeters, a value which is a bit too large, but has the advantage of giving us round numbers. To each of the three grades of brain size shown, illustrative faces of equal length are attached. Facial structure is largely determined by tooth size, so the dentition is shown as decreasing in size in the three grades. The small brains and large teeth of the australopithecine result in an apelike appearance as a simple consequence of the basic geometry of the head. When a sizable face is hafted to a small brain, it necessarily protrudes forward and so gives an overall apelike appearance. The skulls of australopithecines had this characteristic, but their bodies clearly show that they were men. The next grade, that of the pithecanthropines, has an equal-sized face, but with tooth size somewhat reduced, and is hafted to a brain of 1000 cubic centimeters. Without indicating such detailed structures as brow

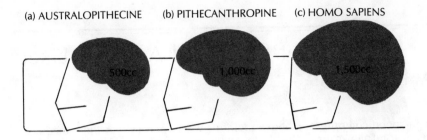

(a) AUSTRALOPITHECINE (b) PITHECANTHROPINE (c) HOMO SAPIENS

500cc 1,000cc 1,500cc

Figure 9-8
Relation of Brain Size to Face

ridges and the like, this diagram is clearly that of a primitive man, well advanced beyond the pongid stage but yet too snouty to be accepted into the modern family of man. Finally, with the same face, but a further reduction in dental size, fixed to a brain of 1500 cubic centimeters, we have reached the modern level of evolution. The enlargement of the brain, combined with dental reduction, allows the face to become subordinated to the brain case and to fit on it so as to give an essentially straight profile. As simple as these diagrams are, they cover the major trends in the evolution of the human head for the last three million years.

Bibliography

BIRDSELL, J. B.
1957   Some Population Problems Involving Pleistocene Man. In Cold Springs Harbor Symposia on Quantitative Biology. Vol. 22, pp. 47–69.
1963   A Review of The Origin of Races by Carleton F. Coon. In Quarterly Review of Biology. Vol. 38, No. 2, pp. 178–85.

COON, CARLETON S.
1962   The Origin of Races. New York: Knopf.

NAPIER, J. R.
1962   The Evolution of the Hand. Scientific American. Vol. 206, No. 1, pp. 56–62.

TINDALE, NORMAN B.
1962   Some Population Changes Among the Kaiadilt People of Bentinck Island, Queensland. In the records of the South Australian Museum, Vol. 14, pp. 259–96.

TOBIAS, PHILIP V.
1967   The Cranium and Maxillary Dentition of Australopithecus (Zinjanthropus) Boisei Olduvai Gorge, Vol. 2. Cambridge, England: Cambridge University Press.

Some Complex Methods of Statistical Analysis Used in Evolutionary Problems

The advent of the modern high-speed computer has been a blessing in many ways, for it allows the rapid solving of problems which could not even be approached in the days of pencil-and-paper solutions. Expectedly it has resulted in a whole series of statistical exercises at high levels of abstraction in which massive amounts of data are manipulated in complex ways and in infinite number of dimensions. At the same time computer "software" includes so many different kinds of programs, available to so many people, that many analyses are run which are inappropriate to the problem, or even to the data. A great many anthropologists using them simply do not have the statistical sophistication to understand the limits of these programs.

Factor analysis was originally devised by psychologists attempting to ascertain the components in their problems. It has been applied, but with no great success, to a few anthropological questions. Multifactorial analyses come in a variety of forms and have the great virtue of handling a great many different aspects of a problem simultaneously and grinding out some seemingly simple answers. The generalized distance of Mahalanobis D^2 is one of the most sophisticated of the newer approaches to complex problems. It takes data in complex forms, solves problems of correlation between traits, and comes up with a single numerical expression of the "distance" between the various populations. It is not to be faulted statistically, nor is it altogether to be trusted biologically. Numerical taxonomy is an overall approach which purports to measure similarities between varieties of organisms. It again can digest enormous amounts of data and reduce them to single numerical expressions. This type of analysis results in certain transformations of data which may on occasion prove illuminating.

From the biological point of view there are a number of implicit assumptions in each of the above methods which may render them inapplicable to problems of evolution. They are generally applied to characteristics measurable on the phenotype of individual organisms. What organisms appear to be on the surface may reflect some of their line of real descent, but their appearance may also be derived from evolutionary changes proceeding in parallel or even converging from originally different configurations. *These methods incorrectly presume that likeness is a measure of relationship.* At higher levels of classification, when due allowance is made for the identification of homologies, this may be true. At the species level, and more particularly within subspecies as is the case in man, this assumption is unjustifiable. It assumes that selection has not been working upon the populations and that drift has no part in the production of their present status. It also tends to ignore, although this is less important, variations in mutational pressure. The method explicitly stands upon the assumption that similarities between populations are produced only by gene flow or through ancestral lines of descent. This position is much too simple to represent evolutionary reality, and so most of the conclusions reached by these kinds of analyses are open to question. On the other hand, it must be admitted, that where the situation is very simple, it may be useful to give a measure of difference or similarity lumped into a single figure for purposes of comparison. But this is a matter of convenience and not a measure of relationship.

One of the advantages of these higher statistical exercises is that a great many kinds of data can be incorporated into a single solution. This produces generalizing power and so is important in that sense. On the other hand, any taxonomically relevant characteristics are certain to be swamped by masses of irrelevant data and thus buried invisibly in the solution.

PROBLEMS
WITH THE
AUSTRALOPITHECINES

The first early, small-brained, and hence apelike men appearing in the fossil record are the australopithecines. Known to range from South Africa to the Red Sea through the Villafranchian or Early Pleistocene, they are among the most abundant fossils known. With brains little bigger than one-third that of modern men, they appear to have been tool users, and at least some of them hunted systematically. They provide an excellent idea of what the first people must have looked like and how they behaved. Yet it is ironic that the time of these fossils indicates they are a side branch of the main line leading to modern men, for there are advanced men living simultaneously with them. Thus human evolution becomes more complex as further discoveries are made. The problem of how the various forms coexisted in the same space has not been totally resolved.

Early Hominid Discoveries

The ramapithecine grade in human evolution is easy to deal with since the evidence is limited to jaws and teeth from less than twenty individuals found in East Africa and India in a time period ranging roughly from 14 to 20 million years ago. There is general agreement that these finds represent a very early type of hominid, primarily based on the dental evidence which shows low-crowned and thick-enameled molars combined with relatively small canines and incisors. The body size of our presumably ancestral form is not known, but it is generally considered that this advanced primate may have stood about three feet high and weighed perhaps 40 to 50 pounds. There is no indication that the bipedal mode of locomotion had been adopted, although the ramapithecines may have moved in this manner more frequently than do the living chimpanzees. Tool using was presumably no more than casual for these early men of the Late Miocene and Early Pliocene. Their environment seems in East Africa and in India to have been riverine forests edging into open grasslands. The most recent find of all, from Pygros in central Greece suggests a more open environment insofar as the accompanying animals can indicate. Among paleoanthropologists there is a general feeling that the ramapithecines are the earliest identifiable hominids and very likely are directly ancestral to us.

After a gap of four or five million years in the fossil record, the australopithecines begin to appear in a spotty way in the fossil record (Figure 10–1). A single tooth from Ngora in Kenya is only identified as hominid, and it is not clear whether it is a ramapithecine or australopithecine. It dates from the time approximately nine million years ago. Then at Lothagam, also in East Africa, was found a portion of a lower jaw with a single molar tooth still embedded in it. It is relatively small in size, shallow in proportions, and it is generally agreed to be the earliest evidence of one of the lightly built australopithecines. From four million years ago onward the record of the australopithecines fills in remarkably well. Teeth and bony fragments of several hundred individuals have been found, almost entirely in South Africa or East Africa. The fossils primarily consist of teeth and jaws, but there are some restorable skulls, fragments of upper and lower limb bones, including some of hands and feet, and perhaps most important of all, three nearly complete pelves.

The greatest number of australopithecines has been found in South Africa and such sites as Taung and Makapan, both excavated by Professor Raymond Dart, and Sterkfontein, Swartkrans, and Kromdraai, where important finds were initially recovered by Dr. Robert Broom and later by his former student, Dr. J. T. Robinson. The East African finds include impressive ones by Dr. Louis Leakey, at Olduvai Gorge, and in more recent years by his son Richard Leakey at East Rudolf. Scattered finds have been made by others at Garusi and Lakes Natron and Baringo. Several teams of field workers have found further australopithecine remains in Lower Pleistocene beds at Omo in southern Ethiopia. Dr. Clark Howell has extensive series of teeth and some jaws which range backward in time from a little less than two million years ago to almost four million years. Recently fossil beds have been found to contain these early men in the Afar triangle where Ethiopia touches the Red Sea.

Figure 10-1
Australopithecine Sites

Discovery of the First Australopithecine

The story of the discovery of the first australopithecine is an interesting one. Its hero is Emeritus Professor Raymond Dart of the Department of Anatomy at the Witwatersrand University in Johannesburg in South Africa. He was a demonstrator of anatomy at University College in London for a few years and then moved permanently to South Africa. Shortly after arriving there, the young anatomist received the skull of a fossil monkey from his only female student. It had been obtained from a limestone quarry near the town of Buckston. Dart arranged for a colleague who was traveling that way to visit the quarry to see what was being found, aside from vast amounts of limestone for a cement kiln. Such mining operations had already pushed the working face of the limestone cliff some 250 feet back into the plateau. By good luck Dart's friend, Dr. Young, arrived on the scene just after a dynamite blast had been fired. It brought down a mass of material, including some containing cave fill deposits, "brown patches" that workmen had long recognized as likely to contain fossil bones. The manager of the lime works sorted through the blasted rubble and handed several blocks of rock containing fossil fragments to Dr. Young. These were shipped to Dart, and thus in November 1924, a new chapter in human prehistory was opened.

Professor Dart easily recognized that his rock specimens contained two more fossil skulls of baboons. But it was the third block of rock which caught his attention. Clearly exposed on one of its surfaces was the cast of a brain larger and more complex than that of any baboon, or even any great ape. It was the first endocranial cast of its kind from the fossil record anywhere. In a matching rock, Dart saw protruding bones which indicated that a face was beneath its surface. After months of careful work with hammer and chisel the specimen was uncovered and found to consist of a virtually complete face, containing the upper teeth and much of a lower jaw with intact dentition. The teeth corresponded in age to those of a six-year-old human child, and thus the potential first Lower Pleistocene ancestor of modern man came to light. Professor Dart made the first formal report of his find in a short article in *Nature* in February 1925. In it he gave a preliminary description of what remained of the skull and its associated brain cast. He claimed that the specimen was important since it represented an extinct race of

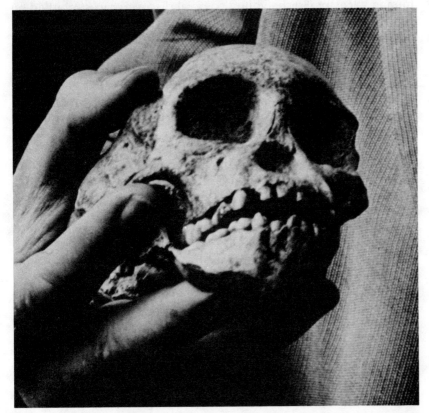

Plate 10-1
Taung Skull

apes intermediate between living anthropoids and man. He named it *Australopithecus africanus,* or the southern ape of Africa. The young anatomist claimed that the Taung child carried its head in an upright position on a better posed body and indeed was capable of bipedal locomotion so that its hands were freed for other functions, including those of offense and defense. In effect, Dart claimed that the Taung specimen was that of a tool user. Since Dart had left behind in London a reputation as an imaginative and possibly even a rather brash young scientist, his preliminary report was poorly received. Professor Arthur Keith, later to be knighted, the leading authority on fossil man, concluded that the Taung specimen represented a pongid related to the chimpanzee and gorilla, although it showed a few peculiar characteristics. Professor Elliot Smith, the great neurologist under whom Dart had previously worked, remained unimpressed with the claims of humanity for the specimen. Dr. W. H. L. Duckworth, an anatomist at Cambridge University, conceded that the Taung child showed some advanced features, but concluded that it resembled the gorilla rather than the chimpanzee. Professor Smith Woodward, paleontologist at the British Museum, doubted that the new fossil had any bearing on the problem of human ancestry. So the young colonial discoverer was rebuffed by four of the most eminent authorities in the English-speaking world, and for some years his find remained in a suspense account. It is a pleasure to report here that Professor Dart never wavered in his convictions, continued his publications, commenced new excavations at the quarry site of Makapan and enriched his collection of australopithecines. In time Dart was able to prove that his original judgment was entirely correct and that the jury of experts was totally wrong.

By diligence and skill, Professor Dart finally cleared the limey matrix from the face and teeth. It then became evident that the Taung teeth were very like those of primitive men and unlike those of the great apes. The reduction in the size of canine teeth produced a rounded palate of parabolic form as in modern man. The child's incisor teeth were of human proportions instead of enlarged as in the gorilla or chimpanzee. The premolars, or bicuspids, were fashioned for grinding rather than shearing as in the pongids. Today it is generally agreed that this and other australopithecines have a dentition very like that of modern man except that the teeth are individually larger in size and the molars in particular are relatively broader.

Absolute dates by the argon-potassium and fission track methods have largely brought the australopithecine fossils of East Africa into a reasonable chronology. But the South African sites remain undated, or rather estimates of their age vary with the winds of scientific fashion. The difficulty in South Africa arises from the fact that the sites are in limestone and no good dating techniques have developed to handle this material. Furthermore, these sites represent caves or broken sink holes or fissures into which animals, including early men, have been deposited more or less accidentally. The majority of the sites have been mined for their limestone, so that the fossils have largely been rescued from the dump heaps of the mines. Only in the last few years have proper works of exploratory excavation really commenced, and the results are not yet in.

Shortly after their discovery the sites containing the larger australopithecines, Swartkrans and Kromdraai, were usually put in the Middle Pleistocene. Those containing the smaller form, Makapan and Sterkfontein, were placed in the Lower Pleistocene. Since a short chronology was then in use, these sites were generally considered to have been a million years old at the earliest and to have lasted perhaps a half million years in time. The estimates of these and later dates have primarily been based upon the faunal

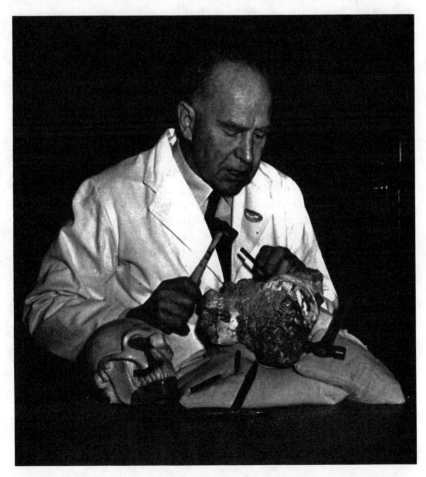

Plate 10-2
Professor Dart developing a portion of an australopithecine skull from cave breccia. The same lump also contains a baboon skull.

contents of the deposits. These bones, generally fragmentary, were frequently misidentified as being the same as living forms, whereas now it is known that many of them belong to extinct species. At the present time it is generally considered that the latest sites are a million or more years old, while the earliest ones extend backward in time to two and a half or even three million years before the present. The latter, then, are earlier than the fossils from Olduvai. It is a little difficult to be certain at this time how much these faunal dates reflect the growing pressures to make the South African sites earlier than most of those in East Africa, and to what degree the new estimates really represent mature judgment based upon the faunal remains. A great variety of dating techniques are now being tried in South Africa on these deposits, and it may be that an absolute chronology in time can be established. It would be most helpful.

Gause's Law or the Competitive Exclusion Principle

Zoologists for many years have held the idea that two closely competing species of animals cannot coexist for long in the same space. The basic idea is simple, since it rests upon the well-demonstrated fact that every species in nature tends to have its own particular array of food resources and other needs, and these in a stable community invariably differ from those of all others. In laboratory experiments it has been demonstrated repeatedly that if two closely related species, whether they be protozoans, or flour beetles, if placed in a limited environment, would shortly result with the extinction of one of the competitors and the triumph and survival of the other. G. F. Gause (1934) was one of these experimenters. But others reached the same conclusion earlier by methods as widely different as observing natural populations of animals on the one hand, and by constructing mathematical models to simulate competition on the other. Even Charles Darwin was well aware of the general principle. Garrett Hardin (1960) has discussed the whole problem of the *competitive exclusion principle* in a useful summary.

The essence of the competitive exclusion principle rests upon the overlap of essential requirements, rather than the close degree of relationship. It is granted that two closely related forms are very likely not to be sufficiently divergent in their preferred food, nesting requirements, and a series of other vital needs, to escape from seriously interfering with each other's survival patterns. But there is another kind of competition between distantly related animals which may prove equally intense. In mainland Australia, it took the imported English red fox less than a century to spread from Melbourne on the south coast to Darwin, some 2,000 miles away on the north coast. Both the fox and the dingo average four pups a year, so the patterns of reproduction in these canids are very similar. Since the dingo did not reach Tasmania, it must have been introduced into Australia after the postglacial eustatic rise of sea level severed Tasmania from the mainland; that is, after about 9,000 years ago. Like the fox, the dingo—by breeding alone—would have crossed the continent from north to south in about 100 years. Its expansion, even in the face of competition from the Tasmanian "wolf," would be virtually instantaneous in geological time.

The dryopithecines covered the same wide range in the Old World as the ramapithecines from which we have descended. The very process of differentiation of these two forms of advanced primates during the basal Miocene radiation must have involved a potential for Gausian competition. The fact that some of the dryopithecines went on to evolve into the living great apes, while the ramapithecines produced the true hominids, of course, stands as direct evidence that somehow they avoided fatal competition in their lifeways. The competitive exclusion principle in this case stands as the best evidence that the two groups must from very early times have differed sufficiently in their vital requirements so as to effectively avoid this type of competition.

Australopithecine Size Variation

One of the points of interest about the australopithecines is that their remains come in two sizes, and the bones, particularly the crania, show very different degrees of robusticity. A few years ago this difference was sometimes said to simply reflect differences in the size of two sexes and to have no further meaning. The proponents of this idea had a very catchy phrase for it; they claimed the two sizes of australopithecines made love, not war. Today the evidence very heavily favors the idea that the australopithecines were in fact two different species, with the lightly built and more delicate one called *Australopithecus africanus,* and the larger form generally known as *Australopithecus robustus.* Since in East Africa the robust form, *Zinjanthropus,* shows even more exaggerated features than that found in South Africa, it is sometimes called *Australopithecus boisei* and so constitutes a third species in the australopithecine complex.

The differences between the two species of australopithecines are worth discussing. *Australopithecus africanus* is estimated to have weighed between 40 and 60 pounds and to have stood a little over or under four feet in stature. The face was dished in, the nasal bones showed no relief, and the teeth, while showing the general proportions of those of modern men, were very much larger (Figure 10–2b). Consequently the jaws which held them were large and the

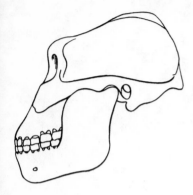

(a) *A. boisei*

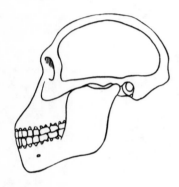

(b) *A. africanus*

Figure 10-2
Variations among
Australopithecines

temporal muscle rose high on the side of the head. Only rarely did those muscles meet and produce a crest down the mid-line of the skull, as in a few males. The gracile australopithecines show a cranial capacity for six specimens at 442 cubic centimeters. Since they were about the same size as the pygmy chimpanzee, and the latter had only 300 cubic centimeters of brain, it is obvious that their brains are about 50 percent larger than the little anthropoid. In both the small and the large australopithecines, there is further evidence that they were essentially bipedally erect, for the spinal cord centers the skull to unite both the brain in the base in a way very much like that in modern man.

Australopithecus robustus and to an even greater degree *Australopithecus boisei* were larger men, probably weighing between 70 and 100 pounds, perhaps standing a foot taller than their gracile relatives (Figure 10-2a). The robust australopithecines, with considerably larger body, show cranial capacities of 530 cubic centimeters for four specimens. This is about what one would expect from their greater body size. Their chewing apparatus was enormous as a consequence of the great tooth size shown in these men. It is peculiar to the robust australopithecines that both premolars became enormously enlarged and changed in both shape and function. Thus instead of three grinding cheek teeth, these australopithecines effectively had five. Their canines and incisors were about the same size as in the gracile australopithecines, but relative to their molars considerably smaller. These proportions are clearly shown in Table 10-1 where it is evident that the East African form *A. boisei* was even more extreme than the robust form from South

TABLE 10-1

Cross-Sectional Areas of Teeth of Four Hominids

Tooth Areas (mm²)	Homo erectus	Australo-pithecus africanus	Australo-pithecus robustus	Australo-pithecus boisei
I_1	79	78	71	80
I_2	66	44	49	52
C	98	91	82	86
PM_1	101	111	140	185
PM_2	95	117	163	212
M_1	150	173	200	269
M_2	147	213	231	361
M_3	116	206	252	335

Ratios of Regional Areas

	Homo erectus	Australopithecus africanus	Australopithecus robustus	Australopithecus boisei
$\dfrac{I_1 + I_2 + C}{M_1 + M_2 + M_3}$	$\dfrac{243}{196} = 124\%$	$\dfrac{213}{228} = 93\%$	$\dfrac{202}{303} = 67\%$	$\dfrac{218}{397} = 55\%$
$\dfrac{I_1 + I_2 + C}{PM_1 + PM_2}$	$\dfrac{243}{413} = 59\%$	$\dfrac{213}{592} = 36\%$	$\dfrac{202}{683} = 30\%$	$\dfrac{218}{965} = 23\%$

Africa in terms of these dental characteristics. The dental differences between the two australopithecines have been overly simplified, but they do indicate that the small ones were omnivores while the robust ones were herbivores. Certainly, both forms would have subsisted on both vegetable foods and meat, but the great grinding surfaces of the robust australopithecines does suggest a greater dependence upon vegetable foods.

As the fossil record now stands both the gracile and the robust australopithecines live overlapping in time for long periods, particularly in East Africa. Thus competition between these two forms would be severe enough, but it is much heightened by the fact that we now know that advanced men lived with them at the same time in East Rudolf. Therefore the competition existed between hominids whose food and other vital requirements must have shown considerable overlap. For the moment we will ignore the more advanced men recorded in East Rudolf and merely consider the coexistence of the two kinds of australopithecines.

Australopithecine Coexistence

There is evidence in favor of the coexistence of two different species of australopithecines in Africa: a small, lightly built form and a larger, robust one. One theory involves the distribution of the gracile and robust forms in time and space. Evidence of their coexistence has been found in the ancient deposits of the Omo Valley. At Olduvai Gorge, only the robust form extends into Bed II. But in South Africa, the various cave deposits do provide more critical evidence. It is generally agreed that the three sites of Taung, Makapan, and Sterkfontein are the earlier, while Swartkrans and Kromdraai are later in time. The three early sites contain a total of about 27 individuals, all of the gracile type. The hypothesis that the differences are due to sex would require that they all be labeled female. The two later sites contain about 38 individuals, all of the robust type, and hence male by the same hypothesis. There are no absolute dates for these South African sites, but they do involve long periods of time, probably hundreds of thousands of years.

How likely is it that only females are represented in the early deposits, while only males occur in the later ones? It is improbable that any separation of sexes should follow this pattern. Indeed, it is so improbable as to count heavily against the theory of sexual differences as being the right explanation. The chance of this improbable event occurring can be expressed numerically. If we assume that males and females occurred in equal numbers in the population as in modern man, and that only about *one-half* of the fossil individuals can be properly labeled as to type, owing either to immaturity or to fragmentary nature, then the chance of finding 13 identifiable females as the total universe in the early cave deposits is equal to $(1/2)^{13}$ or a probability of one chance in 8,192. Applying the same reasoning to the later cave deposits, the chance of finding 19 identifiable males as the only occupants is even less likely. In this instance the chance of such an event is $(1/2)^{19}$, or only one chance in 524,288 trials. Since the earlier and later cave deposits are separate in time, the odds we have just quoted are independent of each other. So, finally, the chance of finding 13 females in early deposits and 19 males in the later deposits is a product of the two probabilities or $(1/2)^{32}$. The chance of finding this distribution of the sexes could be expected in one out of somewhere over 4 billion trials.

As mentioned earlier, the dental characteristics in the two forms of australopithecine differ in significant fashion. This is demonstrated in Table 10–1, where the cross-sectional area of each kind of tooth is given for four different kinds of hominids, beginning with *Homo erectus* and proceeding to the right through small australopithecines, large ones, and the biggest of them all, *Zinjanthropus*. The cross-sectional area is derived by multiplying the length of the crown by its breadth and is expressed in square millimeters. The table begins with the central incisor and proceeds in order around to the third molar. A comparison of cross-sectional areas shows that all three australopithecines differ from the pithecanthropines in possessing relatively smaller canines, larger premolars, and very much larger molars. In the robust australopithecines the latter increase in size from the first molar through the third. But the differences are expressed more strikingly by comparing regional groups of teeth. Taking the most anterior teeth, the two incisors and the canines, and dividing their total cross-sectional area by that of the two premolars, we find that the ratio is 124 percent in the pithecanthropines, while it falls below 100 percent in all of the australopithecines. In the robust form the ratio falls far below this value and climaxes in *Zinjanthropus,* in which it is only 55 percent. For this regional ratio australopithecines are well separated from *Homo erectus* and the robust forms from the gracile ones. These differences primarily arise due to the size of the premolars in the robust form, in which these teeth are changing from their original general shearing function to become functional members of the grinding cheek teeth, the other molars. This trend toward *molarization* has been observed among many herbivores, but characterizes no other early men. The same dental relationship evolved in *Gigantopithecus,* a giant pongid characterized by its great bulk.

A second ratio compares the three foremost teeth of the series with the three posterior cheek teeth, the molars. The reasonable ratio of 59 percent characterizes *Homo erectus*, while all three categories of australopithecines show much smaller values. This tendency climaxes again in *Zinjanthropus,* where the cross-sectional area of the three molars is more than four times as great as for the three anterior teeth. These differences have been recognized by others, and John Robinson (1961) in particular went so far as to claim that because of this molarization of the bicuspid teeth, the robust australopithecines must have been herbivorous in diet. Without accepting that conclusion completely, it is certain that very substantial dental differences separate the robust forms from the lightly built ones.

The dramatic dental shift could have resulted from two rather different kinds of selective pressure. Charles Darwin himself considered that tool-using would free the hands of man's ancestors and so result in a reduction in canine tooth size. Certainly this had already occurred in the ramapithecines. The other hypothesis, put forward by Clifford Jolly (1970), is the so-called "small food" or graminivorous hypothesis. Jolly has suggested that as man's hominid ancestors left the sheltering forest and moved out into the savanna, they would gradually give up their generalized forest diet and become increasingly dependent upon small fruits, seeds, berries, and a wide variety of other kinds of food that come in small pieces. This is exactly what the gelada baboon feeds upon, and it has resulted in some rather interesting anatomical changes compared to other baboons. Along with a variety of changes in jaw structure, this baboon also has relatively small canines and incisors compared to its nearest relatives. These differing ideas are subject to test, for if after ample field discoveries it is certain that the ramapithecines were not habitual tool users, then the Jolly seed-eater hypothesis will look more attractive. On the other hand, there is a certain attraction to the idea that a chimpanzeelike ancestor moved out into the grasslands and by adding more meat to his diet was able to survive on the foodstuffs available there without becoming a confirmed vegetarian living only on small bits and pieces.

How these two closely related australopiths could have coexisted without crippling competition is not clear at this time. But they did, for they occur in the fossil record together for time spans greater than two million years in duration. Their very differences suggest that they must have separated from a common Pliocene ancestry far back in time. On the present evidence both australopithecines may have been tool users. Since pelves of each form indicate an upright and bipedal posture, they must have entered the hominid niche. But it does seem likely that the gracile australopithecines were bipedally more competent, and there have been some clues, particularly in the form of long bones, that the robust form may not have been completely bipedal.

The problem is a fascinating one, of great evolutionary significance, and yet it has recently been pushed into a secondary position by the finds of Richard Leakey at East Rudolf, where advanced men coexisted with at least the robust australopithecine over long periods of time. Therefore, it is now evident that neither of the australopithecines were direct ancestors of more modern kinds of men, at least in the span of time covered by the fossil record. It could be that deep in the Pliocene a form more like the gracile australopithecine did give rise to more advanced types of men.

A General Outline of Australopithecine Behavior

The behavior of these Lower Pleistocene people can be reconstructed in part from the indirect evidence provided by the bones and can be further rounded out with reasonable conjectures based upon good principles. The anatomical evidence that the australopithecines were effectively bipedal in locomotion is indisputable. This conclusion rests upon the nature of the skull, the pelvis, and the feet, and the indirect evidence provided by the reduction of canine teeth. All of the evidence points to the australopithecines' having occupied the human niche for a long time and suggests that they may have been tool makers as well as tool users. Archeology provides confirmation for these estimates.

Early Tools. Roughly chipped stone pebbles used to produce choppers and chopping tools have been known to archeologists in South Africa for many years. This stone industry has been called the Oldowan. The Oldowan industry is known to go back in time to nearly 4 million years in terms of isolated tools. A living floor now being dug at East Rudolf is dated at about 2.6 million years and shows a wide variety of tools. In addition to the easily recognizable choppers and chopping tools, flake tools come in a wide variety of sizes and shapes. Some are very delicate, others are relatively long compared to the breadth, and some have fabricated points. It is not a really primitive stone industry but merely lacks a standardization in its variety of tool types. The complexity of this stoneworking tradition has frankly come as a great surprise to anthropologists and archaeologists. A chopper is shown in Figure 10–3. Stone tools of this type have been found in four of the sites containing australopithecine remains. They are Olduvai Gorge and East Rudolph in East Africa, and Swartkrans and Sterkfontein in South Africa. A so-called living floor in the lower part of Bed I at Olduvai provides the best evidence. Here the skull of a robust type of australopithecine was found together with several leg bones. The remains were associated with chopper and chopping tools and a variety of flakes which had been struck off when trimming the pebble to form the core tools. The stones from which the tools were made had been carried in from some considerable distance, and so

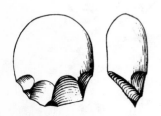

Figure 10-3
Oldowan Chopper

there is no doubt that actual toolmaking occurred on this living floor. Dr. Louis Leakey, in some personal experiments, has demonstrated that a simple chopping tool is adequate for the task of butchering sizable game. If the stone employed is somewhat glassy in nature, the edges of the flakes are razorlike and hence cut flesh very well.

The simple chopping tool functions so well in a variety of ways that some living hunting and gathering people still use them. Green saplings can be cut down and worked into a number of useful forms with a chopper. There is no evidence that the australopithecines made such wooden tools, but technically this would have been within their capability. Most important of all simple wooden tools is the digging stick carried by the women of hunting and gathering people everywhere in their food quest. It consists of nothing more than a sapling four or five feet in length, sharpened at one end, and weighing several pounds. This simple device opens to human beings all of the underground food resources that can be reached by digging to depths of about six feet.

The simple chopper also provides the means for making other wooden tools. These would include straightforward clubs which serve to give men an extra joint to their arm and so speed the blow as well as extend the distance at which it would be struck. Simple thrusting spears are well within the capabilities of manufacture by these stone implements. Finally a simple stick, about two feet in length and slightly curved, becomes an offensive projectile when thrown at game. Australians have been seen to bring down running small game up to distances of 40 yards with these simple throwing weapons. Rotating as it flies, it cuts a swath equal to its length, and its lethal effectiveness is about the equivalent of a modern shotgun. There is virtually no chance of such wooden implements being preserved for a million or more years of time, and so we shall probably never have direct evidences of whether the australopithecines actually made and used them. But they are so simple in nature and so highly effective that it is likely that man invented them at a very early date.

While two of the South African australopithecine sites do show Oldowan tools, the Makapan site has been diligently examined by a competent archeologist who went over twenty thousand stones looking for worked tools and found none that were convincing. But the same site yields a great variety of bones, many of which seemed to have been broken so as to make them into useful tools. In a few cases the edges even show the wear which comes from use. Professor Dart has proposed that the australopithecines there used bones and animal teeth in a wide variety of ways. His claims have not received wide acceptance, but he has passed through this situation before and won his case.

The australopithecines lived too early to be users of fire, but there remains the interesting question of how they dared to inhabit caves in a country noted for its large carnivorous cats. Both lions and leopards do much of their hunting by night, and one authority has suggested that if these early men were like their modern counterparts, their snores would have attracted hunting felines into their bedquarters. We certainly shall never know whether these little men snored, but it may be that they made themselves safe at night by blocking the entrance to their sleeping caves with thorny brush, a method used by modern Africans for similar reasons.

Fossil Evidence of Game Hunting

On one of the lower living floors at the bottom of Olduvai Gorge,* the detailed plot indicates an impressive number of dismembered carcasses of large animals, mostly antelope. Also included were wild cattle, primitive-antlered, short-necked giraffes, very large pigs, and an early form of the elephant. In general, the bones were somewhat scattered, but a few large animal remains were so compactly distributed as to indicate butchering and eating where they fell. This variety of big game, among others not reported here, indicates that the hominids at this time and place were effective hunters and not limited to small and slow forms of game. The techniques by which these large animals were taken are not known, but the fact that the animals were butchered in a limited place, combined with their size, suggests that they were killed there and not carried in sections.

Professor Dart has analyzed the bones from Makapan which represented at least 433 different animals. Ninety-two percent came from antelope ranging from small to very large, but the majority came from moderate-sized forms. In addition, there were the remains of a few fossil horses, giraffes, rhinoceroses, a hippopotamus and several chalicotheres, extinct ungulates equipped with bear-like claws. Other bones represented a few hyenas and porcupines as well as slow game such as water turtles and terrestrial tortoises and a variety of birds and their eggs. Very significant, as we shall see later, were remains from no less than 45 baboons. The presence of these bones in the cave does not prove that their owners were capable of killing them all. Most of the creatures are represented by selected parts, of which more than 80 percent are fragments of the skull. Of the lot only one neck vertebra was present, a fact that suggests to Professor Dart that the aus-

---

*Although *A. boisei* was found on this living floor, it is not clear whether he produced it by his own activities or was eaten there by more advanced hominids such as the habilines who were his contemporaries.

tralopithecines were in a sense headhunters. The curious biased pattern of the bony remains of the food animals suggests that the australopithecines may have spent some of their time scavenging from the kills of lions, leopards, and cheetahs. The combined evidence suggests some competence in hunting big game, together with thrifty scavenging when this was easy.

Hunting Baboons. Professor Dart had noticed depressed fractures in some of the 42 baboon skulls from Taung, Sterkfontein, and Makapan that he examined. Of the total no less than 27 skulls or 64 percent had been fractured by blows from the front. A further number had fractures on the left side of the face, from the front. Dart's baboon skulls gave evidence that their brains had been picked out and eaten, a human custom we shall see again later. The series included adults of both sexes as well as some juveniles. So the total series of skulls showed that more than 80 percent of the baboons associated with the australopithecines had died of violence vented on the head. All but two blows were delivered on the left side, and Dart hypothesizes that the australopithecines had already developed a preference for using the right hand. Only six of the blows appeared to have been delivered from the rear, and so by stealth. The rest suggested frontal assault, a tactic not easy to carry out.

It is difficult to build sound behavioral hypotheses on such data as the battered baboon skulls. Nevertheless, if the figures offered by Dart are correct, a ratio of 14 out of 15 baboons had been killed by what seemed to be a right-handed australopithecine. Among modern human populations six out of seven individuals are right-handed. The differences are no cause for worry, because the samples are small. If the evidence is to be read at face value, the murdered baboons stand as witnesses that handedness had already been evolved among these South African australopithecines, and that then, as now, the right hand was favored for weapon wielding. The chimpanzee shows relatively little preference in terms of handedness. The evidence suggests that specialization in unilateral tool use had been established much earlier than the age of the South African deposits. Thus both the reduction of the canine tooth and the hypothesis about handedness point to the probability that the human niche had been occupied for a long period of time prior to the Lower Pleistocene.

Some anthropologists still doubt that the australopithecines were capable of killing big game and favor the idea that they scavenged from the kills of the big cats. The Olduvai living floor and the baboon kills point in the other direction. The species of baboon found in these deposits is smaller and more generalized than living baboons, but must be viewed as having social habits of the same type. Modern baboons travel in bands ranging from 20 to 200 to provide protection against predation by carnivores. Baboons today are formidable animals whose long canine teeth can inflict serious wounds. A group of dominant male baboons is perfectly capable of driving a leopard from its kill, if the latter is one of their own group. If baboon hunting is hazardous for a leopard, it certainly would be dangerous for little men like the australopithecines. And yet they did systematically secure baboons as food.

Miocene grasslands began to spread extensively in both East and South Africa. The climate as revealed in the South African cave deposits indicates rainfall of about ten inches per year for the earlier deposits, increasing only to about fourteen inches per year in the later ones. This type of climate is dry by any standards and the country approaches desert conditions. The australopithecines obviously were well adjusted to it and so were the local baboons. In East Africa, baboons secure safety for the night by roosting in sizable acacia trees. In South Africa there were no such trees available. Today's baboons tend to hole up for the night in a cave or fissure in the face of cliffs. Baboons have been observed to awaken slowly and to spend a little time rising to a normal level of activity. Since it is inconceivable that the australopithecine could have successfully hunted baboons moving abroad in their large troops in daytime, it seems probable that they chose more favorable circumstances. One can imagine one or more australopithecine men quietly taking their stand outside of the opening of the baboons' night shelter. If the entrance was narrow, the baboons would emerge more or less in single file and so be vulnerable to clubbing. Prudence would suggest that such hunting was not done by single men, but rather in small groups, so that the retreat of the successful hunter could be covered by other unencumbered australopithecines.

The clubs used by the australopithecines to murder baboons have in all likelihood been identified. The depressed fractures of the baboon skulls frequently contain double parallel depressions within their general boundaries. Dart noticed that one of the most frequent bones in the australopithecine deposits was the upper bone of the foreleg of a zebra. Such a humerus, with the two ridges at its far end, made an effective weapon. Many of these bones were broken in midshaft as though from heavy use, and their double ridge ends frequently showed batterings that had occurred before fossilization. The evidence is purely circumstantial, but it seems likely that Dart is correct in this matter, and that they were used as clubs.

The bones of the australopithecines themselves show evidences of violence. Sometimes the top of the skull contains depressed fractures not unlike those found on the baboons. Another time the injury seems to have been inflicted with a round, pointed weapon. Some years ago when talking about the problem with Professor

Dart, I asked him what proportion of the australopithecines he thought had been murdered. "Why, all of them, of course," he replied. The fragmentary nature of their remains, combined with the visible variety of fatal injuries, suggests that his estimate is not unduly exaggerated. The data on the aging of the available series of australopithecines show that more than half of them belong to immature individuals. Other evidence based upon sequence in timing of molar tooth eruption indicates that the australopithecine may have been slow growing, somewhat in the fashion of modern men. These data indicate that growing to adulthood was something of a triumph in terms of survival. No wonder that men living in simple cultures considered that their elders were repositories of wisdom and listened to their words. The available evidence suggests that in addition to some scavenging and big game hunting, these early direct ancestors of ours also indulged in murder and gustatory cannibalism.

Some Speculations About Australopithecine Society.   The social life of the australopithecine is an attractive area for speculation. A good many anthropologists have taken living baboons as models representing the type of social organization which must have been followed by the australopithecines. Much of their reasoning is based on the fact that both primates were ground-dwelling and in the open grasslands of Africa lived amid formidable carnivores. The body size of the australopithecine was sufficiently large, so that they didn't need to fear the smaller carnivorous animals. While it is true that the band structure of baboons may tend to preserve them from undue predation by lions and leopards, they need not be the most realistic model. Let us turn back to Jane Goodall's data on her chimpanzees. Their social units were loosely structured and showed nothing very close to the tight organization of baboon troops. They lived in a relatively open environment but did not act as though they feared leopards and to her knowledge were not attacked by the big cats. There have been speculations that man, and possibly the other ground-living grade apes, have evolved characteristic odors which make them unattractive prey for carnivores. No adequate test has been made of the idea, and so the nature of the factors allowing survival remains in doubt. But in a number of areas even today native people live at peace with dangerous carnivores and are very rarely attacked by them. This seems true of rain forest peoples in Southeast Asia, where tigers are relatively numerous and man's weapons are ineffective. Everywhere the literature indicates that except in the cases of old or injured felines, they are not man-eaters and go their own way.

Density is one of the critical points to be taken into account when speculating about australopithecine social organization. With the assumption of a partially carnivorous diet these early men moved into a niche which limited their numbers. Whereas baboons today may have an effective density of ten animals per square mile in East Africa, the best of human hunters only approached one per square mile. Allowing for the relative inefficiency of australopithecine techniques of hunting and gathering, they would not approach the latter figure. In essence, these early men would have been rare animals in an African grassland environment which carried more meat on the hoof than any other region of the world. Carnivores of all kinds specialize in their hunting and show marked preferences for certain kinds of prey. These are usually animals which are very numerous and which pose very little threat when assaulted. The comparative scarcity of the australopithecine may have given them relative immunity from predation in Pleistocene Africa. When these qualities are combined with weapon use and an increasing intelligence, we may even guess that they could show considerable cunning in terms of living in an environment filled with effective predators. Indeed, we may go back to considerations of energy in the food chain and recall that predators in general do not eat other predators. Thus while we probably shall never define the type of social organization which characterized our distant ancestors, it can be viewed as more likely to have evolved from uni-male units as found in several terrestrial primates.

Because behavior itself does not fossilize, reconstructions of the way people lived in the past probably always underestimate their capabilities. Only in recent years have evidences of types of shelter used by early man been dated earlier than seventy thousand years ago. But to find that early hominids built themselves shelters was unexpected. The Leakeys uncovered an arrangement of stones on the living floor at the bottom of Olduvai Gorge which is best interpreted as a shelter of some type. The site was a stony point reaching out into an ancient lake, and the rocks utilized were obtained from it. They were arranged in a rough semicircle about a dozen feet in diameter around a saucer-shaped area. Totaling some hundreds of individual pieces, this rock pile obviously represented some sort of capital investment in time by these little men. The rock arrangement most probably served as some type of windbreak to moderate the chill of night breezes. Whether it was further surmounted by a wall of thorn brush is, of course, conjectural. The size of this "archeological feature" suggests, if it is compared with the windbreaks used by living Australian Aborigines, that it is of the size to provide sleeping accommodations for a family of four or five rather than a band of 25 or more early men. Perhaps with some further future discovery, in which function can be more accurately identified, it may be possible to make more accurate estimates about the normal group size of the australopithecines.

The conclusion that the australopithecines must have been relatively rare animals requiring a moderate number of square miles to support each one raises the question of why the human niche which they occupied should have been advantageous to enter in the first place.

Miocene pongids, living in forested areas, very likely as densely as today's chimpanzees, must have enjoyed a diet including a wide variety of fruits and vegetable products. Why should such well-entrenched pongids have deserted a good environment to enter a risky life on the open savannas of Africa? Climatic change may hold some of the answers, for in general the Miocene was a period of progressively drier climates in which the grasslands of the world expanded rapidly to replace the retreating forests. It has even been suggested that the human way of life grew out of some dry-opithecine's attempts to continue his forest life. This imaginary model visualizes the ancestors of the ramapithecines scuttling from one forest patch across the grasslands to another in order to continue their accustomed diet in their normal habitat. The idea is not improbable. The availability of slow game and of newly dropped young of the herbivores may have initially provided the increase of meat in the diet which seems to have set man apart from the other hominoids.

In the wet forests of the tropics the seasons pass with no marked changes in the climate. Fruits ripen throughout the year, for even in a single species of tree the fruiting period is not rigidly fixed. Since tropical forests contain an enormous variety of plants, there is always suitable food for the big primates. But in the open country of the grasslands and savannas, seasonality is much more marked, and with the reduced rainfall, vegetable foods become abundantly available for shorter periods of time. Consequently, there may be considerable periods throughout the year when plant foods are relatively scarce. In this type of environment the addition of meat to the diet may increase survival chances even though it condemns the primitive hominid to relative rarity in terms of his density. The partially carnivorous diet of the australopithecine must be considered as an expensive one in energy terms. Nevertheless, by broadening the total range of diet, meat eating may have been a major factor in allowing primitive men to live almost everywhere, so extending the range by broadening their niche.

We assume today that the ramapithecines were ancestral to the much later australopithecines among the early people. Hence it is useful to use the more extensive data on the latter to help reconstruct the missing structure and lifeway of the ramapithecines. Since

Feedback to the Ramapithecines

we know the australopithecines had nearly but not quite completed the evolution of erect posture, it is perfectly reasonable to assume that the ramapithecines living some ten million years earlier were less advanced in this respect. The reduction of their canine teeth suggests that they had entered the human niche, but one might predict that the structure of their pelvis and feet, when recovered, will be less advanced than that of the australopithecines. Perhaps they will prove intermediate structurally between dryopithecines and australopithecines. Since the three largest of the great apes make nests either in the trees or on the ground, it is likely that dryopithecines and ramapithecines would behave in some similar fashion. After all, the circle of rocks on the Olduvai living floor need not have been the first attempt at human shelter. Since the Oldowan stone tools of the australopithecine are so crude, and indeed barely recognizable as man-made, it is unlikely that the much earlier ramapithecines would have formed tools which we can recognize today. Various forces in nature fracture rocks, and a rare one of these will almost look as if it had been fashioned by man. So the earliest stages of stone tool-making will probably remain unidentifiable. Tool use by the ramapithecines may have to rest upon the evidence of its occurrence among chimpanzees on the one hand and the presence of recognizable tools in the hominid deposits on the other. Or alteration, through use, may inform us.

Bibliography

BARTHOLOMEW, GEORGE and J. B. BIRDSELL.
  1953   Ecology and the Protohominids. American Anthropologist. Vol. 55, No. 4, pp. 481–498.
CLARK, W. E. L.
  1967   Man-apes or Ape-men? New York: Holt, Rinehart, and Winston.
DART, RAYMOND A. with DENNIS CRAIG.
  1959   Adventures with the Missing Link. New York: Harper.
GAUSE, G. F.
  1934   The Struggle for Existence. Baltimore: Williams and Wilkins.
HARDIN, GARRETT.
  1960   The Competitive Exclusion Principle. Science. Vol. 131, pp. 1291–1297.

JOLLY, CLIFFORD.
  1970   The Seed-Eaters: a New Model of Hominid Differentiation Based Upon a Baboon Analogy. Man. Vol. 5, pp. 5–26.
ROBINSON, JOHN T.
  1961   The Australopithecines and Their Bearing on the Origin of Man and of Stone Tool-Making. South African Journal of Science. Vol. 57, pp. 3–16.
TOBIAS, PHILIP V.
  1967   The Cranium and Maxillary Dentition of Australopithecus (Zinjanthropus) Boisei Olduvai Gorge, Vol. 2 Cambridge, England: Cambridge University Press.

The Evolution of Increased Reproductive Efficiency

In nature, communities of animals and plants are generally considered to be in a state of equilibrium or balance. This condition refers to the long-term state of balance and does not require that the numbers of a given species of animal or plant remain constant over short periods of time. Indeed, the members of any natural community are constantly changing in number, but if the situation is stable, these variations tend to oscillate gently around an optimum value for each species, which is determined by the nature of the community as a whole. Changes do occur but when the numbers of a species are disturbed, there are forces acting to return them to the normal optimum position. Such systems are called *stable equilibrium systems* in mechanics, *homeostatic systems* in physiology, or *systems involving negative feedback* in communications and related fields.

A stable equilibrium model, Figure 10–4, shows the numbers of a given species of animal in a community of fixed area plotted against time. In the initial stages of the diagram the numbers of the animal oscillate somewhat erratically about a constant optimum denoted by the dotted trend line. Their degree of departure from the optimum is not great, but in each case restoring forces are shown which tend to return the numbers to a proper value for the system. A great variety of forces tends to depress numbers when they are

Figure 10-4
Dynamic Balance in Natural
Communities

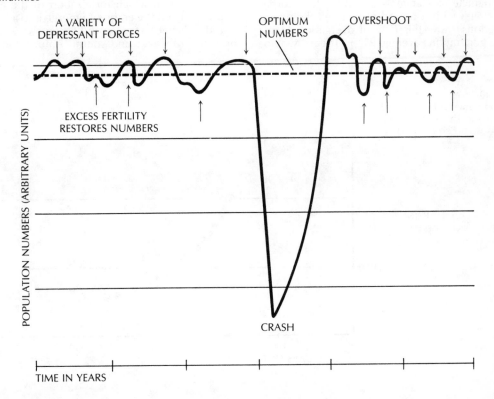

excessive and their nature depends upon the species considered. When numbers fall below the level which the community will support, the built-in excess fertility of the animal suffices to rapidly restore numbers to the values proper for the system. As mentioned earlier, all successful species not only can but tend to produce greater numbers of offspring than a given environment can support.

About midway in the diagram a population crash has been plotted. Perhaps due to drought or disease, the total numbers of the group in the community are much reduced compared to their optimal value. At the bottom of the crash the species is in danger of locally going to extinction simply because of its small numbers. Populations reduced to this extent may be wiped out by a variety of random processes or succumb to directed ones, such as adverse selection. Hence a population crash represents a crucial stage for the group, and it is essential that its numbers be restored to proper values as quickly as possible. As excess fertility builds up group numbers at a geometric rate there is a tendency to overshoot the optimum value briefly until restoring forces depress numbers to a proper level. This particularly characterizes animals of moderate to large size—sheep with a period of one year between generations show this type of overshoot. Man, whose generations average 25 to 30 years, would

also show it under these circumstances. Once the restoring forces reduce numbers to the proper value, and perhaps even a little below, the species again enters the stable, gently oscillating condition of equilibrium. It is important to recognize that *homeostatic conditions in nature do not involve unchanging numbers, but rather oscillations around the level of the most suitable numbers.*

Among the ways in which progressive evolution can be measured is one which evaluates the amount of wastage in reproduction at various levels of organization. If we look at the whole of animal life, the range represented by reproductive activity is so great as to suggest that perhaps fertility is unorganized and even may lie random in its pattern among animals. This view has been argued by a few very eminent scientists, but it is not the case. There is a general principle in evolution which might be stated as follows: *the rate of reproduction of animals tends to directly reflect the risks of life they face at all stages, from egg to reproductive adult, and is selected to maintain stable population numbers under normal conditions, while yet allowing rapid restoration of numbers in crisis.*

In animals in which fertilization of the egg is haphazard and the embryos emerge into a hostile world, it is necessary to produce very large numbers of both eggs and sperm. Among those in

Figure 10-5
Relation between Fertility and Longevity in Three Species of Deer Mouse (*Peromyscus*)

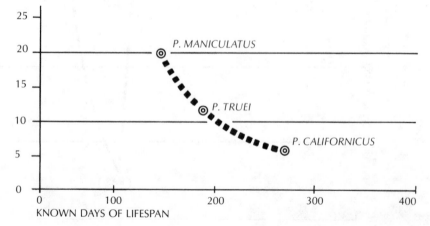

NUMBER OF OFFSPRING PER BREEDING FEMALE PER SEASON

KNOWN DAYS OF LIFESPAN

which internal fertilization increases efficiency but the newborn young receive little or no parental care, the numbers required to maintain the species in balance are reduced but still remain large. In species in which egg fertilization is efficient and the young receive some maternal care in an environment not overly hostile, the numbers needed to maintain a balanced species are much reduced.

The fact that evolutionary processes tend to produce the right relationship between fertility and the species' need for replacement is illustrated in Figure 10–5. There the significant data have been plotted for three different species of the deer mouse, *Peromyscus*. These are closely related species which live in different geographical areas in which the risks of living are sufficiently different so that their average life span varies greatly. Accordingly, they have evolved different rates of reproduction as measured by the number of offspring per breeding female per season. The rate of reproduction is affected both by the number of litters born per year as well as the number of young in each litter. The shortest-lived race, *P. maniculatus,* has a known life-span of 152 days, and each breeding female produces 20 young per season. Females average four litters per year, with five in each. The intermediate species, *P. truei,* has a known life-span extending to 190 days, and each female of the species on the average produces 11.66 young each year. Finally, in long-lived *P. californicus,* which lives in the most equable climate, the life-span rises to 275 days, and the number of offspring per breeding female per season has dropped to 6.21. As the trend in Figure 10–5 indicates, there is a tightly structured relationship between the length of life and the reproductive activity of these three closely related species of deer mice. This example may reasonably be extended to other species of animals on the grounds that excessive fertility, over that needed for the crisis periods when survival is threatened, would be wasteful, and, on the other hand, insufficient fertility would quickly lead to their extinction. Thus reproductive activity in itself represents a variety of factors integrated to produce a stable equilibrium system.

Let us examine a wide variety of animals to relate the magnitude of their reproductive effort with the kind of lives they live. We shall assume in all instances that we are dealing with stable equilibrium systems so that among the bisexual animals in which we are interested, each pair of reproducing adults in their turn will ultimately be replaced by only two of their offspring. If more offspring survived on the average the population would be expanding in numbers and the community be unstable. If their numbers were declining the species would be on the way to extinction and again not in a stable equilibrium situation. The ratio which expresses the reproductive risk of a species can be constructed by taking the total number of gametes or zygotes produced by a female during the average number of reproductive years characteristic of the species. Thus, if a female in species X on the average produces 100 offspring, the *ratio of reproductive risk* is 2 divided by 100, or 1 in 50.

The potential rate of increase among some marine invertebrates which produce free-swimming larval forms is so great as to involve very large numbers. But we are primarily interested in human evolution and so will examine the reproductive risk ratio as it occurs among the vertebrates. The figures used are necessarily approximate, for while egg counts can be taken with reasonable accuracy, the average life-span of the breeding female in nature is a very difficult figure to come by. Even so, the differences among the various groups of vertebrates are sufficiently great so that the trend, when plotted in Figure 10–6, gives clear-cut results.

Let us begin with a sizable marine fish, the codfish. A mature female cod lays as many as six million eggs in a single clutch. If her life expectancy in terms of breeding years is estimated as averaging five years, each female cod on the average will produce 30 million eggs. In a stable oceanic community only two of these will reach their own middle years of effective reproductive adulthood. It is easy to calculate the reproductive risk ratio, which is that 15 million eggs are required to produce one effective adult. This ratio indicates an appalling rate of reproductive waste. But codfish have not inherited the oceans of the Earth and are merely represented in reasonable numbers in their own oceanic communities. Hence it is important to inquire why so many eggs are needed to maintain a population in essentially stable numbers. In the first place, as the female lays her eggs in the continental banks of the north Atlantic, they are externally fertilized by the milt, or sperm, ejected by the male. I know of no figures to indicate what proportion of the eggs escape successful fertilization in this method, but in the open waters of the ocean it must represent a moderate minority. The eggs that do become fertilized are left untended and so immediately become subject to being eaten by a wide variety of predators for whom codfish eggs are tasty morsels. After hatching, the tiny codfish fry are free-living and active but still subject to prolonged and systematic predation by a wide variety of hungry, free-swimming animals. During the entire period of growing to sexual maturity, codfish are subject to the toll taken by a wide variety of marine predators. Since 15 million cod need to be born to provide a single functioning reproductive adult, it would almost appear as though codfish are produced to serve as food for other species. This is not quite true, but it would seem these oceanic fish face circumstances that make their reproductive activities appear vastly inefficient. In fact, they live in a nicely balanced system and hence they need the total number of eggs they lay.

Life in the ocean involves much greater risks for young fish than life in fresh water. The number and kind of predators are reduced in fresh water, hiding places offering protection at all stages of life are more frequent, and so in general mortality figures are much reduced. A few examples will illustrate the point. A large fresh-water trout will lay 25,000 eggs at each spawning period, that is, once a year, and it can be assumed that mature breeding females survive an average four

years in virgin waters without fishermen's pressures. This gives each one an output of 100,000 eggs, of which two must grow to the middle stages of reproductive maturity. Here the reproductive ratio falls to one in 50,000, a very much better figure than that of the oceanic cod. But the eggs of trout in a stream run greater risks than in some other freshwater environments. Let us consider a sunfish, which lays about 1,000 eggs at a time, and assume that each female succeeds in doing this for two years on the average. This gives a total egg output of 2,000, and so a reproductive ratio of 1 in 1,000, or 50 times better than that for the stream trout. Both species are well adjusted in terms of their numbers in the total aquatic communities. In both cases fertilization is external, so that some reproductive inefficiency arises from this cause. But the trout lays her eggs in the gravel shoals of the swift-running stream, whose volume of waters may vary with the vagaries of the weather. Furthermore, there are a variety of fish that love to eat trout eggs and exert their best efforts to do so. Finally, when the young trout are hatched, they must take care of themselves by their inherited and coded behavior in a highly variable and rapidly moving environment. Trout need the number of eggs they lay. Sunfish, on the other hand, live in the still waters

Figure 10-6
Development of Increasing
Reproductive Efficiency

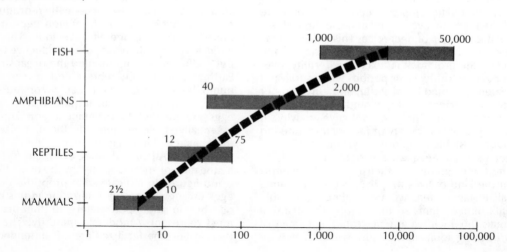

of lakes and spawn in their shallows. They are pugnacious little fish and guard their egg nest aggressively. Both in terms of the stability of their watery environment and in the nature of their elementary parental care of the eggs, the risk ratio of the sunfish is much reduced compared to that of the trout. Even so, a risk ratio involving 1,000 eggs to raise one adult sunfish still seems wasteful from a purely numerical point of view. But it represents the level needed to maintain the numbers of the species stable with a sufficient extra margin of fertility to allow them to survive through environmental crises.

Let us climb the vertebrate scale to the amphibians. These are animals still chained to moist environments to carry out their reproductive activities. While some amphibians lay their eggs in the waters of lakes or streams, others have evolved a wide variety of alternative methods to lower the risk of reproduction. The toad represents one of the amphibians which usually returns to water to lay her eggs. In some species these number 2,000 in each clutch and are laid once a year. If we assume on the average that a female breeder lives for two years, we have a risk ratio of 1 to 2,000 representing the level needed to maintain a constant population. On the other hand, some of the salamanders lay no more than 40 eggs in the clutch. If we assume that breeding females on the average live to produce two annual clutches, the eggs total 80 in number and give a risk ratio of 1 in 40. These figures are very much better than those of the fish discussed above, although those for the toad are more wasteful than those for the nest-guarding sunfish. It is of some interest to see where the greater efficiency arises in the amphibians. First of all, fertilization is more efficient among them since the male positions himself on the female so that his sperm are ejected onto the eggs as they stream from her body. This is still external fertilization, but it must reduce the risk of some eggs not being fertilized. Further, those amphibians which lay their eggs in water generally enclose them in large gelatinous masses of material which tend to deter the smaller predators which would eat them if the eggs were lying free in the water. This kind of protective cloak must reduce the chances of egg destruction considerably. Many of the salamanders lay their eggs in moist places, as in rotting logs, and these semisecret places on land minimize the chance of being discovered. Since amphibians show no cases of well-developed parental care, the improvement

in their reproductive risk ratio must be largely credited to their improved method of external fertilization and to the variety of ways in which the eggs are made more secure until hatching. Once hatched, the larval amphibians are subject to predation, but they are sufficiently numerous to perpetuate the species.

Since the reptiles represent an evolutionary advance over the amphibians, it is to be anticipated that their method of reproduction would be more efficient. Among generalized terrestrial reptiles, there are a number of small lizards which range from six to twelve inches in length, and their average clutch of eggs is about eight. The average female is reproductively effective for three years, and has a total egg output of 24 eggs; thus the reproductive risk of her species would be about 1 to 12. Larger lizards of the monitor type average about 25 eggs per clutch and are presumed to lay six clutches in their lifetime. This would be a total egg production of 150 eggs, giving a reproductive risk ratio of 1 to 75. This represents a considerable advance over the amphibian grade of organization. Certainly a part of this improvement in reproductive efficiency involves internal fertilization, which is universal among the reptiles and insures the fertilization of all of the female's eggs. There is very little or no parental care among reptiles, but in general they choose advantageous places in which to hide their eggs, which must remain dry during their entire incubation. These two features taken together presumably suffice to explain the improvement in reptilian reproductive efficiency as compared to that of the lower vertebrates, the amphibians and the fish.

Mammals are correctly regarded as the apex of evolutionary development. Certainly their reproductive methods are well ahead of the lower vertebrates. If we limit our discussion to placental mammals—those in which the young are carried in the uterus of the mother and receive their nutritional and respirational needs through the placenta attached to the uterine wall—then we do find a very real improvement. Small mammals tend to produce many more young than large, long-lived ones. If we go back to the deer mouse discussed earlier and take the shortest-lived of the three species, then we may equate the 20 young produced during a breeding season with their total output, since the life-span of that mouse was about 120 days. This little mammal gives a risk ratio of 1 in 10, which is rather high but even so is better than most reptiles. A medium-sized mammal, a carnivore such as the coyote, produces four pups per litter, and we can assume that she will live for three reproductive years. This gives a total pup count of 12 and so a reproductive risk of 1 to 6.

Now let us turn to man himself, the supreme mammal. The figures from modern populations in Western civilization are misleading, for those populations are not in a stable equilibrium, but constantly increasing every year. Perhaps the highest fertility recorded among such expanding populations is that among the Hutterites, a sect of thrifty farmers living in western Canada. Careful studies show that among these people the average woman who has completed her child-bearing period has produced 10.7 children. This is a world's record in human fertility and of course results in enormous rates of expansion for the Hutterite population, both absolutely and relative to other human groups.

But for our purposes we must seek an appropriate population remaining constant in number and in equilibrium with their total envi-ronment. The Australian Aborigines are such a people and there are good data on their fertility. Details of their reproduction are discussed in later sections, but here it is enough to indicate that each child must be nursed for a period ranging from three to four years after birth. Since breast feeding does not always prevent a new pregnancy in the nursing mother, some sort of artificial method of spacing is needed. Most desert Aborigines wander ten miles a day in search of food, so that the women, who carry most of the family equipment on their head, can carry and nurse no more than one young child at a time. For these two adequate reasons the Aborigines space the arrival of the children very carefully by resorting to infanticide when necessary. Under such a system the average Australian woman may bear more children, but only attempts to rear about five of the total number. This gives the very low risk ratio of 1 to 2.5. This is as economical a reproductive pattern as will be found among the higher mammals, and so among all of the vertebrates.

CHAPTER 11

MEN
OF THE
LOWER AND
MIDDLE
PLEISTOCENE

The evolution of the early hominids began in the Pliocene. By the Upper Pleistocene modern types of men had become widely established. In the three million years of the Lower and Middle Pleistocene there was a complex transition between those two grades of humanity. Then populations of men showed greater variation by region and in time than before or after, and so different grades of hominids coexisted over long time periods. The story is still incomplete and has recently been revolutionized by Richard Leakey's discovery of ER-1470. No new evolutionary synthesis has yet been achieved.

The Three Grades of Human Evolution

There is a normal tendency in the natural sciences to build systematic interpretations of data and to ground these in the established principles of the field. And this is as it should be. Paleoanthropologists, or those who deal with fossil man, have long wrestled with their data, which is spotty in both time and space. They have generally agreed on a threefold scheme for the evolution of man. The earliest *grade,* or structural level, consisted of the australopithecines, and more particularly the lightly built ones, *Australopithecus africanus.* This level involved a bipedally erect hominid with a small cranial capacity, now known to be considerably less than 500 cc., who was apparently able to both use tools and hunt sizable game. The second structural grade consisted of *Homo erectus,* the kind of fossil man found extensively in most of the Old World in the Middle Pleistocene. Structurally these pithecanthropines were also erect bipeds. Their cranial capacity averaged around 1,000 cc., and they were bigger-bodied than their predecessors. These men were big game hunters and fire-users, and generally their stone tools were more advanced than the preceding australopiths. The better known series of the pithecanthropines come from Java and China, with a few isolated instances in Africa and the Middle East. Finally, the third grade of human evolution, that of *Homo sapiens,* arrived everywhere about 30,000 years ago, and there are troublesome suggestions that they may have been considerably earlier in some regions. In recent years it has been common to include classic Neanderthal man in this category even though morphologically he is very different from living populations and must be classed as a separate subspecies of man.

This threefold scheme for human evolution has served its interpretive purposes well, and indeed much of the fossil data fit into it conveniently. It is even possible to plot the cranial capacities as they are estimated for the fossil skulls of the Pleistocene and to calculate a convincing trend through time. Statistically this is a satisfactory exercise, and the results seem to confirm the threefold scheme. This approach emphasizes that brain capacity increases, consistently in time throughout the Old World during the Pleistocene. It would tend to make human evolution appear as a very regular process with few or no interpretive problems left for the future.

The Problem of ER-1470

Generalized schemes are subject to upset and the one described above is no exception. The new find that makes a different game of the interpretation of fossil man is known as KNM-ER-1470. This remarkable skull which is shown in Plate 11–1 is sufficiently complete so that its general evolutionary status is evident. It is a surprisingly modern-looking man to have come from the Lower Pleistocene and the horizon that is probably 2.9 million years old.

The discovery of ER-1470 culminates several years of field work by Richard Leakey, one of the sons of the late Louis Leakey who worked with such great success at Olduvai Gorge nearby in East Africa. The arid country to the east of Lake Rudolf is forbidding in appearance and difficult for field workers. Yet in the brief space of six years

Plate 11-1
Cranium KNM-ER-1470, a new *Homo* from East Rudolph, from beds 2.6 to 2.9 million years old.

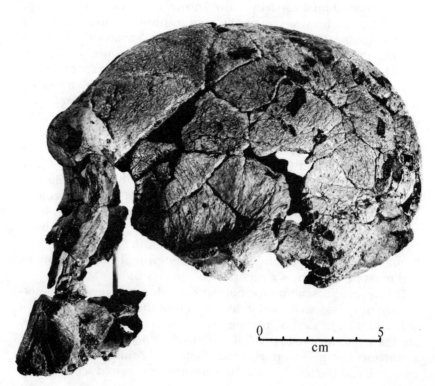

```
0                           5
      cm
```

Richard Leakey and his field workers have found fragments of more than 100 hominids in a relatively small area. Since these largely represent surface collecting, the East Rudolf beds must be considered both the richest and the oldest in which extensive human remains have been found, and many remain to be discovered. Along with the advanced hominid represented by 1470 there is a series of skulls and limb bones of a robust australopithecine very much like *Australopithecus boisei* discovered at the bottom of Bed I in Olduvai Gorge by Louis Leakey. There is clear evidence of long coexistence between the australopithecines and more advanced men in this region, and somehow they avoided Gausian conflict for a period of no less than two million years. This date is in some dispute, but the discovery of even older forms of *Homo* in late 1974 in the Afar region of northeastern Ethiopia at a date older than 3.1 million years tends to validate it. The new finds, made by Carl Johanson, consist of a complete upper jaw, half of an upper jaw, and a half-mandible, all with their rather small teeth intact.

Cranium 1470 is clearly a *Homo* and not an *Australopithecus.* The revised estimate of its cranial capacity of 780 cc. is well outside the range of the australopithecines. The form of this skull is like more modern people rather than the australopiths. It differs from the latter in having its foramen magnum, the place where the spinal cord enters the skull, relatively far forward, and hence its head was better balanced on the vertebral column. The vault is rather smoothly rounded and without marked muscular relief. The brow ridges are relatively small. When viewed from the rear the base of the skull remains relatively broad, as in the australopiths, but the sides of the vault rise almost parallel, approximately as in modern men. The constriction beyond the forehead is moderate compared to the australopithecines. The occipital bone has no crest for the attachment of powerful neck muscles. Unlike the australopithecine the nasal bones show some relief in this face so that it appears less "dished." All of the teeth are missing from the upper jaw, but the palate can be described as shallow, short, and broad.

Coming from a higher level, and so at a somewhat more recent date, is a nearly complete femur with portions of a tibia and fibula. All three bones come from the left side of an individual, and suggest an accidental burial of some kind millions of years ago. The thigh bone is slender, straight shafted, and together with the other two leg bones show characteristics resembling those of modern man. This individual's stature is estimated to have been 5 feet. Many modern hunters are no taller than this.

From the very nature of its characteristics, cranium 1470 does not seem to fit the standard scheme of the three grades of human evolution. But before accepting this find as truly exceptional, it must be carefully tested against known materials. Compared to con-

Plate 11-2
Views of ER-1470

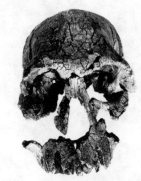

(a) Front View. The reconstruction involves many small pieces which fit together accurately.

(b) Rear View. The sides of the vault are nearly parallel.

(c) Top View. The brow ridges are moderate-sized by modern standards.

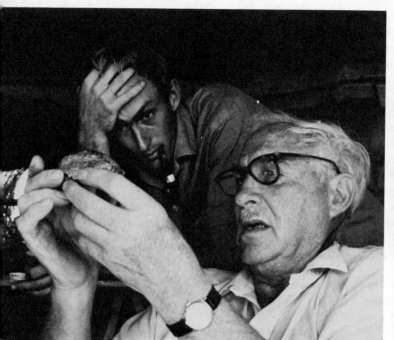

Plate 11-3
Louis Leakey and his son study
the fossil skull of a monkey in
East Rudolph.

Plate 11-4
Richard Leakey with a robust
australopithecine found on the
surface at East Rudolph.

temporary finds, 1470 is very much larger in cranial capacity, much
more modern in general form, and totally out of place at a date
somewhere between 2.6 and 2.9 million years ago. To put it another
way, to find the next human skulls of comparable capacity one must
go to Java where in the Trinil bed two of the *Homo erectus* crania are
smaller than 1470, and five are somewhat larger. These skulls are
dated at least two million years later in prehistory. Further, the older
pithecanthropines of Java, represented by eleven crania of Solo
men, and the pithecanthropines of Chou Kou Tien in China show
some individuals larger than 1470, but none as modern in general
morphology, even though they lived only about 300,000 years ago.
Clearly ER-1470 is a very unusual skull for its surprisingly early date.

   Paleoanthropologists and anatomists frequently test a new find of
fossil man to see whether it falls "within the normal range" of some
other fossil population. The test is seldom more than a simple
quotation of the minimum and maximum values, and if a fossil falls
within these it is often considered to belong to the same group.
Actually this is not a proper test to apply within a single evolving
species, for ranges of dimension overlap enormously between
populations which are different in time or in space. The real
question in the case of 1470 is whether the discovery should be

considered as representative of its population or as an extreme variant within it. This requires some sort of probability distribution, and its discussion is put into a brief supplement at the end of this chapter. The estimates do not indicate that 1470 should be classed with any other contemporary group.

It is worth asking what the sex of cranium 1470 might be and what effect differences in sex would have on the problem. The skull is so modern in its general form that it is very tempting to consider that it is indeed a female. In this case it can be estimated that an average male from the same population would be about 55 centimeters larger, or have a cranial capacity of 835 cc. This places the population from which we have this interesting sample even higher on the evolutionary scale and so creates a greater problem because of the very early date at which it lived. It cannot be made to disappear within the normal range of any other known early human population.

Richard Leakey has provisionally attributed 1470 to the genus *Homo* but has not designated its species name. This is probably a wise procedure. His late father, Louis Leakey, recovered no less than four quite advanced hominids from Olduvai Gorge and labeled them *Homo habilis,* or "Handy Man" because of the initial association with tools. These habilines have not received widespread acceptance among students of fossil man, for there has been a feeling that they are little more than the gracile australopithecines found in East Africa. This conservative view has been based upon the fact that none of the skulls were very complete and so full comparisons with the australopiths were handicapped. Nonetheless estimates of cranial capacities for four habilines averaged 637 cc., or nearly 50 percent more than those for the gracile australopithecines. This difference is certainly significant and it is likely that the habilines will receive wider acceptance in the future. Certainly now that 1470 is available for study, the Olduvai habilines look very much like the same sort of hominid. But there still remains a problem, for Louis Leakey's finds date from about 1.8 to 1.6 millions of years in age and so are about a million years younger than 1470. Furthermore, they average nearly 150 cc. less in cranial capacity. Even though these difficulties remain unresolved at this time, it is tempting to lump all of this group together thus separating them from the australopithecines on the one hand, and from the later pithecanthropines on the other. One of the so-called habilines of less than average cranial capacity has rather large brow ridges, and it has been suggested by some that it represents a transitional stage in evolution toward the pithecanthropines of later time.

Fortunately there is some evidence of the tool-making capabilities of men at this general level in time in Africa. A living floor site has been excavated at East Rudolf by Glynn Isaac and is dated at about 2.6 million years ago. The stone tools are characteristic of the

Oldowan culture but coming as they do from an undisturbed and extensive deposit show greater variety than previously known. In addition to choppers and chopping tools of the core type, there were a great many flake tools in a large variety of shapes, some even so long and narrow as to be classified as blades. The skill of the stone work at this site is surprisingly high. At the bottom of Olduvai Gorge Bed I Louis Leakey found a ring of basaltic stone which has been widely, and no doubt properly, interpreted as a hut ring. Even today in parts of Africa huts have series of stones piled around their bases to give them stability. The inner diameter of the Olduvai stone ring is about eight feet, or size enough to accommodate a small group, perhaps a family, of habilines. With the definite association of the Oldowan industry with 1470 and habilines at Olduvai, the question is reopened as to whether the australopiths, both gracile and robust, also made stone tools in the same tradition. The tools have been found with these smaller-brained men, both in East Africa and in South Africa. There is no reason why simple stone-working could not have been conducted by several kinds of hominids in the same space and at the same time. But with the discovery of 1470 and the validation of the category of the habilines, perhaps the question should be reexamined.

The *Homo Erectus* Grade of Human Evolution

The earliest forms of *Homo erectus* discovered, and still the best known ones, come from the Far East on the island of Java and from North China. They were originally labeled by their discoverers with the genus names *Pithecanthropus* and *Sinanthropus*. In 1950, Ernst Mayr stressed that these Asiatic forms and all later men belonged to a single evolving genus and so should be included in the genus *Homo*. The finds from Java and China were labeled *Homo erectus*, with the additional subspecific labels of *javanensis* and *pekinensis* in order to specify the regions in which they were found. Since the terminology of fossil man has been both changing and confused, we shall use the term pithecanthropine to refer to the grade *Homo erectus* while identifying individual finds by the name of their locality of discovery. Thus the finds from Java can be classed either as *Homo erectus javanensis*, or more generally as pithecanthropines, or yet colloquially as Java man.

The grade of *Homo erectus* is one of the clearly identifiable structural grades in human evolution. The crania from these people are characterized by big brow ridges, low long vaults, pinched occipital outlines, and cranial capacities which average around 1,000 cc., with some as low as 750 cc. and others as large as 1,250 cc. One of the distinguishing marks of the vault is that when seen from the rear, the sides slope inward from the base toward the top, giving the impression of an ill-filled skull. The bones of the vault, face, and limbs are unusually thick for reasons that are not clear at this time. These pithecanthropines primarily occur in the Middle Pleistocene and they seem to have been the dominant form of men in their time. It is an interesting point that they are not as advanced-looking as ER-1470.

The discovery of the original Java man is one of the most romantic episodes in the whole unfolding story of human prehistory. It begins with the famous German naturalist, Ernst Haeckel, speculating about a "missing link" which was conceived of as being a common ancestor to both men and to the living apes. He named his hypothetical form *Pithecanthropus alalus,* or the ape-man without speech. Darwin had suggested Africa as the place to look for human ancestors on the grounds that the chimpanzee and gorilla lived there, but Haeckel thought that Asia would be the more likely place, since to him the skulls of gibbons appeared more "manlike" than those of the other great apes. His ideas inspired a young Netherlander by the name of Eugene Dubois to search for the "missing link." Dubois had studied medicine and had become a teacher of anatomy in Amsterdam. The young doctor gave up his career and signed on as a health officer in the Dutch Colonial Forces as a step toward attaining his goal. Dubois reasoned that of all the Dutch possessions, those of the East Indies would be the most suitable for his search. Accordingly, he asked for assignments in that part of the world and spent several fruitless years in Sumatra seeking early man. In 1889 he obtained a transfer to Java, where his position allowed him to devote much of his time to his quest. By the very next year Dubois had found a series of relics belonging to early Javanese man. His discoveries, which were made in a terrace of the Solo River in central Java near the town of Trinil, included a jaw fragment, some teeth, a complete femur, and most important of all, a skullcap of a very primitive type of man.

By 1894, Dr. Dubois returned to Europe as a celebrity. His published descriptions stirred up controversy which lasted for many years. He and his supporters considered that the first *Homo erectus* was really neither yet a man nor still an ape, but transitional between the two. Other authorities suggested that the finds were those of a true but primitive type of man. The great pathologist, Virchow, who had been wrong about the meaning of the Neanderthal skullcap, was wrong again, for now at an advanced age he proclaimed that these were the finds of a giant type of gibbon. It is ironic that Eugene Dubois, who was so nearly correct in his initial opinion, in later life changed his views to conform with the erroneous ones of Virchow.

Today, thanks to the dedicated efforts of a number of later workers, of whom a Dutch geologist, G. H. R. von Koenigswald, is the most important, the remains of these pithecanthropines now include portions of a total of eight skulls, a number of incomplete mandibles and other fragments, and a large series of teeth. All were found in riverine deposits of volcanic ash and mud flow which resulted from the active volcano on Java in those days. The deposits are divided into an upper series, known as the Trinil beds, and a lower series named the Djetis beds. These beds were formerly considered to be Middle Pleistocene in age, based on the fauna they contained, but the Djetis beds are now known to be much older. These pithecanthropine finds represent the earliest known forms of *Homo erectus,* and so they are of great importance as documents in the story of human evolution.

The remains from the earlier Djetis beds consist of the rear half and upper jaw of the skull of a robust male, the complete vault of a child about two years old, and a number of fragmentary mandibles. The infant's cranial capacity has been estimated at about 700 cubic centimeters, which is much less than a modern child of that age would have. The bones of its vault are rather thin; its forehead rises steeply as in living men. All in all the infant of Modjokerto had a surprisingly modern look about it. But the shape of infant skulls is misleading in many ways, as shown in Figure 11–1. There the age changes are diagrammed for gorillas, pithecanthropines, and modern men. Let us compare the skull of an infant gorilla with that of an adult male. These apes as babies look much more human than they do when grown. Admittedly the face is a little scooped out and the jaw retreating, but these features hardly foreshadow the monster into which the baby will grow. From infancy to adulthood brain size increases a little but not greatly. The most important change involves the eruption of the permanent teeth, which are enormous and require a vast increase in the size of the face. The heavy face and jaw and the ill-balanced position of the skull on the vertebral column require the growth of a great central crest to provide points of attachment for the chewing muscles, and a great flanging of bone around the rear of the skull to support the massive neck muscles which keep the head in proper position. The acceptable infant has grown into something of a nightmare adult. The changes indicate that the fossil skulls of children provide relatively little evidence about the adult populations from which they were drawn.

The modern appearance of the child from Modjokerto results from the fact that we have mentally compared it with the skulls of modern adults. Let us compare it with the skull of an infant of roughly the same age. Now the differences become apparent. The skull of the Javanese infant already shows the beginning of the development of brow ridges, the forehead slopes a little to the rear, and the skull is unusually low in proportion to its length. A modern child has a much more bulbous skull, with bulging forehead and

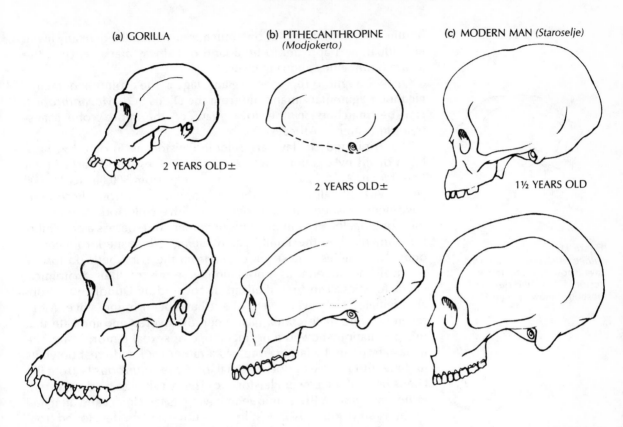

(a) GORILLA

(b) PITHECANTHROPINE
(Modjokerto)

(c) MODERN MAN (Staroselje)

2 YEARS OLD±

2 YEARS OLD±

1½ YEARS OLD

Figure 11-1
Age Changes from Infancy to
Adulthood in Three Evolutionary
Grades of Males

greater height relative to its length. If we consider the adult forms of
these two populations, it can be seen that each child dimly fore-
shadows what it was to grow to be. The restoration of the robust
male from the Djetis beds—in which the infant was also found—is
given to show what it would have grown to be. By modern standards
this adult pithecanthropine is relatively apelike, although not to the
degree shown by the australopithecines. Compared to a modern
fully grown male, the Java man is impressively inferior in terms of a
small brain, to which is hafted a face that is capable of carrying his
large teeth.

The restored male from the Djetis beds represents *Homo erectus*
(Figure 11–2). The vault enclosing the brain is low relative both to its
length and breadth. The maximum breadth of the skull occurs at the
level of the earholes rather than well above them as in modern men.
The brow ridges are very large, and the forehead slopes abruptly
backwards. At the rear of the skull a bar of bone, the nuchal torus,
projects to serve as an area of insertion for the heavy neck muscles.
The lower jaw is heavy in form and shows no chin. This Djetis

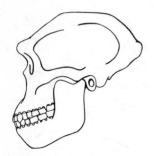

Figure 11-2
Pithecanthropus IV from Djetis, Java, shown as completely restored. Note the missing portions below in Plate 11-5.

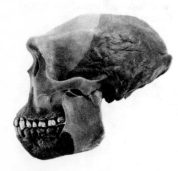

Plate 11-5
Pithecanthropus IV (*robustus*) as Restored by Weidenreich. The lighter portions are plaster.

hominid is a true man, but one much more primitive than any living individual, and appears to be a kind of halfway mark between the australopithecines and ourselves.

Once thought to represent a giant stage in the evolution of man, a gigantic fragmentary mandible from the Djetis bed, *Meganthropus*, can be matched in the lower jaw of the more robust australopithecines in Africa.

The Trinil beds in Java are Middle Pleistocene in date but have been dated by absolute methods for only a restricted part of their time range. Argon potassium datings range from 500,000 to 700,000 years. Virtually all of the pithecanthropine relics from these beds have been washed out and found by native collectors. Not only is their original location in the beds uncertain, but there is a possibility that they reached the Trinil beds by being redeposited from earlier ones. Nonetheless the seven crania from the Trinil beds do form a relatively homogeneous population in terms of their anatomical characteristics. The first skullcap discovered by Dubois was probably female, and several very like it have been found since. More recently some male crania have been discovered and indicate that this population showed a normal range of sexual dimorphism. The males are naturally larger, one has a rather rounded crest down its midline, but in no way does it equal the *Pithecanthropus* IV from the Djetis beds of the Lower Pleistocene. The last skull to be discovered, *Pithecanthropus* VIII, is unique in having its face intact and attached to the vault (Figure 11-3). While it has not yet been developed from the encrusting deposits of lime, its general proportions are such as to make this cranium look more modern, that is, less archaic, than the skullcaps alone seem to indicate. Judging from its general contours, this find is again probably a female, but it does throw rather a different light upon this Middle Pleistocene population.

Most of the Trinil crania have lost their basal portions in such a fashion as to suggest that they were murdered, and then their brains eaten. It has recently been suggested that this portion of the skull is thinner than the rest, and crania rolling down rocky river beds would be expected to lose just these sections. On the other hand, the fact that the faces are missing in all but one indicates that they were probably intentionally removed immediately after death. There are a number of thigh bones from the Trinil beds in the collection gathered by Dr. Dubois. They are straight in shaft, modern in general form, and certainly indicate they belonged to bipedal individuals with upright posture. But the evidence is not as good as it sounds, for these femora have recently undergone reexamination and according to gossip circulating about them, they lack the characteristically thick walls of pithecanthropines. Therefore they may well represent much later men whose thigh bones somehow became mineralized and were recovered in a situation where the native collectors attributed them to weathering out from the Trinil beds. In short they may not belong to *Homo erectus* at all.

Cranial capacity for these Javanese pithecanthropines ranges from as low as 750 cc. to as much as 975 cc. This is not notably so much greater than that of ER-1470 who lived in Africa nearly two million years earlier. When combined with the fact that the African skull is generally more modern in form than these pithecanthropines, an evolutionary problem is posed but not solved.

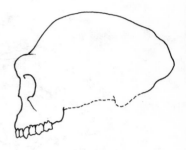

Figure 11-3
Pithecanthropus VIII, from Djetis, Java

An important Middle Pleistocene discovery from South China was found at Lantian in the form of a nearly complete but much crushed skullcap and face. It has been described as an earlier and more archaic form of the population found at Chou Kou Tien. It seems to have been earlier in date than that population, which has been estimated to have existed anywhere between 400,000 and 250,000 years ago. The Lantian skull, with its estimated capacity of about 750 cc., is said to be considerably earlier in time. A primitive mandible, not associated with it and found in another locality, is also attributed to the Middle Pleistocene. It will be recalled that a number of mandibles of *Gigantopithecus* have also been found at this time level in the same general region of China. It would seem that the giant open country ape was a contemporary of pithecanthropine men, no doubt coexisted with some form of *Homo* for many millennia, and perhaps went to extinction at their hand.

The record of fossil man in Java continues into later time and above the Trinil beds lie those known as the Ngandong. These deposits contain a new wave of mammal, apparently coming out of mainland Southeast Asia onto the Sahul Shelf. Most of the fauna are still living, although their modern descendants may have become reduced in size. These deposits have long been called Upper Pleistocene which would have made them anywhere between 10,000 and 200,000 years of age. In the last two years an absolute date has been obtained for them, and it has the very interesting value of 300,000 years plus or minus 300,000 years. The first figure represents the best estimate of the date, while the second figure represents the error on either side which the sample and the laboratory techniques might allow. In short these upper beds of the Solo River might be anywhere between zero and 600,000 years in antiquity, but the best estimate provided by the test is about 300,000 years in age. This puts the men contained in them back into the upper portion of the Middle Pleistocene. They are approximately of the same age as the pithecanthropines of Chou Kou Tien, and this is the population with which they should be compared. With this antiquity Solo man fits better into the chronology than he did previously, but there is no telling at this stage whether he left any direct descendants or merely passed on some genes via hybridization to later populations. At the present time the latter alternative seems the most likely.

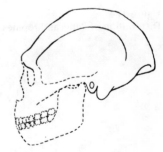

Figure 11-4
Solo Man—Face Reconstructed after Weidenreich

Figure 11-5
Distribution of Pithecanthropines

Men at the pithecanthropine grade of evolution were firmly established, and as far as we now know they were the only human occupants of East Asia in the Middle Pleistocene. Traces of them are found farther to the west, and it is important to try to estimate what the full range of this type of humanity may have been. There are no human fossils from peninsular India for the whole of the Pleistocene; this represents a very important gap in the total record of evolving humanity. Yet farther to the west in a plowed-up field of the Kibbutz Hazorea in Israel were found two types of flint industries, one a Mousterian type that is usually associated with Neanderthal men, the other a Middle Pleistocene fist axe complex. From the same undated deposits came two general types of cranial bones. The one that is of interest was an isolated occiput which in its dimensions and pinched contours fits very neatly into the category of *Homo erectus*.

Turning to Africa, Olduvai Gorge in the upper levels of Bed II yielded a fine skullcap of an exaggerated pithecanthropine type. Its capacity, estimated at about 1,000 cc., is normal enough, but the brow ridges are larger than ever seen before on a human skull (Figure 11–6). There can be no question that this places *Homo erectus* in East Africa in the Middle Pleistocene. In South Africa in the same cave of Swartkrans which yielded large numbers of bones from the robust australopithecines, there occurred two mandibles which were like those of more advanced men. Their teeth were smaller and of the general pithecanthropine type. The mandibles were shallow and the ascending ramus low. They could not have been part of an australopithecine's skull as we know it. Recently the reexamination of a museum collection has established that portions of a face and a frontal bone fit together, although there is no assurance they belong to the same individual. The resulting composite fossil has been classified as *Homo,* but again the subspecies has not been identified and the find is undated. There are differences of opinion as to whether this individual represents some form of habiline or possibly a very small-brained pithecanthropine. The skull is nowhere near as advanced-looking as ER-1470, although it is presumed Middle Pleistocene in date instead of Villafranchian. Tools found in this annex portion of the cave have been classified as developed Oldowan, and include some rather crude fist axes. As it now stands this find hardly clarifies the picture of human evolution in Africa.

From the area near Lake Tchad on the southern edge of the Sahara in Central Africa comes the frontal bone and face of another puzzling human fossil. It has not been reported upon as yet, but published photographs suggest that it is not an australopith, which lived this late in time, nor does it fall in the category of the pithecanthropines. One of the most unusual things about it is that the orbits for the eye are quite large for its moderate sized face. What this means is not certain.

The Mediterranean coast of Africa has provided evidence about Middle Pleistocene men which can only be described as confusing. The remains consist of a few odd bones, a number of more or less complete lower jaws, and some loose teeth. The most important of these are the discoveries at Ternifine near Oran, Algeria. There, in strata in a flooded sandpit, were found many tools which seemed to be of mixed traditions, since there were both Acheulian fist axes as well as many earlier types of choppers and chopping tools and a variety of flakes. The bones include three mandibles and one

Fossils of the Middle Pleistocene in Africa

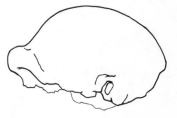

Figure 11-6
Homo erectus skullcap from Bed II, Olduvai Gorge

parietal bone, the latter forming the top and central portion of one side of the skull. The lower jaws vary greatly in size, but all are rather archaic in appearance, with sloping chin regions. One of the moderate-sized mandibles, presumably belonging to a female, has teeth even larger than those of the eastern pithecanthropines. The largest mandible has smaller teeth although they still are of the size generally found in *Homo erectus*. Perhaps the most striking thing about the large jaw is its breadth at the point where it hinges to the base of the skull. This is greater than recorded in any other men, fossil or living, and is correctly taken to indicate that the base of the skull must have been very broad. Some authorities have argued that this points in the direction of the pithecanthropines, but a mandible with bicondylar dimensions nearly as great belonged to a round-headed type of European Neanderthal found in a cave at Krapina. This individual also must have had a broad base to his skull, but certainly was not a pithecanthropine, and is today classed as a *Homo sapiens*. The available data are too scanty to classify correctly these early men in North Africa. But perhaps it is significant that the earliest complete skull from this general region, that of Ir Hout, is that of a Neanderthal. At the moment we cannot state what the detailed relations are between Neanderthal men and the pithecanthropines. Perhaps ultimately it will be shown that the finds at Ternifine and elsewhere in North Africa are those of men ancestral to Neanderthals and may or may not belong to the *Homo erectus* grade.

Fossils from Europe in the Middle Pleistocene

As rich as Europe is in fossil men from the Upper Pleistocene, it is relatively poor in fossils from the Lower and Middle Pleistocene. It is convenient to deal with the finds in terms of their position within this time span. Remembering that the Middle Pleistocene is the time of the pithecanthropine grade of human evolution elsewhere, the primary problem presented by Europe is that there is not one bit of evidence that a *Homo erectus* type of man ever lived there. The famous Heidelberg jaw is the earliest substantial relic of fossil man in Europe, and it is sometimes assigned to the Günz glacial, and at other times to the following interglacial period. This puts it near the base of the Middle Pleistocene. As in all early men, this jaw is chinless. The bony scaffolding for the teeth is enormous in depth, and the ascending ramus, which carries the condyles as well as the attachments of the great temporal muscle, is of unusual breadth. It does not closely resemble the known jaws of any pithecanthropines. The breadth at the condyles, which articulate with the base of the jaw, is great, indicating a broad skull base. But one of the jaws from Krapina, with an identical dimension, is associated with a round-headed type of skull. The Krapina people lived late in the third interglacial or early in Würm I and were rather modern appearing types of Neanderthal man, not pithecanthropines. Further, the teeth in the Heidelberg mandible are not extremely large, being equalled among such big-toothed modern populations as the Australians and the New Caledonians. But they are interesting teeth, in having moderately large pulp cavities, a trait frequently associated with Neanderthal types of man—a point to be discussed in detail later.

The Vértesszöllös Fossil.   A more revealing fossil fragment comes from Vértesszöllös in Hungary. It consists of a single, complete occipital bone of an adult male. Associated with it is a chopper-chopping tool complex previously unknown in Europe. Fortunately, the cultural deposit in which the occupation layer occurs was sealed both above and below by unbroken beds of travertine, so that its date of Mindel II is undoubted, with no chance of intrusion from later layers. In the shorter chronology it is given a date of about 400,000 years, but with the longer one this might rise to 700,000 years. Ordinarily single bones do not provide critical information about the type of the whole man from which they were derived. But this complete occipital bone defines the shape of the rear of the skull, which was totally unlike the vertically compressed occiput of the pithecanthropines. It has a rounded form and open angle found among modern men. It has correctly been identified as belonging to a *Homo sapiens,* and so it is the earliest relic of this most advanced grade of humanity.

One of the most striking features about the occipital bone from Vértesszöllös is its great size. Professor Thoma, in relating its dimensions to cranial capacities, gave 1,516 cc. as the best estimate for its capacity. The lower limit of the estimate, at a 95-percent level of reliability, was 1,405 cc. Of course there is equal probability that it may have been as great as 1,600 cc. The skull must have been very large brained judging by the standards of any modern population. It is the best evidence that Europe at this time was inhabited by some large-brained forms of the grade *Homo sapiens.* Since the rest of the world then was populated by pithecanthropines, it is likely that the man of Vértesszöllös was on the direct ancestral line leading to surviving modern groups.

One of the dimensions which characterizes this occipital bone is the extreme breadth between the two points known as *asterion,* which is formed where the suture bounding the occipital bone unites with that separating the parietal from the temporal bone. It is characteristically found in the pithecanthropines and also in the Neanderthal fossils and their derived hybrids of the Upper Pleistocene. What this great posterior breadth of the skull really means in terms of brain organization is not known, but it does seem to be

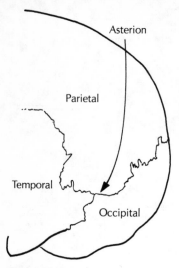

Figure 11-7
Position of Asterion

archaic. All modern populations have a greatly reduced breadth in this region (Figure 11-7).

The site in which this bone was found provides a convenient example as to how sampling accidents may affect our interpretation of prehistory. That portion of the cultural layers which contain the human occiput primarily yielded animal bones of slow game type, including a good many small rodents. On the basis of this limited evidence our early *Homo sapiens* would have been judged incapable of hunting big game. Subsequent seasons of excavation extended the area examined, and, thus, our knowledge of human behavior, for at the far end of the site were recovered the skeletal remains of no less than four cave bears.

The Swanscombe Skull. Proceeding to the time of the Great Interglacial lying between the Mindel and the Riss glacial advances, there are two fossil finds in Europe of great importance. They are the Swanscombe skull from England, and the more complete find from Steinheim on the Rhine River in western Germany. The Swanscombe skull was found in a gravel pit on the Thames River not far from London. The relic consists of three bones, found on separate occasions over a time span of 25 years, and as far apart as 75 feet. It is something of a miracle that all three pieces fit together exactly and beyond doubt came from the same individual, a young adult whom most authorities judge to have been a woman. Sexing can be a tricky business, but the low relief of the bony ridges to which the neck muscles were attached suggests that this individual was a female. In recent years further excavations have proceeded to show that this site at Swanscombe was a living site, although no further human bones have been found to this date. The stone industry is called Acheulian, and along with flakes and a few blades are the common fist axes which characterize it. These occur in a variety of sizes and shapes, with a typical example shown in Figure 11–8. The Acheulian industry occurs at about the same period in Spain at Torralba and Ambroyna, in North Africa in the sandpit at Ternifine, and in Bed II at Olduvai Gorge. It further extends as far eastward as India. At Olduvai, Hazorea in Israel, and possibly at Ternifine the human types associated with this fist axe culture are pithecanthropine. Therefore, it is of interest that in Europe on the Thames the associated skull is assigned to *Homo sapiens,* indicating that tool-making traditions may be more widespread and uniform than the men who produced them.

The skull at Swanscombe must be judged as large brained, with an estimated cranial capacity of about 1,300 cc., as large as that of a modern English female. The vault is very thick boned, but otherwise modern in character. The occiput is well rounded, and the height of the vault is modern. Perhaps the only noteworthy thing about the

Figure 11-8
Acheulian Fist Axe

cranium is that the breadth at asterion is considerably greater than in modern Europeans. This biasterionic breadth was also excessive in the occiput from Vértesszöllös, and as we shall see later, in the man from Fontéchevade. Apparently these very early Europeans, like later classic Neanderthals, had brains broader at this point than do their living descendants. It is not known what this means in terms of brain function. One of the contended points about the skull from Swanscombe is what the missing frontal bone would have shown in the way of brow ridge development. A good many authorities, impressed by the modernity and the sex of the skull, believe that the brow ridges would have been small, as in modern women. Others have rather convincingly suggested reasons why the brow ridges should be reconstructed as larger than in living European women. Perhaps the work now going on in the gravel pit will provide the missing frontal bone and end speculations about it.

The Steinheim Skull.   More complete information about the type of people living in Europe during the Mindel-Riss Interglacial period is provided by the skull discovered at Steinheim, Germany. Since the find was made by professional archeologists, there is no doubt that it belonged to the age of the beds in which it was found. The skull is nearly complete and, most significant of all, retained the greater portion of its face. Once again the base of the skull had been pried open, no doubt for the usual reason. The vault is rounded in the back, has its maximum breadth high on the side, and is apparently both narrower and lower than the skull from Swanscombe. Its capacity has been estimated at about 1,150 cc., a value on the borderline between that of the pithecanthropines and modern man. But in shape Steinheim has little in common with the Far Eastern pithecanthropines and must be placed in the grade of *Homo sapiens* (Plate 11-6).

The sex of this skull is open to some question. Influenced by its rather small capacity, and the evidence of weak neck muscles as shown by their attachments in the base of the occiput, it is frequently classed as female. On the other hand, the forehead shows considerable slope, and terminates in enormous brow ridges which project over the eye sockets in a continuous bar of bone like the visor of a cap. The breadth of the nasal aperture is wide, but this does not help determine sex. Having closely examined many hundreds of Australian aboriginal women and never having seen brow ridges of this size among them, my own bias is to consider that the enormous development of the brow ridges, and the slope of the forehead indicate that the skull from Steinheim probably belonged to a small-bodied male. One other surprising feature is that the upper third molars are much reduced in size, which is surprising for this early date. The teeth show moderately developed pulp cavities

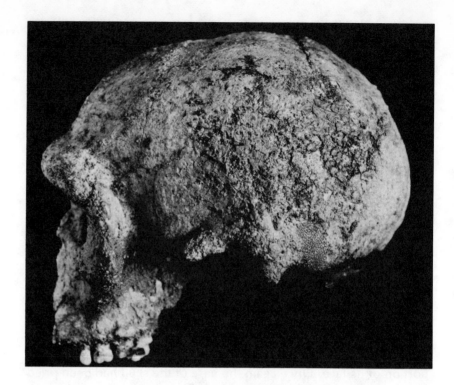

Plate 11-6
Second Interglacial Skull from
Steinheim

and are smaller than those of the average Australian. Irrespective of
these points of disagreement, the important conclusion is that
another Middle Pleistocene skull from Europe proves to be of the
grade *Homo sapiens* rather than *Homo erectus.*

At the end of the Middle Pleistocene, during the third or Rissian
glacial period in Europe, there are a few tantalizing relics of men of
the times. One of the most significant of these comes from Arago on
the French side of the Pyrenees Mountains. The primary fossil
consists of a frontal bone and face, undoubtedly male in sex. While
it has a general Neanderthal appearance, such as occurs in the
classic type at the beginning of the fourth glacial period, there are
some deviations from that type. The forehead is more sloping than
usual, and the constriction behind the brow is more marked. The
brow ridges if anything are a little larger than in the classic
Neanderthal while the face does not protrude as far forward relative
to the skullcap as in the classic type. The face seems to be
dimensionally smaller, and the orbits more rectangular rather than
large and circular in shape. Mandibles found in the deposit include
one of the great size which shows some resemblance to that from
Heidelberg. It would seem that these specimens occupy an intermediate position between *Homo erectus* and Neanderthal man.

From the same glacial period some human relics from Abri Suard give a restorable rear and side of a vault, comparable to that of Swanscombe. The reconstructed vault is rather small in capacity, estimated at 1,063 cc., and is low and broad. The occiput shows a fairly acute angle. Together with the man of Arago some authorities have the feeling that these precursors of true Neanderthal men of the Upper Pleistocene are forms transitional between *Homo erectus* and the classic Neanderthal. Certainly they are considerably smaller brained and in a sense less specialized than Neanderthal man himself. Whether they are anything more than ancestral Neanderthals is hard to tell from their scanty remains. It has been said occasionally that the classic Neanderthal looks like nothing more than a cranially inflated pithecanthropine. But this is an oversimplification. On the other hand, back in time before the classic form of Neanderthal man appear, there must have been a line of ancestors, and one would expect them to be of a lower grade. Whether this is the exact equivalent of the *Homo erectus* grade or represents a regional population passing through a similar grade remains yet to be discovered.

The Evolution of Regional Populations

Since the evolution of large brains is one of the important landmarks in human development the question of when and where modern size was attained is certainly important. It is possible to conceive of great regional populations of men all evolving along the same parallel path at essentially equivalent rates. This model involves an initial radiation of hominids in the Lower Pleistocene followed by long periods of parallel evolution into recent times. One of its interesting provisions is that the great modern blocks of humanity—the Caucasoids, the Negroids, the Mongoloids, and the Veddoids—were reconsidered to have evolved into the grade of *Homo sapiens* independently from regionally separated populations of *Homo erectus*. Such a scheme is shown in Figure 11–9.

An alternative model for depicting human evolution starts with the same initial radiation of men in the Lower Pleistocene and their parallel evolution into the grade of *Pithecanthropus* by Middle Pleistocene times. But then this model deviates from the former in visualizing the evolution of big brain types ancestral to *Homo sapiens* originating in the west and radiating strongly in the Upper Pleistocene to extinguish and replace populations of the other regions. This model has the consequences of deriving all living people from a common ancestor late in evolutionary time and stressing their biological relationship. It overcomes the necessity of having four regional populations of pithecanthropines evolve to the next higher grade more or less simultaneously while at the same time maintaining a great deal of genetic homogeneity. One of the realistic

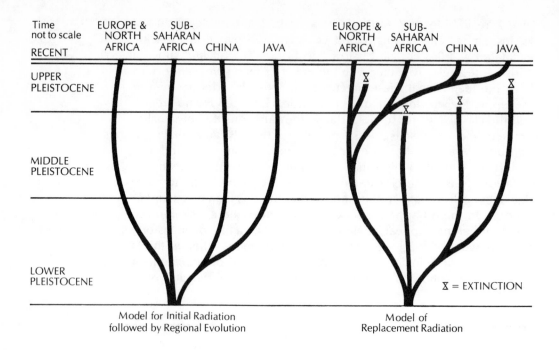

Time not to scale	EUROPE & NORTH AFRICA	SUB-SAHARAN AFRICA	CHINA	JAVA	EUROPE & NORTH AFRICA	SUB-SAHARAN AFRICA	CHINA	JAVA

RECENT

UPPER PLEISTOCENE

MIDDLE PLEISTOCENE

LOWER PLEISTOCENE

X = EXTINCTION

Model for Initial Radiation followed by Regional Evolution

Model of Replacement Radiation

Figure 11-9
Models of Radiation

assumptions in the second model is that evolution did not proceed at the same rate in all major land masses and that any given point in time regional differences were rather considerable. This model fits in very much better with the lessons of long-term evolution and further evidence for it will be discussed in the next chapter.

It should be pointed out that the very early and remarkably advanced skull ER-1470 does not fit into any conventional model for human evolution at this time. Louis Leakey, referring to the habilines from Olduvai Gorge, saw in them the ancestors of modern man with evolution by-passing the grade of *Homo erectus.* Richard Leakey, referring specifically to 1470, had some of the same feelings. This is an attractive idea but supportive evidence is lacking at this time. Instead the very considerable evidence of the Middle Pleistocene suggests that most of the inhabited portions of the Old World were in fact occupied by pithecanthropine grades of humanity. While there is great awkwardness in evolving a *Homo erectus* out of ER-1470, for the moment this must reluctantly be accepted as likely. Anatomically in some ways such an evolutionary stage would seem retrogressive, for in a real sense it postulates that more archaic kinds of men evolved out of a surprisingly advanced form, ER-1470. For the moment evolution remains very complex, and certainly many of its details are poorly understood.

The Pithecanthropines of North China

Behavior is as revealing as are fossil bones. The lifeway of the pithecanthropines can be well reconstructed from a single site. Chou Kou Tien, or the Dragon Hills, lie about forty miles southwest of Peking in North China. The name was derived from the fact that the Chinese considered fossil bones of all kinds to be those of dragons. The fissures and caves shown after excavation in Plate 11-7, provide us with the best information about the behavior of men of the grade of *Homo erectus*. Many years of excavation gave fragmentary remains of more than 40 individuals representing both sexes and a variety of ages. But the majority had not yet reached adulthood, a point which makes the story more compelling. The account of the initial discovery of the bones of these fossil men contains its own drama. Here, as in South Africa, the limestone hills were being blasted and quarried. In 1927 a human tooth was sent to Davidson Black, anatomist at the Union Medical College in Peking. It was the nearly unworn tooth of the lower molar series, so that after being freed from the stony matrix, its characteristics were clearly revealed. Upon examination Professor Black was of the opinion that this tooth could not be referred to any known type of man, living or extinct, and so he took the rash step of claiming it to be a new human genus, *Sinanthropus pekinensis*. With the revision of human nomenclature, this name has now been "sunk" taxonomically, and the find is technically referred to as *Homo erectus pekinensis,* or, colloquially, Peking man. The Chou Kou Tien tooth did have some rather unusual characteristics, but Professor Black was to a degree lucky in that his opinion ultimately proved to be correct.

The excavation of bones from the consolidated deposits of the caves went on for many years. Professor Black died and was succeeded by Dr. Franz Weidenreich, an anatomist of great industry and intellectual distinction. In a series of publications (1943 et seq.), Weidenreich described and interpreted the remains of the more than 40 individuals represented in his collection. The series included 14 skullcaps in generally incomplete condition, 6 separate skull bones, portions of 10 jaws, and a total of 147 teeth. The distribution of skeletal parts in the deposit is biased, for from these nearly four dozen individuals were found only a few fragments of the postcranial skeleton. These included seven fragmentary thigh bones, portions of the shafts of two humeri, one damaged clavicle, and a single hand bone. There was no evidence of intentional burial and, ominously enough, all of the skulls had had their bases pried open,

Some Aspects of Life in the Middle Pleistocene

Plate 11-7
Excavation of one of the deep
bone-bearing fissures at Chou
Kou Tien. The paleontological
laboratory can be seen in the
background.

and some showed head injuries which had occurred in life. Chinese researchers now place the Chou Kou Tien deposits in the opening phases of the Mindel-Riss Interglacial of European chronology, but others place them later in time.

Let us turn first to the anatomy of these Chinese pithecan-thropines. The very limited evidence provided by the thigh bone allows an estimated stature of one man of about 5 feet, 1½ inches. The other fragments suggest the population was short, perhaps with a stature similar to modern Eskimos. Weidenreich's reconstruction of the skull of a female Chinese pithecanthropine (Figure 11–10) shows features typical of the *Homo erectus* grade. For a woman the brow ridges are large and beetling, the vault itself is low and long, and its maximum breadth occurs at the level of the skull base. The occiput is acutely angled and its nuchal crest again reveals the insertion of powerful neck muscles. The face is moderate in its dimension, with moderately low nasal relief, and cheekbones that jut forward and laterally a little in the fashion of living Mongoloids. The lower jaw is heavy but shallow, with a sloping chin region in which nonetheless the beginnings of a modern human type of chin are evident. The teeth are large by modern standards, but do not greatly exceed those of Australian Aborigines in dimensions. Compared to the teeth of australopithecines they have become considerably reduced in size. Cranial capacities as estimated by Weidenreich for five vaults range from 915 cc. to 1,225 cc., an average of about 1,040 cc.

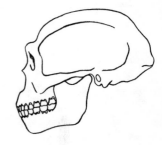

Figure 11-10
Later Pithecanthropine Form (Female) from Chou Kou Tien (based on restorations by Weidenreich)

Authorities agree today that these Chinese forms of early man were closely related to those from the Trinil beds of Java even though the Chinese forms seemingly lived considerably later in time. The Chinese pithecanthropines had better-filled and steeper foreheads, as well as larger brains. Facial regions cannot be compared, since they are generally lacking in the Javanese finds. Weidenreich said, and few disagree with him, that the differences between the two forms were of the size that characterize different races and less than between different species. Therefore, the modern practice of including both populations within the species *Homo erectus* is perfectly compatible with the evidence. If the Chinese men seemed to be somewhat more evolved than the Javanese ones, it is because they lived much later in time. It is fortunate that Dr. Weidenreich described the Peking finds in such great detail and distributed casts of the better preserved specimens to the scientific outer world, for when the Japanese brought war to the mainland of China, the Peking finds were hurriedly packed for shipment away from the war zone. They have not been seen since and were presumably lost with the sinking of some unknown ship.

Deposited within the caves and fissures at Chou Kou Tien was a variety of materials which help us reconstruct the way in which its human inhabitants lived. Their tool kit contained stone, bone, and antlers. The stone tools were largely made out of quartz, a difficult material for chipping, and consisted of a variety of chopping tools and retouched flakes in simple forms. The charcoal in the deposits may be the first evidence of man's use of fire. In the lower layer occupancy seems to have been occasional, as if seasonal in nature, with carnivores being the alternating tenants. But in the upper deposits, evidences of hearths become common and occupancy seems to have been continuous.

The climate of North China at that time was temperate in character and colder than the region is today. The available botanical evidence indicates a covering of pine and beech forests. Since Peking winters are severe today, the question arises as to how early men could have survived them. The ability to use fire was, of course, an important part of man's cultural adaptation to such a climate. But physiological evidence shows that death from freezing would strike down an unclothed man in such winters in less than a half hour's time. These pithecanthropines could not have hunted through North Chinese winters unless they were equipped with reasonably good fur clothing of some kind.

These early men at Chou Kou Tien were accomplished hunters, for the deposits contain the bones of nearly 90 identifiable species of mammals. Their favorite game was a small deer, but, in addition, the bones represent wild boars, bison, horses, gazelles, rhinoceroses, and a giant beaver. Carnivores represented included large hyenas, bears, wolves, tigers, and a saber-toothed cat. It is not certain whether the latter were food animals, or simply occupants at times when man was not in residence. In any case these pithecanthropines had gone beyond the australopithecine level in hunting ability and placed their main reliance upon big game hunting. This is not to say that slow game was ignored, for very few hunters do so, but most of the meat represented animals requiring added skills for their capture. No doubt their diet consisted predominantly of vegetable foods, but very little record is left of its character. The Chou Kou Tien deposits are unusual in that they show hackberry seeds to have been a staple.

The simplicity of the stone tool kit does not explain how these men could hunt big game with regular success. No evidence has been left of projectile points, yet most hunters require one for the capture of big game. Simple wooden spears would have been well within their competence and probably were used. Since the cave represents a base camp, it gives no evidence concerning their use of traps, pits, or possible fire drive to procure their game. At the same time, in distant Spain, an open air kill site does indicate that men of the period successfully did drive big game into bogs to be killed.

Reference was made earlier to the fact that the human bone

deposits were a highly selected series. These deposits involve skulls which were badly broken, with the faces removed and the base opened by breaking away the bone around the foramen magnum. Numerous skull fragments showed depressed fractures received in life. Dr. Weidenreich (1943, p. 190), who examined the question carefully, summarizes the situation in the following words:

My early suggestion still stands, namely: that the strange selection of human bones we are facing in Chou Kou Tien has been made by *Sinanthropus* himself. He hunted his own kin as he hunted other animals and treated all his victims in the same way. Whether he opened the human skulls for ritual or culinary purposes cannot be decided on the basis of the present evidence of his cultural life; but the breaking of the long bones of animals and man alike, apparently for the purpose of removing the marrow, indicates that the latter alternative is the more likely.

Cannibalism

Cannibalism among men occurs in a variety of forms. Ceremonial cannibalism, as it is practiced by some, consists of nothing more than reverently nibbling some cheek meat from a dead uncle. Emergency cannibalism, also known as Arctic, lifeboat, or island cannibalism, occurs among people everywhere when they are reduced to the hard choice of death through starvation or the eating of a companion. Gustatory cannibalism involves the eating of human flesh for the sheer pleasure that it gives. In protein-starved societies this may be understandable. But even where meat is a normal portion of a well-rounded diet, men frequently are thrifty, not letting human protein go to waste. Among Australian Aborigines, mothers have been known to eat the bodies of babies which they were forced to kill since they could not raise them. Among the same people men killed in fights are cooked and then eaten with pleasure. All the evidence points to the consistent practice of cannibalism in Pleistocene times. But Weidenreich's statement that the people of Chou Kou Tien consistently hunted neighboring men is probably an overstatement of the case. No carnivores earn their living by hunting their own kind, for this way of life would lead to extinction. Further, the meat of carnivores is expensive and to obtain it may be dangerous. When it is considered that the deposits at Chou Kou Tien may extend over a period of fifty thousand or more years, then the cannibalized relics of some forty individuals need not represent this practice going on at a very active rate. But the appetites of these pithecanthropines were not selective, as was indicated by the fact that both males and females are represented as adults in these grisly relics, and, even worse, 40 percent of the total were children of various ages. Those were not pleasant times in which to live.

Bibliography

HOCKETT, CHARLES F. and
ROBERT ASCHER.
    1964   The Human Revolution. Current Anthropology. Vol. 5, No. 3, pp. 135–168.
LE GROS, CLARK W.
    1967   The Fossil Evidence for Human Evolution. Chicago: University of Chicago Press.
WEIDENREICH, FRANZ.
    1943   The Skull of Sinanthropus Pekinensis. Paleontologia Siniea. New Sohos D., No. 10, pp. 1–298.

    1945   Giant Early Man from Java and South China. Anthropological Papers of the American Museum of Natural History, New York. Vol. 40, Part 1, pp. 1–134.
    1951   Morphology of Solo Man. Anthropological Papers of the American Museum of Natural History, New York. Vol. 43, Part 3, pp. 203–290.

SUPPLEMENT NO. 14

The Concept of Range of Variation as Applied to ER-1470

New finds such as cranium ER-1470 pose the problem of whether overlapping ranges are indicators of identity or whether statistical distributions are the best evidence of similarity and difference between populations of men.

Let us assume, for there are no very good data to apply, that the standard deviation, or *sigma,* of cranial capacity for the population from which ER-1470 is sampled amounts to 90 cc. If 1470 stands in the middle of the distribution then it can be predicted that plus and minus three sigma would give extreme values of a maximum of 1,050 cc. and a minimum of 510 cc. Since the range of cranial capacity is great in all living human populations, these are reasonable figures for a fossil of this capacity. Let us examine the lower limits first. The gracile australopithecines show an average of 440 cc. for six specimens. The lower predicted value for 1470 is 510 cc., or well above their average value. It is certainly possible that the ranges of the two populations might overlap, but the probability that 1470 belongs with the gracile australopiths is very small. At the other end, the maximum capacity for a population from which 1470 is drawn totals 1,050 cc., and there are a

good many modern individuals, representatives of *Homo sapiens,* with a capacity no greater than this. We can then say that 1470 has a probable range which overlaps with that of modern man, but the statement means very little.

In a normal probability distribution it is known that 68 percent of the values of a population will follow within plus one sigma or minus one sigma. If we assume that 1470 is at the exact average of his group, then more than two-thirds of his population should range in cranial capacity between 690 cc. and 870 cc. A standard deviation of plus two or minus two encompasses 95.5 percent of a normal distribution, and so we can likewise say that the chances of a member of 1470's population being smaller than 600 cc. or greater than 960 is less than 5%. These probabilities give a reasonable estimate of the range one might expect around the cranium found. Of course one could assume that 1470 is an unusually large skull for his population, and then the game is played somewhat differently. If the skull is assumed to be one standard deviation above the average, then the range in size of that population would be from 960 to 600 cc. for a standard deviation of plus or minus two. Such preliminary estimates for capacity alone indicate the improbability of 1470 being any type of australopith, despite its very early age.

Figure 11–11 shows skull A falling within the range of population B in cranial capacity. Where the range alone is used to accept or reject the identity of a new find with an existing population A could belong to population B, but this common method of testing totally ignores the statistical distribution of cranial capacities in population B and by implication in population A.

Turning to Figure 11–12, the probability distribution of population B has been generated as a normal one, and it is immediately seen that skull A fits into the extreme end values of the distribu-

tion; in short it has a low probability of belonging to population B. If it is further assumed that skull A is an average member of its own population, a normal distribution can be generated for it as later and successive finds define its population characteristics. It will be seen that the normal distribution of A greatly overlaps the distribution of B even though the central tendencies or mean values of the two series are far apart. This immediately points out that raw range values do not serve to discriminate proper memberships in different populations. Of course, when new finds fall totally outside of the range of preexisting populations, there is no question that they belong to different populations.

Figure 11-11
A seems to be a part of
Population B

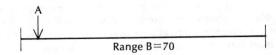

Figure 11-12
The Real Distribution of
Populations A and B

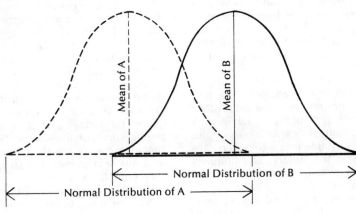

Some Speculations on the Origin of Human Speech

"Chattering like a monkey" was a Victorian suggestion that someone was talking too much. But the order of primates is a noisy group of mammals, and its higher arboreal forms vocalize a great deal. Even so the development of speech among the later hominids is one of the most remarkable aspects of human evolution. The stages of the evolution of speech in human prehistory cannot be reconstructed by direct evidence, but it is possible to suggest the pathway which may have been taken. Charles F. Hockett, a linguist at Cornell University, has provided a series of interesting speculations (Hockett and Ascher, 1964). Most primates live in social groups known as bands, which persist through some time. The collective activities of the members of the band require coordination between its members through at least three forms of communication. Body language involves movement of the body, as well as helpful pushing and prodding. In addition, all of the higher primates communicate effectively through changes in facial expression. Finally, vocal signaling provides communication between members within the band, as well as warnings of proximity to other bands. For most primates the call system is rather simple, containing a repertory of half a dozen or more distinct signals. Each of these refers to a specific situation such as the discovery of food, the presence of danger, a friendly invitation to join the company, or a call which merely says, "I am here." Each of these calls elicits a proper response from other primates within hearing. All of the signals in such a call system are mutually exclusive from the others and contain a simple message which cannot be blended with the others. The linguists technically describe a system with this mutual exclusiveness as *closed*.

In sharp contrast to the closed call system, human language is *open*, allowing us to speak with words that we have never said nor heard before, and so incorporating new blended meanings. One of the limitations of the closed call system is that its signals apply only to the immediate presence. With open language we can and do refer freely to things that are out of sight, belong to the past and future, and may not even exist. The flexibility of the open language results from the arrangement of elementary signaling units which the linguists call *phonemes*. In themselves they have no meaning, but serve to keep sounds distinct and apart. Whereas the signals in the primate call system may well have a basic genetic determination, the utterances of the open language system are primarily learned through repeated observation and conditioning, with only the capacity for speech being genetically determined.

The major problem for the linguist is to explain how a closed call system could have evolved, even given great periods of time, into an open language system. Hockett suggests that early in human evolution the development of bipedal locomotion and the freeing of hands for tool-using may have resulted in the opening of the call system of the early hominids to some degree, and so they took steps in the direction of the development of open language. At its simplest level the opening of the call system seems to require the development of the habit of blending two old calls so as to produce a new one, one with a different meaning. Beginning with a few closed signals, the process of blending rapidly increases the repertory of communicable ideas, but, more importantly, provides the basis for building composite signals out of meaningful parts. Hockett suggests that the opening up of the closed call system by our ancestors required thousands of years of development. In its early form such a system could be called a *prelanguage*. Nothing like such an ancestral system remains in use in the world today, and it is worth noting that among some of the simple living hunters, languages are enormously complicated. In northwest Australia grammatical structure is among the most complex known.

With the conversion of the open call system into a prelanguage, there presumably went hand in hand an increasing innervation of the vocal organs, the tongue and the larynx, and the enrichment of that portion of the cortex which represented the vocal region. This neural evolution is basic to the development of true language. Hockett suggests that the emergence of true language from a closed call system should be conceived of not as a replacement of one sort of system by another, but rather as the growth of a new communication system within the matrix of the old one. Some traces of the old call system are still found in the vocal communications of all men.

The crucial development from a closed signal system into an open language system is believed by Hockett and Ascher to have occurred at the beginning of the Pleistocene, which in today's chronology would be something like three million years ago. This speculation would give to the australopithecines prelanguage at the very least and possibly endow them with true language. A number of anthropologists would disagree with this idea on the grounds that the brain size of these earliest of men shows but relatively little enlargement, still averaging less than 500 cubic centimeters. And at the same time the types of stone tools which they made, the Oldowan tradition, persists over very long periods of time. Indeed, for the entirety of the Lower Pleistocene, a period with a duration in excess of two million years, the stoneworking tradition changed so little as to barely be detectable. At the same time the growth of man's brain proceeded slowly. One would expect that the development of language would be reflected in a more rapid change in technology and to further serve to explain most of the great increase in brain size which we show compared to the australopithecines. For these reasons many students of human evolution would prefer to place the development of language at a later date, possibly early in the Middle Pleistocene.

While Hockett and Ascher appreciated that relatively little was known of the real means of communication among the higher primates, they certainly would not have been prepared for Sarah. Sarah, pictured in Plate 11–8, is a seven-year-old female chimpanzee educated at the University of California at Santa Barbara by psychologist David Premack. This gifted anthropoid has been exposed to a series of carefully controlled experimental situations focused upon the problem of communication between Sarah and her human trainers. Unlike some earlier investigators, who tried to teach chimpanzees to communicate orally as we do, Premack has investigated silent communication. The results could hardly have been anticipated by the most enthusiastic admirers of chimpanzees.

By 1970 Sarah, in little more than two years of training, had developed a working vocabulary of more than 120 words. She not only comprehended their meaning but could combine them to build original sentences. More startling, she used them to ask questions of her own (Figure 11–13).

The impressive thing about Sarah is that she communicated in terms of plastic symbols whose shape and color contained no visual clue as to their real meaning. Initially Sarah learned the meaning of her symbols as a simple memory feat. Once she had command of a reasonable vocabulary, she was taught the meaning of the preposition *on.* In time developing a more complex understanding of syntax, Sarah invented a sentence completion game and invited her trainer to play. Repeated tests show that her accuracy of comprehension was rated between 80 and 90 percent. There is no doubt that Sarah communicated through a silent language with her human trainers. By 1972 Sarah was approaching puberty

Plate 11-8
Seven-year-old Sarah playing word games with a trainer

growing stronger and showing temper tantrums when she made a mistake. She was obviously becoming increasingly dangerous to her teachers and her training had to be stopped.

There were and are a number of gifted chimpanzees involved in the language game. Washoe learned American Sign Language from the Gardeners at the University of Nevada, and her training, although less structured than Sarah's, resulted in her ability to communicate effectively with 130 learned signs. A new chimp, Lana, at the Yerkes Regional Primate Center in Atlanta, Georgia, has been taught to read and write simple but complete sentences in perfect syntax and even punctuates them correctly. She operates a keyboard on a computer console. By the time she had a vocabulary of 50 words Lana began to make her own sentences. There are a further series of young chimpanzees in training, including Peony, Elizabeth, Salome, and just to prove that little boy chimps can learn language too there are Bruno and Booee.

There is now convincing evidence that chimpanzees do have the intellectual capacity of communicating with silent language in one form or another. Apparently in their natural life in the wilds they do use hand gestures frequently, but certainly none of the investigators, including Jane Goodall, realized the linguistic potential of these manlike apes. All hypotheses about the origin of spoken language in our ancestors must now be reworked.

An excellent report on Sarah is contained in "Teaching Languages to an Ape" by Ann James Premack and David Premack in the *Scientific American,* 227, October 1972, 92–99. An overall summary of the status of "talking" chimpanzees is well reported in *Psychology Today,* "Field Report on the State of Apes," Vol. 7, No. 8 (1974), pp. 31–50.

Figure 11-13
One of Sarah's
Sentences

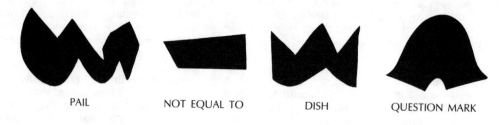

PAIL   NOT EQUAL TO   DISH   QUESTION MARK

CHAPTER 12

THE
COMING OF
MODERN MAN

Modern types of human populations took over the available land masses in the late Pleistocene. The fossil record is best known for Europe, but even there the problems of evolution and replacement are not fully understood. The grade of modern man, Homo sapiens, *contains two differing anatomical types, big-faced, big-browed Neanderthals, and populations akin to living ones, with small faces and brows. Only their big brains place them in the same structural grade. The homeland of living types of modern populations remains unknown. Their appearance in marginal areas such as Australia and probably North America prior to Europe poses real problems which existing data cannot solve.*

Characteristics of Modern Man

Modern man is defined in several ways. First of all the term applies to all living human beings whatever their skin color, hair form, or stature. Living peoples are classed as *Homo sapiens,* and genetic tests show them to be very similar, aside from the frequencies of some variable genes. Further, all modern human populations can, and where chance allows do, interbreed freely with no reduction in fertility. Going back into time, fossils that looked like some of the living populations were classified as *Homo sapiens.* Other fossils, which deviated in visible, and indeed in sometimes spectacular, terms were classified in other taxons. If the differences were not too great they were placed in the genus *Homo.* If, as with the australopithecines, the differences were very great they were given a genus of their own. But this convenient partitioning of fossil forms prevails only as long as there are big gaps in the fossil record and the actual remains come clustered in time and morphology. Since evolution itself represents a continuum of change, it is obvious that classification pressed back into time reaches points where the categories cannot be maintained except by arbitrary decisions. They look good as long as there are gaps in the record.

Until two decades ago the practice was for every new fossil man that looked slightly different from others to be given a special designation. This kind of splitting brought chaos into the field of paleoanthropology. In 1950 a series of eminent taxonomists suggested that Neanderthal man should be classed with other modern men and classified as different only at the subspecies level. The suggestion meant that his taxonomic title would be *Homo sapiens neanderthalensis,* or more simply the Neanderthal subspecies of modern man. This tendency to clump together forms of fossil man which did not differ significantly from each other, compared to the standards used for other animals, greatly simplified the terminology used with regard to human fossils. But the same advantageous clumping brought with it the difficulty that a given grade of man was

Figure 12-1
Relative Size of the Neanderthal
Face Compared to a Modern
Man

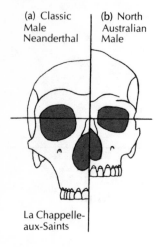

(a) Classic
Male
Neanderthal

(b) North
Australian
Male

La Chappelle-
aux-Saints

so broadened as to include more variable forms; for instance, Neanderthals as well as living types of men were grouped together.

Since people of the third interglacial show characteristics somewhat intermediate between classic Neanderthal man of the early fourth glacial and the types of modern man that succeeded him, it is appropriate at this time to discuss some of the differences between the two forms. Both living man and classic Neanderthal man are included in the same grade of humanity primarily because they both have big brains. There is some evidence that the classic Neanderthal has a bigger brain volume than the modern man of the Upper Paleolithic, and certainly it is bigger than our volume. But Neanderthal man's big brain is packaged differently from ours. As shown in Figure 12–1, classic Neanderthal man has an enormous face, in both its height and breadth. There are immense rounded orbits and a nasal aperture broader than any men before or since. Even in the North Australian shown here the nasal aperture is small, and orbits are generally inclined to be rectangular among living peoples. The Neanderthal vault is low and well rounded in its outlines. By comparison modern men have higher vaults with somewhat flatter sides and a suggestion of a roof-shaped pitch to the top. Seen from the side as in Figure 12–2 the facial differences are again apparent, and added to them is the fact that Neanderthal man tends to have little or no chin, while it is present and positive in more modern types. The vault of the classic Neanderthal man is not

Plate 12-1
Skull of Classic Neanderthal Male
Circeo I. Note limey concretions
deposited by water in burial
grotto.

only low but absolutely and relatively long. His occipital bone protrudes far to the rear, and forms an acute angle in its contours. Plate 12–1 of the classic Neanderthal male, Circeo I, emphasizes these features.

In addition to a low vault, Neanderthal crania show sloping foreheads receding backward from large continuous brow ridges. Their nasal bones are concave, whereas in modern man they are usually straight or even convex. In most instances the face of the Neanderthal is placed well forward relative to the skull and in addition slopes forward to the teeth. This provides a good deal of room for the teeth, which are not excessively large. It will be noticed in turning to Figure 12–2 again that the rear or third molars of Neanderthal man are considerably forward of the broad ascending ramus of the mandible whereas they are partially hidden in modern man. These are substantial differences in form and they help to explain why classic Neanderthal man was once classed as a different species of *Homo.*

Viewed from above Neanderthal man has a maximum skull breadth well to the rear and this is accompanied by a great absolute and relative breadth across the occipital bone from asterion to asterion. This has been commented upon earlier as a characteristic of the pithecanthropines of the Middle Pleistocene and so seems to be an archaic human feature. It will be remembered that other skulls from the same period in Europe not classed with *Homo erectus* show a similar feature. These are Vértesszöllös and Swanscombe.

The teeth of Neanderthals show a tendency toward an enlarged pulp cavity in the molars, a condition known as *taurodontism,* because it occurs in cattle and other grazers. The condition is illustrated in Figure 12–3 where it is shown developed to an extreme degree in molar teeth from Krapina, a Neanderthal site in central Europe, to a moderate degree in the very early mandible of Heidelberg man, and absent in the teeth of modern men. The last condition gives what is called a *cynodont tooth,* also found in dogs and other carnivores. Large pulp cavities like these are not an absolute character in Neanderthals, but their frequency is greater in this population than in others.

Figure 12-2
Skulls of Classic Neanderthal, Modern Man, and Comparison

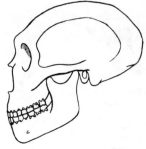

(a) Classic Neanderthal Man

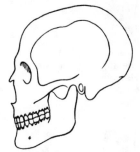

(b) Modern Man

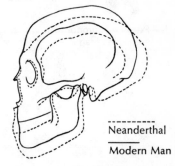

- - - - - - - - - Neanderthal
———————— Modern Man

(c) Comparison of Profiles of Classic Neanderthal and Modern Man

Evidence from the Late Pleistocene of Europe

With the differences between Neanderthal and living types of modern men in mind, it is time to examine the fossil record of man in the third interglacial period, that is, between the Riss and the Würm glacial advances. According to the short chronology in vogue in Europe it might have commenced about 120,000 years ago, whereas the long chronology advocated by this author would place it at about 200,000 years ago.

The limited number of fossil remains that come from the lower

Figure 12-3
Variations in the Size of the Pulp
Cavity in Human Lower Molar
Teeth

(a) Cynodont Tooth
—Modern Man

(b) Moderate Taurodontism
—Heidelberg Man

(c) Extreme Taurodontism
—Krapina Man

part of the Upper Pleistocene tend to be intermediate in their anatomical characteristics and so can neither be called classic Neanderthal nor yet thoroughly modern in their structure. For this reason, many anthropologists have come to view them as the common ancestors of both of the latter populations. In a broad sense this is likely in the long run to be proven true. Two skulls from this period were found just outside Rome in a gravel pit at Saccopastore. Both are generally Neanderthal in their type but deviate from the classic form in some minor ways. Again at Ehringsdorf, in what is now East Germany, another restorable vault resembles the classic type of Neanderthal, save that its forehead rises steeply to considerable height and the maximum height of the vault is placed rather far forward. The cranial capacity of this young woman has been estimated to be as large as 1450 cc., considerably greater than that of a modern woman. Finally, very late in the third interglacial period, or possibly early in the first advance of the Würm, are placed a series of remains from Krapina. The exact number of individuals represented is difficult to estimate, for they were victims of intensive cannibalization. Skillful reconstruction allows the finding that their skulls were broad relative to their length, that is, verging upon what is called roundheadedness. Facial remains are Neanderthal in type, but the people were short and relatively slender in body build, again deviating from the stocky classic model. These three series of finds in general incline more in a Neanderthal direction than toward modern man.

From a cave in France at Fontéchevade come fragmentary remains of two individuals of a different kind. The stratigraphic evidence in the cave makes it clear that they lived during the final interglacial period, and this is confirmed by relative dating tests by the fluorine method. Their importance lies in the fact that we are once again face to face with the remains of people which remind us of Swanscombe and yet earlier Vértesszöllös. One individual is represented by an incomplete skullcap, which nevertheless reveals three points of importance. It came from a big-brained individual whose cranial capacity was about 1465 cc. The shape of the skull was long, low, and relatively broad, with bones as thick as those of Swanscombe. Finally, as in the case of the latter and in Vértesszöllös, the breadth of the lower occiput is great, so that the value of the biasterionic diameter is an important common feature. A few other technical characters heighten the resemblance between Swanscombe and this fossil man.

The fragments from the second individual are seemingly insignificant, and do not exceed two inches in any major dimension. But coming from the critical region above the root of the nose which carries the brow ridges, they provide important evidence. Judged to be adult on technical grounds, this forehead fragment shows no superorbital torus, but rather separate brow ridges as small as in European women of today. Except for the thickness of their cranial bones, the inhabitants of the cave at Fontéchevade could pass as modern Europeans. That point is to be stressed, for with the first advance of ice in the Würm glaciation, the whole of Europe was inhabited by a totally different type, classic Neanderthal man.

As scant as the fossil remains are, they do establish that forms ancestral to classic Neanderthal man in the great glacial period were present in Europe earlier. In general they show features in a less extreme form, and so they have been considered by some to be possible ancestors of both classic Neanderthal man and later *Homo sapiens*. It is common to refer to them as generalized Neanderthaloids. Since their characteristics are intermediate between the two populations it is possible that they could have been ancestral to them. Of course the divergence into two quite separate subspecies would require isolation in space and through time.

The Advent of Classic Neanderthal Man

As the ice sheets moved southward from the Scandinavian highlands for the final thrust, the climate of Europe chilled and the warmth-loving animals disappeared. All of the human remains in Europe belonging to Wurm I are those of classic Neanderthal people and the tools they made were various varieties of the terminal Mousterian stone industry. Throughout the period of their occupancy, which may have been from 40,000 to 70,000 years in duration, the culture remained static and the people themselves seemingly unchanged. Their remains have been found particularly in central and northern France and extending into Germany, where it will be remembered the skullcap of the original Neanderthal man was the first important fossil to reach the view of the scientific world. While most of these folk lived in areas where the glacial climates would have been severe in winter, their remains have also occurred farther to the south. The ritually-used skull of Monte Circeo was found in Italy, and the newest and best preserved of all was found in Greece at Petralona. Their range is extended outside of Europe during the same period by the fossils found at Shanidar in Iraq, at Tabun and Amud in the Near East, and Ir Hout in Morocco, although in the last instance the exact dating is still unclear.

The classic Neanderthal people were the most distinctive population within the species of *Homo sapiens*. In the central portion of their domain, that is, Europe proper, they were characteristically a very short, heavily built, and big-headed people. Their skulls were distinctive in that within their grade they resembled a considerably enlarged *Homo erectus*. But the classic Neanderthal skull differs from that of the pithecanthropines in that it is much better filled,

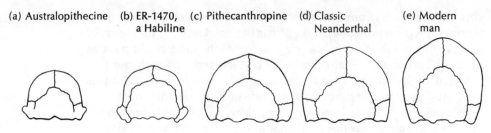

(a) Australopithecine   (b) ER-1470,   (c) Pithecanthropine   (d) Classic   (e) Modern
a Habiline   Neanderthal   man

Figure 12-4
Human Skulls as Seen from the
Rear (to approximate scale)

and the expansion of brain size has brought its maximum breadth
well above the level of the base of the skull. Viewed from the rear,
this skull shows a cylindrical shape with a flattened basal region. The
gracile australopithecine, shown at the extreme left of Figure 12–4, is
more or less half cylindrical, with a flattened basal region and a
maximum breadth at the bottom. In the only reconstruction of a
habiline, the outline is more angular owing to the parietal bosses
which mark a break in the rounded contour of the vault. In the
pithecanthropine, the skull is broadest in the region of the base,
above which the rest lies like an ill-filled half-cylinder with some
angular projections. It is visibly poorly packed with brain, which its
actual capacity reflects. All of the classic Neanderthal skulls, when
viewed from the rear, are seen to conform closely to the cylindrical,
except for the flattening of the base. The maximum breadth lies
considerably above the base, and the skull looks considerably better
filled out than it does among the pithecanthropines. Modern man
has maximum breadth high on the elevated vault. They show a
curious combination of primitive features combined with large
cranial capacities. Their behavior, as we will see later, seems to have
differed very little from that found among living hunters.

One of the surprising aspects of the classic Neanderthal population is the very low range of variation shown. If you have seen one
male skull, so to speak, you have seen them all. They are not only
very much alike in all the aspects of their form, but their dimensions
differ only by fractions of an inch. This uniformity is so striking that a
number of hypotheses have been proposed to account for it. Some
have suggested that this reduction in variability might be an expression of extreme inbreeding. We will see later that this is unlikely.
Others have pointed out that the total Neanderthal population of
Europe was certainly only a few thousand people, and so it might
be that the effect of genetic drift was the causative factor. This is
not a very probable cause of their reduced variability. To some, the
surprising constancy of features and size among the Neanderthals is
a consequence of the enormous selective pressures which they

visualize as operating during an ice advance in Europe. This is a very attractive hypothesis, but it must be remembered that the first advance of the Würm glacier came on gradually, with its intensity increasing over a period of more than 40,000 years. Throughout this considerable time period the population shows the same strange constancy in its characteristics. Additionally, the Neanderthal people of Italy and Greece, both of which were well outside of the stringent effects of the glacier to the north, had as classic Neanderthal features as those found in central France and Germany. The reduced variation in these populations can be noted, but at this time no explanation seems satisfactory.

Mousterian Industry. Whenever the bones of classic Neanderthal man are accompanied by stone tools, they belong to an industry known as the Mousterian. It is basically a flake industry containing various types of scrapers, notched or denticulate flakes, and a few forms that might serve as boring or engraving tools. A few small hand axes are still occasionally associated with the Mousterian. For the first time pointed stone flakes which seem certainly to be spearheads are a part of the culture. As a whole, Mousterian industry seems somewhat advanced over the preceding Acheulian in that a more diversified range of forms is present, but it cannot be claimed that the evolution of man's technology has been greatly speeded. A typical sidescraper is shown in Figure 12–5. Through a series of sophisticated analyses Louis and Sally Binford have suggested that the variations within the Mousterian industry really are a consequence of different types of tool kits used at varying times for specific functions. These include a variety of tools for maintenance activity in keeping weapons and shelters in repair, tool kits involved in the killing and butchering of animals, and others which seem to be concerned with food processing. This type of research will produce more refined analyses of the behavior of prehistoric men.

Figure 12-5
Mousterian Side Scraper

With traces of human shelters going back almost two million years, it is not surprising that Neanderthal men had their own types. With the advancing of the Würm ice sheet from the Scandinavian highlands onto the plains of North Europe, people for the first time took to systematically living in caves. These were modified to make them more suitable for winter occupancy. Cave deposits regularly contain the charcoal from fires concentrated in hearths, indicating that the technique of fire making was now controlled. Hearth fires certainly produced heat, and charred bones indicate that they were also used for cooking. In some carefully dug sites the remains of filled-in post holes suggest that some sort of framework was placed across the mouth of the cave upon which hides could be stretched. These would deflect the wintry wind and at the same time allow the inner chamber's temperatures to be raised somewhat. In the Mousterian horizon of Combe Grenal, Professor François Bordes has recovered bone needles, indicating beyond doubt that classic Neanderthal men made tailored fur clothing. The severity of the periglacial climate would not have permitted survival unless they were capable of making sophisticated clothing.

Cult of Bears.   The Neanderthal hunters were capable of killing any of the big game available to them. These include such formidable animals as extinct forms of elephants and rhinoceroses, giant grazers such as the so-called Irish elk and bison, and such formidable carnivores as cave bears. The latter were related to and the size of living Alaskan brown bears and so when full grown may have weighed up to three-quarters of a ton. In common with some living northern hunters, such as the Ainu, Neanderthal men evolved a bear cult, judging from the way in which they handled the remains of these animals. In a number of excavated caves their skulls have been found placed in seemingly significant ways. But the most remarkable instance comes from Regourdou, in southern France, where a rectangular stone-lined pit contained the skulls of more than 20 cave bears, and was covered by an enormous flat stone slab. Wherever men lived in the same environment with the big brown bear, the animals became woven into their mythology and, of course, were held in high respect. A large bear standing erect on its hind legs would exceed a height of 10 feet, and so, in addition to looking somewhat manlike, presented a commanding appearance. Small wonder that early men armed with simple spears gave them a special position in the world of their ideas.

The cult of bears can be interpreted on the basis of similar behavior among living peoples. But the Neanderthals practiced other types of rites for which there is no modern parallel. In the inner chambers of Guattari cave some sixty miles south of Rome was discovered the skull of a classic Neanderthal man under what are certainly unusual circumstances. The find, known as Circeo I, was almost complete except for two significant mutilations. The right side of the skull and face has been badly damaged by violent blows. The base of the skull has further been mutilated so as to get at the brain in a way that is exactly similar to that practiced by living headhunters in Melanesia. The skull was found in an inner chamber which had not been used as a living site. It lay surrounded by a circle of stones with its base pointed upward. Three bundles of bones of wild cattle, red deer, and pigs were placed in the chamber in calculated clusters. From this evidence, Professor Blanc (1961) hypothesized that this skull represented a sacrificial victim who had been killed by a heavy blow on the temporal area of the skull as is done among present-day headhunters. The victim was beheaded and the skull mutilated outside of the cave, since there is no trace of either the rest of the skeleton or the fragments of the skull inside. The skull was then brought into the inner chamber and placed in an honored position within the circle of stones. This position suggests that it may have been used as a cup. Blanc considered the animal remains as perhaps analogous to the modern Mediterranean practice of sacrificing the bull, the ram, and the pig. In this instance occurring some 40,000 to 50,000 years ago the deer stands as a precursor for the modern ram. Professor Blanc's reconstruction may seem a little fanciful, but it is as convincing as any other which has been offered.

Human Burials. Big-brained Mousterian man provides the first evidence for intentional human burials. Whereas the relics of earlier men have been found in deposits whose character suggests that they have been victims first of violent death and subsequently cannibalism, most of the finds of Neanderthal man have been recovered from intentional burials. Sometimes these are single, sometimes multiple, in such form as to suggest a family cemetery. In most cases, the bodies are accompanied by burial offerings in one form or another. A flint tool kit is frequently placed with a dead man. Food offerings seem to have been common and may take the form of charred bones which are reconstructed as having once been a cooked joint of game. One young Neanderthal child was buried with the horns from four of the great mountain goats, the ibex. But the most interesting burial of all was found in Shanidar cave in the Zagros Mountains of northern Iraq. It represents a man about 30 years of age whose skull had been badly crushed, as if by a falling rock from the roof of the great cave. And he was buried on a bed of flowers and covered with a blanket of them. Pollen analysis showed the abundant remains of blue grape hyacinths and bachelor buttons, hollyhocks, and golden ragwort both above and below the skeleton. One can almost visualize his family and friends going out into the hills to pick their arms full of floral tributes to be buried with this man. Shanidar I was remarkable in another way. He was born with a withered right arm which was amputated above the elbow in life and healed successfully. This demonstrates a surprising skill in surgery in a Pleistocene society and, more importantly, shows that a handicapped individual was able to survive. Presumably, it means that the society contained some sense of altruism, enough to provide the essentials of food and shelter to its injured members. The fact that he was also honored with his burial blankets of flowers suggests that he was an individual of some importance. In modern parallels, he may have been one of the men practicing magic to mediate between his people and the supernatural.

The practice of intentional burial of the dead is the first direct evidence for the evolution of religion in the human record. Among living primitives, burial is always associated with a belief in life after death. Like the ancient Neanderthals, modern people often provide offerings to assist the dead in reaching the next world, and tool kits appropriate to the individual's role in life are frequently placed in the grave. While no reconstruction of prehistoric behavior can ever be absolutely certain, the probabilities all favor the hypothesis that the Neanderthal people had a belief in life after death, and consequently had formulated a system of religious beliefs to go with it. Despite their brutal features they were people of some sensitivity and perceptiveness. Therefore, it will be worth tracing their course in time to see what became of them.

The Appearance of Modern Men in Europe

Some 40,000 years ago, at the end of the first ice advance of the complex Würm glaciation, classic Neanderthal populations disappeared from Europe. Their going remains something of a mystery, for while cave deposits show that Europe continued to be inhabited, there were no further human skeletons recovered during an interval of about 15,000 years, that is, until 25,000 years ago, although stone tools continue to be present throughout the time gap. Then a totally different man is found, a *Homo sapiens*, so like modern living Europeans that there is no doubt that he is an early Caucasoid.

As the second advance of the Würm glacial cycle began, with its even colder climate, the new occupants of Europe lived there in greater numbers than Neanderthals previously had. Since the archeologists themselves have not fully unraveled the so-called Upper Paleolithic cultures of the area, there is no need to discuss them in great detail. There seem to be two different early traditions in stoneworking, the Perigordian and the Aurignacian. Both appear about 35,000 years ago and continue for some fifteen millennia. They are followed by the Solutrean, a specialized stone industry which lasts from 20,000 to 17,000 years ago. The final Upper Paleolithic culture, the Magdalenian, persists for the next 5,000 years down to 12,000 years ago. The most important shift in the making of stone tools is to the production of what are called blades, that is, flakes which are long and narrow, struck off prepared cores by special techniques (Figure 12–6). The methods used include indirect percussion, in which a punch made of bone or antler is positioned against the flint and then heavily struck with a hammer stone. This simple method made possible a much greater variety of tool types. Whereas the broad, short, and frequently thick flakes struck from flint with a blow from a hammer stone allowed some variation in the form of the final tool produced, the new elegant blade allowed many more types to be fabricated. These included a variety of projectile points and knives much more effective than their Mous-

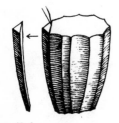
Figure 12-6
Upper Paleolithic Blade and Prismatic Core

terian counterparts. A series of new types of scrapers appears, some held in the hand, and others undoubtedly hafted in wooden handles. Burins, or bone engravers, begin to appear in considerable number, and ultimately in no less than half a hundred identifiable types. In the Solutrean the pressure chipping of flint makes its first appearance, and long, leaf-shaped points were produced. They are so long and fragile that, like the similar points made by northwestern Australians today, they probably broke every time they were thrown at game. Since the latter produce their elegant pressure-chipped points largely as a demonstration of skill and carry them as a matter of show, it seems likely that the same human motives carried the Solutrean flint workers well beyond the point of diminishing utility. With the advent of the Magdalenian, flintworking returned to sensible proportions, and an increasing use of bone, as in barbed harpoon points, is noted. As a result of this bone work, the varied burins attained their maximum development because they were necessary in working this material.

The Upper Paleolithic flint industries represent a rather rapid upswing in the evolution of technology and help account for the increasing population. But it is not certain where the new people came from or what the changes in industrial tradition mean. It may be that the technical changes in tool-making are as much an expression of style cycling as they are of the appearance of new peoples from outside Europe. The details of the story will no doubt emerge in time.

It has been estimated that when classic Neanderthal populations occupied central France, their numbers may have been as few as 2,000 persons at a given time. By the end of the Upper Paleolithic, the Magdalenians, who specialized in hunting reindeer on the upland tundra plain, may have numbered many times as many people. The changes in stone technology are not sufficient to explain the much greater efficiency in exploiting the environment that this shift in densities must imply. True, the character of both the country and the climate shifted as the glacial age progressed. Paradoxical as it may seem, the near Arctic tundra which replaced the northern type of woodlands of Neanderthal times may have carried much more game in the last phase of the Würm glacier. For the mixed bag of big game hunted by the people of Mousterian culture was now replaced by reindeer which, as we know from the closely related caribou of North America today, lived in enormous herds.

Certainly the Magdalenian hunters became specialists in procuring reindeer. How they hunted we do not know. Some of their summer reindeer hunting sites cover as much as five acres, while it has been estimated that the almost continuous string of rock shelters along the rivers of central France in the Dordogne may have housed communities of 300, 600, or even 900 people. The reliability

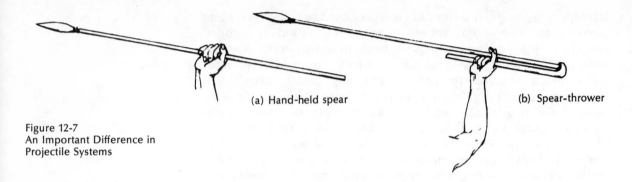

(a) Hand-held spear   (b) Spear-thrower

Figure 12-7
An Important Difference in
Projectile Systems

of reindeer as a source of food must have played a large role in sustaining such sizable communities. The earlier invention of the spear-thrower, which serves to add an extra joint to the human arm, meant that a new and more efficient projectile system had come into being (Figure 12–7).

Such large communities of hunters, as are suggested by the archeological remains, probably have no real parallel among living hunting people. Perhaps the nearest might be found among the salmon fishers of the northwest coast of North America. But even without the latter to serve as a model, it is clear that communities of this size require the instituting of new forms of social control to make life in them bearable. No doubt society began to be more complicated, and perhaps even stratified. The successful hunting of herds of big game, as shown by the buffalo hunters of the plains of North America, requires strong leadership if the exercise is to be efficient. The reconstruction of Upper Paleolithic life is difficult for the very reason that we have no real living models upon which to base it.

Housing among Upper Paleolithic peoples involved some sort of movable lightweight summer tent, presumably constructed of saplings with a covering of hides. Winter dwellings were situated in the rock shelters and caves in protected river valleys. Here again there is evidence that the open mouths may have been covered by a wooden scaffold across which hides could be stretched. Other evidence suggests the summer tents may have been pitched in rows within the shelters. Improvements in housing over those demonstrated by Neanderthal men are not great enough by themselves to explain the great increase in human numbers.

Aurignacian hunters living farther to the east in central Europe and southern Russia specialized in hunting the woolly mammoth, which required a sizable task force of able-bodied men. Excavations at Dolni Věstonice have shown that their houses, more or less circular in shape, held several hearths, and so sheltered several families. Since little wood was available in the tundralike country, mammoth bones were one of the major structural materials in the houses. The people of Dolni Věstonice practiced some of the practical arts, where their interest centered in game animals and women. Carvings both on the flat and in the round on mammoth ivory depicted females, the so-called "Venuses" of the Upper Paleolithic. More interesting, the people at this site were the first known to have fired clay to make pottery figures. These again include both fat women and a variety of game animals such as mammoths, horses, rhinoceroses, and the well-executed head of a carnivore, the great cave lion. Pottery containers were not among the items made, although the settled way of life would have made them useful in many ways.

Upper Paleolithic Art

Art as practiced by hunting peoples takes many forms. It is art in the eyes of modern critics, but for the people who made it, it probably was magical and decorative. We know from such people as the living Australian Aborigines that much of this so-called art is transitory in nature, for such things as paintings on bark, designs of eagle-down feathers stuck to the human body by blood, and great designs as ground paintings all rapidly disappear after the ceremony is finished. Cave art as such does exist in Australia, both painted and engraved. It does not occur in all regions and shows its maximum flowering in the northern tropical regions.

We know nothing of the transitory art of the Upper Paleolithic hunters, but a considerable amount concerning their cave art. The surviving examples have been found mostly in those regions where people at this time lived in caves, the Pyrenees Mountains which separate Spain from France, and the central region of France itself. During the earlier period, the techniques are relatively simple and consist of animals painted in a single color, or perhaps engraved into the stone face of the cave. No paintings are specifically identifiable with the short-lived Solutrean occupation, but with the advent of the Magdalenian culture the techniques of cave art show advances both in execution and in the use of several colors in a single design. The final polychrome animals done in yellow ocher, red ocher, and charcoal are effective in a design sense. Six examples chosen from the best paintings in the early linear forms are given in Figure 12–8. In western Europe the designs are executed in a bold naturalistic fashion which simplifies the forms of the animals but at the same time expresses their structural essence very well. They express the kind of elegance found in Chinese ink paintings. Art critics agree that the designs are very well executed but do not call it great art. Much of this art is rendered in a simplified naturalistic manner. But distortion may be purposefully used. Compare the lifelike woolly rhinoceros (b) with the exaggerated menacing one

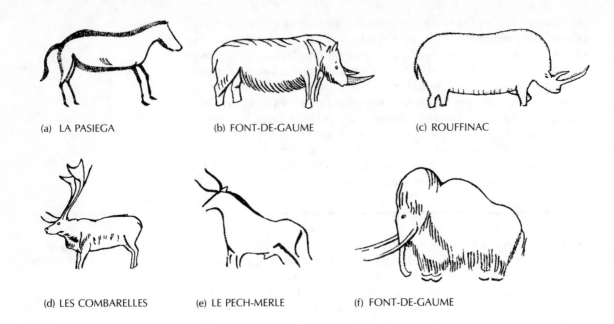

(a) LA PASIEGA

(b) FONT-DE-GAUME

(c) ROUFFINAC

(d) LES COMBARELLES

(e) LE PECH-MERLE

(f) FONT-DE-GAUME

Figure 12-8
Cave Art of the Upper Paleolithic

(c). Since judgments in art are very subjective, it is wise to return to the actual scene before forming opinions. The designs, whether engraved or painted, occur in the total darkness of distant chambers sometimes as much as a half mile away from the cave entrance. They were executed high on walls, on ceilings, and in other positions that must have required scaffolds to reach. Since they were done with no light better than a pine torch, the artists deserve credit. Some of the great animal figures are as much as 18 feet in length. Frequently the same walls are used over and again, with new animals executed over older designs. In spite of the fact that very few instances show animals being speared or dying at the hands of the hunters, there is general agreement that these well-executed designs involved a form of hunting magic. The exact details of the ritual can never be reconstructed, and it may be that the designs were drawn to insure success in a hunt, or perhaps as a kind of offering to celebrate success. In Europe, female figures showing exaggerated secondary sexual traits were carved in ivory, presumably as fertility fetishes (Plate 12–2). Other movable art includes very skillful engraving upon bone and stone.

The Ancestors of Modern European Peoples

Classic Neanderthal populations existed unchanged until the end of Würm I, when they seem to have disappeared rapidly and nearly completely. The next people found in the archeological deposits of caves in Europe are completely modern and differ from living

Europeans only in being bigger brained, possessing larger teeth and so more robust facial bones. Their bodies were like those of modern peoples, that is, relatively lightly built compared to the Neanderthals, and they ranged in stature from short to moderately tall. There seems to have been a considerable difference in the size of the two sexes. A study revealed that these Upper Paleolithic peoples, frequently referred to as the Cro-Magnon type, showed a rather remarkable uniformity. Their cranial measurements were no more variable than those of a large series of Londoners who were buried in a single late medieval plague pit.

In appearance, however, these new Upper Paleolithic people seem to vary more than the measurements indicate. The famous skull of Chancelade from the Dordogne in the late part of the period had a number of features which led early anatomists to classify as Eskimolike. These include a kind of keeling running down the top of the vault, a flaring of the cheekbone, and some changes in palate and jaws. It is true these are found in Arctic peoples such as the Eskimo, but they also have arisen rather recently among the Europeans who inhabited Iceland since medieval times. These Eskimo-like characteristics are apparently not genetic in nature and represent modifications of the external form resulting from life in near Arctic conditions. The man at Chancelade was a big-nosed European, living in the final maximum of the last glacial period, and was not related to the Mongoloid peoples of Asia.

Some of the earliest remains of Upper Paleolithic people were found on the Mediterranean coast of France at Monaco in the Grimaldi cave. A mature woman and a teen-age boy were buried together. It has often been speculated that they represented a mother and son who died at the same time, but, of course, there is no way of testing this idea. Their characteristics are of special interest. They are totally modern in type, showing no traits resembling those of the Neanderthal people. In fact, it was first claimed that they showed some Negroid characteristics, since the lower segments of their limbs were relatively long compared to the upper ones, and their somewhat crushed faces seemed to indicate a little facial protrusion. The consensus of opinion today includes them within the normal range of variation of the European Caucasoid population of the times.

The sudden appearance of modern peoples in Europe raises evolutionary problems. Virtually all authorities agree that they represent a wave of incoming people who replaced the Neanderthal inhabitants. One or two anthropologists persist in the opinion that classic Neanderthal man evolved into modern Europeans, in spite of the fact that there is no bony evidence to confirm this hypothesis. The problem is of particular interest since it involves our two primary evolutionary models for man, the ongoing regional evolution model as opposed to the replacement radiation model.

Plate 12-2
The Upper Paleolithic "Venus" from Lespugue. Note the exaggeration of portions of body and suppression of extremities.

Europe provides the best data for testing which of the models is the most appropriate. People lived continuously throughout the whole of the fourth glacial period in that region, for the stone tools persist throughout it. The classic Mousterian culture is abruptly replaced in the cave deposits by the more specialized blade industries of the Upper Paleolithic peoples. A few types of tools which occur in the earlier cultures persist into the later, suggesting only enough contact between the two peoples for the transmission of some ideas. But the two styles of flint-making are so different that the replacement of flint-working techniques cannot be doubted.

Human remains are few and scattered in time so that a gap in the data appears at the critical time of replacement. In Europe no burials have yet been found which represent the important time gap between 40,000 years and 25,000 years ago. This interval largely represented a mild oscillation between two peaks of the Würm ice advances and so the region was perfectly fit for human habitation. This period of ignorance has allowed an occasional anthropologist to speculate that classic Neanderthal man somehow evolved with enormous speed into modern European types of man.

On the Mechanisms of Population Replacement

It is probable that no human population had displaced another without some genes being transferred between them. This is frequently a one-way matter, with the women of the displaced being absorbed in the successful population and so producing hybrid children. This is true today where European colonial populations have dispossessed native peoples. In the Pleistocene it is probable that larger genetic contributions were involved. Among those anthropologists who view the Neanderthals proceeding to extinction under the assault of more modern people, there is a general consensus that Neanderthal genes are still present in all of the living populations of Europe. Estimates usually range from about 5 to 15 percent in the total modern gene pool.

The recent expansion of Europeans into territories of native peoples was made possible primarily by a superior technology, which involved firearms as the primary referee. Some advances in technology are probably involved in the European replacement in the Late Pleistocene. While classic Neanderthal men could hunt cave bears successfully, and there are stone points of a type that would have hafted to spears, it is likely that these were primarily used for thrusting and were at best awkwardly thrown. The Upper Paleolithic people, with their much more advanced tool kit, perfected the spear-thrower (Figure 12–7) which added many yards to the distance that a spear could be accurately thrown. This more advanced projectile system may have played a role in the displacement of Neanderthal men.

The caves in the Dordogne region of south central France were occupied by classic Neanderthal families as well as by the modern kinds of men who succeeded them. The archeologists agree that the incoming Paleolithic people became much more numerous, as judged by the amount of cultural rubbish they left behind. If our ancestors had merely learned how to hunt in better organized groups, the difference would have supported a greater population density. And it is well known that numbers are very important in determining success in conflict. Finally, data from living peoples indicate that superior forms of social organization may play a deciding role in the outcome of population contests. If a people are so organized that any member may successfully call upon many related families for assistance, then they can focus numerically powerful fighting bands quickly at the required point. If, on the other hand, the social bonds between people are weak, and each family or band lives more or less to itself, then it must succumb to better organized forces. It is probable that all of these factors played their role in the displacement and in ultimate extinction of the Neanderthals of Europe and elsewhere.

Late Pleistocene Problems in Africa are Numerous

Evolutionary problems and paradoxes persist in sub-Saharan Africa into the Late Pleistocene. For instance, there are skulls of striking uniformity from four different sites extending 2,500 miles from East Africa in the North, to the Cape of Good Hope in the South, which stand together as a group. The most complete and best known of these is the so-called Rhodesian man. The most southerly is from Saldanha, discovered only a few miles north of Cape Town. The others consist of a fragmentary skull from Florisbad and a restored skullcap from Lake Eyassi in East Africa. The reconstruction of the latter is somewhat doubtful, since it consisted of a great number of small pieces which did not fit together tightly. The important point is that these four fossil finds in a number of characteristics verge on the pithecanthropine grade of humanity, and that they apparently lived late in time, some of them no more than 30,000 or 40,000 years ago.

The skull of Rhodesian man is the most impressive, not only because of its completeness, but because it is a record breaker in several aspects. Its brow ridges are among the largest ever recorded. The height of its upper face is the greatest ever found in man. As indicated in Figure 12–9, the vault is low and long, with the rear portion acutely angled, in a way that differs in detail from the Far Eastern pithecanthropines. Cranial capacity is 1,250 cc., which is high for the pithecanthropine grade, but low for a male *Homo sapiens*. Had his brain capacity been no more than 1,100 cc., Rhodesian man would have been classified as *Homo erectus,* so this

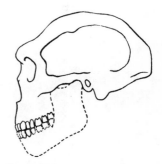

Figure 12-9
Rhodesian Man—Mandible Reconstructed

find is difficult even from a taxonomic point of view. It is estimated that Saldanha man had a cranial capacity almost as large, namely 1,225 cc.

Aside from a somewhat larger cranial capacity, Rhodesian man closely resembles the skullcap from Bed II of Olduvai Gorge known as Chellean man. In spite of a time difference of perhaps one million years, both have similar and enormous brow ridges, skull proportions are generally alike, and there seems little doubt that they represent the same lineage, which seems to have evolved very slowly. In form this descent group differs markedly from living Negroids and so cannot have contributed importantly to their evolution.

The picture in Africa is complicated by the fact that perfectly modern appearing fossil men are known from about the same time period as that in which the Rhodesian group has been placed. The earliest of these consist of four fragmentary skullcaps, found by Louis Leakey many years ago at Kanjera in East Africa. This was his first important discovery and unfortunately it has never been properly dated. There is general agreement among workers that the Kanjera finds belong in the Upper Pleistocene, but they could be as old as 100,000 years or as young as 30,000 years in age. The problem takes on its importance from the fact that they are small browed, and aside from some thickness of the bones in the vault, perfectly modern in appearance. They would in fact be good candidates for the ancestors of the living Blacks of Africa. The problem arises from the fact they seem to have lived along with or perhaps earlier than the Rhodesian types of man.

Leakey's son, Richard, a few years ago recovered two more restorable skullcaps from the Upper Pleistocene deposits of the Omo Valley in southern Ethiopia. Preliminary publications indicate that they are large-brained, rugged, and modern in type. Carbon-14 dating gives them an age older than 35,000 years. Again their time position is awkwardly close to that of the Rhodesian series. The simplest solution for this situation would arise from future finds and better dating, which indicate that the men of the Omo and Kanjera type lived a little later than the derived pithecanthropines and caused them to go to extinction. But even if such a simple replacement radiation should be confirmed, there would remain considerable time overlap, and so a contemporaneous period. It remains a little disturbing that such derived pithecanthropine types as the Rhodesian group should in any case survive so late. The problems of Gausian competition seem to occur throughout the Pleistocene of Africa.

North Africa and the Middle East provide too loose a pattern to confirm either of the evolutionary models. Throughout the first part of the fourth glacial advance, this entire region seems to have been occupied by Neanderthals of near classic type. These occur from Jebel Ir Hout in Morocco, through the caves at Tabun near Mount Carmel and at Amud in Israel, to as far east as Shanidar in Iraq. Throughout this region modern men appear before the end of the Pleistocene, but the dates are so poor and the evidence so scanty as to favor neither model. The cave known as Mugaret-el-Skhul, situated close to Tabun at Mount Carmel in Israel, provides a special problem of its own. Among the nearly one dozen human burials excavated there, two adult skulls were complete enough to allow reliable restorations. Both deviate markedly from the classic Neanderthal form and come closer to some of the early and big tooth ancestors of modern Europeans. They are dated at about 36,000 years before the present. Their brow ridges are still large, but not of the Neanderthal size, and are divided instead of continuous in form. Faces are somewhat reduced in dimension, and have lost the inflated look of the classic men of Neanderthal. Limb bones are evidence that these people were rather slenderly built, that is, modern in their proportions. All in all they show a general level of modernity combined with some dilute Neanderthal characteristics. This has led some authorities to suggest that they represent a transitional population resulting from hybridization between a classic Neanderthal population situated somewhere to the south of Mount Carmel and fairly modern types of Europeans which are usually placed to the north, and often in the steppes of southern Russia. A few anthropologists have suggested that the Skhul population was sampled at a time when it was actually evolving from classic Neanderthal to modern form. This would be very rapid evolution, and no explanation is offered as to what would bring it about. Even so these skeletons can be used to substantiate either of our models, depending upon one's basic point of view.

The Advent of Modern Man in the Hinterlands

One way of evaluating the significance of totally modern types of people in Europe is to see when they occur elsewhere and in marginal areas. It will be recalled that the earliest skeletal material of modern populations dates no farther back than 25,000 years in Europe. The arrival may have been somewhat earlier, since the Upper Paleolithic blade industries appear on the scene in France about 35,000 years ago. This date would represent the approximate time at which Neanderthal man was replaced by the ancestors of modern Europeans. Both, of course, fall into the taxonomic category of *Homo sapiens.*

Turning toward Southeast Asia, a find from the cave at Niah in Borneo is of a delicately formed type of modern man. Its carbon-14 dating of 40,000 years ago now seems generally accepted, although there were initial questions about its validity. A part of the change of attitude about the Bornean find results from discoveries in Australia. It will be recalled that at Mungo Lake in western New South Wales an extremely modern type of individual has been dated at about 26,000 years before the present. In the same fossil dune a series of artifacts and hearths go back to approximately 38,000 years. At nearby Arumpo lunette, another fossil dune, a well-defined midden was found late in 1974 and dated at 36,000 years before the present. Since these discoveries come from a single limited area which happens to be favorable for their initial occupancy and subsequent discovery, there is no reason to believe that these are the earliest modern men to have arrived in Australia. But it is significant that these skeletal finds are earlier than those of any modern man in Europe and reach back into the time when the Neanderthal man held sway there. Workers in Australia now have the feeling that the first occupants may have reached their continent fifty or sixty thousand years ago or even slightly earlier, as the sea level lowered during the fourth glacial period.

The situation in North and South America is somewhat similar. Good radioactive dating on perfectly modern-looking Indian skulls go back to at least 27,000 years. If the amino acid method of dating developed by Bada proves secure, there are indications that modern types reached the New World about 50,000 years ago. Thus two marginal areas in terms of the central Eurasian land mass show modern types of men arriving earlier than they did in Europe itself. Furthermore, in both Australia and North America the early occupants are very like those who live there in modern times, and show no archaic features. The situation again raises the question of where the fairly modern types of man first evolved.

The New World and Australia can be ruled out as their homelands. Africa which has such a complex and long-standing record of fossil man does not answer this question clearly. The arrival of modern types of man is late in Europe. The Middle East, like Europe, has an early fourth glacial population of Neanderthal men, and so this does not seem to be the likely spot. India gives us no fossil documents about human evolution. The question remains unanswered and will remain so until new discoveries are made. But the fact that populations had already differentiated regionally into their modern form by at least 40,000 years ago is a point of compelling interest.

How Rates of Evolution Help to Define the Problem

Not all populations of an organism evolve at the same rate, and so it is possible in some cases to examine the rates of evolution and to help distinguish various populations. In Figure 12–10 are plotted the cranial capacities of the most important fossil men of the Pleistocene. The graph must be viewed as approximate for not only have

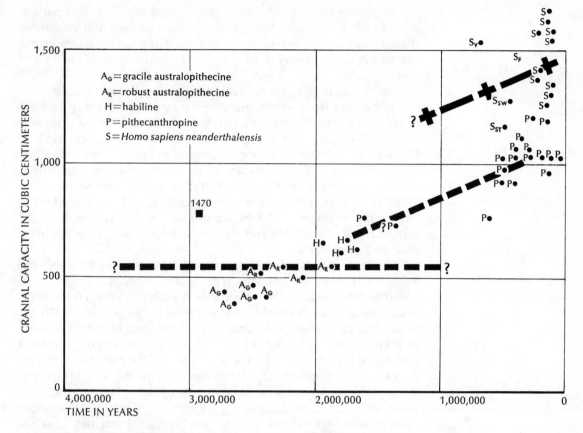

Figure 12-10
Evolution Based on Cranial Capacities

the difficulties of chronology been mentioned previously, but actually cranial capacities vary a good deal depending upon who takes them and by what methods. The volumes of the skulls of the early Africans are probably the most consistent for they have recently been examined by one worker using modern and acceptable techniques. The fossils of the Middle and Late Pleistocene have been estimated less carefully by a variety of workers and so are less reliable.

One of the most striking aspects of the graph is the position of the robust australopithecines and their constancy through time. Four individuals give an average capacity of 517 cc. This type survived into the Middle Pleistocene, and at least two finds are approximately one million years old. Louis Leakey claimed to have found fragments of this type of man as late as a half a million years old in Olduvai Gorge, but dates in that site have been changing rapidly in recent years and his estimate is perhaps suspect. The general morphology of this hominid seems to have changed very little through time, although the latest members seem even more extreme in form than the earlier ones from South Africa. Since there can be little doubt that the robust and gracile australopithecines had a common ancestor somewhere back in the Pliocene, perhaps even the middle of that period, it looks very much as though this robust lineage did not show changes in cranial capacity through time. For this characteristic the rate of evolution seems to have approximated zero over millions of years.

Cranial capacities are available for six of the gracile australopithecines from South Africa, and they give an average capacity of 442 cc. This is well below that of the robust australopithecines and so sets them apart. The datings of these fossils are so uncertain that estimates of rates of evolution among them are avoided. But if it is assumed that the South African forms are between two and three million years in age, and that the East African habilines evolved from them, then a line with positive slope could have been placed upon this graph to connect the two populations. But so much doubt surround both the dating and the relationship that this has not been done in Figure 12–10.

The greatest factor for uncertainty lies in the capacity and the position in time of ER-1470. It is earlier than the gracile australopithecines and has almost double the capacity. It is a full million years earlier than the habilines of Olduvai Gorge, but is 150 cc. larger than their average volume. If the individual found is a female, as seems likely, then the differences in capacity are even greater than those just quoted. ER-1470 is a "flyer" in the pattern of data on the rest of the graph. It is both too early and too big to fit into any of the patterns revealed by other clusters of data. This is a point of some significance, for by ignoring 1470 it would be possible to show by statistical methods of analysis that human capacities have increased consistently through time. This was done several years ago by Lestrel and Read (1973) in an article which showed that central tendencies determine this type of analysis, and outlying pieces of data become buried under general trends. It must be pointed out that 1470 was not known at the time of their analysis. But data which do not fit the general trends in evolution sometimes are the most revealing, and must not be ignored for mere economy of hypothesis.

The pithecanthropine volumes cluster around 1,000 cc. and occur rather late in time. They would show no time trend were it not for two early forms, OH-12 from Olduvai Gorge and *Pithecanthropus* IV from the Djetis bed of Java. If these two examples are representative of their time period, then a trend line can be drawn as indicated in the figure. It seems to originate above the position of the habilines. Note the long period of time between 1.4 and .6 millions of years ago where the data are lacking. Still one point is clear: the pithecanthropines of the Far East, between 200,000 and 300,000 years ago, as represented by the populations of Chou Kou Tien in China and the Solo series from Java, are conspiciously smaller in capacity than Europeans living in the same time period. There is only one point of overlap, where the largest of the Solo men exceeds the capacity of the smallest of the European series, Steinheim. Beyond this all the pithecanthropines fall well below and the Europeans well above the point of range overlap.

The sapient men of Europe are represented in the earliest portion of their time span by Vértesszöllös, Steinheim, Swanscombe, and Fontéchevade. The rest of the data are derived from six male Neanderthals who show an average capacity of 1,551 cc. and four female Neanderthals with an average capacity of 1,339 cc. All of the Upper Paleolithic finds could also have been plotted and would have reinforced this distribution pattern, but obscured its clarity. It is enough to say that the classic Neanderthals and the modern men who succeeded them firmly fixed the most recent end of the trend line estimated here. Europe has since at least the Middle Pleistocene been populated by big-brained humans who seem to have been well advance of other known populations who were their contemporaries. This suggests that the model favoring the relatively late spread of *Homo sapiens* with the absorption and extinction of archaic populations elsewhere in the Old World may be the correct one.

Obviously the whole study of fossil man continues in a state of flux and new finds no doubt will alter the interpretations of our evolution. It would be of particular interest to obtain further material bearing on the problem of 1470 from the region of East Rudolf. So rich is this area that one can be reasonably certain that the finds will come to hand in the next few years. Beyond that, Lower Pleistocene fossils from Europe and the Mediterranean area are needed to fill in the background of the evolution of the big-brained people of that area in later times. We still do not know in a convincing way exactly where modern types of men did evolve, so that further work in southern Russia and the Middle East are clearly necessary. Since India has yielded no fossil men from the Pleistocene, this deficit is a particular handicap to efforts which try to coordinate the evolution of men in the Far East with those in the land masses to the west. But interest in this whole project remains high, new data will continuously come to light, and our interpretations of human evolution will necessarily change with them.

Bibliography

BLANC, ALBERTO C.
  1961   Some Evidence for the Ideolo-
         gies of Early Man. In Social Life
         of Early Man, Viking Fund Publi-
         cations in Anthropology, No.
         31, New York [Editor, S. L.
         Washburn], pp. 119–136.
DEEVEY, EDWARD S., JR.
  1960   The Human Population. Scien-
         tific American. Vol. 23, No.3,
         pp. 195–204.

LESTREL, PETE E. and
DWIGHT W. READ.
  1973   "Hominid Cranial Capacity Ver-
         sus Time: A Regression Ap-
         proach." Journal of Human Evo-
         lution. Vol. 2, pp. 405–411.

SUPPLEMENT NO. 16

The Petralona Caper

The uncertainties of chronology and the flux of anatomical opinion concerning fossil man are well demonstrated by the case of the skull from Petralona in central Greece. This is a large cranium, complete on the left side but with some damage on the right and with the mandible missing. It was found shortly before 1960 together with a number of animal bones by six amateurs. The fauna primarily consisted of cold-loving animals, and since the skull in its general characteristics approximated the classic Neanderthal type, it was considered to have been from a Würm I deposit and to belong to that branch of humanity.

It has not yet been technically published but already some very different opinions are appearing in print. A paleontologist has studied the fauna, and claims that only part of it belongs to a cold climate, and some elements are clearly associated with a warm climate. This could be any one of the three interglacial periods of Europe. The investigator rather favored the first interglacial period, which is the very end of the Lower Pleistocene.

With this tentative date in mind, some students of fossil man have revised the classification of the Petralona skull so that instead of being a very good representative of classic Neanderthal man,

it is now called a *Homo erectus.* This is a major change in evolutionary grade and it is clearly derived from chronological considerations rather than anatomical criteria. One of the justifications for putting this cranium in a pithecanthropine category is that the walls of the vault seen from the rear are said to slope inward or converge as they proceed upward. This is a pithecanthropine trait, but I once studied under a Harvard professor of archaeology who had the same characteristic to a notable degree. I doubt that he would have welcomed being called a pithecanthropine Bostonian. In actuality most of the characteristics of this skull resemble those of classic Neanderthals, particularly if one allows for a normal range of variation in that population. It would be interesting to learn by what evidence the skull is associated with the so-called warm fauna and how massive the data are that it is the first interglacial from which they are derived, rather than the second or even the third. In short with the evidence originating in a cave dug by amateurs, many questions remain unanswered, and it is perhaps best to consider that the skull was late in the Pleistocene. If it ultimately should prove to be early, then it would stand as evidence that large-brained Europeans go back into the Villafranchian, a point of considerable interest.

The Human Numbers Game

The reproductive pressure of human fertility can bring monstrous results. Until the last 10,000 years, which ushered in the agricultural revolution, men lived in a more or less balanced condition with the forces in their environment. Their numbers did not increase because infanticide was practiced, death rates immediately after birth were very high, and life was both difficult and dangerous. But in recent millennia man has fallen out of balance with natural forces, as a consequence of his progressive technological achievements. We are just now beginning to realize the danger of unchecked human population growth, and have not yet provided the difficult and necessary solutions.

Long-term perspectives frequently place problems in a better context. Accordingly, we shall estimate the growth of human population numbers from a very early period in man's evolution. This method is subject to error, but of such relative insignificance that the major features of the curve we are about to plot will show no great changes. A model of this type was first developed by Edward Deevey (1960), an ecologist at Yale University. I have altered the values he used for his early periods, since more recent data suggest changes. During the thousands of millennia of human prehistory there are, of course, no good data on the numbers of men at any instant of time. Consequently, we must use information derived from living hunters and project it backward, using transforming assumptions that seem reasonable.

Let us begin with a time period 3,000,000 years ago. The men who lived then were various kinds of early hominids, including the two species of australopithecines, primarily inhabiting Africa. It is likely that they had already radiated to tropical regions elsewhere in the Old World, and the southern regions of Europe and Asia can be included as their probable range. If we assume that they found Africa about 70 percent inhabitable, they would have occupied 8.3 million square miles of that continent. If it is further assumed that only 10 percent of Europe and Asia could be included in their range, then the former continent adds .4 million square miles and the latter 1.7 million square miles to the range available to them. Their total range is then put at the sum of these three figures or 10.4 million square miles.

The second stage of the problem involves estimating how many square miles were required to support each of these archaic hominids, that is, to imagine a reasonable density figure. Since most of these early men were concentrated in Africa, a continent unusually rich in animal biomass, I have chosen the figure of 25 square miles to support each person. This is very much less than the one square mile achieved by hunters in good regions elsewhere. It is considerably better than the densities recorded for Australians living in the worst desert, supported primarily by little lizards, or for the Caribou Eskimo living in the Arctic tundra. All in all, it seems like a reasonable figure, and so we shall use it. If the total range estimated as available is divided by 25, we have an estimated population of 415,000 individuals at a time after the initial radiation and when they are considered to be in balance with their environment. For our purposes it really is not very important whether the number was actually ten times as great or only one-tenth of this estimation model. In the Middle Pleistocene, the time of *Homo erectus,* the evidence of stone tools suggests that these more advanced men inhabited most of Africa, part of southern Europe, and an extended range in southern and eastern Asia. If we estimate that they occupied 90 percent of Africa, and 20 percent of each of Europe and Asia in the more suitable portions, then their total range can be calculated at about 14.8 million square miles. With the advances in technology which had been achieved, this higher grade of human being had expanded its livable range by a little more than 40 percent. By the same token we must assume that they were able to live in greater numbers in a given region than their precursors. Modern Australians, taking good country with bad, require 12 square miles on the average for each individual. One must consider that Australia was a rather unfavored continent for hunters, but at the same time allow for the fact that the Aborigines were considerably more advanced in their technology than were the pithecanthropines, judging from the archeological record. Therefore, a figure of 15 square miles to support each Middle Pleistocene man, as an average throughout their range, seems reasonable. Using this value and dividing it into their estimated range, the population can be estimated as being very close to one million persons. The date used for the Middle Pleistocene is somewhere around 500,000 years.

During the Upper Paleolithic, or about 20,000 years ago, man had been able to expand into considerably more stringent climates, for his cultural advances included both tailored clothing and housing capable of sheltering him through near-Arctic winters. If we now assume that he was still capable of living in 90 percent of Africa, 40 percent of Europe, and 60 percent of Asia, his range becomes extended to 22.2 million square miles. It may be assumed that in tropical and subtropical climates these modern types of men were on the average characterized by higher densities than those found among recent Australians. On the other hand, they had penetrated into the sub-Arctic regions of the Old World, and while they may have been very much better off than the Caribou Eskimo, still their overall density in the cold areas must have been reduced. Therefore, I have assumed that throughout the entirety of their range it required 10 miles on the average to support a single person. Combining the area of their range and the assumed density, we have an estimate of 2,200,000 people for the Late Upper Paleolithic.

Over the 3,000,000 years' time span approximately represented by the above estimate, human numbers had been increasing very slowly. For the first two and a half million years they did little more than double. This would represent the total human population of the world increasing at a rate of one person every five years. This rate of change is so slow that no one then would be aware of its existence, and in fact for any reasonable time span all of those hunting populations may be considered constant in numbers and in equilibrium with the forces in their environment. From Middle to Upper Pleistocene times, a period of 500,000 years, numbers again more than doubled. Obviously the rate of increase has picked up, but it is still barely measurable. For now the sum total of humanity would be increasing about three persons every year. Even though the rate has visibly quickened, if we plotted the results on a piece of graph paper, the line would appear horizontal.

Once man gained some control over his own food supply, through the invention of agriculture and the domestication of chosen animals, a new future opened up. It was a world in which increasing population numbers became possible because the food supply was increasable. It amounted to a revolutionary transformation in economics. Whereas the best country previously would support no more than one hunter per square mile, then even primitive forms of agriculture provided enough food for 100 persons per square mile. Today, in very favored areas, but still using primitive technology, farmers may reach a density of 800 to 1,000 persons per square mile.

Choosing a period of 4,000 years ago, when agriculture had spread to most parts of the Old World, and then was independently invented but not spread so widely in the New World, we can make estimates of what the population may have totaled. The problem is more complex, for major portions of each area still remained in the hands of hunting and gathering people. Only the more favored regions were taken over by agriculturalists. While agriculture was in existence in both hemispheres about 8,000 years ago, its spread from its areas of discovery required time, and its practice was less efficient in the outer and marginal areas. For this reason I have assumed that throughout the inhabitable range of the period, densities averaged out to about one square mile to support each person. This is, of course, much lower than the number of people supported by agriculture in good areas, but on the other hand, it is certainly much higher than the density to be found among hunting people by and large. It seems to be a good compromise estimate. The area occupied by man has been calculated as follows: 90 percent of Africa, Europe, and North and South America was assumed occupied; but 80 percent of Asia was included in the estimate, owing to the extensive Arctic regions; while 70 percent of Indonesia was included in the estimate. Australia was occupied much earlier than this, but its area has been treated so as to add no more than 300,000 people to the total world population. In effect, the model assumes that man occupied 42.8 million square miles, and this, with the density assumed, gives a population estimate of about 43 million people.

Let us turn to Figure 12–11, which begins with the estimate of 43,000,000 people alive 4,000 years ago and then proceeds down into modern and even future times. I have plotted populations as estimated in 1936 by Carr-Saunders for 1650 A.D., 1750 A.D., and 1850 A.D. The values, respectively, are 545,000,000, 728,000,000, and 1,171,000,000. They are plotted in the appropriate place on the right-hand side of the chart and are important because they control the sharp point of bending in the curve of population increase. The United Nations provides an estimate of 2,406,000,000 people alive in 1950. In 1975 the world estimates approximated 4,000,000,000 people, and it is estimated by a series of projections that by 2000 A.D. there will be between six and eight billion people alive on this planet.

All of the time period covered by this figure is postagricultural, so that the control of the secure food supply does not alone explain the change in rate from a very slight increase in population to the geometrical increase evident now. About two centuries ago the industrial revolution dawned in England and rapidly spread to the other countries in the Western world. Even today this revolution is making further inroads on previously untouched peoples. Until recently the underdeveloped nations were consumers of industrial goods but produced none. Now the methods of this revolution, particularly as they apply to light industry, are rapidly being exported to peoples everywhere.

In recent decades an equally critical change has occurred with the export of Western medical practices to nations which had lacked them. This humanitarian gesture saves lives on a short-term basis, but increases the magnitude of many problems. By drastically cutting infant mortality, as well as risks of death at all later ages, these nations will see a rise in life expectancy from about an average of 30 years to that approaching our own of 70 years. Thus in a generation the mortality figures will go from Pleistocene to modern values. Unfortunately, food-producing technologies are not so easily exportable, and the underdeveloped countries everywhere are faced with the prospect of their food supplies growing much less rapidly than their population numbers. A good many students of the problem predict in the decade of the 1970s that gigantic famines will begin to occur, and the death toll will number many millions of people. The ideas of Malthus were correct but were merely expressed two centuries too early.

Figure 12-11
Increase in Human Numbers for
Last 4000 Years

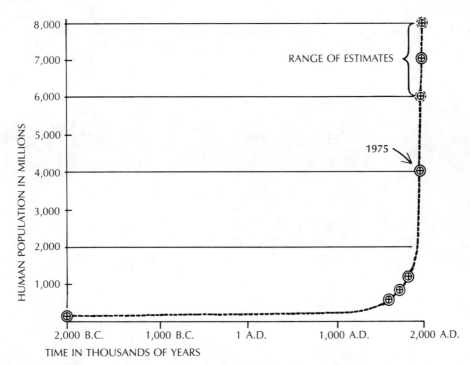

The curve of population growth is accurate enough so that within the next generation all the problems it produces must in effect be solved and under control. Remembering that our numbers will double in the next 30 years, it is prudent to ask what this will mean in terms of problems other than mere food supply. Men everywhere are asking for more meat to eat, but in the future it will dwindle instead. Estimates have been projected to indicate that whereas every American consumed 186 pounds of meat in 1970, in the year 2000 this will be reduced to 123 pounds. In less fortunate countries it, of course, will be much smaller in value. Throughout the world today there is a trend to ask for more energy per person to lighten the tasks of life. This primarily is derived by burning fossil fuel in the form of oil and coal. The problem of air pollution will not consist of how to live in a few of the world's great cities, but will irreversibly alter the very composition of the atmosphere and waters of planet Earth, and consequently change the climates of the world in ways that are difficult to predict. The feeding of more men will require the use of more pesticides, and already our entire planet has been contaminated with a single one, DDT. Urban populations are demanding more and more packaged foods, and their containers, together with man's offal, pose a rising problem of disposal. Certainly the world is moving toward more rapid transportation, and each new advance seems to involve higher noise levels. Noise is a form of pollution of the environment with serious repercussions upon human health. This is a doleful picture and we have not even considered the problem of disposing of radioactive residues and the heat pollution from nuclear power plants, nor even glanced in the direction of nuclear warfare.

The problem of the next few decades can be dramatized (as if it needed it) by a simple calculation which goes as follows. The present annual rate of increase in our world population is about 2.5 percent. If this rate were projected back to the time of Christ, about 2,000 years ago, and allowed to operate constantly, the world today would be totally uninhabitable, for in this model there would be 160 people for every square foot of land surface on this planet. Imagine life with 159 people standing on your shoulders.

CHAPTER **13**

THE
STRUCTURE OF
SIMPLE HUMAN
POPULATIONS

Man is the animal which best illustrates that behavior evolves as structure evolves. Bipedalism and tool-using combined with an enormous expansion in brain size, resulted in complex, largely conditioned human behavior. Evolved behavior consists of two general types: deep-lying behavior which is coded in the DNA of the species, and conditioned behavior which results from the learning experiences of the individual. These two types of behavior are interwoven and difficult to distinguish.

One of the striking things about Pleistocene man was that with a life expectancy of about three decades, one-half of his total life was devoted to growing up and learning how to live successfully. So adaptive has the human species been that before the end of the Pleistocene man learned to live in every environment available to him. Successful adaptation requires different kinds of technology and various types of behavior. The range is well demonstrated among living hunters and gatherers: the Aborigines of Australia, the Shoshoni of the Great Basin of the United States, and the Caribou Eskimo of Central Canada. Such peoples show a high degree of environmental determinism, both in their technologies and in their societies.

Various Types of Behavior

It has long been customary to speak of instinctive, or genetically-determined, behavior as opposed to conditioned, or learned, behavior. This polar concept is no longer valid. It is known that all types of behavior are influenced both by the genetic constitution of the individual and by the environment in which the behavior occurs. The visible evolved behavior is the product of the interaction of these two basic sources, each of which contains a number of variables. It is more profitable to consider the development of each form of behavior and its stability in the life cycle of the individual. Behavior that develops early in the individual and shows little variation in different environments is usually called "genetically coded" behavior. But because even coded behavior is subject to genetic variation, it may vary slightly from individual to individual although stable in each. If behavior is subject to modification through learning experiences, it is usually called "learned behavior." In its development this kind of behavior can be channeled in many directions and will vary greatly from individual to individual. Its variety is as great as the individual learning experiences and as the diversity of the environments in which the learning occurs.

Apparently many of the multicellular soft-body invertebrates are capable of some learning from experience, and so they become conditioned by their life experiences. Flatworms are of considerable use in experimental work, just because these unimpressive and simple creatures can modify their behavior under repeated stimuli. No doubt their learning is based on genetically-determined substructures of considerable stability. The social insects manifest highly structured forms of behavior which hold their societies together, and yet the element of learning is present. Much of the behavior of the honeybee in the hive is highly structured, and so presumably coded, but in the foraging for pollen and honey they readily learn the approximate location of rich sources from the messages brought back and transmitted by their scouts.

Some of the solitary wasps provide a useful example. Each of these little animals is born from an egg which has been attached to a paralyzed caterpillar. The infant grub eats this store of provisions provided by its unseen mother, in time transforms into a fully fledged wasp, and emerges from the burrow prepared for life with a store of coded behavior. A newly hatched female wasp manages to live, to mate with a male of her kind, to reproduce her own eggs and provide them with the necessary food resources so as to ensure species survival into the next generation. Her attitude toward the responses of other insects, including all wasps, is coded. She must not respond to the courting advances of other related species, but she must accurately recognize the mating pattern offered by a male of her kind. Then copulation must proceed effectively so as to ensure the internal fertilization of her eggs. Obviously in these vital functions there is no room for trial and error learning as occurs among higher animals like ourselves.

Once fertilized, our wasp must choose the right place in which to build her burrow, in soil of the proper consistency, and construct it to the proper dimensions. Her innate drive then leaves her to identify the right kind of food, usually a caterpillar, which she paralyzes by stinging. Dragging the inert victim to her burrow, she carefully places it in the bottom. If she judges it to provide enough store of energy, she lays a single egg in the burrow and closes its mouth. This complicated chain of events is repeated until her egg supply is exhausted.

Behavioral investigations show that the solitary wasp is coded so that each successive action serves as a stimulus for the next, and so she acts out the meaning of her life in the form of chain behavior. If any of her actions are interfered with, as when a human investigator removes the paralyzed caterpillar from her burrow, the wasp is incapable of altering her behavior to accommodate the disaster. Instead she lays an egg in the empty burrow, and seals it just as though nothing had happened. On the other hand, she can fly away from each sucessive burrow a considerable distance and return to it unerringly with her paralyzed victim. Experiments in the field have shown that while the bulk of this little animal's behavior is tightly coded, individual insects can learn and so do develop some conditioned behavior. Presumably some female wasps make mistakes and so their offspring fail to survive. This kind of elimination serves to maintain the accuracy of the coded behavior intact through time, and yet to improve it in some individuals with adaptive conditioning.

With few exceptions the young of both amphibians and reptiles hatch unattended by either parent, and so depend for immediate survival upon their own coded behavior. Wastage among the young is great but the survivors in time profit from learning and so complicate their behavior to attain greater adaptive success and individual longevity. Brains are not greatly elaborated in these forms, since in general their behavior is not complicated enough to require advanced neural structures. Still the sensory organs of some reptiles are noteworthy. In the pit vipers, of which the rattlesnake is a familiar form, a pair of special facial pits lie on either side of the muzzle, between the nostrils and the eyes. Equipped with a rich supply of sensory nerve endings they allow the detection of warm or cold objects and record the intrusion of such in something like $1/100$ of a second. This provides a very special kind of responsiveness to those snakes which feed on warm-blooded animals. Like most other snakes and reptiles, the pit vipers also have another special sensing instrument known as Jacobson's organ. This consists of a pair of cavities, one on each side of the inside of the mouth. They are heavily supplied with nerve endings like those which produce the smell in the nose. The snake's vibrating split tongue takes up odorous particles from the ground or even out of the air itself and transfers them to the openings of Jacobson's organ, thus giving the animal a high degree of awareness of its immediate environment. A rattlesnake finds its prey some distance away by virtually molecular tracing of its path. One further adaptation in this serpent is that its poison begins to predigest the victim while it is still alive. This is highly adaptive, for during the period of torpor in which the snake is gorged with his victim and virtually defenseless, the length of the time of hazard is cut down by this action of the poison. Of course for a man struck by a rattler, this is uncomfortable knowledge.

Learned Behavior Among the Mammals

Mammals have evolved more complex behavior than any of the other vertebrates for the simple reason that the mammary gland insures that the young will remain associated with their mother for a considerable period of time, an interval in which she can and does instruct them. The duration of this intimate learning period varies with both size and type of mammal. Among the smaller forms, like rodents, the period of exposure at the mother's breast may be but a week or so before the young are weaned and dispersed to seek their own destinies. Since the rate of reproduction is very high among these little creatures, the loss of many juveniles can be afforded. Among larger mammals the number born in a given litter is reduced, and, among the large advanced mammals, it is usually one at a time. Among herbivores it is important that the young be able to move on with their mothers, so as to escape the danger of prowling predators. Such animals are born long-legged and within a few hours after birth have achieved sufficient muscular coordination to be able to run effectively. This early competence is necessary for sheer survival.

Young carnivores are frequently born in dens and in any case are well protected by their mother. They are relatively helpless for some weeks after birth and do not even open their eyes for a considerable period. This early helplessness among the newborn of our domestic dogs and cats is well known. Cats as pure predators provide well-known examples of teaching, which clearly involves conditioned patterns of learning. When the kittens reach the right age, their mother brings in a slightly injured bird or mouse for the kittens to play with. She stands by to make sure that the victim of this heartless lesson does not escape, and she actively encourages the kittens to assault the poor animal. The lesson serves to instruct the kittens in some of the rudiments of killing and to further stimulate their interest in the necessary action. Not all cats become equally good killers upon growing up, which indicates that there may be variation both in the genetic code and in the conditioning process.

The relationship between body size, brain size, and rate of maturity is interesting. Larger animals take longer to grow to maturity than their smaller relatives, and Rensch (1960) claims that experiments demonstrate that the larger animals are also more intelligent. It is not certain how much of this is concerned with the length of time required to grow big bodies and how much of it involves an extended period of time to condition more complex brains. The young of small wildcats mature and become independent at the end of a year, whereas those of the great cats take much longer. Tiger cubs do not learn to kill successfully on their own until they are about three years of age. George Schaller (1967) has seen a mother tiger sit by complacently while three of her nearly grown cubs demonstrated incompetence in trying to kill a young water buffalo. They showed no initial flair in their task and did not even seem to know how to approach the victim for a proper kill. In time, and with their mother's help, they learn to hunt by themselves and disperse to find their own territories. Supervised learning seems an important part in the conditioning of most mammals.

Growing Up Among the Primates

The primate order is characterized by relatively large brains, which provide a basis for very complex learned behavior. Beginning with moderate-sized animals such as the baboon, the young remain immature for a period of at least four years. Among the man-sized apes, such as the chimpanzees and gorillas, the subadult state is lengthened even further, for their young cannot be said to enter a mature state prior to reaching eight years of age. The process culminates in man. Limiting the discussion to simple living hunters, the young of both sexes do not become economically self-sufficient until they reach about the fifteenth year of life. Aboriginal Australian boys spend most of their time perfecting their future skills by throwing imitation spears, clubs, and boomerangs. They are diligent in learning their tracking, and I have seen six-year-olds successfully track a little lizard across rough country. Little girls practice the skills necessary for feminine economics and very early in life learn the art of grinding seeds and collecting foods. Such skills are essential for survival, and require long learning periods for perfection.

Man is an advanced primate with complex learned behavior, therefore it is of some interest to see whether remnants of coded behavior in our species can be identified. It would be characterized by extreme developmental stability and would manifest itself in almost all environments. There seem to be three manifestations of this type of behavior among newly born human infants. First, all babies cry and each mother recognizes her own infant's cry within two or three days, even under crowded conditions as in modern hospitals. Thus there is an acoustical bond between the newborn infant and its proper mother. Second, all normal babies have rooting and suckling behavior, which of course ensures nutritional security when the mother is available. The mere presence of the breast stimulates this basic response. Finally, the smiling of a baby is a programmed expression which the average normal baby shows at about six weeks of age. It is dependent upon the maturation of the neocortex. The fact that it is found in babies born both blind and deaf shows that it is genetically determined and completely stable in its development. Smiling solicits caretaker response, which is of course important for the baby's survival. While people live primarily by learned behavior later in life, these three forms of programmed behavior are essential for survival early in life.

Human Life Expectancy

Fifteen years is a long time to invest in the process of learning and growing up, especially among people whose normal life expectancy does not much exceed an average of 30 years. It is the price exacted to condition that complex instrument, our brain. Relatively early death has characterized men throughout the Pleistocene, and still holds among populations not yet reached by Western medicine. Table 13–1 gives the data on the death of ancient man as estimated from the age of the individuals whose burial had been excavated. The techniques of aging used by Vallois (1961) are sufficiently

TABLE 13–1

Ages at Death from Burials in Pleistocene and Early Recent Periods (Modified from Vallois, 1960)

	Series	Percent of Neanderthal	Percent of Upper Paleolithic	Percent of Mesolithic
SUBADULT	Number	39	76	71
	Age in years			
	0–11	38.5 } 48.8	38.2 } 54.0	29.5 } 38.0
	12–20	10.3	15.8	8.5
ADULT	21–30	15.4	19.7	49.3
	31–40	25.6 } 51.2	14.5 } 46.0	8.5 } 62.0
	41–50	7.7	9.2	1.4
	51–60	2.5	2.6	2.8
	TOTALS	100.0	100.0	100.0

Ages at Death by Sex (after Vallois, 1961)

Ages		21–30	31–X	Totals
Series				
Neanderthal	Male	1	11	12 } 20
	Female	5	3	8
Upper Paleolithic	Male	5	15	20 } 35
	Female	10	5	15
Mesolithic	Male	15	7	22 } 44
	Female	20	2	22

TABLE 13–2

Ages at Death in Pecos Pueblo Cemetery for 1,772 Burials (after Kidder, 1958)

Age	Percent	
0–1 month	18.2 } 24.8	
1 month– 3 years	6.6	
4–12 "	6.8	} 45.8
13–17 "	8.2 } 21.0	
18–20 "	6.0	
21–35 "	43.5	
36–55 "	10.7	} 54.2
56– X "	0.0	
	100.0	100.0

reliable for these purposes. The remains of 39 Neanderthal men were almost equally divided between immature and mature individuals. It will be recalled that earlier figures were quoted for the australopithecines of Africa, and they showed almost exactly the same division. In Upper Paleolithic man, 76 individuals show an even higher proportion of immature among the burials. For the Mesolithic, during the first few thousand years after the end of the glacial period, the proportion of early deaths seems to fall somewhat, but the figures may be misleading.

If we look for a moment at Table 13–2, which records the ages at death for 1,772 burials recovered by archeologist Alfred Kidder (1958) from the cemetery of Pecos Pueblo in New Mexico, which span several centuries of time, we find that subadult death totals almost reach the same frequency as adult burials. This table is interesting in that it provides evidence of the very high rate of mortality of the newborn, for almost one in five died before reaching the age of one month, even among sedentary, gardening Indians. Such high neonatal rates are perfectly normal for people living under simple economic conditions. Returning to the data of Vallois, it is likely that the very young infants are underrepresented in these burials and so missing from his data. Very few infant Neanderthals have been found, and the youngest is estimated to have been two and one-half years old. Those facts would suggest that something like 25 percent of the infant deaths which occurred in the Pleistocene are missing from Vallois's table. If proper corrections are made for these missing babies, deaths among the subadults would come closer to 60 percent than the 50 percent recorded. During the Pleistocene people lived precariously, and early death was a commonplace thing. It was something of a triumph to reach the age of 20, for one-half or more failed to do so. The life of hunting people was hard enough without trying to live in a region that was drastically influenced by glaciers. The early death of infants must have been a regular event. Among the infants spared from infanticide by the aboriginal Australians, it is estimated that about 50 percent die from natural causes during the first year of life.

In the lower part of Table 13–1 Vallois analyzed the age at death by sex. In all three groups, the Neanderthal, the Upper Paleolithic, and the Mesolithic, women died more commonly than men between the ages of 21 and 30. After that age, deaths among the men predominate. Vallois suggests that the higher death rate among the young women must have been connected with the hazards of childbearing. He is certainly correct. The process of human birth is dangerous. Among most mammals the true birth canal is of ample size to allow for the easy passage of the full-term fetus. Even among our closest relatives, the great apes, birth is achieved with no great difficulty. But among men evolutionary specialization has resulted in a rapid increase in brain size. In our populations, irrespective of status, body build, or economics, the head of the full-term human fetus fits tightly in its mother's birth canal. Without the loosening of the pelvic ligaments and joints through the preparatory action of the hormone prolactin, birth would be even more difficult than it is. Anyone who has seen a large number of new babies in the maternity hospital will have noticed how many of them show visible deformations of their skull as the result of the difficult passage through their mother's pelvis. Fortunately, the effects of this rough trip are temporary and seem to produce no ill effects in brain function in later life.

This birth struggle between mother and infant must have been a rather recent human development. Small-brained australopithecines presumably had ample-sized pelves to pass their young without difficulty. At the pithecanthropine level, brains were still only two-thirds as large as ours today, whereas the bodies of these early men were comparable in size. This again suggests that they would have no great difficulties in normal birth. But with the evolution of the grade *Homo sapiens* there must have been a period of stringent selection to make the fetal head size concordant with the passage in the mother's pelvis. If a child's head is too great to pass through the pelvis, both infant and mother will die in the effort. Good statistics provided by one medical mission at the Fly River on the coast of southern New Guinea indicated that without the intervention of Western doctors, about three percent of the observed 700-odd births would have resulted in killing both mother and child. The odds for death in each childbirth were about one in 30 for the average woman in this simple society. In recent or Late Pleistocene times, if the average woman conceived as many as ten children, her chances of dying in childbirth rose greatly. The data of Vallois suggest that this was a fact.

With life expectancy approximating three decades for men and women in the Pleistocene, very few children knew their grandparents, and many children were orphaned. If a person survived to reach the age of 50 years in a hunting society, it represented a personal triumph. He had not only won the battle of life, but had amassed more skill and knowledge than most individuals had time to achieve. These advantages were, of course, translated into simple social power. Elders were not only respected; they were even listened to.

The present situation in the United States is a very different one, for males born today have an average life expectancy of a little less than 70 years, while females exceed that value by a few years. Both of these calculations include the infant mortality of early years. Although such figures represent a kind of statistical abstraction, they do provide a basis for comparison. Very few children today are orphaned by losing both parents. It is commonplace that they will know all or most of their grandparents, and may even find a few great-grandparents surviving. Birth is no longer a hazardous process, for with X-rays and Caesarean deliveries few infants are lost. After birth the drugs and skills that go with Western medicine save many individuals from risks of early death. In the developed countries today a substantial minority of people have a good chance of exceeding the biblical age of three score and ten. The sum of all wisdom is no longer invested in the elders of the society, and with a technology which changes more rapidly each decade, the gap between generations becomes increasingly widened.

Relating People to Their Land

Density is the ratio which expresses the number of individuals per unit of land area. It is the most important single numerical expression describing the success of man in his ecological niche in a given country. Under our system of measurement we speak of density in terms of the number of individuals per square mile where densities are high, or for hunting societies the number of square miles required to support a single individual. People are counted irrespective of age or sex. Other types of measurement are useful in the understanding of human evolution, but none tells as well as density how a given region supports its inhabitants. The ratio relates man to nature directly.

Some groups of living hunters serve as models for the interpretation of the way men lived in the Pleistocene. Better than stones and bones, they allow us to observe their complex learned behavior, to examine their social organization, to count the frequency of the genes in their population pools, and so to look deeply into their living systems. People lived under a variety of climates in the Pleistocene and do today. Not all have evolved the same techniques of tool-making and using, but the overall variety of hunters of any period is not great. With care it is possible to match groups of living hunters with their Pleistocene counterparts with some reliability. Among people who live by hunting and foraging, the range of variation in their density is illuminating. From the worst desert in Australia, where it requires 75 square miles to support one person, the densities increase until in a number of places hunters live in favored environments in which two square miles will support each person. The home range of a family of five, therefore, varies from 375 square miles down to only 10 square miles to provide the energy resources needed for their support. The worst environment inhabited by living hunters is found in the Arctic, where the Caribou Eskimo of north central Canada needs more than 125 square miles for each individual, or a total of 625 for the average family of five.

When men live so thinly scattered on their land, there are both social and biological repercussions.

With a few dubious exceptions, no modern hunting people living in an inland environment have a density exceeding one person per square mile. On islands and seacoasts where marine resources are available, this number is exceeded. But this seemingly low density appears to represent a kind of ceiling for this type of lifeway. By contrast simple gardeners and agriculturists may reach very high densities without elaborate technologies. In India in favored regions the population may exceed 800 persons per square mile and be supported by local food production. The biology of crowding and the evolution of social control are much altered under these circumstances. But among hunters people are generally well scattered and for the greater part of the year live in very small social groups.

Three living groups of hunting and foraging people seem especially useful as examples to help us understand how men lived in the Middle and Late Pleistocene times. The Shoshoni Indians of the Great Basin of western United States were primarily foragers, eating all of the available food plants of their desert environment and the few animals that they could kill. The Australian Aborigines live in a greater variety of environments and have more animals to hunt, but like the Shoshoni may be called generalized in that they eat anything and everything which their environment allows them to take with their simple techniques. The Caribou Eskimo, conversely, are specialists to the extent that caribou and fish provide the vast bulk of their energy intake, and their techniques have been sharpened and specialized to obtain these animals. As people living in the Arctic they provide more suitable models for the hunters of the Upper Paleolithic of Europe, who lived near the edge of the continental glacial masses to the north. None of the living peoples serve as completely accurate models for those in the past, but they do provide much useful information.

The Shoshoni of the Great Basin

The Shoshoni-speaking tribes of the arid Great Basin of North America were economically among the simplest people on Earth. They lived in an unusually stringent environment consisting of the interior drainage system lying between the Sierra Nevada range on the west and extending to the Wasatch Mountains of Utah on the east. They ranged from the Colorado River in the south to the Columbia River in the north. Their domain consisted of minor mountain ranges running north and south and separated by desert valley floors. The latter averaged about 5,000 feet in altitude, with rainfall ranging from only five to twenty inches a year. The ground cover was primarily sagebrush and greasewood. The few streams

Plate 13-1
Two Shoshoni-speaking Paiute
Women, about to set out on a
seed-gathering trip. The
winnowing baskets held in their
hands and the storage containers
on their backs are examples of
the specialized baskets
developed for the seed-gathering
economy.

originating high on the mountain slopes ended in the valleys in
saline marshes or lakes. Animal food resources were scarce, with
jackrabbits and pronghorn antelope providing skins and meat.
Most protein was derived from such lesser game as mice, rats,
gophers, small reptiles, and such insects as ants, their eggs, and the
larvas of the flies which bred in the salt lakes. Consequently,
vegetable foods were very important, and they ranged from the
seeds of a variety of small annual plants to the all-important nuts of
the piñon pine. Plant seeds were distributed erratically each year in
the paths of thunderstorms that crossed the country. The piñon nuts
were derived from trees growing between 6,000 and 9,000 feet,
which bore prolifically at very irregular times and places. The erratic
nature of the small seeds as well as the piñon nuts required the
Shoshoni to live opportunistically and to move where food could be
found.

This type of environment required a very special kind of human adaptation. Since food resources occurred in random and un-predictable ways, there was no advantage in creating fixed territories. The population was sparse, and their food gathering was more efficient when they broke down into small biological families which scattered to wherever food was available. The Shoshoni had no large permanent social units or villages. They occasionally came together for short periods of time to drive rabbits into long nets. These were the big events of almost every year, for the rabbits bred prolifically. A much rarer drive was for antelope, whose population took much longer to recover from the kill.

Families were bound together by the exchange of their women in marriage. Fecundity was controlled through infanticide. Regional groups of families had some knowledge of each other as the result of the random contacts which their foraging produced. Families wintered in separate houses erected in the piñon grove which provided good harvest that particular year. Under such circumstances it is remarkable that the Shoshoni still consisted of tribes whose primary unity was a common language.

The technology of these gatherers was simple, except that the emphasis upon the collecting and processing of small seeds resulted in the elaboration of baskets in which to collect and hold such items. The bow and arrow were present, but unimportant since big game was scarce. The bulbs and greens of growing plants provided seasonal food in addition to the seeds. This type of cultural adaptation in which the family is the basic unit of society is rare among gathering peoples of today, but it may have been more common in the Pleistocene. This way of life requires a minimum of social elaboration, for the family as the basic unit is adequate to rear the children and provide the most efficient unit for exploiting resources. This kind of human adaptation must go back a long way into the Pleistocene, for it requires no more than the modern kind of sexual relationship in which women have no period for breeding, but are sexually attractive and available throughout the year. Such people leave so little evidence archeologically that their presence in the Pleistocene is difficult to detect.

The Australian Aborigines as Models of Pleistocene Man

Australia is the smallest of the continents, generally the most arid, and climatically and topographically the most homogeneous. Even so it exhibits a considerable range of climatic variation. Its mean annual rainfall goes from a minimum of 4 to more than 160 inches. Most of the country is low-lying and is lacking in sharp relief except along the east coast. One-half of the continent, the central interior, has less than 10 inches of rainfall annually and so is to be classed as desert. At the time of contact with colonial Europeans, the entire

Plate 13-2
These aboriginal women from the Western Desert are preparing digging sticks from a mulga shrub before beginning their daily gathering activity. Note the last-born children under foot.

aboriginal population totaled about 300,000 people living by hunting, fishing, and foraging. It was the only place in the world where men at this level of economy did not have neighboring agriculturists to apply pressures upon them. For these reasons the Australians serve as useful models of earlier men living under a similar range of climates. Of course, they provide no information about human adaptation in the Arctic.

Aboriginal technology was simple but sufficient. A wide range of stoneworking techniques coexisted down into modern times. Tools ranged all the way from the simple choppers similar to those found in the Oldowan culture of early Africa up through a variety of more sophisticated implements made by percussion methods, even microliths. In the northwest, elegant bifacial points were made by pressure flaking. The spear-thrower was present and bows were absent. Boomerangs of various types, nonreturning for use against marsupials, and returning forms against flocks of birds, were used throughout most of the continent. The wild dingo was tamed by most groups, but the dog was not important in the pursuit of food animals. Women used digging sticks to obtain root vegetables and collected various types of fruits and berries. Their contribution amounted to about 70 percent of the food eaten by a family. Men contributed big game using the spear-thrower, and their contribution averaged about 30 percent of the total intake. The basic economic posture of the Australians was essentially Upper Paleolithic. By depriving them of their spear-thrower and reducing them to a spear thrown from the hand, we can in our imagination even let them serve as models for men throughout much of the Middle Pleistocene.

The adaptability of human behavior is well demonstrated by the fact that the Australians, using essentially similar tool kits, were able to exploit with equal effectiveness every variety of environment found within the continent. If the tools remained unchanged, their extractive behavior was suitably modified to fit each local situation. In one portion of the western desert several tribes lived in a region almost totally lacking marsupials in any form. Their primary protein consisted of lizards only a few inches long and usually weighing no more than an ounce. The role of men in the daily food quest consisted of going out into the sandhills, tracking these little reptiles to their hidden lairs, and crushing them with their heels. The successful hunter returned home with 20, 30, or perhaps 40 little lizards whose heads had been tucked under his hair-string girdle. Spears were not carried in the hunt because they were only used against men. At the other end of the scale the early occupants of Australia lived with giant forms of marsupials and birds. They apparently drove to extinction kangaroos standing eight or nine feet high, giant emus, and the ponderous diprotodon, roughly the size of a rhinoceros. A man armed with a spear-thrower could bring down such big game.

Vegetable food ranged in size from minute grass seeds, much exploited in the dry areas, through a wide variety of underground roots and bulbs, and included all the fruits and berries in each region. In a small patch of climax rain forest in the northeast Queensland most of the food was gathered high in the canopy, a hundred or more feet off the ground. The diminutive Aborigines living in that region used a section of vine to hold themselves against the straight trunks of the giant trees and by shifting it rhythmically ran vertically up the smooth bole almost as rapidly as they could on the ground. This particular technique was used nowhere else in the world but was invented there to aid in rain forest foraging.

One of the most important lessons taught by the Australians is that among such generalized hunters the diet is so broad as to include all of the energy sources available to them. Captain Grey

(1841), one of the earliest explorers to penetrate the west coast of Australia, listed the categories of food items utilized by the tribes in the extreme southwestern corner of the continent. Their diet contained 17 broad categories of animal foods and 10 of plant foods.

A. Animal foods
  (1) 6 types of kangaroos
  (2) 5 marsupials somewhat smaller than rabbits
  (3) 2 species of opossum
  (4) 9 species of marsupial rats and mice
  (5) 1 type of dingo
  (6) 1 type of whale
  (7) 2 species of seals
  (8) Birds of every kind, including emus and wild turkeys
  (9) 3 types of turtles
  (10) 11 kinds of frogs
  (11) 7 types of iguanas and lizards
  (12) 8 sorts of snakes
  (13) Eggs of every species of bird and lizard
  (14) 29 kinds of fish
  (15) All saltwater shellfish except oysters
  (16) 4 kinds of freshwater shellfish
  (17) 4 kinds of grubs

B. Plant foods
  (1) 29 kinds of roots
  (2) 4 kinds of fruit
  (3) 2 species of cycad nuts
  (4) 2 other types of nuts
  (5) Seeds of several species of leguminous plants
  (6) 2 kinds of mesembranthemum
  (7) 7 types of fungi
  (8) 4 types of gum
  (9) 2 kinds of manna or acacia gum
  (10) Flowers of several species of Banksia

This food inventory is so exhaustive as to indicate that very few food resources remained unexploited if they could provide enough energy to warrant their collection. Even Captain Grey's statement that the natives did not eat oysters may be open to some question, for it may have reflected a seasonal avoidance such as we practice. Certainly in other parts of Australia the rock oyster is a delicious shellfish and much relished by Aborigines and Whites alike. Other careful observers in Australia have also provided food inventories of the same exhaustive nature as the one listed above. Generalized hunters there and on other continents will eat anything available.

The organization of Australian society is more complex than that found among Shoshoni. Among them, as everywhere else, the basic biological family serves as the unit of procreation and economic exploitation. Aboriginal families spent a considerable portion of the time apart from others, whether living in the desert or in the better watered coastal regions. This reflects the fact that the family is the most efficient economic unit under most circumstances. The largest unit in Australian society is not a social one, but is a linguistic one, the *dialectal tribe*. As among the Shoshoni, its existence depends upon the pattern of face-to-face interactions, which operate in such a way as to keep language uniform among groups of neighboring families. The analysis of existing data (Birdsell, 1953) shows that the Australian tribe tends in a statistical sense to approximate 500 persons. This same number characterizes the size of such tribes among a variety of other hunting and collecting people. If during a sequence of good years a tribal population should breed up to perhaps 1,500 persons, there is a tendency for it to fragment as the increase in numbers allows. linguistic diversity to spring up. On the other hand, should the tribal population be reduced to perhaps 200 individuals, they become exposed to increased language pressure from their neighbors and may be absorbed by one of them. Travel by foot allows distance to become an isolating factor among Australians and of course all the Pleistocene people. Forces tend to break up tribes that become too large numerically, as well as to eliminate those whose numbers become too small for linguistic survival. The dialectal tribe represents a system maintained in a stable equilibrium.

An important social and territorial unit exists between the family and the dialectal tribe. It is the *patrilineal band* or horde. It is characteristic of all Australian groups and is found in most other generalized hunters except the Shoshoni. Such local groups consist in essence of a male line of descent, grandfathers, fathers, and sons who own and inherit the food rights to a tract of land whose boundaries are well defined. Associated with this patrilineal descent group are the women who marry into it, as well as the daughters not yet married out of it. Marriage rules are *exogamous,* that is, matches are arranged with the women of neighboring bands. The men remain in residence in their band territory, where their new wives join them. Daughters are given in exchange and go to live in the territories of their new husbands.

Precontact bands observed in Australia by N. B. Tindale averaged 25 persons of both sexes and all ages. Still functioning bands observed by Lorna Marshall (1960) among the Kalahari Bushmen in South Africa and by B. J. Williams among the hunting Birhor of North India both showed averages of 25 to 26 persons. The evidence from these three generalized hunters suggests that band size is remarkably stable and again operates within a balanced system. This somewhat magical number of 25 persons seems to be the optimum size of the local group which can maintain itself under a variety of social pressures, including the arranging of exchanges of daughters as wives. It may also represent the number of adult men comprising the best size work group among generalized hunters. Where the resources become richer, as along seacoasts and rivers due to marine food supplies, the Australian bands frequently range from 50 to 100 persons. These ecological circumstances allow people to live more densely and they choose to do so where possible. All such societies are so small that there exists a strong tendency to gather in larger numbers when food supplies allow, in order to minimize the boredom of seeing the same faces year after year.

The exogamous, patrilineal, landholding band represents a widespread and effective type of social organization. It may reasonably be projected well back into the Pleistocene since it requires nothing more complex than local territoriality, the necessity to travel on foot, to speak, and the desire to marry outside of one's own local group to minimize sexual rivalries. Archeological evidence is already suggestive that men of the Middle Pleistocene often lived in local groups of this size. The two important sites of Ambroyna and Torrelba in central Spain, dating from the end of the Middle Pleistocene, suggest this size of groups at kill sites.

The reader may wonder why any men remained living in a hostile environment when physically they could migrate to a better one. The answer is that *local group territoriality acts as a spacing mechanism to distribute men in the best ways throughout the area in which a living can be earned.* It provides opportunities for the greatest number of people by scattering them throughout the total environment without overcrowding. In this sense it serves to safeguard the survival chances of the species. *Territoriality characterizes virtually all terrestrial vertebrates—reptiles, birds and mammals.* It does not consist of ownership of real estate, in our modern sense, but rather of the establishment of rights to utilize the food products of a given area. Territoriality is established both by real and by symbolic fighting. Birds sing to post their territory and intruders of their own species are driven off. A sense of territoriality operates to even out physical differences. A small dog will successfully drive a much larger one out of its own yard or territory. The larger animal senses that it is an intruder and so leaves when the smaller dog yaps his objections and outrage. This accommodation is usually achieved with no real fighting. Among men like the Australians, uninvited trespassers were speedily speared, but of course on most occasions friends from outside the local group were invited in to share a temporary surplus of food and to engage in social and ceremonial events. The Australians show a tendency to arrange their local group territories so as to maximize the reliability of food support base.

Land is usually divided so that it provides reasonably reliable food for all the seasons of the year. Even in the most arid stretches of the Australian desert, the patrilineal band remains the functioning social unit. If a severe drought requires a band to leave its area, it seeks temporary hospitality among other friendly groups, usually those with which it has by custom exchanged daughters. When the crisis is over, the strong feelings of territorial attachment and local ceremonial sanction invariably draw the displaced band speedily back to its own land. The patrilineal local group is a successful spacing mechanism that must have prevailed in many regions far back into the Pleistocene. Where exceptions occurred, there would have been good ecological reasons for them.

Density Variations in Aboriginal Australia

The fact that the Aborigines eat everything edible in their environment makes it evident that their numbers must somehow be closely related to the amount of food available. The relationship is orderly and mathematically rigorous. After many years of experience in the field, Norman B. Tindale (1940) published a map containing the boundaries of 400 tribes in Australia. The fact that desert tribes occupy large areas and those in better favored environments only need small ones made it clear that the environment was influencing Australian populations. Some years later a mathematical analysis of Tindale's data (Birdsell, 1953) showed that in this arid continent a single environmental variable, *mean annual rainfall,* sufficed to relate men to their land with mathematical precision. The data were fitted well by an equation of logarithmic form which need not be given here. The observations fitted this curve sufficiently well to give a coefficient of curvilinear correlation of 0.81, which is high for two unrelated factors occurring in nature. The shape of the relationship is shown in Figure 13–1.

The curve allows reasonably accurate predictions of how many square miles of land are necessary to support one person anywhere in Australia. One of the striking features of this relationship is that human densities are continuously related to changes in rainfall, and some portions of the curve respond very sensitively to differences in it. In going from the most arid region with but four inches of rain annually to one in which 20 inches falls, an increase that is fivefold, the land needed to support one Aborigine is reduced about twelve-fold. In this part of the curve very slight differences in rainfall greatly affect aboriginal densities. Above 40 inches annually, human densities increase only very slowly. In the eucalyptus woodlands, in the 40-inch rainfall region, both the plants and animals are very different from those which grow in the climax rain forest, which receives more than 100 inches a year. Yet in passing from one type of

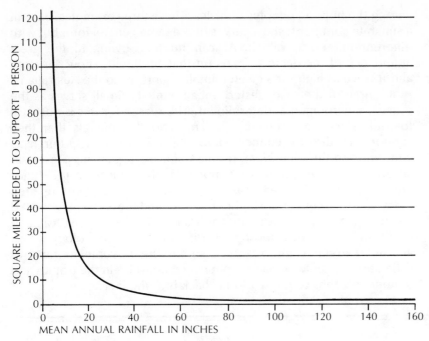

Figure 13-1
Relation between Men and Land
in Aboriginal Australia

environment to the other, changes in aboriginal density are both slight and continuous. Our curve indicates that in all environments in Australia the Aborigines were pressing very close to the maximum density which the country could support on a long-term basis.

Ever since the original work in 1953, it has been evident that the relation of men to their land in aboriginal Australia is more complicated than the simple relation of density to rainfall. In 1973 I had the opportunity to spend another year in Australia examining this problem at a more complex level. The preliminary results at this time indicate that the correlation between aboriginal densities and 65 environmental variables, many of which are related to rainfall, gives a coefficient of curvilinear correlation of 0.96. This spectacular relationship can best be understood by assuming a model in which the correlation equals 1.00. In such a case aboriginal people would be spaced entirely in terms of the environmental variables. Since the area occupied by a tribe correlates exactly with the numbers in the tribe, it follows that each tribal population would rigorously number 500 persons. The correlation of 0.96 is so close to the maximum possible value that it indicates a virtually rigorous environmental determinism for these people and nearly as tightly a self-regulating population system of numbers.

Similar relationships have been found by graduate students of mine for the Shoshoni of the Great Basin in North America and among the hunting peoples of Africa. These three examples, each in a separate continent, show curves of the same general form but with different numerical values. African hunters, owing to the great abundance of the game animals on that continent, show densities about twice as high for a given rainfall regime as do the Australians. In all three of the cases tested, mean annual rainfall serves as the primary environmental determinant. The reason for this is not hard to find, since the growing of plants and the animals that they support are directly connected to utilizable rainfall. Other facts contribute to the total *biota,* or mass of living organisms present in an area, but their influence is not enough to disguise the major impact of rainfall. These cases make it clear that hunting men everywhere, except in the higher latitudes and the Arctic, are related to their land by the amount of rainfall in a mathematical and predictable fashion. Allowing for differences in technology which are not extremely important, it should be possible in time to estimate accurately human numbers and densities in the tropics and temperate zones of the world far back into the Pleistocene.

Man in Balance with Nature

The rigorous way in which the environment determines human densities among hunters, and consequently their numbers, shows hunters everywhere at a given time existing in a dynamic state of balance. The balance is not steady, but constantly shifting. If one local group increases its numbers and succeeds in maintaining them, then another one nearby will be diminished through the action of compensating forces. Overpopulation frequently results in pressure which leads to conflict between groups, and the casualties that occur may be the releasing force. When numbers become excessive, they will be knocked back by local starvation. The men of an expanding group may find difficulty in obtaining women as wives, and so that group suffers from reduced birth rates in the next generation. All of these dislocations tend to produce more personal stress and more intergroup quarreling. Natural forces act in many ways to prune off surplus numbers both in men and among other animals.

The Role of Systematic Infanticide in Human Evolution

People tend to manage their affairs to their own best advantage. This optimizing principle works among simple hunters. One of the obvious ways in which people can maintain their own numbers at the best level involves the murdering of excess, or awkwardly spaced, children. The motivation for the killing of the newborn most frequently arises from the mother's inability to properly raise it. But there is some evidence to indicate that group policy is involved in the mother's decision-making. To understand such antisocial behavior, we must examine its causes in detail.

Genealogical Background for Aboriginal Infanticide

During 1953 and 1954 Tindale and I engaged in field work around the great Western Desert of Australia. In the genealogies we collected, 194 referred to the aboriginal matings prior to the time of contact with Whites. A carefully taken genealogy provides much information, including the number of children born to a couple, their sequence, sex, and as many other matters as the investigators may choose to inquire about. Genealogies inevitably contain some systematic errors involving human forgetfulness. Some infants that died very young are omitted, and individuals who in rare cases move away from their band tend to be omitted. One of the errors in the genealogies taken involves the fact that children who did not live into adulthood tend to be forgotten. But, by and large, genealogies provide the only way of reconstructing the numbers of people who lived in the past among groups who have no written records.

Sex Ratios. These genealogical data were analyzed in terms of families arranged according to the number of surviving children. The *sex ratio* was then calculated among the offspring of the precontact mating by dividing the number of male children by the number of females. In ordinary populations this value at birth runs close to 100 percent, indicating that about as many boys are born as girls. The Australian data are shown in Figure 13–2. In families of all sizes boys outnumber girls to a degree that is well outside of errors resulting from sample size. In precontact families in which one child grew to adulthood there were 186 men for 100 women. In the more moderate-sized families in which two and three children survived, the sex ratio was reduced to about 130 percent. Among the larger families in which four children grew up, there were 185 males for each 100 female children. Finally, in the largest families, in which five children were remembered, the sex ratio rose above 260 percent. Taking all the families together, the ratio of male children surviving to females was 153 percent.

Methods of analysis that need not be detailed here show that these deviations in sex ratio were the consequence of human behavior rather than the result of deviant biology. The high sex ratios clearly reflect family planning, and exhibit a marked preference for male children. This preference became exaggerated in families which were either smaller or larger than the average. The motivation for preferential male survival can be found in the cultural

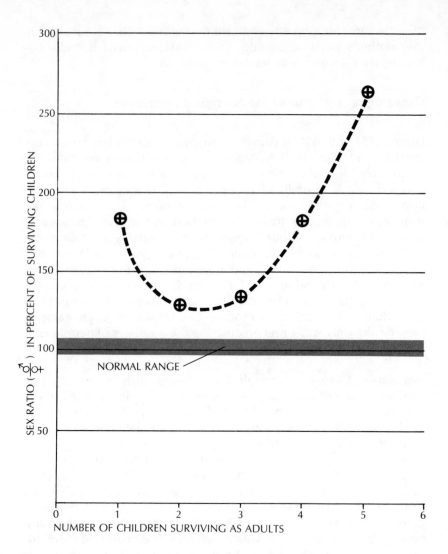

Figure 13-2
Sex Ratios as an Indicator of
Preferential Female Infanticide
among Precontact Australian
Matings

realities of Australian life. Men are the hunters and so provide the
meat. They are the fighters and so defend the band. They hold the
territorial rights to their country and so live in patrilineal descent
groups. The fact that parents will grow old in the territory of their
sons certainly provides motivation to have and save male children.
Custom is such that the parents continue to receive the best
portions of the animals killed by their active adult sons. Altogether,
Australian society is male-dominated, and so there are good functional reasons why, where the numbers of children must be limited,
males should be preferentially spared.

Factors in Decisions for Infanticide

Evidence not detailed here points to the fact that in precontact times, prior to disruption by Whites, the Australians killed up to one-half of the total number of children born. Three major factors are involved in this seemingly cruel practice. Two of the limitations involved the mother's inability to care for all the children to which she could give birth. The third factor is more subtle and involves conscious population planning.

Physical Factors. As an introduction to the subject, let us consider the fertility patterns of human females. During the period of sexual competence the mature woman ovulates once each 28-day lunar cycle, and so for a short period thereafter she is capable of conception. This cycle of ovulation continues unless interrupted by pregnancy. In all mammals but man the nursing demands of the newborn inhibit further ovulation until the lastborn is effectively weaned. The regulating mechanism is hormonal and serves to efficiently space births so that new conceptions do not occur until the lastborn are out of the way. Wild mammals never need to practice infanticide.

Present-day women follow the basic mammalian pattern but do so imperfectly. The available medical information indicates that human females are not always able to suppress the resumption of ovulation while still nursing the last child. The ability to suppress ovulation is somewhat variable in different women, with some following the proper mammalian pattern perfectly and others conceiving a new child within one month after the birth of the last. The pattern of distribution between these two extremes is medically unknown. In most simple societies, including both those of hunters and primitive agriculturists, children are weaned only after the third year, and before the fourth. In hunter groups women cannot nurse two children simultaneously, so that where birth spacing is improper the newborn must be killed for the benefit of the older child.

There is a second factor which may spread the interval of spacing of births beyond the fourth year. Among hunting peoples camps may be moved almost every day, and the ability of a mother to transport young children acts as a second bottleneck. Even when the band or family remained camped in the same spot for some time, the woman's economic role required her to gather food for some hours each day and at distances as much as three to four miles from the campsite. Most mothers can sling a three-year-old across the small of their backs and still walk along with their digging sticks and containers. But no mother can effectively carry two children simultaneously in this fashion. Good precontact evidence among the

Australians indicates that on longer journeys children as old as five or six years may have to be carried a part of the time by their already overburdened mother. This transportation factor adds further years to the ideal spacing pattern of births.

Group Fertility Ratios.   A third and final act involved in infanticide is concerned with the reality that the fertility of a group is determined by the number of its adult women, rather than by its adult men. Undoubtedly, one able-bodied male could keep ten women continuously pregnant. This would produce the same number of births as if the group consisted of ten men and ten women. But if we can imagine a local group consisting of ten men and only one woman, the birth rate would necessarily be only ten percent of the former examples. *The number of women determines the rate of fertility.* Most Australian data were not obtained by sufficiently careful investigators to demonstrate that conscious population control is a part of the rationale for infanticide. But one capable professional woman doing research on the problem has found that in her North Australian community one of the senior women accompanied the young mother into the bush for the act of birth, and had an important role in deciding whether the newborn child should be killed or spared. The presence of this elder strongly suggests that this painful decision-making also included the group interest as a major component. Presumably there would be times when the mother might be able to rear another child, but it would not be to the best interests of the local group to let it live, especially if the birth involved a girl. Simple hunters are perfectly capable of this kind of direct population control, for they, too, recognize that it is the number of women which controls the rate of reproduction.

The fact that the mother of a newborn baby can kill it directly after birth of course represents murder of the most fearful sort in terms of our Western ethics. The magnitude of the crime is further compounded because the mother is flooded with the hormone prolactin at that time, which conditions her psychologically to love the newborn child fiercely. But sheer necessity overrides ethics and maternal feelings. Our behavior is sufficiently complicated that we might expect some sort of rationalization to have been developed to help the mother overcome her normal guilt feelings after the killing of her newborn. In fact, the Australians have developed a suitable rationale. Throughout Australia, there existed a systems of *totems*, that is, spirits which all of the Aborigines associated with some features in the landscape. The totem spirit usually involves animals, birds, or economically important plants. The spiritual essence of a totem is situated at a focal point, frequently a water hole, and it is from this totem spot the spirit of each newborn child enters the body of the woman who is to bear it. After death, the spirits of all

Aborigines return to the same totemic spot from which they originated. This is one of the very important factors which relates these people spiritually to their country. Consequently, a woman forced to kill her newborn baby realizes that the loss is not a spiritual one, for the spirit of the dead child returns to the totem center and likely will come again to enter her body the next time she becomes pregnant. This emphasis upon the spiritual value, as opposed to the bodily importance, undoubtedly operates to relieve the conscience of the prolactinized mother at the time she commits the murder. Other peoples in other times have committed infanticide and not all had developed similar kinds of emotional compensation.

How men can have multiple wives when they outnumber the women is explained in Supplement No. 18.

The Problem of the Specialized Hunters

Men become generalized hunters when food resources are varied and scattered. Where one or two food resources are sufficiently abundant, hunters prefer to tie their economies to them. The buffalo hunters of the Great Plains of North America illustrate the point. The herds of buffalo seemed to present so inexhaustible a supply of meat that when the Indians obtained the horse indirectly from Spanish sources, many of them migrated out onto the plains to become full-time parasites on the herds of buffalo. They became the human equivalents of the buffalo wolves which had lived off the young, aging, and sick of the buffalo herds for thousands of years. The biomass presented by these herds was such that several hundred Indians could meet in various places during the summer to conduct the great buffalo hunts. Such gatherings naturally required the development of new forms of social control, and so during the summer months their societies became complex compared to those of other hunters. During the winter months they broke up into small groups to live out the bad season, and often died of starvation after exhausting their cured buffalo meat.

In California several species of oak trees provided another specialized food base in the great annual crop of acorns. Throughout the interior of the region the Indians became specialized acorn gatherers, with most of their diet and economy revolving about that food. It enabled them to live in large, stable communities involving 100 or more people for most of the year. They varied their diet with other vegetable foods, such meat as they could obtain, and fish and honey when available. Their settlements were as large as those of primitive agriculturists, and so they, too, had to devise new means of ordering society to make life bearable in such relatively large gatherings.

The northwest coast of Canada provided for another type of

specialized hunting existence. There the rivers were seasonally filled with great runs of three or more kinds of salmon and the inshore waters of the Pacific provided a great variety of other fish. Indians living primarily at the mouths of these rivers perfected their fishing techniques so that they could live in substantial groups, again in villages numbering 100 or more people. Their life was sufficiently sedentary so that they erected substantial communal houses built out of beams and planks of split red cedar. The wealth of their food supply provided them with the luxury of stratified classes, the privilege of waging war, and the ostentatious destruction of real property.

The Upper Paleolithic communities of Europe represented yet another kind of specialized hunting society. Although the climate was near Arctic, the sparse tundra vegetation of the plateaus supported a seasonally great biomass of game animals. Reindeer, woolly mammoths, and a variety of other big game traveled in herds through regular annual movements. The hunters of the period were relatively well off in the caves and the rock shelters along the bottom of the river valleys. Archeologists are convinced that they must have lived in communities of several hundred persons. Since the climb to the tundra plateaus was only a matter of a few hundred feet, they had a large supply of game accessible nearby. It is probable that their communities also required the development of rather elaborate forms of social control in order to make them manageable. Certainly their social organization must have been as complicated as that of the modern buffalo hunters, salmon eaters, and acorn gatherers. No Arctic peoples today serve as models to represent the Late Pleistocene type of economy and society.

The Caribou Eskimo

Most of the Eskimo living along the Arctic coast subsist primarily upon sea mammals such as seals, walruses, and the smaller whales. No people in the Pleistocene are known to have lived in this fashion. But in the so-called Barren Grounds of north central Canada we find a people who provide some information applicable to the hunters of the last glacial advance in Europe. They are appropriately called the Caribou Eskimo, for it is the caribou which provides the real staple of their economy. The Barren Grounds are typical Arctic tundra upon which no trees grow, and all the vegetation is low and prostrate in the form of mosses, lichens, and dwarf willows. The country is monotonously flat and extremely cold in winter. During 1922–1923 these Eskimo were studied by Kaj Birket-Smith, who found them in virtually a precontact condition. He counted 432 Caribou Eskimo who were divided into four dialectal subtribes. Allowing for a few people not counted by direct census methods, the total tribal population would once more approximate 500 persons. The area they lived in was greater than that of all the New England states, but they were so sparsely scattered that about 125 square miles was required to support each person. This is the lowest recorded density known for modern man.

The caribou, which are the New World representatives of the reindeer, spend their winters on the northern fringe of the coniferous forests lying to the south and in the summer move in great herds out into the tundra, remaining there until fall. The Eskimo hunt them and must, in the late months of the fall, obtain enough meat to see them through the winter. Occasionally they fail, owing to the reindeer migrating by different trails, and then whole groups die of famine. The Eskimo hunt the caribou with bows and arrows which can shoot accurately only to distances of 50 or 60 feet. This is in no way superior to a spear projected by the spear-thrower in the hands of a competent hunter. Consequently, the Eskimo hunter uses a variety of methods to bring himself close enough to be sure of killing his game. It is advantageous to hunt during snowstorms. They stalk the animals with caribou horns as a disguising headdress. They drive them into lakes where they can be dispatched with lances by men in skin kayaks. Some of these hunting methods may resemble those of the Late Pleistocene hunters of Europe.

A good many anthropologists have reconstructed Upper Paleolithic life around the idea that the hunters followed the herds of animals as they moved through their seasonal round. The evidence from the Barren Grounds disputes this. The Caribou Eskimo wait for the herds to come and ambush them at suitable points. After the caribou have left for the winter, there always remain behind some stragglers of the herd, and these provide additional meat for the worst part of the year. These classic hunters of the caribou reside in their own fixed territories and do not trail the moving herds. It is very likely that our European ancestors followed similar practices.

The clothing and shelter of these people again throw some light on what must have been necessary for life in late glacial Europe. Like all Eskimo, they wear tailored clothing, but of a somewhat primitive pattern. In the deep months of the winter they live in snow igloos, some of which have skin linings to make their temperatures more comfortable. During other months of the year they live in skin-covered circular tents which are transportable when broken down. There is archeological evidence for similar kinds of shelters in Europe during the Upper Paleolithic. Unlike the maritime Eskimo, these people do not have seal oil to fuel their lamps, and what little heat they have for cooking comes from twigs of bushes collected by their women. Economically they stand as impoverished models of the kind of life that must have been possible around the edges of the last glacier to advance into Europe.

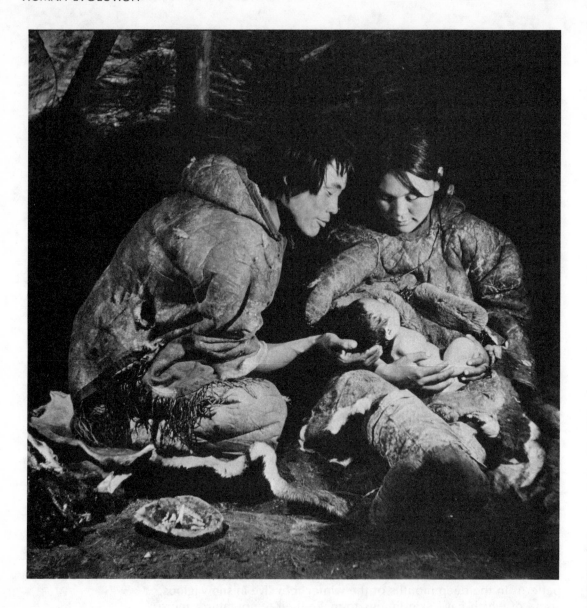

Plate 13-3
This Caribou Eskimo Family demonstrates that the human family can be beautiful even on the harsh tundra of the Barren Grounds.

Birket-Smith classified these people by age and sex. Only two categories of age were provided, immature and grown people. For the four subtribes of the Caribou Eskimo the 230 adults showed the surprisingly low sex ratio of 79.8 percent. In effect, there were only 80 grown men for 100 women. Since the men are the exclusive producers of food in this economy, these figures obviously mean that some of them had two wives, and they must have been the superior hunters to support them. Turning to the subadults, Birket-Smith found the unnaturally high ratio of 145.1 percent for the 201 juveniles involved. In all four subtribes the separate figures follow the same pattern, so that it is certain we are not dealing with accidents of sampling in these distorted sex ratios.

The anthropologist who collected these data merely reported that infanticide had previously existed but did not comment upon these unusual sex ratios. With the insight provided by the Australians, it is possible to give a more detailed interpretation. The high sex ratio among the immature can only mean that infanticide was preferentially practiced upon females. The difference involved the death of at least one out of every three girl babies, over and above the children of both sexes who were killed to provide the necessary spacing. Various maritime Eskimo show similar sex ratios among their immature members. The very low sex ratio found among the adult Caribou Eskimo indicates that one-half or more of the boys die in manhood. Males are lost primarily through accidents and the dangers that go with hunting in the Arctic. Death by freezing, assault by animals, and accidents leading to drowning all take their toll. The interesting point about this Caribou Eskimo census is that it provides definite evidence for population planning from generation to generation. There is no other explanation for the high sex ratio among the immature other than that infanticide was responsible for it. The low sex ratio among the adults, which ranges only between 72.3 and 83.3 percent for the four subtribes, testifies to the success of their population planning. These values represent a balanced state in which a minority of superior hunters could take over widows and children of the men who died. What a pity that Birket-Smith did not press his research in this direction, for it must be that the Caribou Eskimo had very explicit ideas on population control as a necessary aspect of survival in their rigorous environment.

Conclusions from Living Hunters

The Shoshoni, Australians, and Caribou Eskimo live in two different continents and under very different circumstances. The animals they hunt and the plants they eat all vary. Their climates range from the wet tropics through deserts to the Arctic. But all three groups show certain features in common that suggest that they may be projected back into the Middle and Late Pleistocene period. In each instance, the biological family is the basic unit of society. In all three cases, the largest recognizable unit is one of linguistic character, the dialectal tribe. Where no political authority is present, as it is in our model populations, the dialectal tribe approximates 500 individuals. Territoriality is manifest in all three groups, but in differing forms. Among the Shoshoni food resources are claimed by the simple right of first arrival, and because of the erratic nature of their distribution, territories are not bounded. Among the Caribou Eskimo a regional territory is well defined, but the migrating caribou allow no finer subdivisions in claimed ownership. Among the generalized hunters in Australia, band territories are rigorously defined and defended. This type of behavior is found among most other generalized hunting peoples. In all three cases infanticide is practiced systematically so as to space children to allow them to be raised by their mothers. Where the data are well known, infanticide is preferentially female. This bias results from the fact that hunting societies are male-oriented, and, among the Eskimo, hunters are the only primary producers. On the other hand, women represent the medium through which fertility can be controlled. Society seems to have taken part in the planning of population numbers. These uniformities may reasonably be considered applicable to hunting and foraging men everywhere, except where local ecology suggests changes in the adaptive pattern of behavior.

Bibliography

BIRDSELL, JOSEPH B.
1953 Some Environmental and Cultural Factors Influencing the Structuring of Australian Aboriginal Populations. The American Naturalist. Vol. 87, No. 834, pp. 171–207.

BIRKET-SMITH, KAJ.
1929 The Caribou Eskimo. Report of the Fifth Thule Expedition, Vol. 5, parts 1 and 2. Copenhagen.

GREY, GEORGE.
1841 Journals of Two Expeditions of Discovery in Northwest and Western Australia During the Years 1837, 38 and 39, 2 vols. London: T. and W. Boone.

KIDDER, ALFRED V.
1958 Pecos, New Mexico: Archaeological Notes. Phillips Academy, Andover. Robert S. Peabody Foundation for Archaeology. Papers, Vol. 5.

MARSHALL, LORNA.
1960 Kung Bushman Bands. Africa, Vol. 30, No. 4, pp. 325–355.

RENSCH, BERNHARD.
1960 Evolution Above the Species Level. Columbia Biological Series. New York: Columbia University Press.

SCHALLER, GOERGE B.
1967 The Deer and the Tiger. Chicago: University of Chicago Press.

STEWARD, JULIAN H.
1938 Basin-Plateau Aboriginal Sociopolitical Groups. Smithsonian Institution, Bureau of American Ethnology, Bulletin 120.

TINDALE, NORMAN B.
1940 Distribution of Australian Aboriginal Tribes: A Field Survey. Transactions of the Royal Society of South Australia, Vol. 64, pp. 140–231.

VALLOIS, HENRI V.
1961 The Social Life of Early Man: The Evidence of Skeletons. In Social Life of Early Man. Edited by S. L. Washburn. New York: Viking Fund Publications in Anthropology, No. 31, pp. 214–235.

How Do Those Old Men Get All the Girls?

The discussion of preferential female infanticide in the body of the preceding chapter indicated that the Australians, some Eskimo, and no doubt many other peoples controlled the sex of the children saved at birth in such a way as to keep about 150 boys for every 100 girls. Yet in most of the same societies in which all men marry, a minority of the men have two or more wives. Among the precontact desert Australians 15 percent of the men had multiple wives, ranging in numbers from two to five. The problem, of course, is how to reconcile an excess of males with the practice of *polygyny,* or the marriage of one man to more than one woman.

In spite of the excess of males, this type of marriage pattern can be achieved. Figure 13–3 represents a model tribal *population pyramid* in which both sexes have been arranged by their numbers in each given age group. For our purposes, ten-year intervals have been taken as the suitable group size. In populations this small, chance variations affect each age group, so that

the steps in the pyramid are uneven. A sex ratio of 150 has been used, based upon infanticide data. It follows that the 20 or so bands that make up the dialectal tribe will each show much greater irregularities by age and sex than does the full tribe of 500 persons. Such chance deviations pose difficult problems in practice. For example, a given band may need two women as wives for its unmarried young men. The operations of chance are such that in a band population of 25 there might not be two girls of marriageable age to give in exchange. With a system of *exogamous* marriage, the brides must be provided by an outside band, but with a strict understanding that they are on loan, and an equal number of women will be provided by the first band when in time they have two girls of marriageable age. It is processes of this kind which very likely operate to make the average band number of 25 one which is difficult to fall below and still maintain social responsibilities. In practice in Australia and elsewhere, the bookkeeping accounts of women given and

Figure 13–3
Model of a Precontact Australian
Population Pyramid

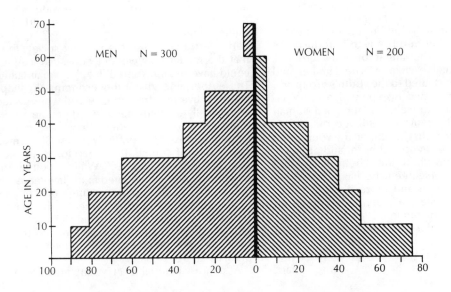

owed are kept very rigorously in the minds of the old men, and a full return is expected when circumstances allow. The giving and taking of women in marital exchange between bands is, of course, one of the mechanisms which originate and maintain friendly relations. The balance in terms of women is very seldom even, and so bands have various social debts and credits with surrounding groups.

To return to our basic problem, which is the difficulty of providing a plural number of wives when the men are more numerous, the solution depends upon the way in which marriages are contracted. Among Australians the problem is solved by deferring the age at which men can marry. The system works in the following fashion. Young men pass through a ceremony and are usually fully initiated by the age of 16. Their society now requires that they demonstrate that they are economically self-supporting before they are considered eligible for marriage, a perfectly reasonable test. So for two or three years the young initiated men live in bachelor camps by themselves and away from the married couples. Once a youth has demonstrated that he can provide his own food, he becomes eligible for marriage. A suitable relative, frequently his mother's brother, then takes him on a tour of the country and introduces him to the camps of people all around his own territory. Sometimes these "grand tours" in the desert may last for several years and cover many hundreds of miles. One of the primary purposes of this pilgrimage is to show the young man the surrounding country, to familiarize him with its totemic spots and water holes, and so widen his geographical horizon should he have to move out in case of drought or famine. But another purpose is to introduce him socially to men who might owe his group a woman in marital exchange. The families and bands who are indebted to the youth's group are, of course, known, and his sponsor exerts considerable pressure to make certain that a woman owed to them be returned. But the rub in this whole procedure is that in general the young man is promised a woman not yet born, and probably not even conceived. If this "betrothal" takes place when he is eighteen, he has at least one year to await a birth and more likely a longer period. For if the wife of a man who has promised his next daughter to him in fact conceives a male child, his wait is prolonged. Even after a promised female baby is born, and spared in terms of the spacing problem, our young bachelor faces several other risks before he takes a wife. Infant

mortality is high, and the little girl may die from a variety of diseases. If he is very fortunate, he has spread his risk by having two or more men promise their daughters to him in marriage. But in any case he must wait a long time for his "fiancee" to grow up.

The age of marriage for girls in Australia is at about the age of puberty, or perhaps even a few years earlier. This means the young bride may be somewhere between 11 and 14 years of age. If he is very fortunate, the man may, at the age of 30 years, obtain a wife, but with preferential female infanticide the hazards of health and life, he is more likely to receive his young bride after he has reached an age of 40 years. This is approaching old age in aboriginal society, and so the old men do get the young girls. Figure 13–3, which is not exact, has 35 men between the ages of 30 and 40 years. At the same time, there are 50 girls in the marriageable range from 10 to 20 years of age. There are enough girls for the old men.

But a man may not remain a bachelor until he receives his promised bride, for widows of the very old men are immediately allotted to younger men who stand in the correct social relationship. Life for hunting peoples is hard, and the women particularly show the signs of age early in life. Under traditional conditions the Australian woman of 35 looks more like 70. So a hand-me-down bride is no great luxury, but if she is healthy, she may be something of an economic asset, for women provide about 70 percent of the total food in Australia. In any case she is a woman. So the society has invented an interesting scheme which works. The old widows go to the young men and the young, nubile girls go to the older men.

The restraints are so slight in our society today that the reader may wonder how such a system could have been invented, let alone maintained. These Aborigines, like other economically simple peoples, live in a traditional world where values are virtually unchanging. The old people are the repositories of knowledge, they have demonstrated success by merely living for 30 or more years, and, of course, they keep forces of social control "in their pockets." While a young Australian may be fully initiated in his late teens, the inner world of the ceremonial life is only opened to him gradually and in installments. The spiritual values of his culture are revealed to him at such a rate that he is not fully acquainted with the full sweep of the ceremony until he has reached perhaps an age of 50.

At a more material level, young men are under

a rigorous set of food taboos which provide that they must give the choicest parts of the various animals they kill to their older relatives. These food distribution rules are exactly defined, so that if a young man brings a kangaroo carcass into camp, one hind leg goes to a certain relative, another to a different one, and so on as the carcass is butchered, and he ends up with the worst part of the animal for his own meal. These rules are submitted to, for there is no place to go if you leave your own band, except to neighboring bands who live by the same dogma. But the system has its points, and presumably the young men see this, for they make certain that the able-bodied men feed their elder relatives—parents, uncles, and the like—who are no longer able to hunt vigorously. It is a kind of old age insurance, and the young men know in time they will profit from it. This is one of the reasons that male children are preferred.

Australian society, and that of many other living hunting peoples, is dominated by males, and particularly by the old men. They have designed the system of individual rewards in such a way that young men are placed under constant constraints until they themselves reach old age, and so become the decision-making group. Naturally the old men have arranged that they should get the young girls. The devices by which the young men are maintained as tame young animals obviously include forms of supernatural sanction, threats to withhold a promised wife, group approval, and the knowledge that they will have their turn someday, if they live long enough. This is a stable type of society, and it is well adapted to the hunters, whose lives are ordered by a whole array of unwritten laws, which prescribe certain punishments for certain offenses. The young men in general accept this traditional pattern of life. And of course even in such small societies, real rebels can be speared in the back if they prove too troublesome. Living within the system carries the further reward that as a man grows older he can trade his daughters for more young wives.

CHAPTER 14

A VARIETY
OF MATING
FACTORS THAT
STRUCTURE
HUMAN
POPULATIONS

The way in which individuals choose to mate, or marry, may be their own affair. But different patterns of mating have consequences that may genetically affect the structure of populations. The genetic makeup of the individuals that comprise the population may be influenced by the mating pattern of their parents. Systems of mating may be essentially random, show preferential seeking or avoidance of certain kinds of mates, or involve inbreeding. Each of these variations has the potential for different consequences. The mating patterns alone do not produce evolutionary change, although they may set the stage for it. Fortunately, the deviations in mating patterns which can occur in theory seldom are found in real populations.

The term *random mating* suggests complete sexual promiscuity and could be achieved this way. But matings may be entirely random in strictly monogamous societies. Mating patterns are random when every individual of one sex of marriageable age has an *equal opportunity* of marrying any other mature individual of the opposite sex. In essence it simply means that mating patterns are without any bias. In random mating in an ordinary population, a man marries a woman who is, as a rule, unrelated to him. But at the same time a sister, or even several female cousins, might be included among the marriageable women. Since there is no bias in the choice of a wife, a man by pure chance might very well mate with one of these closely related female relatives. Such a mating would constitute genetic inbreeding, and the random pattern of mating would predict it to occur as a rare event. But even in a real population, most of the eligible young women will be unrelated to the man making the choice and generally will come from the more distant parts of the area in which he is wife hunting. The importance of the random mating pattern is that it is easy to handle mathematically, and so it provides a base line from which deviations occurring in other mating systems can be detected.

The Nature of Random Mating

Factors Affecting Random Mating

The idea of random mating is simple enough at the level of mathematical probabilities, but does not strictly apply to real human populations. All human societies prohibit marriage between close biological relatives, that is, have an *incest taboo*. This universal law causes real marriages to deviate somewhat from those predicted by the random pattern of mating. On the other hand, men everywhere have their courting activity limited by distance. When men walked to do their courting, their radius of action was probably limited to the investment of an hour's time going and another returning, or a radius of about four miles. With a horse to ride, a comparable radius of courtship perhaps rose to six miles. The invention of the bicycle might have increased courting distance enough so that all women within a ten-mile radius would be potential spouses. Today, with the automobile, the radius of effective courtship has certainly risen to 30 miles. But in all these instances, women outside of the cruising radius were not actual candidates for marriage with the given man, and so distance acted as an *isolating factor*. In effect, this meant that many women in distant portions of the community were not available. Some years ago sociologists investigated the pattern of marriage in a modern city and found that on the average couples had lived about a mile from each other prior to marriage. Despite those who like to think that marriages are made in heaven, these figures suggest that proximity is what makes for marriage. Men are socially prohibited from marrying their close female relatives and can only court at distances which are practical, so these two deviations from random mating tend to cancel each other, with the consequence that matings in most societies are, in effect, random.

The mathematical model of random mating differs from the pattern practiced by men in that mathematicians assume no overlap between generations. In some short-lived animals this condition exists. But for long-lived animals like ourselves, generations overlap in practice in a confusing fashion. In a given family line parents can always be distinguished from their own children, but parents and children are not so cleanly separated in the population at large. The point may be illustrated with a simple example. Let us assume that a man of 45, who already has three children by his first wife, is widowed, and later marries a girl of 20. Agewise, the new wife belongs to the generation of his other children, but the act of marriage raises her nevertheless to the status of coparent with him. When they have children, it will be difficult to relate them generationwise to the children produced by his first wife. A more important cause of the blurring of generations in real populations arises from the fact that childbirth is a continuing process, and so each year many new young men and women reach the age of marriage. Obviously, in actual human populations generations flow on continuously, and do not consist of alternating sets. Fortunately, this makes very little difference in evolutionary problems.

The Hardy-Weinberg Law

The most important mathematical use of the concept of random mating is to project the frequency of genes in a total population into the phenotypes of the individuals produced in such a mating pattern. The population is the basic unit in evolution, and the individuals who make it up consist of nothing more than transient packages containing a certain combination of genes. Each individual has received his unique combination by the gametic contributions from each parent. While it may seem humanistically degrading to so view ourselves, nonetheless all evolutionary processes are conceived of in terms of the population and the genes contained in its so-called *pool.* There are various ways of counting the genes in each individual sampled in a population, and it may be by direct methods or by mathematical estimation. But there is only one way of taking frequencies of genes in the population pool and predicting how they will be combined in separate individuals. This is achieved by the Hardy-Weinberg equation, which was published simultaneously in 1908 by Hardy, a British mathematician, and Weinberg, a German physician.

A binomial equation is nothing more complex than an equation which has two terms on its left-hand side. The Hardy-Weinberg equation is such a one. Let us start with a simple genetic system in which two allelic genes, A and B, occur at a single locus. Their relationship can be numerically written as:

$$A + B = 1$$

This simple statement merely says that the sum of the two alleles equals unity. This binomial equation is a very simple one. Since the genes exist in populations in different frequencies, let us rewrite this equation introducing the frequency P for the gene A, and the frequency Q for the gene B. Our elaborated binomial equation now reads as follows:

$$PA + QB = 1$$

Now we have indicated what genes occur in the pool of the population and specified the frequencies for each. It should be evident that the sum of the frequencies also equals one, so that P + Q = 1.

The conversion of gene frequencies in the population pool into the combination of genes in all the individuals in the population requires a mathematical way of drawing genes, two at a time, and at random, from the pool. This is easily achieved by squaring both sides of our basic equation.

$$(PA + QB)^2 = 1^2$$

In practice this can be achieved by simply multiplying each side by itself, so that PA + QB is multiplied by PA + QB and 1 is multiplied by 1.

$$PA + QB \qquad\qquad = 1$$
$$PA + QB \qquad\qquad = 1$$

$$P^2AA + PQAB$$
$$\qquad\qquad PQAB + Q^2BB$$

$$P^2AA + 2PQAB + Q^2BB = 1$$

The product of this multiplication now gives the genotypes of all the kinds of individuals possible in this population and the frequencies of each. Thus, homozygous A individuals, that is, those with the genotype AA, occur with the frequency of P^2. Homozygous B individuals, that is, those who are BB above, occur with the frequency Q^2. And the heterozygous individuals, those of the genotype AB, occur with the frequency of 2PQ. It is easy to derive these same results quickly by saying the first term of the binomial is squared, then the second term is squared, so producing the frequencies of both homozygous kinds of individuals, while the heterozygotes are obtained by taking twice the product of the two terms in the equation.

An example becomes appropriate at this stage. Let us assume that we have a population in which the gene A has a frequency of .3 and the gene B a frequency of .7, so that the sum of their frequencies is equal to 1.0. Now substitute these frequencies into the Hardy-Weinberg equation and it reads as follows:

$$(.3A + .7B)^2 = 1^2$$

If we multiply this out, the population contains the following genotypes:

$$.09AA + .42AB + .49BB = 1$$

In other words, the *individuals* of this gene pool consisted of nine percent homozygous A's, 42 percent heterozygous AB's, and 49 percent homozygous B's. The mathematical reason for the squaring of the binomial equation is that man, like most other animals, is diploid, with chromosome sets in pairs, and has two genes at each locus. The random combinations of the genes at a locus is mathematically achieved by squaring the binomial equation. Some organisms are not diploid, but we are not concerned with them here.

How to Determine the Frequency of Genes in a Population

The Hardy-Weinberg equation allows us to go from the frequencies of genes in a population pool to the genotypes of the individuals which make it up. But the reverse process must be undertaken first. As an example, let us take the population figures used above and from them calculate the gene frequencies existing at the AB locus.

The equation reads .09AA + .42AB + .49BB = 1.00. Let us convert this into a percentage basis by multiplying both sides by 100. We can now consider that the entire series consists of 100 individuals, of whom nine are homozygous for the gene A, 42 are heterozygous AB, and 49 are homozygous for the gene B. To determine the frequencies of these two genes in the population pool, we in essence count the genes in each individual, remembering that every person contains two at this locus. Thus the nine persons who are homozygous A contain 9 × 2 or 18 A genes. The only other A genes found in the population occur among the heterozygous individuals, of whom there are 42. Since each of them contains only one A gene, their contribution to the frequency of A is 42 × 1, or 42 A genes. By adding together genes found in the homozygous individuals to those occurring in the heterozygous ones, we have a total of 60 A genes in the population of 100 people. By similar calculations, the 49 homozygous B people give us 49 × 2, or 98 B genes, while the 42 heterozygous members give us 42 × 1, or an additional 42 B genes. Adding them together, the sample population contains a total of 140 B genes. The sum of the two alleles, 60 A genes plus 140 B genes, gives a total of 200, which is, of course, the correct number to be found in a sample of 100 diploid persons.

These gene counts can be converted to gene pool frequencies by dividing them by 200. This divisor represents a return from a base of 200 to one of 100, and then by dividing by 100 we get two frequencies, which add to unity. Thus the frequency of the A genes becomes $^{60}/_{200}$ or .3, and the frequency of the B genes becomes $^{140}/_{200}$ or .7. If our calculations have been accurate, the sum of the two gene frequencies should equal 1.0 and, of course, .3 + .7 do.

This simple, direct method of counting can be followed whenever all the genotypes of the individuals are known. This is true of allelic series of two, three, or more genes insofar as all the heterozygous genotypes can be identified and so distinguished from the homozygous ones. In the preceding example, the Hardy-Weinberg equation had two terms, one representing each gene. If the allelic series consists of three genes, then the equation will contain three terms. But in all cases the term containing the gene frequencies is squared, since the genes are being distributed *in pairs* to individuals on a random mathematical basis. Methods of calculating gene frequencies in cases in which one or more of the genes in allelic series is recessive, and hence unidentifiable in the heterozygous form, can be found in a number of modern textbooks on serology, such as Race and Sanger (1964). The reader may test this more complex situation on the O-A-B locus. Assume that a population contains O = .6, A = .3, and B = .1; insert these frequencies into a Hardy-Weinberg trinomial and square the equation. The results indicate at the genotypic level the makeup of each individual. By remembering the rule of dominance that both A and B are dominant over recessive O, the phenotypic frequency of all the individuals in the population can be determined. The six classes of genotypes should be reduced to four phenotypes.

Deviations from Random Mating Systems

Since the Hardy-Weinberg equation provides us with a distribution of genotypes reflecting random mating conditions, it can be used to test to see whether mating in real populations proceeds in a mathematically random fashion. *The Hardy-Weinberg prediction will hold good only if no evolutionary forces are changing the frequencies of the genes in the pool, and if mating in the pool proceeds at random.* In practice there are a number of ways in which the mating system may be biased and so not operate strictly in a random fashion. Deviations from a random mating system fall into two major categories: (a) patterns of mating which involve a preference or bias are called *assortative mating systems;* and (b) marriages between genetically closely related relatives are called *inbreeding mating systems.*

Assortative Mating Systems

Men marry or mate for so many reasons that they cannot all be listed. At one end of motivation it may be the way light strikes a girl's hair, the sweetness of her smile, well-turned legs, or high, firm breasts. Reasons less connected with physical appearance may involve conservation or acquisition of wealth through marriage, the fact that a man has got a woman pregnant, or merely that a man has reached an age when society expects him to be married. In some societies no real choice in a marriage partner is involved, for a man is promised one or more future wives before they are conceived, as in some regions in Australia. In any society both the patterns and the motivations are complex. Our purpose here is to see if there is any overall bias in the end produced by marriage patterns. If so, it will fall under the category of assortative mating. If patterns of mating either allow or encourage marriage between physically alike partners, then the system involves *positive assortative mating.* On the other hand, if the net result is to encourage marriage between people physically unalike, then the situation involves *negative assortative mating.* Both pose a potential shift in evolution, if accompanied by differential selection of various phenotypes.

Positive Assortative Mating

In positive assortative mating people marry others who are physically like themselves. It is the physical externals that count, not the internal and hidden biochemistry. Who cares whether a girl has the blood group B? If this type of mating preference is consistently displayed, then later generations will differ from the Hardy-Weinberg prediction in that there will in time be more individuals of the homozygous phenotypes, and fewer representing the heterozygous ones. Before examining the theoretical basis for this deviation from random mating, let us consider our own values. Mating practices are influenced by cultural values. It was claimed some forty years ago by a lighthearted author, Anita Loos, that gentlemen preferred blondes. If this were true, it would introduce some bias in mating preferences. Today there certainly is a tendency to tamper with the genetically determined color of one's hair, by bleaching it by one means or another. Apparently our society still prefers the type of depigmentation that is common in northwest Europe. For my part, while growing up I thought I preferred blondes, but when it came time to marry, I chose an elegant brunette. Only a statistically and cosmetically controlled survey could really show what goes on in our own society at this time.

The variety of reactions attached to stature in the opposite sex provides a less difficult problem, since it cannot be cosmetically altered. Most of us have observed certain preferential patterns involved in both dating and mating. Tall women certainly prefer tall men, preferably ones taller than themselves. Short men search diligently to find yet shorter women whom they hope to be able to physically dominate. For undoubtedly our present sexual values are still based upon the ancient idea that men should be stronger, and usually taller, than the women they marry. But in between these extremes are other kinds of matings that tend to muddy the bias in the system. Tall men have no very fixed ideas about stature in the women they marry, for they can dominate them at any size level. Indeed, they seem to show a kind of perverse preference in frequently choosing to marry little women. Short-statured women are equally uncommitted, for they can pick males of any size range and still be physically subordinate. This type of mating preference is perfectly testable and a real example will be given shortly.

The consequences of various kinds of mating systems can be investigated mathematically. Sewall Wright (1921) has been an evolutionary pioneer in this area. Before examining his conclusions, let us set up the kind of hypothetical population that he is dealing with. Assume once more that we are dealing with two allelic genes, A and B, at a single locus. In this instance it is convenient to make them equal in value, both with a frequency of 0.5. We can now set up a Hardy-Weinberg relationship as follows:

$$(.5A + .5B)^2 = 1$$
or when expanded
$$.25AA + .50AB + .25BB = 1.00$$
When multiplied by 100 on both sides, it becomes
$$25\%AA + 50\%AB + 25\%BB = 100\%$$

Given this sort of Hardy-Weinberg equilibrium population, it is now proper to ask what would happen with given various types of positive assortative mating. First of all, let us assume that every individual marries someone like themselves, so that the assortative mating is positive and perfect. This mating system produces drastic alteration and at the end of only ten generations of practice, the frequency of the heterozygotes is reduced to 1.0 percent. The equation for the population then becomes:

$$49.5\%AA + 1.0\%AB + 49.5\%BB = 100\%$$

This type of mating produces the virtual extinction of heterozygous individuals in the population in a very short time. But if we were to complicate our mathematical model and make it closer to real populations by assuming that a number of genetic loci are involved in the mating system, our results become less impressive. Under the same conditions of perfect assortative mating, with two pairs of factors the proportion of heterozygous individuals remaining in the population after ten generations rises to 8.8 percent, while with four pairs of factors the value rises to 23.3 percent. Obviously in actual situations the reduction in heterozygosity is not as rapid in genetically complex situations as it is in models in which we look at only one locus at a time.

This model errs in a very serious way. It assumes that everyone marries a like-looking person 100 percent of the time, whereas in life this occurs in only a fraction of the marriages. If we assume for ten generations that the practice of positive assortative mating is reduced to 50 percent of the marriages, then heterozygosity is reduced to 33.3 percent for a single pair of genes, and remains as high as 44.4 percent with four loci involved. We may conclude that in real life situations, in which positive assortative matings occur only as a minority of the total marriages, that the reduction of heterozygosity is both slow in time and small in magnitude. Conversely, the increase in homozygosity in individuals shows similar rates of change (Table 14–1).

Let us now look to an actual field study on a modern population. James Spuhler (1962) studied approximately 1,000 persons in Ann Arbor, Michigan, during the years 1951–1954. Of the total, about three-quarters represented married couples. His problem was to determine how much assortative mating was practiced in a population and whether it had any significant effects. His method involved measuring 43 different aspects of the body, including the various lengths, breadths, and circumferences. Of this series 14 traits or 42.6 percent gave no evidence of positive assortative mating among the married couples. This left a residue of 57.4 percent in which some degree of positive assortative mating was evident. Among the 29 of the measurements so involved, five did not differ at a statistically acceptable level of significance, but 24 did. Thus only 24 of the 43 measurements used gave evidence of positive assortative mating in the Ann Arbor series.

Spuhler proceeded to analyze the meaning of assortative mating in his series. He noted that many of the measurements taken were highly correlated with each other, so that they did not represent independent biological entities. Most of the measures which did show assortative mating involved body length, body breadth, and body circumferences. Reduced to a simple statement, this meant that there was a positive tendency of people in Ann Arbor to marry others of about the same size, making allowance for sexual differences. Briefly, big people married each other, as did small ones. We have already concluded that this is not an unexpected finding. The investigator then analyzed his material to see if this bias in the mating system had produced any evolutionary consequences, that is, changes in the gene pool of the population. Evolutionary changes will result if there are different degrees of fertility among the different types of couples involved in the bias. Using very careful measures of fertility, he found that there was no significant association between reproduction and the pattern of assortative mating. He therefore concluded that in his series the mating biases changed the distribution of the phenotypes with regard to size inheritance, but had no lasting effect through changing the content of the gene pool. Since preferential mating is probably practiced in more extreme form in modern American societies than in any other contemporary ones, due to culturally manipulated values, and certainly in any past ones, it is fair to conclude that positive assortative mating has little effect in an evolutionary sense.

One of the primary purposes of examining reproductive behavior in modern societies is to project the findings back into Pleistocene times. Judging from all the living hunting and gathering peoples, marriage is primarily an economic union and relatively little freedom of choice is involved. Such societies are small, and certain obligations reflecting the exchange of sisters and daughters in previous generations must be honored. In a number of hunting societies there are more men than women, and to obtain a wife is something of an achievement. It is safe to infer that positive assortative mating plays almost no role in the marriage systems of such people. Remembering that unless it is perfectly practiced it has very little effect, we may conclude that it has been of no importance in past times.

Among more sophisticated societies, bias in the mating systems may affect small numbers of the elite. One of the most likely examples is involved in modern Polynesia, where the chiefly rulers of society (usually called kings and queens in our version) have rather consistently been men and women of almost gigantic size. The high chiefs are not infrequently six feet, six inches tall and 300 or more pounds in weight. The women in these royal families are built on the same generous lines. Although anthropologists have not investigated this particular problem, it seems evident that preferential mating has been practiced to breed up chiefly lineages characterized by impressive size (See Plate 18–5). Another unsolved problem involves the professional wrestlers of Japan. They, too, are men exceeding six feet in height, and some 300 to 400 pounds in weight. Some of this mass is obviously blubber, but they are also enormously strong individuals. For such formidable men to come out of island stock of small people again suggests some type of biased mating. It would be interesting to research this problem, to see how much of the impressive end product is a matter of biased marriages, and how much it may involve special diet fed during the growing period. Very likely both factors are involved.

Negative Assortative Mating

Negative assortative mating is a biased mating system in which unlike individuals marry each other. This again refers to characters which can be seen in the exterior phenotype of individuals. Wright (1921) also explored the consequences of this system of mating. For a single pair of genes, his models showed that a number of heterozygous individuals would increase from a predicted 50 percent to 66.7 percent at the end of ten generations if this type of mating was perfectly practiced. For reasons that don't concern us here, this value of a 16.7 percent increase in heterozygosity is stable and does not increase in further generations within the population. In this system, too, if the model is elaborated to include more than a single locus, then the percentage of heterozygous individuals falls off. Dealing with four pairs of genes, it increases to only 53.3 percent in ten generations, compared to the original predicted value of 50 percent. These examples show that negative assortative mating produces even less change in the distributions of phenotypes in the population which practices it than does the positive form of bias. We may conclude that if this system is practiced imperfectly, meaning that all people do not marry their opposites, that the proportion of individuals who would show an increase in heterozygosity can be ignored.

Research has not revealed any negative assortative mating in human societies today. Curt Stern (1973) suspects that it may be a practice among redheaded individuals in our population. He thinks that marriages between redheads are less frequent than might be expected from their numbers. This may be true, but the definition of red hair is a little tricky, for it not only ranges from the lightest of strawberry blondes, through the fiery reds, but into the auburns, so dark as to merge into ordinary dark browns. Furthermore, the fact that in many red-haired individuals color darkens with age further confounds the issue. Nevertheless, it is a somehow attractive idea that redheaded people avoid each other. Those of us who are not gifted with this pigmentation tend to find them perhaps more than ordinarily attractive individuals. Certainly, red hair is one of the gaudiest of natural decorations found in man.

Under these circumstances, one would not expect to find negative assortative mating practiced either by living hunting peoples, or their Pleistocene ancestors. And yet there is a very interesting case which suggests that under some special circumstances it may have been done. An early observer reports for the Wirangu tribe, on the southern coast of Australia, that marriages were arranged by elders in a way which would tend to produce a system of negative assortative mating. These people lived on the margin of the great Central Desert belt where blondeness in childhood was a very common characteristic among the dark-skinned Aborigines. In the Wirangu tribe blondeness occurred only in a minority of children. Our early observer reported that parents took hair color into consideration in arranging marriages between their children. If their own son was dark-haired, they tried to obtain the promise of a girl from a mother who was blonde-haired as a child. If we presume that the unborn girl would likely be blonde in infancy, this practice amounts to negative assortative mating. The reverse type of marriage would be arranged by the parents who were themselves blonde infants. It is unlikely that the practice was carried out perfectly, that is, in 100 percent of the marriages, so it probably produced no evolutionary changes in the gene pool. Nonetheless, it remains an interesting example.

Systems of Mating Involving Inbreeding

Inbreeding involves the mating of persons having ancestors in common. For practical purposes their relationship need not be traced beyond the great-grandparental generation. Many simple societies claim that they prefer marriages between first cousins or second cousins. In virtually all cases, these relationships are only carrying the label of cousin in a classifying system of kinship, and actually these people are not close relations by descent. In such systems, and they are common in Australia, matings are usually arranged to minimize the possibility of involving real biological cousins of either the first or second degree.

Inbreeding results in depressing heterozygosity and increasing homozygosity compared to the predictions of the Hardy-Weinberg equation. Thus it would operate somewhat in the same fashion as positive assortative mating. Theoretically, closely related individuals are genetically more alike, and hence more alike in appearance,

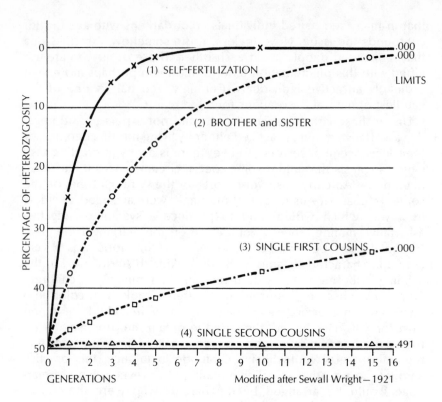

Figure 14–1
The Way in Which Inbreeding
Systems Reduce Heterozygosity

than those chosen at random from the population. Again, we must turn to Sewall Wright (1921) for the definitive investigations of inbreeding systems. The result of his mathematical modeling is shown in Figure 14–1. He has assumed a situation in which two allelic genes are present, each with a frequency of 0.5. Since he was exploring the consequences of inbreeding systems on a wide basis, he commences with the situation in which self-fertilization is the method and ends with a system in which single second cousins always marry each other. The figure reveals that self-fertilization reduces heterozygosity from the initial 50 percent to 1.6 percent in only five generations, and to no heterozygosity by the tenth generation. This is a very rapid decrease in time and in theory leaves us with populations in which all individuals are homozygous for one or the other gene originally present. Naturally, it cannot occur in bisexually reproducing animals but selfing is fairly common among plants.

Brother-sister systems of inbreeding are theoretically possible and are followed from time to time by animal breeders. The amount of heterozygosity diminishes less rapidly than in selfing and still amounts to 5.7 percent at the end of ten generations in which it has been practiced perfectly. This type of inbreeding has never been systematically pursued in man, although an occasional case of brother-sister incest may turn up among isolated communities anywhere on Earth. Since it is not a population practice, it is of no concern (Table 14–1).

TABLE 14–1

Effects of Various Kinds of Inbreeding
Percentage of Heterozygosis

Generation	Self-Fertilization	Brother-Sister	Single First Cousin	Single Second Cousin
0	.500	.500	.500	.500
1	.250	.375	.469	.491
2	.125	.312	.453	.491
3	.062	.250	.439	.491
4	.031	.203	.427	.491
5	.016	.164	.416	.491
10	.000	.057	.374	.491
15	.000	.020	.344	.491
	.000	.000	.000	.491

The inbreeding system which involves continuous and perfect mating between first cousins is theoretically possible in man. Single first cousins have one pair of grandparents in common, that is, a total of six instead of the usual eight. Thus they have a probability of receiving some of the same genes from the common grandparent. The decrease in the frequency of heterozygosity is very much less rapid here than in the previous cases, and by the end of the tenth generation in which the custom was practiced in all cases, the percent of heterozygous individuals was 37.4 instead of the predicted 50.0. If continued for an infinite length of time, it would ultimately extinguish all heterozygosity in the population practicing it.

Finally, we have the situation in which the inbreeding system involves the perfect and continuing mating between single second cousins. These are individuals who have one pair of great-grandparents in common, that is, a total of 14 instead of the usual 16. The consequences of this type of inbreeding are interesting, for in the first generation heterozygosity is reduced from 50.0 percent to 49.1 percent of the total, and then remains constant at this figure forever, even though the mating system is perfectly practiced. In small populations the reduction would be so slight as probably to escape our notice.

Wright makes it abundantly clear that his predictive models are only exploratory in nature, and their conclusions may not fit real populations. There are no people who practice any of the inbred marriage patterns perfectly, and he warned that any deviations from the pattern of 100 percent correct mating causes a rapid deterioration in the predicted results, namely a reduction in heterozygosity.

Secondly, his models were constructed in such a way that none of the evolutionary processes were allowed to operate, since he assumed they did not interfere. If there were any differences in fitness between the two kinds of homozygotes and the heterozygote type of individual, then Wright's predictions would not hold. In real populations there is considerable evidence to indicate that heterozygous individuals do in general have somewhat higher fitness, and this may result in discarding all ideas about the consequences of inbreeding practices.

Models such as Wright's must be tested against the reality of nature in order to determine their worth. There are a number of populations which have been competently studied which allow us to do this. A *coefficient of inbreeding,* F, has been devised to provide a symbol to express the consequences predicted in inbreeding. The value of F merely indicates the amount of deviation in the heterozygotes from the value predicted by the Hardy-Weinberg expansion. Theoretically it varies in value from 0.0 to 1.0, the latter case indicating complete inbreeding and the total extinguishing of heterozygosity in the population. In actual practice coefficients of inbreeding involve only low values, and there is some doubt that these represent an actual situation. In the first place, even in random mating systems, there are a few marriages between inbred individuals. But the coefficient does not take this into account. Secondly, the coefficient is calculated on the probability that individuals will be homozygous, based upon their line of descent. If in fact the homozygotes have a general fitness inferior to the heterozygotes, the real population will not correspond to the calculations.

Let us test the real effect of inbreeding on some living populations. Patterns of reproduction which involve self-fertilization, or selfing, provide a rigorous type of test. Allard and others (1968) have investigated the problem in some wild grasses in which seed production results from selfing in all but a fraction of one percent of the grains. This is a nearly perfect system of selfing, that is, self-fertilization, and in theory should produce total homozygosity in each local population. Field results, instead, show that each little patch of grass maintains a high degree of genetic variability. Apparently the populations of this grass have adjusted their gene pools to suit each little local aspect of the microenvironment. This achievement, combined with the observable genetic variability, indicates that selfing has had little effect under these extreme conditions in reducing genetic variability. The polar differences between the ideal mathematical model and the real local population are caused by a series of differing factors which result in the maintenance of heterozygosity and the depression of homozygosity. In these populations the most stringent form of inbreeding has produced no visible diminution in genetic variability.

Man is a mammal and so it is suitable to turn to evidence from the lower mammals. Laboratory strains of the white rat have been perpetuated by brother-sister mating for very many generations. The goal of this type of inbreeding is to eliminate all heterozygosity, and so to present to laboratory researchers, who wish no genetic variables to interfere with their procedures, strains of rats which presumably have become genetically homogeneous, that is, *isogenic.* In recent years some investigators have questioned whether long continued and perfect brother-sister mating does achieve this desired goal. They have found that in general, while each strain investigated shows a marked reduction in overall genetic variability, none of them has totally lost it. In effect, the strains of rats which should have become isogenic many generations earlier have failed to do so. Combined with these investigations are the commonplace observations that inbred lines of animals tend to become reduced in vigor and to lose their fertility. Some are so biologically depleted of vigor that they can hardly be maintained under the best of laboratory conditions. The two sets of ideas may be put together to consider that the continued survivorship of inbred lines of rats depends upon individuals who chance to retain more heterozygosity genetically than they should in theory. If an inbred female rat produces four offspring in a litter, one or two of these may die before reaching sexual maturity. If such deaths are purely random in terms of genetic makeup, then there is no selection. But if the death differential occurs among the more homozygous individuals, then the reduction in heterozygosity in the line is not proceeding according to plan. The evidence indicates that inbred lines retain more heterozygosity than they should in theory.

The evidence of inbreeding in man should now be examined. Close inbreeding may in some families produce unfit offspring, but this depends upon the genetic endowment of the individual parents, and biological deterioration does not always result. It is worth digressing to examine what may well be the most extreme instance ever to occur in our species. In mid-nineteenth century a drover came to Australia from the Punjab, a province in northwest India, to manage the camels that were being imported for transportation in the Central Desert. He and many other immigrants like him were known in Australia as "Afghans." Our particular Afghan drove his camels out of the town of Oodnadatta into the southern portion of the desert. There he took to bed (for to say married would probably be presumptive) a full-blooded aboriginal woman who produced a number of children for him. The offspring were *F-1,* or *first generation hybrids* between the "Afghan" and the Australian woman. These children, of course, were genetically one-half Afghan and one-half aboriginal. The number produced in this generation is not certain, but the old Afghan, Jack Abdullah, took one of his daugh-

Plate 14–1
This normal appearing man is 7/8 Afghan from his great-grandfather, Jack Abdullah, and 1/8 Aboriginal from his great-grandmother.

ters, by the name of Augusta, and began producing another family whose genetic composition was three-quarters Afghan and one-quarter Aboriginal. Our Afghan was a remarkable man, and he eventually took to bed one of the girls of this generation, his own granddaughter by the name of Annie, and she produced for him no less than eight children, six boys and two girls. This interesting family genetically was seven-eighths Afghan and only one-eighth aboriginal. This is an instance of double father-daughter incest and, to my knowledge, the only one which has ever been recorded. Theoretically, this kind of extreme inbreeding should have resulted in signs of physical deterioration in the great-grandchildren of the camel drover. Six of the eight great-grandchildren survived into adulthood and showed lower infant mortality than most families living under normal conditions in the center of Australia.

As interesting as the Afghan caper is, the real test for the effects of inbreeding in man must be sought in populations rather than in single families. In recent years a number of investigators have examined inbreeding under a variety of circumstances. Wherever the coefficient of inbreeding has been calculated, whether it be among isolated groups of American Indians or Indian castes whose members are required to marry within their own general group, F is always low. Usually its value is less than 0.01, and the highest calculated reaches 0.074 percent. But it is to be remembered that the coefficient is calculated from the matings recorded in genealogies and does not actually measure the degree of reduction of heterozygosity that is attained in the real population.

Some number of years ago Norman B. Tindale and I investigated a unique population living on one of the islands in the Bass Straits separating Tasmania from southern Australia. This population had originated when a number of English and American sealers and whalers kidnapped aboriginal women both from the island and the mainland. The founding generation totaled only 21 persons. During the first years of the colony, which subsisted on sealing, the hybrid children born were not permitted to live, since that would interfere with the women's work of curing the seal hides required by the White men. During the 20-year interval prior to 1835 only six children were spared. But as time passed, the population increased at a rapid rate, and Tindale (1953) showed through his genealogical analysis that the gene pool of the population theoretically consisted of $^{22}/_{64}$ Tasmanian, $^{6}/_{64}$ Australian Aboriginal, and $^{36}/_{64}$ European in its genetic derivation. Its number exceeded 400 persons.

This little island experiment is of interest, for the population started with a very small founder stock, remained isolated, and so should have been an ideal instance of inbreeding in man. Today more than 90 percent of the surviving individuals have one of four names as their surname, Brown, Everett, Mansell or Maynard. These

Plate 14–2
This F-2 Tasmanian man had two White grandfathers and two Aboriginal grandmothers. His general appearance is that of a European.

names are, of course, derived from the original White sealers and suggest that the community is highly inbred. One of my former students, Robert Littlewood, analyzed the genealogies of the group and calculated the coefficient of inbreeding. The value of F was surprisingly low, amounting to only .023. This theoretical estimation of inbreeding suggests that even in their isolation, these hybrid islanders avoided close marriages insofar as they could. My own blood group data based upon 80 adults gave the following results:

TABLE 14-2

Phenotypes of Hybrid Islanders	M	MN	N
Observed	13.8 mm	48.7 mn	37.5 nn
Calculated	14.59mm	47.22mn	38.19nn
Difference		1.48mn	

[Calculations $(.382m + .618n)^2 = 1 = .1459mm + .4722\ mn + .3819\ nn$]

The phenotypic frequency of the blood group MN was shown as the observed frequencies above. The calculated frequencies were obtained from deriving the actual frequencies of the genes in this population, which were .382m and .618n. When these values are squared, the calculated frequencies above are obtained. The expectation is that this small isolated community would show a considerable reduction in the heterozygous phenotype MN. In actuality, our sample shows an excess of almost 1.5 percent in the heterozygous category. This result is not statistically significant, owing to the small size of the sample, but since it is in the opposite direction from that expected if inbreeding was to reduce heterozygosity, it does suggest that this hybrid population is maintaining normal genetic heterozygosity in spite of its breeding history.

On the island of Tristan da Cunha, in the South Atlantic Ocean, a population of less than 300 persons, showed some degree of inbreeding, with a coefficient, F = .04. But at three blood loci, D. F. Roberts (1967) found substantially more heterozygotes than would have been expected in even a random mating system. The data from selfing plants, inbred lines of mammals, and human populations all suggest that in natural populations inbreeding results in little reduction in expected genetic heterozygosity. Hence from the point of view of evolutionary process we may ignore it.

On the other hand, under certain circumstances, students of the genetics of families properly regard inbred marriages as important, and as potentially harmful. This different view arises from the fact that biologically handicapped children are a humanistic concern in our society, whereas in simpler human populations they died early and were not missed. An experienced geneticist can make predictions about the probability of producing a biologically handicapped child. He must know the genetic nature of the defect, and whether or not it is in the familial background of the married couple. Let us take a clear-cut example in which a young couple have already produced one child suffering from albinism, a defect in the production of normal pigment, or *melanin*. The fact that they have such a child demonstrates that both parents carry one recessive gene for this trait. The defect is produced when two recessive genes of albinism occur together in an individual, or in the homozygous state. To understand how predictions are possible, let us return to examine the process of meiosis that goes on in both of the parents. In the father the locus involving this defect in pigmentation has both a normal and an albinitic gene. His sperm will then contain the normal gene in one-half of the cases, and the abnormal one in the other half. Exactly the same processes go on in the mother, except that any given ovum has a probability of one-half of either being normal or abnormal. With random fertilization going on, the chance that a given child will be abnormal, that is, an albino, is the product of the independent chances of both the sperm and the ovum carrying the albinitic gene. This probability is $1/2 \times 1/2$ or $1/4$. Thus the genetic counselor can predict that in a family which has already produced one homozygous recessive albino, the chance of any future child being the same is $1/4$. Where genetic defects are known to be caused by a single gene, whether dominant or recessive, this type of prediction is possible. Two individuals are more apt to carry this same type of harmful gene if they have an ancestor in common. Grandparents in common occur in such cases and so first cousin marriages are prone to produce a higher frequency of genetically defective children than are random matings. This is the source of the geneticist's concern about inbreeding in our own population. It is a humanistic problem rather than one of overall evolution.

Bibliography

ALLARD, R. W., S. K. JAIN and P. L. WORKMAN.
- 1968 The Genetics of Inbreeding Populations. Advances in Genetics. Vol. 14, pp. 55–131.

RACE, R. R. and RUTH SANGER.
- 1964 Blood Groups in Man. 2nd Edition. Springfield: Charles C Thomas.

ROBERTS, DEREK F.
- 1967 The Development of Inbreeding in an Island Population. Ciência e Cultura. Vol. 19, pp. 78–84.

SPUHLER, JAMES N.
- 1962 Empirical Studies on Quantitative Human Genetics. In The Use of Vital Statistics for Genetics and Radiation Studies. New York: United Nations.

STERN, CURT.
- 1973 Principles of Human Genetics. 2nd Edition. San Francisco: W. H. Freeman.

TINDALE, NORMAN B.
- 1953 Growth of a People: Formation and Development of a Hybrid Aboriginal and White Stock on the Islands of Bass Strait, Tasmania, 1815–1949. Records of the Queen Victoria Museum (Launceston) Tasmania. Vol. 2, pp. 1–64.

WRIGHT, SEWALL.
- 1921 Systems of Mating. Genetics. Vol. 6, No. 2, pp. 111–178.

CHAPTER 15

THE FORCES
OF EVOLUTION

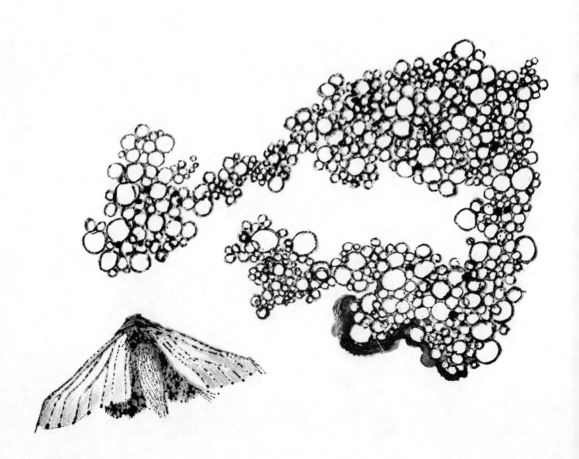

The process of evolution consists of continuing changes within a natural system. The causes of such changes, called the forces of evolution, are discussed in this chapter. Four such forces are currently recognized, and each operates in changing the frequency of genes in the pool of a population. These forces are mutation, selection, gene flow or hybridization, and random genetic drift. The last occurs in two forms, one as founder effect and the other as intergenerational drift. In most natural populations three or even four of these forces may operate together at most times. But the changes produced in such cases are of such complexity that even a computer has difficulty digesting them. For the purposes of simple explanation it is necessary to discuss each force separately, always remembering, however, that they operate together.

The Problem of Dynamics as Opposed to Description

Many readers will be reluctant to turn away from the fascinating sweep of the fossil record of the evolution of life to study the nature of the forces which have produced this record. The evidence pieced together from teeth and bones is not static, for it changes constantly in time. But it fails to reveal *how* and *why* the observed changes have come about. The evolutionary forces are best investigated on living populations. A number of animals with short lives, economical to feed and manage, have genetic material in a form that facilitates study. Among them the fruit fly, of the genus *Drosophila,* is perhaps the most suitable. In the hands of gifted experimenters this little insect has allowed the testing of most of the ideas which originated with such great mathematical biologists as R. A. Fisher and J. B. S. Haldane in England and with Sewall Wright in the United States. Their mathematical predictions about the ways in which evolution operates required confirmation on living organisms. The fruit fly among insects, mice among mammals, and corn among plants have been particularly useful here. But the results of experimental research must then be projected onto natural populations and tested there. Just as mathematical predictions often err in resting upon overly simple assumptions, so the findings produced by experimental laboratory research sometimes suffer because they do not apply to real life. More and more frequently researchers who look to natural populations, including those of man, find aspects of evolutionary processes which were not predicted by either the mathematical or the experimental methods, since both of these are simpler than life. Man is an especially suitable subject to study in populations in nature, for we know a great deal about his genetics, his behavior, and his social organization. Above all he can talk and so recount his life experiences to the interested investigator. Some of these results will be sketched in the next few chapters.

The Forces of Evolution

An evolutionary force changes the frequencies of the genes in a population pool. It is analogous to a force in physics in which an object remains at rest unless acted upon by a mechanical force or continues in uniform motion until resisted by a force. Evolutionary forces manifest themselves by the genetical changes they produce in populations. The various types of mating patterns discussed in the last chapter, including both types of assortative mating and inbreeding, are not considered to be evolutionary forces, even though they redistribute the frequency of genotypes in individuals; acting alone, they produce no changes in gene frequencies in the gene pool.

The four major forces of evolution—*mutation*, *selection*, *gene flow* or *hybridization*, and *genetic drift*—are divided into primary forces and secondary forces. The first two constitute primary forces in that they are basic to the process of organic evolution. The latter two are secondary forces, since acting alone they could not produce prolonged evolutionary change. There is an extensive literature covering the four forces at the level of mathematical modeling, in terms of experimental research, and a few scanty records in terms of investigations of natural populations. Our ideas about the relative importance of each of these forces is certain to change somewhat in the years ahead. In the following section the forces will be discussed briefly, with somewhat more extensive presentations in the supplements which follow the end of this chapter. As a word of warning it should be noted that a few writers have tacked on other categories to their evolutionary forces. Some consider that isolation between populations is a force. Strictly speaking, it is a condition in the environment that directly and by itself produces no evolutionary changes of a genetic nature. More reasonably, there is experimental evidence of the probable existence of a fifth force, *meiotic drive*, but it has not been demonstrated as yet in man. It changes gene frequency through systematic distortions which occur in meiosis. As a consequence, the expected ratios of segregating genes in both sperm and ova can be much biased. It has been shown to occur in both mice and fruit flies, but its causes are not understood. Until it is shown to affect man, we can safely disregard it.

Mutation as an Evolutionary Force

Mutations consist of abrupt changes in the genetic code in body and sex cells. The latter are inheritable in the descendants of the individual who has the mutation. Mutations are roughly classed in two broad categories: point mutations and chromosomal rearrangements. Point mutations, those which involve very limited changes in the code, may result from change in a single base pair in the set of instructions. For example, one base pair may be replaced by a different pair. Or adjoining base pairs may be reversed in their sequence. These seem to be the smallest units of mutation, yet they do produce considerable changes in individuals. It is known that most of the abnormal hemoglobins have been produced as a mutation by the change in a single base pair in the DNA code.

Mutational changes of the second type consist of rearrangements of considerable sections of the code. External inputs of energy sometimes produce breaks in the chromosome which reknit in a variety of new positions. These include changes such as inversions, translocations, and deletions of sections. Examples are given in Figure 15–1. An *inversion* (Figure 15–1b) consists essentially of a

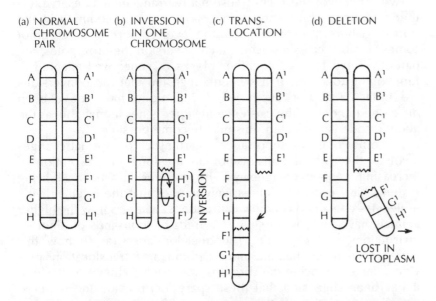

(a) NORMAL CHROMOSOME PAIR   (b) INVERSION IN ONE CHROMOSOME   (c) TRANSLOCATION   (d) DELETION

Figure 15–1
Types of Chromosomal Aberrations

broken fragment of one chromosome healing in an inverted position to the section from which it was torn. This results in an inverted arrangement of the loci on the chromosome represented. The sections of the code have not been changed in their overall total content, but merely in the order in which they occur. Yet what is known as *position effect* results in the inverted chromosome showing somewhat different biochemical activity than the normal arrangement. It would seem that the activity of each section of coded instruction is dependent to some degree upon the actions of the neighboring sections of code and the sequence in which the instructions are followed. Figure 15–1c gives an instance of *translocation*. Here a portion of one chromosome has broken off and attached itself to the other chromosome of the pair. Again, the detailed sequence of these pairs in a given stretch of code is not changed, but the instructions are now adjacent to different neighbors, and so position effect produces a change. The translocated loci are now reduplicated. *Deletion* (Figure 15–1d) occurs when a broken section of chromosome fails to fuse to another chromosome and so simply becomes lost in the cytoplasm of the cell. In this instance the chromosome which has lost a terminal portion no longer has the full code of instructions, and if this occurs in a gamete, the individual it produces may not live. Very small deletions may not produce visible results, but if any appreciable section of the code is missing, the individual probably will not develop normally. Presumably most individuals conceived suffering from deletions are lost as spontaneous miscarriages in the early stages of fetal life.

Two aspects of these chromosomal rearrangements deserve further attention. Chromosomal inversions are very common in the fruit fly, where, if heterozygous, the inverted portions of chromosomes inhibit crossing-over. Such inversion heterozygotes apparently serve to keep together blocks of genes well adapted in functioning together. It has been suggested that these inversions are, in fact, a kind of "super gene." Techniques have not yet been developed to detect the existence of super genes in man, but as the genetic code is translated, this may become possible.

The other point of importance regarding chromosomal changes involves the role of translocation in the long-term evolution of increasing genetic complexity. The manner in which all living systems have evolved progressively implies that the original living molecules contained very few instructions, whereas now the higher animals have tens or possibly hundreds of thousands of different instructions in their code. The question arises as to how this increase has been attained. Since chromosomal translocations provide some extra loci to the chromosome to which they are attached, it has been suggested that these spare loci provide the required opportunity. In their initial generations of existence, the extra loci would be duplicates and essentially functionless. But at the same time they are sites at which mutations can occur without harming normal code functioning. Over long periods of time the extra loci may evolve new sets of instructions which add to the fitness of the species, and, of course, increase its genetic complexity. In this sense, chromosomal translocation may have a very important role in overall evolution.

The Importance of Mutations in Evolution

As Charles Darwin noted many years ago, the individuals in natural populations vary among themselves. He considered that some were more fit than others and so would leave more surviving offspring. The reverse of this position is that the less fit individuals, being eliminated from a population, cause a slow decrease in its basic variability. This constant selective erosion of biological variability would lead to evolutionary stagnation and species extinction. It is evident that one of the primary roles of mutation is to maintain variability in spite of the constant drain which tends to reduce it. In this sense it is a necessary process. From this long-term point of view, mutations are beneficial.

Mutations operate as an evolutionary force when they change the frequencies in a gene pool. This involves the idea that mutational changes are not unique but are *patterned.* A mutation which has occurred once may occur again and again at intervals. Experiments have further demonstrated that mutational changes are *reversible.* If gene A has mutated to gene A', then it follows that gene A' sometimes mutates back to gene A. But the rate of mutational change is not necessarily constant in both directions. If gene A can mutate more frequently to gene A' than the reverse mutation occurs, then there will be a *net mutational pressure,* converting A genes into A' genes. In this sense, hypothetically, mutational pressures can modify population gene frequencies. On the other hand, rates of mutation for any given locus are known to be very low. In man, dominant mutations allow the calculations of the rate of occurrence. Six such gene mutations show rates of mutation which are a little above or a little below one mutant gene per hundred thousand gametes per locus per generation. In our terms the gene A may mutate to gene A' only once a generation in 1,000,000 ova or sperm. This is a very low rate and so mutational pressure at best would be imperceptible in local populations. For this reason, we may conclude that while mutations have been important as a base for increasing complexity in species, and while they serve to maintain variability against eroding factors which produce a diminution in species, nonetheless they do not actively produce directly detectable evolutionary change. Favorable mutations occur occasionally and by providing greater fitness to individuals would increase slowly in time. But the rate of change is so minimal that it would escape detection in real populations in nature. Supplement No. 19 provides some further useful details about mutations and man's genetic load.

Selection as a Directing Evolutionary Force

Darwin recognized that selection directed most of the changes evident in the record of organic evolution, and he even framed the idea of *differential effective fertility.* The concept of differential effective fertility revolves about the way in which the parents in one generation contribute their genes to the next generation. It involves not only the number of children born to them, but whether or not these are effectively raised until they in turn reach their full reproductive competence. A woman who gives birth to only two children, both of whom survive into their third or fourth decades of life, is a more effective contributor than the woman who bears seven offspring but raises only one to maturity. The effectiveness of differential fertility is measured statistically by the number of children who reach the midpoint of their own reproductive age. Differential fertility is the essence of selection, but it must be effective fertility.

Differences in fertility which occur irrespective of the genotypes of parents do not constitute selection. For example, deaths arising by accident, which are not related to the biological makeup of the victim, are not acting selectively. But if the pattern of death is correlated with biological makeup, then the differences in effective fertility which it induces do constitute selection. In man, differential effective fertility works in a wide variety of ways, but individual fitness in our species is so complex a matter it is difficult to show selection in action. For that reason it is convenient to turn to some of the lower forms of animal life for the best demonstrations of the process in action.

The Example of Industrial Melanism in Moths

England passed from a state of handcraft production into the beginnings of the industrial revolution little more than two centuries ago. The introduction of machine power involved the burning of fossil fuel, largely coal. Consequently, the westerly winds spread the black plumes of smoke to the east over most of the English countryside. Downwind all vegetation became covered with a fallout of fine soot. Some plants survived in this changed environment, others disappeared. The green-gray lichens which covered the trunks of many gray-barked trees could not survive and disappeared. The tree trunks themselves became blackened with grime and so totally changed in color.

This change in the English environment was for the worse, and benefited neither man nor beast. The effects of this type of pollution on human populations is real but difficult to document. On some of the lesser animals the changing environment necessitated strong selective forces which have been studied. The best example consists of a relatively insignificant insect, the peppered moth (*Biston betularia*). The nature of its life cycle makes it a classic example for studying the process of selection. In preindustrial days this moth at rest displayed a mottled gray upper wing speckled with areas of black. The color and design provided almost perfect camouflage effect against the lichen-covered trunks of unspoiled trees. But in 1840 a black form of the moth was noted at Manchester on the west

Plate 15–1
Dark and light forms of the peppered moth on soot darkened bark

side of the English Midlands. Although this moth has but one generation a year, by 1895 about 98 percent of its population in that area were of the black form. This is a very rapid shift in time, and it has been calculated that the dark or melanic form must have had an advantage of about 50 percent over the gray form to have produced the observed change. A single dominant gene produces the black form. Plate 15–1 shows both dark and light forms.

The details of this evolutionary shift have been thoroughly documented by a British naturalist, H. B. D. Kettlewell, who devoted more than 20 years of his life to the study of natural populations of this moth. He devised and conducted ingenious experiments which involved the breeding of both color forms of the moth, marking them on their underside so that they would be inconspicuous, releasing them in a natural woods, and recapturing them by traps. Thus he was able to count survival rates for the light and dark forms. The primary agency in eliminating these moths was found to be birds of several species which hunted by sight. In one experiment, in isolated woods which had not been subject to industrial pollution, the investigator released equal numbers of the gray and the black forms of moths. He observed the capture of 190 specimens by birds who took the dark, and hence conspicuous, form more than six times as often as the gray and camouflaged type. Reversing his experiment in a polluted forest, he again placed equal numbers of the two types, and this time observed that the conspicuous normal form was captured by birds three times as commonly as the dark and here inconspicuous form.

Kettlewell was a thoroughgoing researcher and did not rest with his demonstration that bird predation played an important part in the evolutionary change of his moths. He suspected that differences in viability might be involved in the rapid upsurge in numbers of the dark moth. By experiment he was able to demonstrate that the black form did have a greater fitness, in addition to its being less conspicuous on black tree trunks. He demonstrated that during the course of the two centuries in which the melanic moths had become the normal type in industrially affected areas, other evolutionary changes had occurred beyond those affecting color. These involved faint differences in the overall pattern of color between moths taken in preindustrial England as compared to the modern forms. He concluded, and other geneticists agree with him, that while the primary thrust of the force of selection was to favor the now inconspicuous dark color, other genes in the population pools of these moths had made their own shifts to produce an overall gene pool better adapted to the new circumstances. The idea that an evolutionary shift in the frequency of one gene might have impact upon others is covered by the concept of *genetic coadaptation,* which in essence considers that all genes affect the functioning of all others to some degree and that changes in frequencies in one gene will have repercussions on the frequencies of others. The process may be viewed as one producing an optimal mixture of genes in the newly adapted pool.

The story of industrial melanism does not end with this example and has been well told by E. B. Ford (1964). The peppered moth was not the only one affected by soot pollution in England. No less than 80 other species of moths reacted to the change in the environment in much the same fashion. Similar types of selective change affecting many dozens of species of moths have been observed in the manufacturing regions of western Europe. In these cases the nature of the selective stress is visible, and the response is constant among many different species. The same is true in New York City, Detroit, and Philadelphia, and no less than 100 species are already affected in the Pittsburgh district alone. These examples are particularly useful because of the demonstration that changes in color do affect survival and hence differential effective fertility through predation by birds. The fact that the example involved more complex genetic interactions than this simple predator-prey relationship does not mar its essential simplicity. It is a classic case of what is called *transient polymorphism.* This phenomenon involves a shift in selective pressure so that the gray allele which was once predominant in its frequency is in time replaced by its alternate form, the dominant black allele. Since the phenomenon occurs relatively rapidly, it is called transient in character. Prior to the industrial age in England the gray form was predominant, with a rare minority of the black forms present. This was a stable or *balanced condition of polymorphism* which remained in effect until the clouds of smoke began to destroy the natural environment. After several centuries of transient polymorphism this species has presumably now attained a new level of balanced polymorphism in which the black moths are the common ones and the gray individuals rare. With this evolutionary shift in the gene pool, new levels of genetic coadaptation must be presumed.

Some General Features of Selection

Examples of natural selection influencing human populations are not as clear as the industrial melanism of moths. For that reason examples will be given in the next chapter involving geographical variation in human populations. Before examining the human data, some of the general aspects of evolution should be considered. There is no doubt that all species alive today, and all which lived in the past, have been subject to selective action in manifold ways. All of the visible changes which the evolutionary record provides have largely been molded by differential effective fertility in one or another of its complex forms. The equations used by the mathematical evolutionists to explore the action of selection necessarily rest upon assumptions much simpler than those found in populations in nature, thus distorting reality. If we are limited to thinking of selection acting upon locus X or Y, or even both, we are missing the point. *Selection really operates upon individuals as a whole.* A man does not die in a single locus, nor fail to produce viable offspring in another. Selection acts upon both men and fruit flies as whole individuals, and the results of differential effective fertility are manifest in population terms by changes in the gene pool.

At this point it is appropriate to consider how much genetic variability man shows as a species and what may be assigned to each of us as individuals. This problem has been explored by Richard Lewontin (1967) by a simple but ingenious approach. Plotting the historical dates of the discovery of new blood group systems, he was able to estimate what proportion of loci in man was generally variable, or polymorphic, in that it could maintain two or more genes. The common blood group loci belong to this category. The other category consisted of rare genetic variants, the so-called family antigens, which have also been discovered at a fairly frequent rate by blood bank workers and those concerned with blood transfusions. The investigators considered that these rare antigens served as markers for genetic loci which were essentially homogeneous, or unvarying. The very infrequent occurrence of the family antigens supports this view. At the present, the ratio between the discovery of the variable and the stable loci has reached an essentially constant value. The investigators conclude that about 33 percent, or one-third, of the total loci in man are apt to be variable, judging from the blood group data. This is a basic estimate of population variability. Translated into terms of individual variability, the estimate is reduced to about 12 percent of the loci as being heterozygous in each of us. The opposite of this situation is that about two-thirds of the loci are nonvariable in a population sense, while seven-eighths are homozygous in the average individual. Other approaches give similar results.

These estimates concerning genetic variability in man raise the question of why these two types of loci should exist. There is a consensus of opinion that the nonvariable loci essentially carry the genetic instructions which make each species unique and different from all others. These loci presumably are so vital to the normal functioning of the individual that few or no mutationally induced variants can be tolerated in the genetic code. This interpretation would label them as the basic and important sections of the code. The variable loci, on the other hand, can be considered in a sense as being less critical in the growth and functioning of each individual, since minor genetic differences are tolerable there. But at the same time it must be pointed out that heterozygosity seems to confer increased fitness upon those individuals who have it, and so the variable loci seem to serve this additional function.

The existence of two types of loci has an interesting bearing on the question of human races. All men seem to have the same nonvarying loci in common, that is, the sections of the code that determine the specific characteristics of our species. It therefore follows that since races differ genetically at the variable loci, racial traits are in a sense to be viewed as biologically rather trivial. And yet racial variation reflects processes of selection acting upon populations to adapt them to their own environment. These different estimates are not in conflict with each other, but should be viewed in their proper setting.

The Intensity of Selection

The force of evolutionary pressure is measured by numerical differences in terms of differential effective fertility. Let us use a simple example as a case. Suppose the individuals of a population can be sorted into two kinds of phenotypes, A and B, differing in a single characteristic. Let us further assume that they are born in equal numbers in the population and that for every 1000 A's who reach effective maturity there are only 999 B's. In this case the difference in fitness between phenotype B and A is one in a thousand. This is the coefficient of selection, and it is written as $s = 0.001$. This is a seemingly insignificant difference in fitness, and yet the mathematical evolutionists have been able to show that it is quite large enough to make evolutionary changes of the kind seen in the fossil record, since in populations involving large numbers it is sufficient to produce steady and consistent changes in the gene pool through time.

Research on natural populations has in recent decades revealed that coefficients of selection may be very much higher in value than indicated above. Cases have been recorded in which the coefficient has been as high as 0.1, 0.3, and even perhaps 0.5. Selective pressures of this magnitude would be expected to produce very rapid evolutionary changes, such as those involved in the industrial melanism of moths. They are more likely to be found in the cases of short-lived animals, such as insects, rather than in such long-lived mammals as man. But it is important to recognize that this range of variation in selective pressures acting upon normal characteristics can be expected in the course of the evolution of natural populations.

The Direction of Selection

The record of organic evolution shows many trends with changes proceeding in a given direction. Many of the terrestrial mammals in the Cenozoic show steady increases in size and mobility. In a general way the grazing and browsing animals follow this pattern, and so do the carnivores that eat them. We are apt to conclude from examples like this that a selective force tends to be constant and to produce changes in a given direction. The story is more complicated than this.

Fruit flies and humans provide an interesting contrast, even though the basic genetic processes operate on both species. The interval between generations in the fruit fly is a matter of a dozen or more days, so that during the course of a year in a suitable climate there may be as many as 30 generations. The examination of natural populations of fruit flies in the mountains of Southern California reveal a repeated pattern of evolutionary change each year. This has been traced primarily through inversions in chromosomes which are easy to identify in this insect. Starting early in the year, one inversion may be common, while another is rare. As the season advances and the weather warms, the first inversion tends to be less common, and the rare one increases in frequency. Finally in the fall the process may be reversed. Obviously, different types of selective forces are operating upon these fruit fly populations at different times of the year.

People, together with other long-lived animals, are subject to a rather different system of selective pressures. If we assume that the interval between human generations is about 30 years, then it is obvious that the forces of selection cannot act upon us on a day-to-day and week-by-week basis. For evolutionary success, man must be able to survive under a great variety of conditions and, in fact, does so. Because learned behavior has allowed him to live in so many different types of environments, seasonal changes are usually such as to have little effect upon human fitness. Exceptions occur in the Arctic, where man is under the stress of cold much of the year, and perhaps in the tropics, where the stress of disease may vary seasonally. But by and large we may consider that different types of selection operate upon long-lived species compared to those whose lives are to be measured in days. For this reason the inversions which occur in chromosomes may be less important, or of different importance, in man than in *Drosophila*.

Disease as a Selective Agent

For many years after their discovery human blood groups were considered to be neutral in nature, that is, not subject to selective pressures. This view was naive, for as a general principle the variable loci are now presumed to be maintained in that condition by selection that slightly favors the heterozygous individual and thus maintains both alleles in the system. In recent years medical researchers have uncovered a number of associations between diseases and blood group types which suggest rather specific kinds of selective forces associated with them. No less than eight different studies conducted on large numbers of individuals in this country and the countries of northwest Europe have shown that individuals with the blood group O are slightly more susceptible to such diseases as gastric ulcers and duodenal ulcers than are those individuals with blood groups A, B, or AB. But a comparable number of investigations has shown that individuals with the blood group A are slightly more prone to develop cancer of the stomach than are those with other kinds of blood types. A small number of studies also indicates that the blood group A type of person has a statistically greater chance of developing pernicious anemia than do other phenotypes. Such studies are in their infancy, and the association between a blood group type and a given disease only deviates slightly from random expectation. With the passage of time we can anticipate more detailed and more convincing investigations. There is no need for anyone to feel predestined to fall victim to a given disease, but perhaps in the future preventive medical practice may take such associations into account.

The internal parasites that plague us evolve with us. The well-adjusted parasite evolves to a condition in which it does not kill its host but merely uses it as a resource upon which to live. By the same token the host develops buffering mechanisms to reduce the severity of the parasite's effect upon it. Given enough time, the mutual evolution of host and parasite results in a kind of tolerant living together. If the host population has not in past time been exposed to a given parasite, the first exposure may be disastrous to the host. When expanding colonial Europeans carried their own mild diseases into new continents in North and South America, as well as Australia, the results were initially disastrous. Diseases as mild as measles and whooping cough became killers when loosed on men who had had no previous exposure to them. The necessary evolutionary give and take came only after Indians and Australians had had a number of generations of exposure to the new parasites.

In the United States we as host animals have developed immunity to a surprising number of diseases. The Asiatic strain of influenza, which has come through this country in recent years, was not expected by health authorities in 1970 to have as serious effects as earlier. Exposure to this particular strain of virus has built up biochemical defenses in the bodies of most of us. More serious diseases, such as paralytic poliomyelitis, have been with us long enough so that a degree of natural immunity has evolved. While individuals who have no natural defenses may suffer crippling and permanent paralysis, many others are infected and show no more serious symptoms than those of a cold, and possibly a stiff neck. Populations maintain themselves by constant battle with the parasite they are host to and change in terms of the deaths caused.

It is occasionally questioned whether men today, insulated by their complex culture and well-guarded by their medical science, are still undergoing selective changes. It is reasoned that we have progressed so far in our technology that we should not be subject to the so-called laws of nature. The idea is attractive but untrue. The direction of selection has changed dramatically with the rise of modern civilization, and it may be that gross structural changes are not evolving at a rapid pace today. But the query can be answered in a definitive way by asking if all human deaths at this time occur in random terms with regard to individual genetic makeup. We have just discussed the relationship between inherited biology and resistance to disease, and it is clear that in the past, today, and no doubt in all of future time each unique individual will respond somewhat differently to the invasion of parasites. Early death does contribute to a reduction in effective fertility, so selection is operative. If differences in effective fertility are totally unrelated to genetic makeup, then selection would cease to operate. There is no possibility of this state of affairs coming into being. The operation of selection in cities is referred to in Supplement No. 20.

Gene Flow or Hybridization as an Evolutionary Force

Most populations, both small and large numerically, differ in the frequencies of the genes in their pool. Almost universally, when adjacent such populations exchange genes. The method may involve a reciprocal exchange of women, a lopsided exchange of women, or simply sexual access to women of the neighboring group. In any case, the genes from one group are transmitted to another, and since they differ in their initial frequency they produce changes in the gene pool which receives them. In all cases the net effect is the reduction of the difference between the population pools, provided no other evolutionary processes override it.

These reductions in gene frequencies have a leveling effect. In populations based upon simple economy, and with limited foot mobility, the rate of gene flow is restricted. In these instances if the population structure allows for founder effect drift and if selective pressures are at least moderate, then gene flow may be overridden and may do little more than gradually spread advantageous combinations of genes through the general series of populations. In aboriginal Australia this seems to be the case. There adjacent dialectal tribes exchange their women at a rate so that about 14 percent of all the marriages involve a woman brought in from another tribe. This frequency is not sufficient to smooth out genetic differences between breeding populations and so their genetic differences remain marked and seem to arise rapidly between adjacent groups of tribes. Examples of this will be shown by blood group data in a later chapter.

On the other hand, populations may exist under conditions which stimulate frequent subdivision, or budding off. This type of population is often found to be expanding numerically. Certain South American Indian tribes and a number of areas in highland New Guinea are characterized by this kind of population dynamics. Among them founder effect drift tends to override the other evolutionary forces, and even though gene flow occurs, it does not level out the general gene frequencies of population groups.

Man is the most far-flung species among large living mammals. Wherever members of distant populations have been brought together, they have produced hybrid offspring with no reduction in their fertility. This genetic test demonstrates the unity of all living men as members of a single variable species. There is some theoretical question as to whether men in the Pleistocene maintained their unity through constant genetic exchange over very great distances, or whether our present-day unity is a consequence of the fact that no human populations are far enough removed in time from the early ancestral stock of *Homo sapiens*. The more recently settled portions of the Earth, such as the New World, the outer islands of the Pacific Ocean, and Australasia, have been penetrated relatively recently and they are not the problem. The question involves whether gene flow could have maintained species integrity in the major land masses of the Old World, Eurasia, and Africa. There are a number of geographical bottlenecks through which genetic exchange would be slow, but there are no points at which they would be cut off. If gene flow had been unimpeded between all sections of the Old World, the evolution of regional differences would have been prevented by genetic swamping. But distance does act as an isolating factor, and when combined with land forms either as narrow gates or mountain ranges, genetic exchange can be slowed enough to the point where distinctly different regional populations can be evolved by local forces of selection. Both the fossil record and living peoples today show that differentiation did occur, but the magnitude of these differences in man has always been at the level of subspecific or racial differences.

In the chapters on fossil man the question was raised whether or not some of the abrupt changes in population were achieved by physical displacement. One can develop complex models testing gene flow against replacement as a process. But the conclusions based upon them concern us more than their actual working. Imagine two adjacent populations which differ much as does modern man from Neanderthal man. Further assume that their relationships are friendly and as a token of this fact they exchange daughters with each other at a regular and even reciprocal rate. After many generations both populations will have become exactly one-half Neanderthal and one-half modern, and this is the terminal condition of equilibrium. Gene flow proceeding under these terms with other evolutionary forces assumed absent tends to produce intermediacy in all the characteristics that show genetic differences.

Now, on the other hand, take adjacent populations of modern men and Neanderthals, and assume that their relations are hostile. If we grant the modern men a considerable superiority in weapon systems and in strategy, in relatively few generations they can totally replace their Neanderthal neighbors. If in the process they have followed human temptation and saved a few young Neanderthal girls for their young men, then the territory of both groups will come to be occupied by essentially modern populations in which there survives a very small minority of Neanderthal genes derived from the captive women. Population replacement operates more rapidly than gene flow and tends to preserve the gene pool of the dominant group with less alteration. These two models have been stated in very simple terms, but they can be extended to cover the problem of the sudden appearance of modern types of man in Europe, Africa, and Southeast Asia about 40,000 years ago, coupled with the nearly simultaneous disappearance of less advanced kinds of men in the same areas.

Problems of Hybridization in the Modern World

Gene flow is the only evolutionary force which allows for simple calculations under certain circumstances. There is a series of computations which predict the kinds of changes produced by hybridization. The method allows estimates as to the future composition of a hybrid population where the genetic characteristics and proportions of the parents contributing to it are known. Other versions allow the reconstruction of the genetic composition of an extinct parental group under certain circumstances, or the determination of relative contributions which the parents make to hybrid populations. But all of these methods of estimation depend upon two simplifying assumptions. In the first place, the populations involved must be so numerous that sampling errors can be ruled out. In the second place, the process of hybridization must be conducted with sufficient speed so that no selective forces will have time to act and so change the calculations. Such conditions exist only where sizable populations have been brought into contact with masses of incoming migrants.

This kind of situation exists in Australia where, in the last few centuries, invading British colonists have come to dominate the country and to displace its aboriginal owners. Today there are about 12,000,000 Australians of European descent. Of the original 300,000 Aborigines who once inhabited their country, no more than 25,000 exist today. But, as always, extensive hybridization has gone on between the two populations, and so there are an additional 100,000 or so persons of mixed European and aboriginal ancestry. Assuming that all three groups have the same effective fertility, and continue to do so, in a few centuries the total population of the continent will become evenly blended and consist of many millions of persons whose gene pool consists of a little more than 99 percent of genes derived from European ancestors, and a little less than one percent derived from the Australian Aborigines. It is very unlikely that any type of social action in the future can prevent this complete hybridization from occurring.

The problem in the United States is one of differing proportions. Today our population totals a little more than 200,000,000 persons, of whom more than 20,000,000 are of ultimate African ancestry. The methods of calculation do show that the modern American Black is on the average about one-quarter White in his genetic makeup and three-quarters African. American Negroes today are a hybrid population. There has always been hybridization between Whites and Blacks in this country, and no doubt it will continue. Much of the White component in the modern American Black was derived during plantation days when slaves were treated as animals, and the pretty young women served in rotation in their master's bed. Since slavery was abolished, the rate of genetic exchange between Whites and Blacks has been somewhat diminished. But it still continues, and in spite of the rising sentiment for Black nationalism, undoubtedly will continue into future generations. Over and above actual interracial unions, there has been a steady disappearance of very light-skinned hybrids into the White population. The rate of this "passing" is unknown, and probably unknowable, since those who have successfully passed certainly will not advertise the fact. But the existence of this type of gene flow results in the steady incorporation of a small minority of Negro genes in the matrix of the predominant White population. If we again assume that barriers to interracial marriages will slowly succumb, then after long periods of time, certainly involving several centuries, the population of this country will be finely blended and the gene pool will have a content of which about 90 percent was derived from colonial White Americans and 10 percent from Black slave ancestors.

The genetic consequences of the ultimate blending of European and African genes in a single national pool will, in general, produce a population intermediate in characteristics but very heavily weighted toward the European parental side of the mixture. Brunette complexions will become commoner and very fair ones rare. Honey-toned skin color will appear more frequently. In general, hair form will show more waviness than it does today, but tight spiral curls will be unusual. In a sense, the extreme differences of form and pigment which exist in the parental populations will disappear as the two different racial codes are blended.

Predictions about the future of mankind are always risky, but the action of gene flow allows a few to be made with safety. Wherever men and women of different racial stocks have met, they have produced hybrid children. This is not an assumption but a well-documented fact. Granted that this behavior will continue in the future, the real problem revolves about the mobility of later generations. We have every reason to believe that our technological progress will make it possible to move more people further and faster as time passes. The problems of distance on our planet will diminish. Racial intolerance is visibly receding. Under these circumstances it seems reasonable to predict that interracial blending will progressively increase. On the other hand, there will always remain sizable blocks of people tied to the land and not moving about in jet-set fashion. Today there are about 800,000,000 people in Red China, all of the Mongoloid race. For political reasons, the Red Chinese travel abroad only in small numbers, and while abroad, are controlled by a tight ideological rein. It is conceivable that restriction upon that population, which is one-quarter of the total population of the planet Earth, may inhibit the total blending of the world's population. But aside from special situations like this, it seems clear that where possible, their genetic characteristics will blend.

Random Genetic Drift as an Evolutionary Force

The joker in the pack is the evolutionary force that is called *random genetic drift*. It is also called the Sewall Wright effect, after the man who first recognized it as a contributing factor in causing evolutionary change. He successfully resisted the efforts of such English mathematicians as R. A. Fisher to minimize its significance. In essence, it consists of nothing more than errors in sampling in genetic events in small populations. This will be explained in detail below. It is a mysterious type of force, since its direction of action can never be predicted, and only its magnitude can be estimated. For this reason, it is hard to comprehend and even more difficult to explain. Finally, it introduces the operations of chance into an evolutionary system of forces which, in the main, proceed in direct and straightforward fashion.

At this point it is necessary to consider the meaning of simple *chance* or *probability*. In statistics it expresses the likelihood of an event occurring. In the flipping of a coin we recognize that the chance of a head occurring is one of two possible events; it is expressed as the probability, $P = 1/2$. Since the chance of a tail being thrown is exactly the same, the sum of the two events, $P = 1/2 + 1/2 = 1$, indicates that one or the other will occur with certainty. Given this situation, we are accustomed to think that since the probability of heads occurring is one out of two possibilities, that in a series of tosses heads should appear one-half of the time. This is based upon application of the statistical prediction to very large samples. To put it another way: if a coin was thrown 10,000 times, heads would appear in approximately 5,000 cases, that is, close to the expected 50 percent.

But if we convert statistical probability into the events that happen in small samples, quite a different system prevails. Given a coin thrown into the air six separate times, what are our expectations? To guess that three heads will occur in the series is statistically naive, for there are no less than seven different kinds of probable combinations of heads and tails. Actually, there are very small probabilities that either six heads or six tails may occur occasionally. Let us see how often six heads might be expected if six throws are repeated as a series of continuing events. The chance that a head will occur in the first throw is $P = \frac{1}{2}$. But as a famous French mathematician expressed it, "Chance has no memory." Therefore, the chance that the head will occur on the second throw is also $P = \frac{1}{2}$. And the same is true for all six throws. This gives the cumulative chance of six heads being thrown one after the other as $P = (\frac{1}{2})^6$ or one chance in 64 such series of six-throw events. Obviously an equal chance occurs that six tails may be thrown one after the other. This kind of probability is expressed by a simple binomial equation.

$$(\tfrac{1}{2}H + \tfrac{1}{2}T)^6 = 1$$

When expanded, this equation becomes:

$$\tfrac{1}{64} H^6 + \tfrac{6}{64} H^5T + \tfrac{15}{64}H^4T^2 + \tfrac{20}{64} H^3T^3$$
$$+ \tfrac{15}{64} H^2T^4 + \tfrac{6}{64} HT^5 + \tfrac{1}{64} T^6 = 1$$

This rather formidable series of figures tells exactly what probabilities there are for any combination of heads and tails to occur together in throws of six. For example, the term H^3T^3 will occur in 20 out of 64 such separate series of throws. This is the expectation of three heads and three tails, but it occurs in less than one-third of the total series. The other terms give the statistical predictions for each combination of heads and tails. For example, a series of six throws of the coin would yield one head and five tails in $\frac{6}{64}$ of the total series. Without belaboring the point, we can conclude that in more than two-thirds of the repeated series, the combination of heads and tails deviates from the expectation of one-half for each. These deviations constitute sampling error in that they do not agree with what is to be expected in very large samples.

The Effect of Population Size upon Drift

The unit of evolution is the population, as opposed to the individuals that comprise it. But overall population numbers are not the direct expression we seek in evaluating drift, and we must turn instead to the size of the *effective breeding population*. This is always much smaller than the total population census, and under most situations approximates one-third of the real head count. Where effective breeding populations are very large, sampling error or drift would matter but little. But under most circumstances, and certainly during the whole of the Pleistocene, man lived in small, effective breeding populations, and so drift must have been a constant occurrence.

Let us approximately calculate the size of the effective breeding population among a people like the Australian Aborigines who stand as reasonable social models for Pleistocene men. Among them the dialectal tribe statistically tends to approximate 500 persons and as it has been shown (Birdsell, 1950), it represents the total breeding population. A very large majority of its members marry within the tribe, and only about 15 percent of the marriages involve women from outside of its boundaries. Let us take a specific tribe such as the Kokata who live in the northern portion of the Great Western Desert. In traditional or precontact times, the tribal census would reveal that about one-half of the population is of adult age, and the other half subadult. Obviously, the effective breeders are among the married portion of the population and so at most would number 250 persons. Among the adults would be some old men and a greater number of aging women who had passed the stage of effective reproduction. These must be numerically discarded. At the same time, some of the adult natives who had children would not succeed in raising them to their adulthood in turn to become effective breeders. This is a difficult figure to arrive at, but it requires a further reduction in the estimation of the size of the effective breeding population. There are several other factors of a technical nature to be taken into account, and we are left with an estimate of about 165 to 170 persons as a first numerical approximation of the effective breeders in the tribe. This is a small enough sample statistically for us to expect that sampling error or drift would occur in its procreative pattern, and as will be shown in the next chapter, it does.

Modern populations are seen as enormous masses of people who aggregate together largely in urban centers. Under such circumstances it would be expected that the effective breeding sizes of these populations would be numerically large. This would be true if all members within a city mated at random, but they do not. The mating patterns which go on in urban centers are influenced by religious preference, race prejudice, economic differences, preferences for marriage among individuals of the same educational attainment, and finally by the ever present fact that distance tends to isolate. When all of these and other variables are taken into account, a city of several million persons may actually consist in an abstract sense of a large number of different breeding populations, each numbering a few thousand individuals rather than millions. Estimates of this sort have been made in a few cases, and in a middle-sized French city the value is about 2,500. This is a much larger effective breeding population than the one just calculated for Australia, but it would still be subject to a slight degree to sampling error, and so drift may still be operating among us today. The chances of identifying its action are remote, for the rapid rate of increase in our personal mobility blurs the picture and, indeed, tends to produce increasingly larger breeding units. Still, it is fair to say in theory that very likely no populations are exempt from drift, but the magnitude would be relatively large among economically simple people and vanishingly small among modern Westernized urban groups.

Founder Effect

Founder effect is a special form of random genetic drift which is sometimes also called *boatload drift*. Its importance in evolution has been strongly emphasized by Ernst Mayr (1966). Founder effect involves a seed stock population, sometimes the remnants left over from catastrophic reduction, possibly a migrant group moving into a space empty of human beings. The colonizing of offshore islands for the first time has usually been achieved by a single boatload of voyagers. Among simpler people the founding group may have been a single family on a raft. But the concept is not limited to the colonizing of islands, for populations on the mainland frequently go through this type of genetic bottleneck, in which their numbers are much reduced by such environmental stresses as drought, famine, or the onset of mortal diseases. For example, in desert Australia, if the Kokata tribe suffered from two or three years of severe drought, and it certainly has had such episodes at various times in the past, the tribal population might have been reduced to perhaps 100 individuals. The survivors would represent effective breeding populations very much smaller than those previously calculated. Its real size would depend upon the number of adult fertile women still alive and capable of reproduction. The size of the Kokata effective breeding population under these circumstances might be reduced to 40 or 50. This type of population fluctuation has very important consequences, for if such events occur at a fairly regular rate, then the size of the effective breeding population remains closer to the minimum value than to the normal value. This introduces a further kind of uncertainty, or *indeterminacy,* into our problem, for there is no possible way of directly determining how many times in the past thousand years or so the Kokata tribe has been subjected to this sort of drastic reduction. All we know is that they live in a desert area in which rainfall is highly irregular and famine resulting from drought must have occurred many times. But since we cannot calculate the number of times such events happen, we cannot fix the real size of their effective breeding population. Perhaps it is fair to say that it is closer to 100 persons than the 165 to 170 persons first calculated. In human prehistory environmental disasters must have left many living spaces empty. In time they were always recolonized. The hardy people who moved back into these temporarily empty areas were usually few in number and probably consisted of a few related families. These became the founding stock of a new population, and, of course, they constituted a very tight genetic bottleneck in the terms we are considering.

Let us take a small but finite population, perhaps numbering 200 individuals, and draw from them at random a "boatload" or handful of their members. If we take ten people from the parental population, there is every assurance that they will deviate genetically from the parental population in the gene frequencies at many of their loci. For each individual is genetically unique and so no one or small group of them can perfectly reconstitute the frequency of the genes found in the population from which they were drawn. Their gene pool differs from the parental gene pool in a large number of ways that are impossible to determine, for the process is a statistically chance one. If a hundred different "boatloads" of the same size were drawn in turn from the parental population, each one would have its own unique gene pool. For this reason the direction of genetic change due to drift cannot be determined. An estimate of the magnitude of change can be made by the proper statistical procedure, if the gene frequencies and numbers of persons in the parental population are known, together with the number in the "boatload." Examples in nature are difficult to identify and study, for the experiment must be caught at its very beginning, or else the population produced by the founding stock must remain isolated.

An Experiment in Nature

The northern coast of Australia is deeply indented by the Gulf of Carpentaria. The southeastern corner of the gulf contains two sizable islands, Bentinck and Mornington (Figure 15-2). Both were a continuous part of the mainland less than 10,000 years ago. At that time during the termination of the last glaciation, rising seawaters returned to normal and cut them off. Mornington Island is inhabited by the Lardiil tribe of Aborigines who maintain some contact through a series of stepping-stone islets with the Janggal tribe of adjacent Bayley Point on the mainland. The Kaiadilt inhabited Bentinck Island until a series of natural disasters forced their exit in

TABLE 15-1

Blood Group Frequencies at Three Loci of the Kaiadilt and the Lardiil

	Number	a	b	o	m	n	R^1	R^2	R^0	R^z
Kaiadilt	42	0	.24	.76	.50	.50	.52	.06	.42	0
Lardiil	67	.08	0	.92	.52	.48	.66	.26	.07	.02

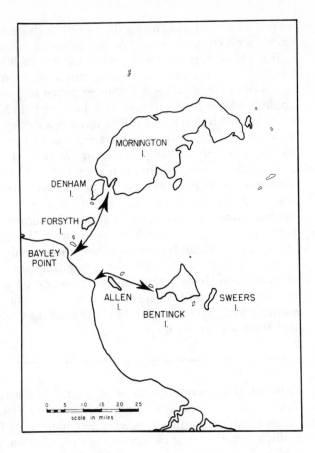

Figure 15–2
The Domain of the Island
Kaiadilt and Lardill Tribes

1947. These Aborigines lived in a completely isolated world of their own and do not remember contacts with either the mainlanders or the Mornington Islanders. All three of these tribes in this limited area are closely related and yet show major genetic differences. Table 15–1 shows the blood group frequencies at three loci for these two island populations.

The isolated Kaiadilt show gene frequency values totally outside of the range for Aborigines in the rest of Australia. They have a very high value of .244 of the b gene, approaching the world's highest values and totally outside of the Australian range. This same gene is absent among the Lardiil. The a gene shows a moderate value among the Lardiil, but is totally absent from the Bentinck Islanders. The m and n genes are close to .50 for both populations, a value generally higher than found in north Australia. But the most startling differences occur at the rhesus locus. The Kaiadilt have no less than .417 frequency for the gene R^0, a value found elsewhere only among African Negroes. The Lardiil have only one-sixth as much, a value characteristic for Australian Aborigines in general. The gene R^2

attains a value of .260 among the Lardiil, high for Australians, while it shows a much lower value for the Bentinck Islanders.

The deviant gene frequencies of the Kaiadilt provide a fine illustration of founder effect drift, not in terms of boatloads, but in terms of raft loads. The available evidence suggests that the Kaiadilt could not have reached their island earlier than 3,500 years ago, and more probably reached it only a few centuries ago. There are about 18 miles of shallow waters separating them from the mainland. Even so the crossing is dangerous and Norman B. Tindale, who has studied these people, found that 50 percent of the rafting natives who tried to reach Allen Island about ten miles away toward the mainland, died by drowning. Thus the passage from the mainland to Bentinck was dangerous, and it may be presumed that only a few raft loads of Aborigines reached the island. The genetic differences shown by the Kaiadilt from their neighbors are the result of this type of genetic bottleneck. Of broader importance is the fact that these circumstances have created major genetic differences between two closely related and adjacent groups of people. The data provide warning that gene frequencies alone offer no adequate evidence for real relationships by descent.

Intergenerational Drift

Drift of the founder type is dramatic but erratic in character. *Intergenerational drift* is a type of sampling error which goes on continuously in populations whose effective breeding size is numerically small. Under such circumstances, children have a gene pool which always differs from that of their parents to some degree. Each parent contributes but one-half of his genetic content to each child, and just what the contribution will consist of is determined by the random processes that operate in meiosis. Which sperm will fertilize the available ovum is a further random type of event. Consequently, each child differs from both of its parents and is genetically unique. This process in itself is sufficient to cause intergenerational drift even if every couple always produced two offspring which survived to breeding effectiveness in a population which was constant in numbers.

But matings differ in their contribution to the next generation and this introduces further distortion into the process. Figure 15–3 shows a frequency distribution of the 194 precontact matings previously referred to in the discussion of infanticide. The distribution looks erratic but in fact it very closely approximates what mathematicians call a Poisson distribution, one which commonly occurs in the reproductive processes of natural populations. The data are broken down into different kinds of families in terms of the number of their offspring which grow to adulthood. As might be expected in a numerically constant population,* the most frequent type of family is one which raises two offspring to maturity. But the other families are skewed around this central type. If we analyze the data according to the relative contribution of the different types of family, we reach an important conclusion. As shown in Table 15–2,

TABLE 15-2

Adult Survivorship of Children Among Precontact Australian Aboriginal Families

Size of Sibship	Number of Families	Percent of Total Families		Number of Adult Children	Percent of Total Children	
0	5	2.6		0	0.0	
1	40	20.5	49.4	40	7.8	27.6
2	51	26.3		102	19.8	
3	44	22.7		132	25.7	
4	37	19.1		148	28.8	
5	13	6.7	50.6	65	12.6	72.4
6	1	.5		6	1.2	
7	3	1.6		21	4.1	
TOTAL	194	100.0	100.0	514	100.0	100.0

families whose adult offspring range from none through two constitute 49.4 percent of the total number. Families producing between three and seven children are little more than half the total, or 50.6 percent. But the reproductive success of these two groups of families is totally different. Of the total number of children produced, families which are low in fertility, with children ranging from none through two, contribute only 27.6 percent of the total of children, while the big families produce 72.4 percent.

This type of differential fertility further distorts the deviation in the gene pool of children as compared to that of their parents. Each type of family is genetically unique as a combination, and, accordingly, each of the categories plotted in Figure 15–3 will differ in its genetic content. When one-half of a population is responsible for approximately three-quarters of the next generation, this further biases the genetic deviations manifest in the children.

This kind of drift, occurring between generations, goes on throughout the period during which the population remains small in effective breeding size. The deviations produced cannot be predicted and may result in a very considerable divergence in time. While the magnitude of drift can be estimated between each generation, the direction in which the next generation's gene pool deviates cannot be determined. Consequently, some very strange

---

*The ratio of children to parents is 514/388 = 1.33, which results primarily from the fact that all offspring had not reached reproductive years.

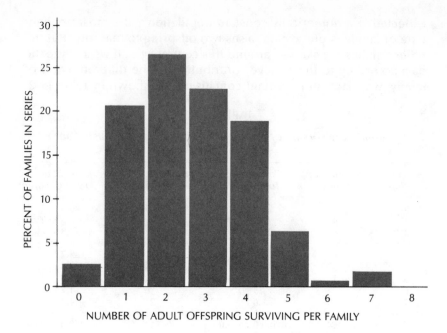

Figure 15–3
Frequency Distribution of 194
Precontact Matings

things happen in small populations. A gene which has a low frequency in the pool is very apt to disappear from it altogether after some generations have passed. This will happen if the new families which possess the gene fail to raise any offspring successfully. On the other hand, by pure chance processes, a rare gene, if imbedded in a reproductively successful family, might rise to surprising heights over a period of time, if consistently favored by chance. As everyone knows, Lady Luck is very fickle and so are the results of drift. It is real, it has considerable importance, but it introduces a kind of unpredictability into the evolution of populations in which it operates. It operated strongly in man until very recent years, when urbanization dampened its effect.

Genetic Coadaptation

The earlier sections of this chapter have given brief discussions of the four evolutionary forces, one by one. Each in its own right is sufficiently complex so that this is necessary for understanding. But, of course, there are no real situations in nature in which one evolutionary force acts alone. In all actual populations, mutation and selection in manifest forms are going on at all times. Gene flow and drift also operate in those situations which favor them. It is probably fair to say that at most times all four processes are acting together. Such a situation is too complex both for naturalistic mathematical modeling and for our comprehension.

Mathematical models of evolution necessarily include the force of mutation in their construction. In real populations, we fortunately can ignore its immediate results. Mutation seldom, if ever, acts to change the gene frequencies of small populations in a direct sense. For the most part, this force can be thought of as ultimately restoring variability lost through selective processes, but not intruding visibly upon the scene. At the same time, the effects of gene flow can be held constant, for evidence from Australia which will be presented in a later chapter indicates that it is overridden there by the combined forces of selection and random genetic drift. This allows us to focus upon two of the four evolutionary forces, selection and drift.

Leading evolutionists have pointed out for some years that selection operates on an internal basis within the gene pool, as well as a force applied upon it by the exterior environment. This is advanced genetic dogma which rests upon several well-demonstrated facts. All genes are visualized as manifesting *pleiotropy* to some extent, that is, contributing to the formation of more than one character. Laboratory geneticists tend to focus upon the most visible manifestation of the action of a single gene, but their investigations also show that genes have manifold effects which ramify throughout the phenotype. They are, in a sense, acting cooperatively to bring a given character to its full normal expression. Under these circumstances it follows that some genes cooperate with the rest of the genes in the individual better than do others. The genes that cooperate well have been called "good mixers." There is a general feeling among geneticists today that probably all of the genes at the variable loci are concerned with a kind of give and take between themselves in terms of their biochemical functions. The most fit individual is visualized as being the product of a genotype in which the genes so operate as to produce an optimum overall functioning. The idea that the genes within an individual may function at various levels of cooperation can be projected directly to the gene pool of a population. When for any reason the frequency of genes is changed within the pool, a period of coadjustment follows, producing what we call *genetic coadaptation*. This is a perfectly reasonable situation but one which has enormous evolutionary consequences.

Let us first look at the classic experiment demonstrating genetic coadaptation in the fruit fly. It was conducted by Dobzhansky and Pavlovsky (1957) and is worth summarizing here in some detail. The experimenters began by choosing two populations of a species of fruit fly, one from Texas and one from California. This determined that the stocks would have many genetic differences. The experimenters bred members of opposite sexes from both of the two geographical populations to provide a mixed and genetically hetero-geneous base stock for the experiment. Their procedure was designed to test the effect of founder drift in experimental popula-tions. It demonstrated that it was important and further revealed that the populations were affected by genetic coadaptation.

The experiment was divided into two halves. In the first, ten different replicative stocks of fruit fly, 4,000 individuals each, were randomly picked from the mixed stocks. Because of the method of breeding, each started with an identical frequency of certain designated inversions, of which the marker inversion was called Pike's Peak, since it was discovered on that mountain. This chromo-somal inversion is rather rare in natural populations. All ten experi-mental cases show a reduction in frequency from 50 percent to values between 43 and 28 percent after the first four months. Populations were allowed to continue for a total of 17 months, and the frequency of the Pike's Peak inversion was further reduced to a range of 20 to 34 percent. The results are shown on the left-hand side of Figure 15–4. Each population behaved differently, for each contained a different frequency of the variable genes. During the period of the investigation these were undergoing the process of mutual adjustment and reaching new *new levels of genetic equilibri-um* as reflected in the marker inversion.

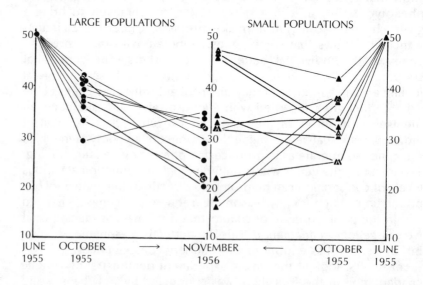

Figure 15–4
Genetic Coadaptation
in the Fruit Fly (from
Dobzhansky and Parlovsky,
1957)

But the second part of the experiment was the more important one. In this the investigators chose from the same basic hybrid populations of fly, ten new and separate populations, each of which originated with only 20 founders. Each population was thus put through a tight genetic bottleneck. At the end of the first four months the frequency of the Pike's Peak inversion had fallen in all populations, showing a range of from 25 to 42 percent. But, more significantly, at the end of 17 months, the range in these populations had expanded to 15 to 47 percent. Having passed through the initial bottleneck, these populations went through a much more active process of genetic coadaptation, and ended up at various levels of equilibrium which differed widely from each other. Their range was more than twice as great as that of the ten populations which started with 4,000 individuals each. These results are shown on the right-hand side of Figure 15–4.

This experiment demonstrates that founder drift and a variety of selective forces act cooperatively and that one does not proceed to the exclusion of the other. The kind of sampling errors which occur in small populations initially changes the frequencies of their gene pools by chance. But thereafter the process of genetic coadaptation operates so as to produce the best balanced combination in the long run. In Dobzhansky's experiment with the ten small populations, no less than four of them ended with higher frequencies for the Pike's Peak inversion than they had at the four-month time level. These represent very dramatic oscillations in the frequency of the marker chromosome. *The experiment summarized above explains why evolution contains an indeterminate component and so does not allow complete predictability.* When the contents of a gene pool are disturbed, there is no way of telling where it will find its new equilibrium position. Since drift, especially in the form of founder effect, produces an unpredictable variety of disturbances, the consequence is that no one can predict the end product. *Drift pulls the trigger and selective forces then come into action to create a new level of equilibrium in the gene pool.*

In a later chapter evidence will be presented in the form of Australian aboriginal blood group data to indicate the same type of interplay of evolutionary forces, in short, genetic coadaptation.

Bibliography

BIRDSELL, JOSEPH B.
1950   Some Implications of the Genetical Concept of Race in Terms of Spatial Analysis. Cold Spring Harbor Symposia on Quantitative Biology. Vol. 15, pp. 259–314.

CALHOUN, JOHN.
1962   Population Density and Social Pathology. Scientific American. Vol. 206, No. 2, pp. 139–148.

DOBZHANSKY, THEODOSIUS.
1970   Genetics of the Evolutionary Process. New York: Columbia University Press.

DOBZHANSKY, THEODOSIUS and OLGA PAVLOVSKY.
1957   An Experimental Study of Interaction between Genetic Drift and Natural Selection. Evolution. Vol. 9, No. 3, pp. 311–329.

FORD, E. B.
1964   Ecological Genetics. London: Methuen & Co. Ltd.

LEWONTIN, R. C.
1967   An Estimate of Average Heterozygosity in Man. American Journal of Human Genetics. Vol. 19, pp. 681–685.

MAYR, ERNST.
1966   Animal Species and Evolution. Cambridge: Harvard University Press.

SIMMONS, R. T., NORMAN B. TINDALE, and JOSEPH B. BIRDSELL.
1962   A Blood Group Genetical Survey in Australian Aborigines of Bentinck, Mornington and Forsyth Islands, Gulf of Carpentaria. American Journal of Physical Anthropology. Vol. 20, No. 3, pp. 303–320.

More Information on Mutations

Chapter 15 contains bare-bones information on mutation, only to the degree that it is needed to understand evolutionary processes. Other points of interest are whether mutations are generally harmful or beneficial, how they are produced in nature, and whether the rates of evolution within a species are subject to natural controls. These points are briefly covered here.

Mutations are usually detected in stocks of laboratory animals or plants, then confirmed by breeding experiments which show them to be something genetically new in the population. It follows that the inventory of recognized mutations is heavily biased in favor of those that produce large visible change, as opposed to the ones whose effect upon the phenotype is insignificant. This fact influences our judgment as to whether mutations are generally beneficial or harmful. Major alterations in the genetic code are apt to produce dislocations in the smooth functioning of the total organism. Geneticists generally agree that the small and insignificant mutations constitute the real building blocks by which adaptive changes within a species occur in time. Under these circumstances, one could predict that most mutations recognized in the laboratory will prove harmful to the animals that have them. And this is generally true. Most mutations are harmful in the sense that they lower the fitness of the mutant individual. The lowering of viability may be detectable only on a statistical basis, or it may be dramatic. The latter kinds of mutations are divided into *semilethal* mutations which usually cause death before the individual reaches the reproductive age, and *lethal* mutations which cause death prior to birth or shortly thereafter. These categories are only roughly defined and have no sharp boundaries.

Before proper judgment can be made about the effects of a given mutation upon the individuals who bear it, the mutation must be examined in a variety of reasonable environments. Mutations which lower fitness under conditions normal for a population may prove neutral under other environmental circumstances, and even superior to the normal gene under some environmental conditions. This is illustrated by a mutation in the freshwater flea (*Daphnia longiaspina*) which has been extensively studied in the labo-

ratory. One mutant type flea thrived and grew rapidly at water temperatures about 10° F. higher than those in which the general population did best. When animals of the mutant strain were placed in a normal environment, none of their young survived. The mutation obviously affected thermal tolerances. If such a mutant should occur in the population under its normal environmental conditions, it could not reproduce and so would disappear. Its reproductive fitness would be zero. But in an environment in which the water temperature was warm, such mutant individuals would become increasingly more fit, until with a rise of 10° F. they would become a successful type of animal with a reproductive efficiency of 100 percent. At the same time, the normal *Daphnia* population would itself have gone to extinction. Clearly, this mutation affecting thermal tolerance is disadvantageous under normal temperatures, but becomes highly advantageous when temperatures rise substantially. How can such a mutation be judged?

The example of the mutation in *Daphnia* points to the fact that judgments must be made in terms of individuals, populations, and specific environments. At the normal temperature the mutant *Daphnia* could not reproduce. Therefore, the mutation was disadvantageous under these circumstances. At the same time, the population had the advantage of possessing potential variability which would allow it to adapt to increasing temperatures. It is fair to say that the handicap to the individual constituted an advantage to the population. Populations can afford the loss of mutant individual members in a given environment insofar as they represent a potential for population adaptation if the environment changes. In this sense, mutations are a kind of tax levied on the population in order to ensure its evolutionary plasticity.

It was mentioned earlier that the rearrangements of the genetic code are a result of some form of energy input. H. J. Müller received a Nobel prize in biology for his discovery that mutations could be induced by bombarding fruit flies with X rays. This not only confirmed the idea that mutations resulted from energy inputs, but provided laboratory experimenters with a means of artificially inducing them. In recent decades it has been determined that a variety of other energy sources produce mutations. In cold-blooded animals, like the fruit fly, a temperature rise of about 15° F. is sufficient to increase the spontaneous rate of mutation two or three times. This increased heat input can be viewed as increasing the molecular activity of DNA, and so fostering mutational changes.

Müller's findings that X rays caused mutations led other scientists to test whether other forms of natural radiation might be *mutagens.* Ultraviolet light was found to produce an increase in the rate of mutation in small animals like the fruit fly which the rays could penetrate. It could not affect the sex cells of large mammals like ourselves. On the other hand, cosmic rays coming in from outer space with very high energies and velocities can cause mutations in us, for they traverse the whole of our body without our knowing it. One of the interesting points about cosmic rays is that people living in high-rise concrete structures are more subject to mutation from cosmic rays than naked Australians sleeping out under the desert sky. For cosmic rays become more efficient causes of mutation if they are slowed down, as they are when they pass through a number of floors of concrete reinforced with steel.

But it has been calculated that all of these natural sources of mutation do not suffice to account for the observed rates of mutation seen in natural populations. Obviously, some other form of energy must be involved. The first clues came in the 1930s when a series of investigators discovered that chemicals such as ammonia, copper sulphate, among others increased the rate of mutation in populations of fruit flies. But the first powerful chemical mutagen, mustard gas, demonstrated the point more forcefully a decade later. This poisonous gas produced a rate of mutation in fruit flies comparable to that induced by X ray exposure. Since that time, many other chemicals have been tested and a surprising variety have been found to cause mutations. They include some apparently very bland chemicals such as boric acid. These varied tests suggest that chemicals are involved as mutagens in perhaps a majority of cases occurring in nature. It is even likely that the chemicals we produce in our own bodies induce mutations under some circumstances. Totally unanswered is the question as to whether the drugs with which man is prone to doctor himself also raise the rate of mutations in the cells of individuals. Since human beings cannot be used on an experimental basis, laboratory animals such as the white rat must be depended

upon to give us approximate answers to this type of question.

Geneticists are fond of talking about the *mutational load.* By this they mean the mutations normally carried in a population in the form of recessive genes. Calculations indicate that every individual carries a moderate number of semilethal or lethal recessive genes. Under these circumstances, it is important to avoid marrying an individual who is closely related to oneself biologically to minimize the chance of bringing together two such genes in offspring. Research has indicated that some of these genes which in a homozygous form produce death are actually advantageous when in heterozygous condition. Thus, the abnormal hemoglobin produced by the sickle-cell gene only kills when homozygous; when heterozygous it actually benefits individuals living in an environment in which falciparum malaria is prevalent.

It is obviously too early to make hard judgments upon the meaning of the invisible genetic load in man. On the other hand, some ideas can be rejected out of hand. In the early part of the twentieth century there arose a movement known as *eugenics.* Moving prematurely with the new genetical knowledge of the day, its promoters claimed that all genetically unfit individuals should be prevented from reproducing, presumably by sterilization. It is ironic that today we know that if such a program would be carried out, all human beings necessarily would have to be sterilized. The answer to improving the biological qualities of the human population obviously lies in other directions.

One of the most interesting questions about the process of mutation involves whether or not it is subject to evolutionary change. One might speculate that the rate of mutation should be controlled by adaptive pressures since otherwise many populations would come to have too high a rate of mutation, and so lose too many individuals for its well-being. On the other hand, populations in which the rate of mutation was too low to allow adaptive shifts would tend to go to extinction. The problem is subtle, but the evidence consistently favors the idea that processes of selection work upon the mutational rate itself. If we compare the known rates of mutations in fruit flies and man and figure them in terms of the number of mutations *per locus per gamete per generation,* then these two very different animals have rates strikingly alike. In spite of the fact that the intergenerational time among fruit flies is about twenty days, whereas that in man is something close to 30 years, the rates when expressed in the proper form are surprisingly close. The same is true of other animals where rates have been calculated. These data suggest that rates of mutation in natural populations are set by a series of forces which tend to produce the best or optimal rates. The exact mechanisms are not yet known, but in the laboratory populations of fruit flies genes have been isolated in a number of instances which either depress or increase the normal rate of mutation. Since the way in which they alter the rate of mutation varies over a considerable range, and because they have been found on different chromosomes, it is reasonable to presume that a rather wide variety of genes cooperates in determining the overall rate of mutation in a population. Under these circumstances, the rate can be selected by natural forces to produce optimum consequences. This type of selection very likely works upon breeding populations, allowing survival of those in which the rates of mutation are optimal while others whose rates are unsuitable proceed to extinction.

Man's increasing use of radioactive materials poses a real problem in terms of his future genetic load. If we totally ignore the likelihood of atomic warfare in the future, radioactivity still poses a real threat. The danger takes two forms. The first is represented by the increased radioactivity in the atmosphere which has resulted from test explosions of atomic and hydrogen bombs to date. The debris of these explosions has raised the radioactive level of the Northern Hemisphere to a very considerable degree, and in time will circulate throughout the entire atmosphere of our planet. Radioactive fall-out products are particularly serious in the Arctic, for the Russians have been carrying on a number of their tests in that zone. It has been found that the lichens on an Arctic tundra have absorbed so much radioactivity that the reindeer in Europe and the caribou in North America, which feed upon them extensively, have become loaded with radioactive products themselves. The Laplanders in Northern Sweden who feed upon these animals now contain 127 times the amount of radioactive cesium that occurs in the bodies of Swedes living to the south. The results of this unintentional experiment upon men have not yet been investigated. But there is no doubt that the rate of mutation in such populations will be increased and other body damage due to radioactivity, such as a rise in cancer rates, is certain to follow. We have not yet learned all the consequences of our tampering with our species and its planetary environment.

Even the peaceful uses of the atom carry their threat to man's biological welfare and future. There is no doubt that in time power must be produced by nuclear fuels. But in the past the Atomic Energy Commission has been extremely casual in the restraints it imposed upon atomic reactors. One of the problems laid in the fact that the AEC promoted the building of atomic reactors, and at the same time constituted the authority which controlled them. A bureaucracy has no conscience, and so this single organization could not perform both functions with equal fairness. Even without the catastrophic meltdown of the uranium fuel of a reactor, there is radioactive contamination of the adjoining atmosphere and hence ultimately of the Earth's atmosphere. It is very doubtful that the present level of control for venting the reactor is satisfactory from a biological point of view. Thoughtful men have suggested that a 10-year moratorium should be enacted before any more atomic reactors are constructed. In this time interval existing reactors could be closely scrutinized to see what may be their real effect upon us.

One of the real dangers of man's increasing use of radioactive fuels is the fact that as a natural population he cannot change his own species' rate of mutation by decree. As our atmosphere becomes more polluted with radioactive products, man's genetic load must necessarily be increased. This will produce more and more individuals suffering from serious genetic defects and consequently a higher wastage of human life. Given a few hundred thousand years, no doubt our species could accommodate to the new level of radioactivity by evolving lower natural rates of mutation. But this amount of time is not left to us.

Selection and City Life

Selection is an evolutionary process of the utmost importance, for it has a major role in most genetic changes in time. It operates in a variety of ways whose total effect is expressed in terms of differential effective fertility, as stressed earlier in Chapter 15. A variety of factors can contribute to differences in the way in which individuals hand on their genetic materials to the next generation. Early death, a refusal to marry or procreate, carelessness in the raising of children so they do not reach maturity, and a variety of other causes can all contribute to differences in effective fertility.

Crowded cities arose for the first time on this planet about 7000 years ago in the fertile valleys of Mesopotamia. Since that time cities have steadily increased in importance, and continue to do so today. In the United States about 70 percent of the total population of 200 million people live in aggregations large enough to be called cities. In Australia, a country known essentially for its primary produce, almost 80 percent of its 13 million citizens live in cities. Quite literally the back country—which produces the animal and mineral products upon which Australia is largely dependent—is becoming uninhabited. As more people crowd into the great cities of the world, the latter become less fit for human beings to live in. As the reader is well aware, the cores of cities are becoming impoverished ghettoes, and at this time there is considerable doubt whether they can be restored to a decent state in which man can live a healthy life. What has been called "pathological togetherness" will no doubt increase as the world's population doubles in the next 25 to 30 years.

It is important to understand that no selection is operative unless differences in effective fertility are related to genetic differences in the people involved. Let us use a gruesome example to illustrate the point. Let us presume a small hydrogen bomb of five megatons is dropped upon the center of Manhattan Island. The so-called "lethal area" extends in a radius of four miles all around the point of impact. Within the 50 square miles represented by the "lethal area," presumably all of the individuals in the central portion will be killed, if not vaporized. This occurs irrespective of their genotype, and so selection is not acting upon them. They have merely been annihilated. Away from the center, an increasing number of survivors will be encountered. Most of these will have survived owing to the accident of their position at the time of the explosion. But as we reach the outer portion of the circle, survivorship becomes more common, and the possibility of selection now arises. If in the outer stretches of the devastated area some people survive because they are genetically equipped to better withstand radiation and blast effects than are others, then selection is going on. The point of this grisly example is that all deaths do not involve selection, only those in which genetic differences contribute to survivorship. Russia now has warheads of the equivalent force of 25 megatons, whose "lethal area" extends in a radius of 12 miles about the point of impact, and encloses slightly more than 450 square miles of devastation. This unpleasant subject is elaborated by Rathjens and Kisgiakowsky (1970), *Scientific American,* Vol. 22, No. 1, pp. 19–29.

To return to normal city living, census figures indicate that the people residing in most cities usually do not reproduce themselves. Their rate of reproduction is insufficient to keep up city numbers, so that the growth of these central metropolitan areas depends upon recruitment of masses of people from suburban and rural areas. Whether this situation has any genetic consequences is not known, for there is no reason yet to believe that urban dwellers of the same race differ biologically from country dwellers. But the fact does raise an important point, namely, that cities are not good places in which to raise children. The total environment of the city is a crowded one, the living space of families within it is constricted, and all of the needs of children are expensive to fulfill. Small wonder that city dwellers tend to restrict the number of children that they bear. Nor is it difficult to comprehend why so many people leave the cities for more open spaces in the suburbs as soon as they can afford to.

Aside from the inconveniences of city living, there are a number of hazards which certainly do contain the potential of acting as selective agents. Let us first consider smog. Medical research has not yet fully established the consequences of this type of atmospheric pollution. But in the city of Los Angeles it is said that an autopsy surgeon can closely estimate the length of time a person has lived in that city from the condition of the cadaver's lungs. Normal healthy lungs are a bright pink in color. Those of the citizenry of Los Angeles range from brown to black. Smog undoubtedly contains cancer-producing compounds, an effect confirmed by experiments with animals. Individual men and women differ considerably in their susceptibility to these cancer-producing agents. Without knowing any of the genetic differences involved, it is safe to say that smog is acting as a selective agent upon human populations living in cities. Even the human fetus growing in a smog-bound environment is deprived of its normal growth potential, for they average more than a half-pound lighter than those born in a clean environment.This handicap results in greater post-natal mortality and may operate throughout life. The smog also contains carbon monoxide (CO), a compound which limits the amount of oxygen which can be bonded to the red corpuscles, and so distributed to the body. Slight amounts of carbon monoxide impair our perceptions, and cause passing illness. In greater amounts it kills. People undoubtedly differ somewhat in their susceptibility to these effects. Finally, it has been shown that the asbestos incorporated in automobile brake linings fills the air around freeways with tiny fragmented fibers. These enter the lungs and in time can cause serious illness. Altogether it can be predicted that air pollutants of one or another kind in cities will so affect the lives of men as to select against those who do not have adequate genetic defenses. Given enough time, the same processes will tend to evolve smog-resistant men.

City planners visualize for the future great buildings 1000 stories high, which contain within their own structure the residential units, the work offices, stores and the entertainment needed for city living. It is even proclaimed that one will never have to leave one's own building for anything short of medical emergencies. Even today cities are crowded and crowding produces stresses. It is characteristic of city dwellers that they insulate themselves from social interactions with their fellow man by a cloak of anonymity, whereas the residents of a small town may know everyone in it, and see them frequently. The latter situation does not overload our social capabilities. I have asked students in a number of classes to ask their parents how many other couples they are sufficiently friendly with to see

on an average of one or more times a month. These urban parents average 2.1 other couples as friends, a striking document for the intimacies (or lack thereof) bred by city living.

The effects of crowding can be studied more carefully among the experimental mammals. It has been known for a long time that rodents and related animals such as rabbits and hares are prone to sudden death under crowded conditions. Where too many animals are put into a restricted space, they begin to die without a mark visible on their body. The phenomenon has been intensively studied during recent years, and it has been determined that the animals die as a consequence of adrenal gland overactivity. The crowding produces more face-to-face interactions than the animals are equipped to handle, so their adrenal activity goes irreversibly upward and they die as a consequence. Autopsies on the dead show enlarged adrenal glands, accompanied by a number of other changes, including deterioration in brain cells.

John Calhoun (1962) has designed a series of ingenious experiments to examine some of these problems among rats. One experiment demonstrated that wild brown rats, placed in a one-half acre enclosure with an unlimited supply of food, simply failed to increase at the expected rate. At the end of the experiment, the cage contained no more than 3 percent of the number expected if reproduction had been conducted successfully. The investigator concluded that the population of rats was rigidly limited by the number of suitable nesting places within the compound. In another experiment he found that under overcrowded conditions only the most dominant rats could protect their females securely enough for them to raise a litter of young. The less dominant rats did not reproduce successfully and formed gangs of what he called "juvenile delinquents," which acted, in the social terms of rat life, destructively. Crowding produces biological stress, and there is reason to believe that individual rats and men differ genetically in their capacity to resist it. Accordingly, it acts as a selective agency.

Noise pollution is one of the further stresses occurring in urban life. Noise levels are scientifically measured in terms of a unit, the *decibel*, which is arranged in a logarithmic scale of increasing intensity. Noise levels which ordinary individuals can tolerate for long periods of time scale between 20 and 50 decibels. Ordinary city automobile traffic averages 95 decibels, a level at which permanent damage to hearing can be sustained. Rock and roll sessions have recorded as much as 175 decibels, and their addicts at ages as young as 14 years have already suffered permanent impairment to their sense of hearing. I do not know whether hearing specialists have recorded the decibel levels of 10 television sets, all tuned too loudly, blaring their unsynchronized noises into the single inner courtyard of the modern apartment building. But stress is there and people react differently to it. Some will become nervous and irritable, and possibly neurotic. Others may seek refuge in smoking, drinking, or becoming prone to acts of violence. But most people exposed to this level of noise will certainly think twice about having children. Since all known characteristics in man show genetic variability, it is extremely likely that people differ in their resistance to high noise levels and so in their responses. Selection surely operates in genetically significant terms in this sort of urban situation. It is quite possible that given enough generations the genotype of city dwellers may be modified to the forces of selection to produce a breed of city human beings. Indeed, this is almost necessary, for all population projections into the future indicate that our living is going to become more and more crowded, and so more highly urbanized.

The interested reader may consult John Calhoun (1962), "Population Density and Social Pathology," *Scientific American,* Vol. 206, No. 2, pp. 139–148.

CHAPTER 16

GEOGRAPHICAL VARIATION IN GENETICALLY SIMPLE AND INVISIBLE CHARACTERISTICS

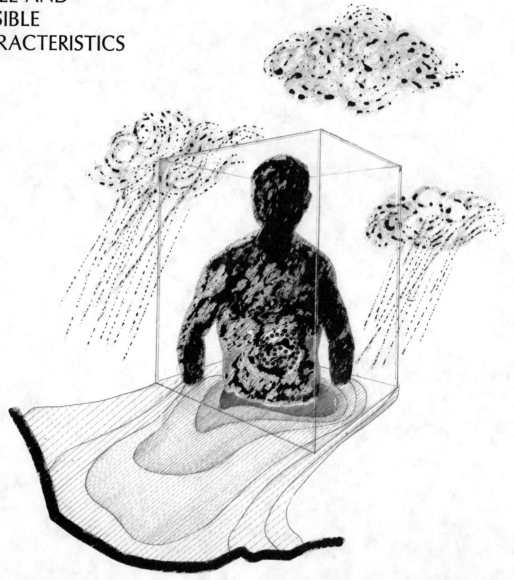

Most of the genetically simple polymorphisms which occur commonly in man are not visible in the surface phenotype but effect invisible biochemical characters, frequently those connected with blood group properties. Their distribution in space varies, and in most cases they present a pattern known as a cline. The occurrence of clines is presumptive evidence for the operation of selection, unless gene flow can be shown to have intervened. As an example of this, we will look at a number of the abnormalities of the human red corpuscle which have evolved to act as buffers against falciparum malaria. Different genetic variants have arisen through mutation in different parts of the Old World, and selection has carried them to heights where they give a balanced genetic protection to many populations. The so-called normal blood group factors also show clinal distributions, and this is the only substantial evidence that they are under selective control. The present world pattern of blood group genes is likely a product of genetic coadaptation, in which founder effect drift starts a local change which is then quickly taken to new positions of gene pool balance by local forms of selection.

On the Dimensions of Time and Space

There are many situations in which changes in time can be measured directly. Geological deposits provide this opportunity where dating methods are sufficiently accurate. With very short-lived animals and plants, the investigator can also follow evolutionary changes in time in life over a fair number of generations. The fact that the fruit fly, *Drosophila,* matures in less than a fortnight allows the laboratory geneticist to run through perhaps 20 generations a year. On human subjects the same number of generations would involve 500 to 600 years. As opposed to studying experiments on laboratory populations, there is the alternative possibility of studying experiments in nature. In past generations, evolutionary events may have occurred in populations now living that provide a suitable model to examine how and why such events happened. Such studies at the present time usually involve the investigator's spending a few consecutive years in the field. If the populations are correctly chosen, they may—in effect—present *evolution caught in the act.* This results from the fact that the differences found in populations distributed in space, as opposed to time, may give direct reflection of events occurring in past times. Such studies are within the framework of *microevolution,* since they are dealing with evolutionary change and process occurring on a short-term time scale. Under such suitable circumstances the dimension of space may indirectly reflect the happenings of the dimension of time. To put it another way, *time may be converted into space.*

The Case of the Tasmanian Brush Possum

A simple but classic example of how variations in space lead to an understanding of evolutionary process is provided by the Tasmanian brush possum (*Trichosurus vulpecula*). This animal is not intrinsically important except for the fur obtained by trappers. Because it contributes valuably to the Tasmanian fur industry, which keeps detailed data on skin, this animal has been studied twice. The initial paper by J. Pearson (1938) has been followed by a more recent study by E. R. Guiler and D. M. Banks (1958). Julian Huxley has been much interested in this example and assisted both investigators.

This arboreal, nocturnal marsupial occurs in two color phases in Tasmania, a normal gray-furred animal and a black, or melanic, one. This is not an unusual occurrence, but the Tasmanian possum is of evolutionary interest since the black phase is present in very high frequencies on the west coast of the island, and these diminish toward the east. The distribution of the frequencies of the melanic possum is regular in space, and so it provides an interesting case for analysis.

The raw data are massive, consisting of nearly 60,000 skins recorded in 1949 and 1952, in terms of fur color and locality of catch. This allowed the investigator to plot the frequency of the black form at more than 200 different stations scattered rather generally over the island. The extreme southwestern corner of Tasmania is uninhabited owing to the impenetrable evergreen beech forests. There are no data from this region, but this is unimportant since the brush possum does not live there in any case. The results are shown in Figure 16–1.

The distribution map of coat color in this animal shows regular gradation, with the gray form absent or reaching no more than 25 percent of the total along the entire west coast of Tasmania. Next, to the east there occurred a relatively narrow belt in which the grays ranged from 25 to 50 percent of the population. The remainder of the island has populations of possum in which the black phase has fallen below 50 percent, and in a tongue extending in from the upper portion of the east coast almost to the center of the island the blacks are further reduced and in some districts disappear altogether. The overall distribution is regular, its general accuracy is assured from the large number of skins upon which it is based, and so the remaining problem is to read its evolutionary meaning.

No breeding tests have been run to determine that black fur results from a single dominant gene, but the high frequencies in the west suggest this. Furthermore, there are many analogues to suggest that melanic forms usually are due to a single dominant gene for blackness. These include the hamster, in which distribution of a black form in southern Russia has been well investigated. Most of

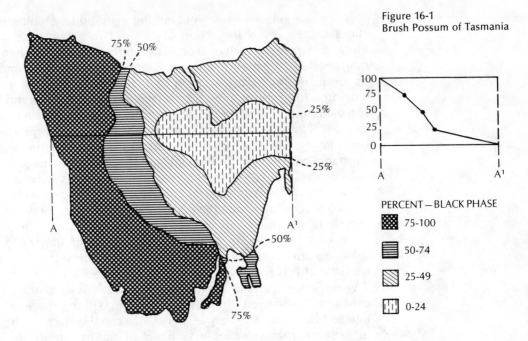

Figure 16-1
Brush Possum of Tasmania

PERCENT — BLACK PHASE

▓	75-100
▤	50-74
▨	25-49
▥	0-24

the moths involved within industrial melanism owe their dark color to a single dominant gene. For reasons that are not clear, this seems to be a general pattern. The fact that there are no intermediate forms between the black and gray-furred possums indicates that the gene seems to be completely rather than partially dominant.

The distribution of the black phenotype in Tasmania is mapped in terms of contour lines which correspond in a general way to the lines of equal elevation on topographic maps, or the lines of equal rainfall or barometric pressure on weather maps. Because they are here mapped as differences in the phenotype, they are best called *isophenes.* Had they been mapped in terms of the frequencies of the gene, they would represent *isogenes,* or points of equal gene frequency. The map really represents a sloping and somewhat complicated surface which reflects the differing frequencies in the black-furred form of marsupial. The surface itself is called a *cline* and is to be visualized much like a contour map. Clines are a useful way of expressing traits which vary continuously in space. A *transect* or section through a cline, as from A to A¹ in Figure 16–1, reveals that the surface of the cline does not slope equally at every point. The higher and lower values show a rather low gradient of change, while the two intermediate values, between 75 and 25 percent of the black form, show a very steep slope. Obviously something is operating there to make the rate of change greater in these populations. But the map in itself does not tell us what forces are acting.

There is a general agreement among evolutionary biologists that clines usually reveal the action of one of two evolutionary forces. Mutational pressure can be discounted since it cannot operate at the rate needed to produce a cline. Random genetic drift cannot be important in the final formation of a clinal gradient since its effect would be to produce a mosaic surface of uneven values and not to show smooth and continuous gradients. We are left with selection and gene flow as the causal factors involved in cline production. Gene flow could produce a cline such as the brush possum shows in Tasmania but only if hundreds of thousands or perhaps even a few million black-furred possums were dumped on the west coast of the island. Having established themselves, their genes would then diffuse progressively into the populations of normally gray possums. This idea can be ruled out as both ridiculous and in practice impossible. It must be concluded that the evolutionary force of selection operating in unknown ways has produced the distribution we have observed.

When investigations are extended to the human characters and genes that show continuous variation in space, there is much intellectual profit. The very existence of a clinal surface can be taken to indicate that selection has been operating on the series of populations included within the cline, for the number of clines which can be attributed to the action of gene flow alone is vanishingly small. Therefore, the analysis of clines provides a quick and easy identification of those characters which are subject to selective forces. The fact that the forces themselves are not directly identified by the cline nor their intensity measured is beside the point. The existence of the cline opens the way for higher powered investigations to determine both the agency affecting differential fitness and the rate at which it acts. If this situation in space is compared with the identification of selection operating in time on man, its great advantages are immediately seen. Rates of selection, save for lethal or semilethal characteristics, are slow enough in general so that they do not destroy the Hardy-Weinberg equilibrium in the population being studied. In short, the deviations they produce from equilibrium are usually not detectable. The fact that man has an intergenerational span of about three decades further complicates the process. For the study of the selective forces acting upon normal genetic characters in man, space offers great advantages compared to the dimension of time.

To return to the brush possum in Tasmania, the heightened fitness of the black-furred form has been studied to some degree. Guiler and Banks (1958) examined one of the major dimensions of weather, mean annual rainfall, and found that high rainfall was associated with high frequencies of black coat color. When rainfall exceeded 40 inches per year, the country was covered with dense eucalyptus forests, and in them the black form occurred with a frequency of 50 percent or more. The association demonstrated was convincing. A few little pockets of exception were noted, and it was clear that where the forests had been cut out, the black possums were reduced in frequency irrespective of the amount of rainfall. Obviously there is some factor operating in the more densely forested regions which makes the black fur color more fit for survival. The animal is not subject to heavy predation, owing to its nocturnal and arboreal habits. A few young may be taken by owls but otherwise predators treat it very gently. Since the Aborigines took these animals out of their high tree dens by daylight, without catching sight of them, and modern white trappers collect them sight unseen by snares, it is unlikely that human predation has anything to do with coat color differences. It would appear that the dark fur color is associated with some greater biological vitality which manifests itself in regions of dense forest and high rainfall. The factors involved are unknown.

The spread of melanism in the brush possum in Tasmania is quite likely a rather recent evolutionary episode. Tasmania was firmly connected to the mainland of southern Australia until the sea level rose from its lowered glacial level and reached its present height about 9,000 years ago. Tasmanian and Victorian possums should have been very much alike. It is known that the black form occurs occasionally on the mainland but does not seem to increase in numbers enough to become locally visible. The results suggest that an occasional melanic mutant in Tasmania did find itself more fit than the normal gray form in the wetter area and with the passage of time established itself there as the regionally predominant color phase. It may well be that evolution occurred in the genetic background of the population and that the heightened fitness is not due to the black gene alone. The example is suggestive but certainly needs further investigation.

The Distribution of Some Genetically Simple Characteristics in Man

Of the tens of thousands of variable loci in man, those involving allelic differences in the hereditary characteristics of various blood group components are the best known and the most revealing. The red corpuscles, or *erythrocytes* as they are known technically, provide the greatest amount of data. *Hemoglobin* has been subjected to a number of mutations which have been incorporated in the gene pools of various populations in the tropics of the Old World. Some of these have produced changes in the genetic code for hemoglobin, except in which one important one, the thalassemia complex, involves various types of blocks in the production of normal hemoglobin. The *haptoglobins* are substances in the serum of the blood that combine with free hemoglobin to prevent them from passing through the capillary network of the kidney and there doing damage because of the large size of the molecules involved. The serum of the blood also contains the *transferrins* which carry iron ions to and from the red marrow. No less than 12 molecular varieties of human transferrin have been identified at this time. Various other proteins such as the group specific component (Gc) and the Gm and Inv serum group are also polymorphic in man. A protein with the horrendous name glucose-6-phosphate dehydrogenase (G6PD) is also found in the red cell. In recent years some 30 or more polymorphisms have been discovered in the white blood cell, or *leukocyte.* It is not surprising that a series of similar genetic variants have been identified in the great apes, which of course are closely related to man.

But the most variable group of loci involves the *antigens* of the red blood cells. These are large protein molecules to which various sugar molecules are attached. They react with antibodies already present in the serum. Their action is to cause the clotting or *agglutination* of the red corpuscle. The surface of the red corpuscle is covered with a wide variety of known antigenic loci, now numbering 30 or more in all. Some of these are found rarely, and are called family antigens, but others are present as common polymorphisms in most human populations and so are of greater importance. An excellent discussion of these and other variable loci in the blood is contained in J. Buettner-Janusch (1966), and the reader may turn to this source for a more complex discussion than will be offered here. In this chapter we are concerned primarily with the important alterations in human hemoglobin and the better known of the blood group antigen systems. These have been intensively studied, so that it is now possible to plot their variable frequencies over much of the globe. For this reason they serve as models for our inquiry into their evolutionary meaning. The persistence of polymorphic characteristics of this type suggests that they serve an important function in adjusting populations to their environments. They may provide information as to how evolution proceeds in some of its detailed processes.

The Impact of Malaria upon Human Populations

It has been known for a long time that swampy environments caused much ill health in man. Centuries ago it was thought that swamps gave off bad air, and indeed some do smell from decaying vegetation and an unpleasant gas known as hydrogen sulfide. It was believed that bad air produced illness, specifically, the disease which came to be called malaria. Today we know that malaria is a complex of diseases in which parasites attack our red corpuscles. The intensity of malaria ranges from relatively mild to severe. The

Figure 16-2
Present Distribution of Falciparum Malaria in the Old World

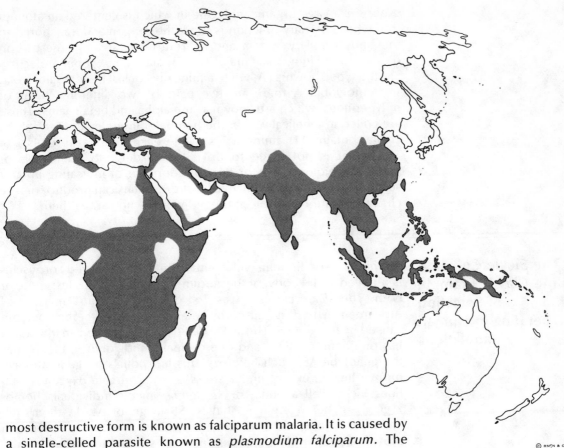

© RMCN & CO.

most destructive form is known as falciparum malaria. It is caused by a single-celled parasite known as *plasmodium falciparum*. The disease is widespread and common throughout most of the wet tropics of the world, extending in some regions into the subtropics (Figure 16–2). This parasite is now known to be carried by a number of mosquitoes of the genus *Anopheles*. In regions where the disease is prevalent, infection begins in infancy and reinfection occurs more or less continuously. Malaria of this kind is more than debilitating; it frequently causes death, with its terminal form known as blackwater fever, since the urine emerges black from the destruction of red blood cells. A disease of this kind causes a heavy drain upon the human population subjected to it, and it might be expected that the population would evolve genetic characteristics which buffer the effect of the parasites.

On the Nature of Human Hemoglobin

All the primates contain hemoglobins which are composed of four *heme* groups and *globin*. The latter is a protein made up of different proportions of the 20 common amino acids. Each of the heme groups is a large organic molecule in which is centered an atom of iron whose primary function is to fix the oxygen and transport it to the body capillary system and on return to carry out the carbon dioxide to the lungs. The majority of human hemoglobin molecules are the type known as A, while a minority consist of hemoglobin A_2. In hemoglobin A there are two pairs of two different chains of polypeptides, which are known as the alpha and beta chains. These are coiled in a helical way in the hemoglobin molecule. Each alpha chain contains 141 amino acids, while each beta chain totals 146. This point is not made to drive the reader into the details of molecular biology, but for the sole purpose of indicating in later sections how very small changes in these chains can produce drastic changes in the condition of the individual who carries them.

The Sickle Cell Trait as the Classic Example of Selection Operating in Human Populations

Plate 16-1
Red Corpuscles of a Sickler, showing various stages of blood cell distortion under reduced oxygen conditions.

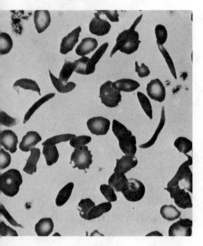

Anemia is a condition in which either the number of red corpuscles is reduced in the body, or the amount of hemoglobin is lessened, or both. The *sickle cell trait* was first discovered in an American of African ancestry. It is caused by an abnormality of the hemoglobin caused by the gene S. The sickle cell trait occurs when an individual has one hemoglobin S and one normal hemoglobin A, that is, has the genotype AS. Such heterozygous individuals, in general, have normal health, except under stress of considerable oxygen reduction. Sickle cell anemia, or *sicklemia,* affects individuals homozygous for hemoglobin S, that is, SS in genotype. In them, the anemia is severe, and few, if any, live to reproductive adulthood. In acute cases of sicklemia under conditions of oxygen reduction the red corpuscles assume a sickle or crescent shape as shown in Plate 16–1. These sickle cells may collect in the spleen, which then enlarges and undergoes fibrosis. There are a series of other consequences. The clumping of the sickle cells may interfere with circulation by causing local failures in the blood supply, produce heart failure, pneumonia, paralysis, or kidney failure among its other manifestations. Since the sickle cells tend to be destroyed more easily, so producing the anemia, this produces an overactivity in the marrow of the bone, which overgrows, and may even result in what is quaintly called a "tower skull."

Hemoglobin S and some 20 other abnormal hemoglobins all seem to be produced by very simple changes in the genetic code, involving a shift in a single base pair of peptide unit. In hemoglobin S, an amino acid known as valine replaces one known as glutamic acid in the beta chains. The DNA codings of these two amino acids are GUU, and GUA, respectively. Thus the replacement of the terminal U in the triplet unit in the code by A is what produces hemoglobin S and all of the manifestations of the disease. The

Figure 16-3
Present Distribution of Hemoglobin S (Sickle Cell Trait)

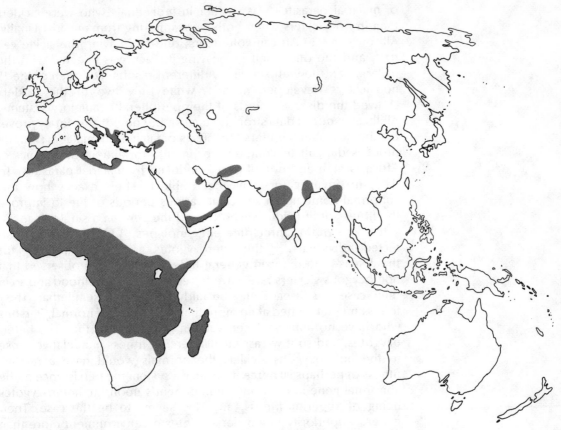

© RMⓈN & CO.

bodily consequences are remarkably large for the small scale of the change in the code.

The evolutionary significance of the sickle cell trait is that the gene S is widespread in the populations of the Old World tropics and in some reaches high frequencies. It occurs throughout most of tropical and subtropical Africa, and the disease ranges up to phenotypic frequencies of 40 percent. Slightly lower frequencies occur in populations in south India, Turkey, and Greece, and it is present elsewhere as shown in Figure 16–3. Why an anemia should be so common in so widespread an area is the evolutionary question. A. C. Allison (1954a, b) has presented three different kinds of evidence to support the hypothesis that hemoglobin S in a heterozygous condition offers some buffering against the inroads of falciparum malaria. In Africa the distribution of the disease closely parallels that of falciparum malaria at its highest levels of intensity. In populations where the disease is severe, the frequency of sicklemia is high. A nearby population, living on higher, drier land, may suffer less from the parasite and in general shows a lower frequency of the disease. Allison further found that the actual count of malarial parasites was smaller in individuals who were sicklers than in nonsicklers when both were suffering from malaria. Finally, Allison used 30 African volunteers, one-half having the sickle cell trait, and the other half not having it. Each was injected with the malaria parasite with startlingly different results. Fourteen of the 15 nonsicklers developed malaria, while only two of the sicklers showed the disease in tests of their peripheral circulatory systems. All three types of data strongly suggest, but do not absolutely prove, that hemoglobin S buffers the effects of falciparum malaria. Definite proof is difficult to come by, for infant mortality in the tropics of Africa is high and not all of it is induced by malarial parasites. To date, no investigators have been able to show exactly how this abnormal hemoglobin serves at various periods of life to improve the fitness of individuals who carried the gene in a single dose.

It is known that three different genotypes, *AA* representing unaffected individuals, *AS* the heterozygotes, and *SS* homozygotes for the disease, do differ in general fitness. It has been observed that homozygous sicklers rarely live to reproductive adulthood and so in their case the disease may be fairly considered semilethal. Their fitness might be rated at no more than 10 percent of normal. Persons who have no abnormal hemoglobin S are less fit than the heterozygotes, and so if we assign 100 percent fitness in a relative sense to the heterozygous sickler, the normals would have a relative fitness of perhaps 90 percent. In such a system the persistence of the abnormal gene in the population depends upon the heterozygotes being of superior fitness, and this seems to be the case. Their general superiority in a malarially infested environment more than counterbalances the constant loss of homozygous individuals who were born in each generation. Let us look at the situation and use the Hardy-Weinberg expansion to demonstrate how it works. We shall assume that the gene determining hemoglobin S has a frequency of .3, while its normal allele, A, has a value of .7.

$$(.7A + .3S)^2 = 1$$
$$.49AA + .42AS + .09SS = 1$$

relative fitness   90%   100%   10%

The above values for fitness are only approximate, but they serve to indicate that the system is in essential balance, for the semilethal condition of the homozygous sicklers is balanced by a reduction in the lowered fitness of the homozygous normals. Such a system is called *balanced polymorphism,* and it depends upon the heterozygotes being more fit than either kind of homozygote. The sickle cell trait is not the only one offering some immunities against the inroads of malaria, and some of these are discussed next.

Some Further Abnormal Human Hemoglobins

Hemoglobin S is the most widely distributed of the abnormal hemoglobins in man, but two others are of some importance in a population sense. They are hemoglobin C and hemoglobin E. Both are found in populations which already have some genetic buffering against falciparum malaria. Both of these abnormal hemoglobins have single base pair changes in their beta chains. The genes which produce abnormal hemoglobins C and E are alleles of the gene producing the abnormal hemoglobin S.

The full range of these two subsidiary hemoglobins is not known at this time. Abnormal hemoglobin C occurs in limited distributions in Africa, which range from the south coast of the Mediterranean through West Africa and some apparently isolated pockets almost at the bottom of the continent. The West African distribution is the best studied, and there it occurs in the same populations with hemoglobin S and other types of genes which presumably help to buffer the effects of malaria. In general, hemoglobin S is a more common gene in the pool of populations than is hemoglobin C. It is hypothesized, but by no means yet proven, that the sickle cell trait is a better buffer against malaria than the abnormal hemoglobin C. Hence in time it would be expected that hemoglobin S would replace hemoglobin C in populations in which both are present. Nevertheless, there is no doubt that having both adds some general fitness to the population prior to the rise of hemoglobin S to its full equilibrium value.

Hemoglobin E has a distribution which also is incompletely known at this time. It is most common in the Far East, in Southeast Asia, and the offshore islands of Sumatra and Java. It also is known to occur at least as pockets in eastern Pakistan, a portion of Ceylon, south of the Caspian Sea, and in southern Turkey. Very likely its distribution will be extended as it is investigated further. Falciparum malaria is present in all of these regions, and they each also contain other buffers against the disease. So once more we find two different genetic abnormalities of hemoglobin prevalent in malarially infested regions. These associations are suggestive that selection there favors the survival of mutations of this type, and in time selection lifts them to visible frequencies (see Figure 16–5).

The Distribution of Thalassemia

Thalassemia, which also goes under the name of Cooley's anemia or the Mediterranean anemia, is not caused by an abnormal hemoglobin but by a block in the synthesis of normal hemoglobin.

Figure 16-4
Present Distribution of Thalassemia Complex

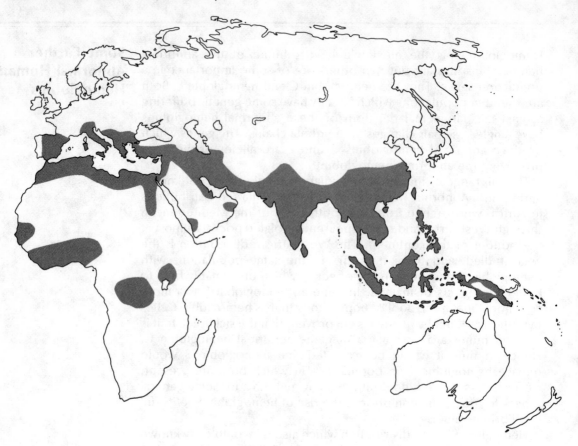

© RMSN & CO

Formerly thought to consist of a simple biallelic system, it has recently been recognized as occurring in a variety of independent forms, so its genetics are not understood. Its clinical manifestations occur in two phenotype forms, thalassemia major and thalassemia minor. In its major form the disease is severe, and individuals suffering from it survive only if they receive repeated and regular blood transfusions. The spleen, liver, and red marrow cells become much enlarged, accompanied by a reduction in the size of the red cells and a decrease in their count. White cells are abnormal in both numbers and size. As a consequence, the affected individual suffers from extreme jaundice and anemia, and the condition must be considered semilethal.

The frequency of thalassemia (Figure 16–4) varies in different populations, but its general distribution coincides closely with that of the falciparum form of malaria. Even in Africa, where hemoglobins S and C are fairly widespread, thalassemia also appears in widely spread regions. As the name suggests, it occurs in all of the countries bordering the Mediterranean Sea and extends in a wide belt from the eastern end of that body of water through southern Asia out into Indonesia and into New Guinea. It is more widespread in Asia than any other gene conferring buffering effects upon malaria. Once thought to be determined by a gene allelic to that producing the sickle cell trait, this trait is now considered to be genetically complex.

G6PD Deficiency as a Malarial Buffer

The red blood cells contain a great deal of protein in the form of hemoglobin, but they also contain as a minority component glucose-6-phosphate dehydrogenase (G6PD). This protein acts as an enzyme furthering chemical processes within the red cell. It was discovered more or less by accident when American soldiers of ultimate African and Asiatic descent were given an antimalarial drug known as primaquinine, a modern substitute for the old therapy with quinine. The new drug produced a mild state of hemolysis, or destruction of red corpuscles. Investigations showed that these affected individuals were deficient in G6PD. The trait is inherited and is an X-linked trait, and it appears to be incompletely dominant. It has many distinct variants, and it has revealed itself under other and curious conditions, involving one of the broad beans (*Vicia fava*). The fava bean is much eaten by people living in the Mediterranean area and G6PD-deficient individuals in these populations also suffer from a hemolytic condition resulting from eating these beans, or even smelling the pollen when they are in bloom. It is widespread in the Mediterranean region, in parts of West and South Africa, through much of the Middle East, and it occurs again in India, Southeast Asia, Indonesia, and New Guinea. Its occurrence is well associated with the distribution of falciparum malaria, which suggests but does not prove that this enzyme deficiency offers the individuals who possess it some immunity to the inroads of the falciparum parasite.

The Evolutionary Lesson Taught by the Distribution of the Red Cell Anomalies

The discussion of the biochemical abnormalities of the red cell given above provide a number of points of interest at the level of hypotheses, if not of proof certain. The microorganism, the falciparum parasite, infects the red blood cells in man. Each of the five characteristics we have discussed alters the biochemical composition of the red cell. Each one occurs in the region where falciparum malaria is endemic and so seems closely associated with the presence of the disease. In spite of the lack of a complete series of demonstrations to show how these changes in the genetic code give greater fitness to infected individuals, it may be concluded that they do so.

The theory of mutations leads to the expectation that changes in the genetic code will occur more or less at random with regard to any given population's need for change. It may be concluded that the one base pair change in both beta chains which has produced hemoglobin S has occurred repeatedly and in a variety of populations. Its presence in both India and Africa suggests this conclusion. Presumably, many populations outside of this malarial zone have also had the mutation, but among them it did not have selective advantage. In the same way, the other changes in the composition of the red cell seem to have been produced by a large but unknown number of mutations, with survival occurring only where selection favored the change. Mutations that are neutral or disadvantageous are rapidly eliminated. Favorable mutations will in time become incorporated in the gene pool of the population that bears them.

The occurrence of no less than five favorable mutations within the

Figure 16-5
Comparison of the Distribution of Falciparum Malaria with Five Genetic Buffers

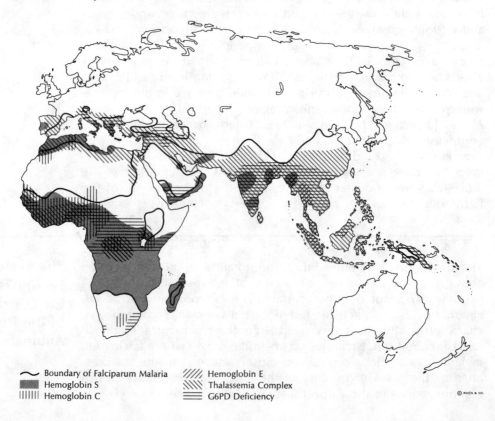

~ Boundary of Falciparum Malaria
■ Hemoglobin S
|||||| Hemoglobin C
//// Hemoglobin E
\\\\ Thalassemia Complex
≡ G6PD Deficiency

malarial belt in the Old World suggests that alterations in the genetic code are of sufficiently wide variety to meet most imaginable needs. No less than four separate changes in the red cell have occurred in both West Africa and Southeast Asia. In the former region, the abnormal hemoglobins S and C are found, together with the genes producing thalassemia and G6PD-deficiency. In Southeast Asia, thalassemia seems very widespread, and hemoglobin E and G6PD-deficiency are also present. In a few pockets, hemoglobin S seems to add to the battery of malarial buffers. It is almost as though we could conclude that wherever fitness is seriously lowered in a species, some genetic alteration will arise which will assist in minimizing its effect. The record to date seems to make that point.

When the five buffers against malaria are plotted on the same map (Figure 16–5), their distribution almost completely covers the regions in which malaria occurs.

The situation in the New World is today something of an anomaly. Malaria is present in the subtropics and tropics of the hemisphere. None of the known buffers have been identified there, but interestingly enough there is a blood group antigen, the *Diego factor,* which is widely distributed in the same area that malaria occurs. The association is very nearly complete in terms of overlap. Furthermore, the frequencies of the Diego gene rise to a peak in areas in which malaria should be presumed the most severe. As yet, the situation has not been adequately investigated, and so it is premature to claim that the Diego gene also acts as a buffer against malaria. But it would not prove surprising if this should prove to be the case.

Evidence from the Regional Distribution of Normal Human Blood Groups

The pattern of the distribution of genetic malarial buffers does not provide clines, for the investigators have frankly been more concerned with determining the range of each variant than in studying how the frequencies change in the gene pools of different populations. For this reason, they do not stand as examples of clinal structures, even though the fact that powerful forces of selection are involved implies that clines would emerge with the proper data. Research in the distribution of the better known and normal human blood groups does provide examples of clinal structures. As mapped in this chapter, they refer to the gene frequency of native populations rather than to the dominant population which may today largely represent immigrants from northwestern Europe. For example, in Australia the clines reflect the gene frequencies found in Aborigines, and those of European descent are excluded from the data. For the purpose of the investigation is to show the nature of biological variation before colonialism disturbed the original scene.

Geographical Variation at the A-B-O Blood Group Locus

The genetic variants at this locus became known earliest and hence have been tested in the greatest number of populations. In a general way it is now possible to plot the distributional frequency of the three major genes, _a_, _b_, and _o_, throughout the inhabited world. The record is more accurate in some areas than others. For example, for many years the Russians have frowned upon blood group research, so variations in the vast regions under their control are poorly represented. In other places, the original inhabitants have essentially passed to extinction. There are no good data for the blood groups of Indians living in the eastern and central portions of the United States, for no unmixed populations remain. Hybrid descendants provide a little evidence, but it is usually ambiguous. South America is shown as though it had been effectively sampled, which is not true. But this is not a serious misrepresentation, for the genes _a_ and _b_ seem to have been totally absent in the aboriginal populations of that continent, allowing a generalized picture which has a high likelihood of being correct. With these comments and reservations the maps do show what they are intended to, namely, that the blood groups at this locus have a worldwide distribution which is clinal in nature.

The distribution of the blood group gene _a_ is shown in Figure 16–6. Its frequencies range from .00 in the southern part of the New World to peak values greater than .50. The distribution of the latter is interesting. One peak value occurs among a group of Indians of whom the Blackfoot tribe is a central member. It and the related surrounding tribes show the world's maximum value for the gene _a_, and the values descend systematically from the central peak. This is an unexpected occurrence among peoples who in the rest of North and South America generally show low values for the gene. It, of course, does not mean that the Blackfoot Indians are unrelated to their neighbors, but rather that a very interesting but unknown evolutionary event has occurred among them. The other peak value for the gene _a_ occurs in Australia. The south central portion of that continent shows values characteristically in excess of .40. But in the Western Desert is a single tribe, the Mandjiljara, whose frequency for this gene reaches .53. They stand alone and their immediate neighbors show a much lower set of values for the same allele.

Some preliminary conclusions can be drawn from the distribution of the gene _a_. Its frequency is under selective control since it cannot be presumed to have originated through massive gene flow. The fact that the clinal surfaces are quite regular gives no other possible evolutionary explanation. On the other hand, the forces of selection which are operating upon this gene are totally unknown, for it is very unlikely that the present distribution involves diseases as the

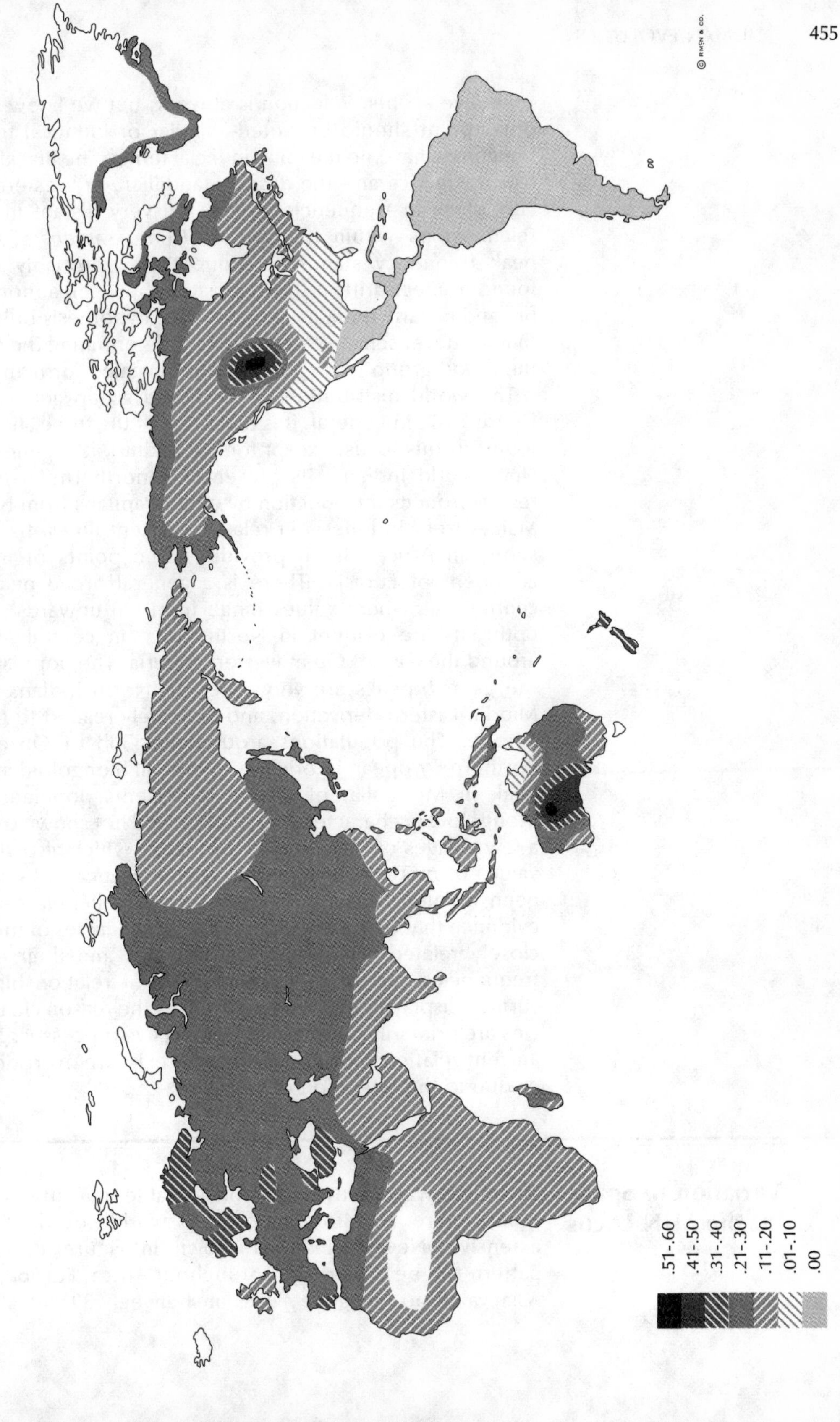

Figure 16-6
Distribution Frequency of Gene *a*

.51-.60
.41-.50
.31-.40
.21-.30
.11-.20
.01-.10
.00

formative agents. Selection is at work, but we know not how. One other point should be noted—similar or identical frequencies for the gene _a_ have no meaning in racial terms. The Blackfoot Indians of North America and the desert Mandjiljara of Western Australia are very close in frequencies for _a_, but very distant in terms of real relationships within the species _Homo sapiens._ The Mandjiljara peak frequency stands so alone that it probably resulted from founder effect drift. If the association between abnormal hemoglobin and malaria remains at the level of hypothesis only, it is obvious that whatever selective pressures are determining the frequencies of the blood group gene _a_ will not be revealed for many years.

The world distribution of the blood group gene _b_ is shown in Figure 16–7. In general, it is the rarest of the three alleles commonly found in this locus. Except for the Eskimo, it is generally absent in New World Indians. Its presence in northern Australia primarily results from its introduction by either Papuans from New Guinea or Malays from Indonesia in relatively recent times. Its distribution is spotty in Africa. But it provides some points of interest in the continent of Eurasia. There is a general broad peak of value in central Asia where values range from .20 upwards. Three further upthrusts are evident in North India, in central Mongolia, and around the Gulf of Ob in western Siberia. The populations showing the gene _b_ peaks are very different. North Indians are largely of Middle Eastern derivation, and ultimately related to the peoples of Europe. The populations around the Gulf of Ob are essentially northern European in origin with a small Mongoloid increment. The peak in Mongolia, of course, represents populations extremely Mongoloid in characteristics. Finally, and not shown on the map, are a few villages of Ainu in northern Japan which also show very high values of _b_. They represent an ancient Caucasoid strain which has been isolated in eastern Asia for many millennia. Again, we have evidence that peoples with the same frequencies of the gene are not closely related. The presence of genes and their equivalents in frequencies cannot serve as markers of relationships. The clinal surface displayed by the gene _b_ makes the reason clear; its frequencies are primarily determined by selective processes. Thus whatever ancient relationships of descent may exist are overridden by changes due to the operation of selection.

Variation in Space at the M-N Locus

This locus was discovered a considerable time after the ABO blood groups were, so information on its worldwide distribution is less extensive. Nevertheless, as shown in Figure 16–8, the general pattern can be discerned. Throughout Africa, Europe, and much of Asia, values for the gene _n_ oscillate between .30 and .60, with a trend

Figure 16-7
Distribution Frequency of Gene _b_

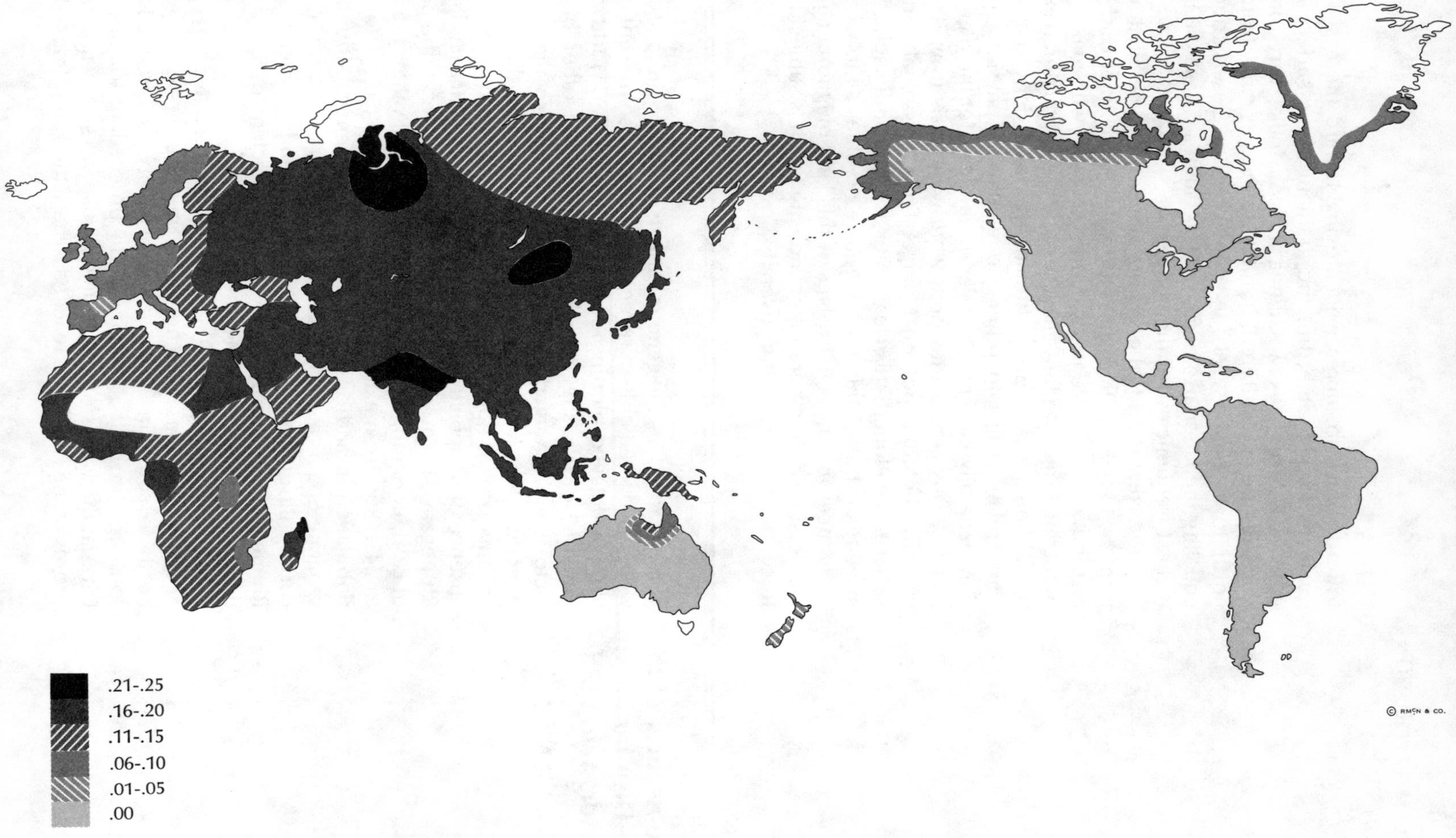

.21–.25
.16–.20
.11–.15
.06–.10
.01–.05
.00

toward values to increase in moving toward the east. These frequencies form a cline, but the gradient of the surface is gradual. When we turn to New Guinea and Australia, very different frequencies are encountered. In the latter, values range from .50 in the north to as high as .98 for some Western Desert tribes. Values equaling the latter also occur in New Guinea. These are the world's highest frequencies for the gene _n_.

In the New World, the gene _n_ values are generally low, ranging from .50 to below .10. Again, North and South America show a clinal distribution of the frequencies for the gene, with maximum values occurring in a broad belt on either side of the equator. It is of interest that although the American Indians are clearly derived from Asia and show relationships to other Mongoloids in physical appearance, they generally have much lower frequencies for the _n_ gene than do Asiatics. This gene varies on a worldwide basis from .00 to .98, or virtually the total potential range. From the clinal nature of the surfaces developed, it is clear that it is again under selective control in general, but at this time we have no clues as to what kinds of differences in fitness it involves. If this kind of ignorance is frustrating, let it be remembered that knowing that it is subject to selective forces is a net gain in our overall knowledge.

Some Aspects of the Distribution of Factors Within the Rhesus Locus

The Rh factors, as the genetic variants at the Rhesus locus are called, were discovered only in recent decades, and so present data do not allow a detailed plotting of their distribution in worldwide terms. What is known suggests that they, like the other blood group factors, are arranged in clinal patterns, and so their frequencies directly reflect the action of selection. One of the better known variants at this locus is called Rh negative, and represents the absence of the positive antigens C, D, and E. Hence it is labeled _cde_. Rh negative is known to reach its highest values in the populations of Europe. Among those populations its usual values are around .35 to .45. But its maximum value, exceeding .50, is found in the population in and on both sides of the Pyrenees Mountains which separate Spain from France. This is primarily an area in which the curious Basque language is still spoken, and the high frequency of Rh negative is associated with these ancient people. The world distribution is shown in Figure 16–9.

The farther away from the Pyrenees peak that we move, the lower become the values of Rh negative. They decline steadily, going from north to south in Africa, and have very low frequencies at the southern tip of that continent. In proceeding to the east across Eurasia, the values of _cde_ again diminish in a steady fashion until the shores of the Pacific are reached, where the gene almost disappears.

Figure 16-8
Distribution Frequency of Gene _n_

.81-1.00
.61-.80
.41-.60
.21-.40
.00-.20

Rh negative is totally absent from all the known populations of North and South America, the outer Pacific Islands, New Guinea, and Australia. If the data do not allow close plotting, it is nonetheless very evident that the distribution of Rh negative is once more clinal in nature.

The Rh negative factor is known to be subject to adverse selection, for populations in which it is high suffer a measurable loss of infants due to the hemolytic jaundice which goes by the name of *erythroblastosis fetalis.* This condition arises as a biochemical incompatibility between fetus and mother and has been briefly discussed in Supplement No. 5. But the world distribution of Rh negative can hardly depend upon this selective pressure alone, for then it would have been eliminated from Europeans as it has from other continental populations. At best, the situation allows the conclusion that this Rh factor is subject to a complex of selective pressures, in which maternal incompatibility is only one and in some areas not the most important.

Where the other types of Rh factor can be plotted—there are four other common ones—they, too, show that their distribution is essentially clinal in structure, and hence they are adaptive in nature. There are about another dozen blood group loci for which even less data are available. But all indicate that variation in the frequencies of their allele again falls into the same type of clinal structuring. The variable loci exist in some form or other of balanced polymorphism, and the frequency of their genetic variants depends upon the complex of selective pressures existing in the local environment. They are not evolutionarily neutral but apparently respond very sensitively to directed evolutionary pressures.

Lessons from the Geographical Distribution of the Blood Group Genes

Years ago, students of human evolution played a game which involved describing regional populations of man and attempting to classify them into a racial scheme which presumably would reflect the relationship, or the pattern of the descent of their lineages. This was a purely descriptive exercise and perfectly normal for the early phases of investigation. It is significant in retrospect that no scheme for regional analysis has ever proven acceptable, and the various classifications which have been produced carry no scientific authority. Where classifications have no consensus among the experts in the field, they are, of course, bankrupt.

Today, in the study of human evolution, the goals are to identify the nature of processes involved, and, if possible, to measure the rate at which they proceed. This is a difficult goal and at this time no one has succeeded in demonstrating for man the nature of selective

Figure 16-9
Distribution of Rh Negative (cde) Chromosomal Frequencies

.41–.50
.31–.40
.21–.30
.11–.20
.01–.10
.00

forces that act upon a single balanced polymorphic locus, nor in determining the pressures which they exert in any single population. Once that has been done successfully, the much more extensive task of determining the coefficient of selection at that locus in a wide variety of populations awaits the investigator as a long-term challenge. But in this decade of the twentieth century it is sufficient to point out that the investigation of process, as opposed to the sterile exercise of classification, is the real goal of evolutionary anthropology.

The blood group clinal maps presented earlier provide three important conclusions about the nature of ongoing evolutionary process in human populations. Very briefly, each one demonstrates that the gene frequency examined varies continuously in space both within continents and throughout the world. Secondly, the fact that the distributions are clinal in nature demonstrates that they must have been produced by selective pressure since gene flow can be ruled out. Lastly, the clines shown earlier are each independent entities, unique in configuration, and produced by the action of different selective forces.

Gene Frequency. The blood group genes chosen for illustration here are characteristic in that all other genetic variants of a simple nature which have been investigated to date show (or promise to show when data become more complete) a similar type of continuous variation in space. Therefore, it may be anticipated that they will usually develop clinal surfaces and that this is a normal expectation where gene frequencies are involved. This apparently simple conclusion has important consequences. It means that *nowhere among natural human populations, as they existed prior to colonial expansion, are there any changes in frequency abrupt enough to act as boundaries defining racial populations.* While members of both the public and of professional scientists may detect regional differences in human beings, neither can define a boundary to separate the central tendencies thus identified. Regional populations of man grade continuously into each other. This single fact is sufficient to explain why racial classifications are unscientific insofar as they attempt to set boundaries around the populations thus defined. This is an important conclusion, for it again demonstrates the unity of all living mankind.

Blood Group Clines. Earlier it was stated that clines for genetically determined characteristics are indicators of the prior action of selection or gene flow, or possibly both acting together. In none of our cases is there any possibility that gene flow has played an important role in establishing the surfaces of the cline. There are a few regions of the world where early mass migrations occurred, as in the peopling of North and South America, but these took place long enough ago so that selection has had ample opportunity to work upon populations since. We may conclude dogmatically that the blood group clines shown are the consequences of selective action, if of unknown types, and unmeasured intensities. If we cannot analyze them in the detailed way that will be ultimately possible, they do serve to point out that these genetic traits are molded by selection and so are adaptive in nature. This has very important consequences, for it means that none of the genetically variable characteristics considered can serve as markers of relationship, racial or long-term. Any analysis which purports to measure relationships on the basis of similar gene frequencies at any of the variable loci can be dismissed. This warning should be repeated, for there are many professional anthropologists and population biologists who still ignore it. The point may be put in another fashion. If a gene is presumed to trace relationship because it is present at about the same frequency in two groups, it must not be subject to the changes imposed by varying selective forces. Indeed, it must be neutral and immune to selective pressures. This requirement in itself is paradoxical, for the genes at the variable loci all seem to be subject to adaptive processes. *Genes do not serve as race markers.*

Independency.   The third point made by our blood group clinal maps is that each reveals itself as an independent creation and neither corresponds nor shows any systematic relationship with the others. This is a general principle, for it occurs in virtually all examples. It carries the very interesting moral that while all of us live, reproduce, and die as individual whole systems, the evolutionary changes that go on in the gene pool of a population proceed locus by locus. Apparently each genetic variant has its frequency determined in local environments by selective pressures acting upon it there. This fact is so significant that an anthropologist, Frank B. Livingston (1962), has proclaimed that clines are chaos. He is, of course, referring to the fact that when one clinal pattern is overlaid upon others, the only correspondences they show are those occurring by chance. Thus *it is impossible to build the description of a population in racial terms by defining it through the frequencies of its variable genes.* This fact does not allow the conclusion that regional differences do not exist, but merely that they cannot be neatly defined in scientific terms. The point is important, for it further emphasizes the futility of attempting the classification of populations, as opposed to studying the processes which have produced regional differences. Thus the study of the variable genes in man has moved biological anthropology on to new and more important problems.

Aboriginal Australia as an Experiment in Nature

In the laboratory an experiment essentially consists of testing a hypothesis formulated by the investigator. A well-designed experiment eliminates as many confusing factors as possible. An experiment is satisfactory if its results reveal the relationships which exist between two variables, with no interference from others. It is the ability of the experimentalist to control the procedures in the laboratory which bring important results when the investigation is finished.

The blood group data presented for the populations of the world suffer from a number of disadvantages. Certainly they have been collected in a helter-skelter fashion, primarily to provide descriptive material concerning the fluctuations of genes in different populations. Very seldom have these investigators designed their surveys so as to illuminate the workings of microevolution. The general results further suffer from the fact that they were largely collected by serologists, scientific workers who know a great deal about blood chemistry but almost nothing about evolution. They have turned up and identified a large number of previously unknown genetic variants but little more in terms of evolutionary biology.

In recent years, students of evolution have recognized that the basic unit in their studies must be the breeding population. For it is under these conditions that evolutionary processes reveal themselves most clearly. Among the thousands of investigations which have provided the data upon which the worldwide clines of the blood groups have been based, virtually none has identified the breeding populations they have investigated. In a few cases this is known because the population samples were small in size and lived in isolation.

As a consequence of these defects, the worldwide clines for blood group frequency are too generalized, the extremes of high and low values have been blurred by the mixing of different breeding populations, and in some instances the slope of the cline is wrong, as well as the values it is presumed to express. The situation may be likened to one in which an investigator looks at a distant heavenly body through a telescope with insufficient power of magnification, and lenses blurred with the fingerprints of children who have been eating sticky candy. Obviously, fine details cannot be seen. The world maps show none of the detailed relief evident in aboriginal Australia.

The Aborigines of Australia at the time of colonial contact existed as a simple and magnificent experiment in nature. The 300,000 inhabitants of the continent were divided into an estimated 585 dialectal tribes, each of which spoke its own homogeneous dialect. These tribes are the smallest breeding populations on the continent, and so each represents one of the basic units of evolution there. Numerically the Australian tribes ranged from several hundred to several thousand individuals, but they showed (Birdsell, 1953, 1975) a central statistical tendency to number about 500 or slightly fewer persons. This meant that the *size of the effective breeding population* was of the magnitude of 150 to 200 persons, values small enough to allow random genetic drift to operate. Therefore, they offer an unexcelled opportunity to see whether that curious evolutionary force is trivial or is really of some importance in general evolution.

Australia varies from uttermost desert to lush climax tropical rain forest, thus providing a great variety of climates. Its inhabitants all live on a hunting and collecting stage of economy, and so the picture was not complicated by the presence of either gardening or herding. Throughout this wide range of environments the Aborigines everywhere used very similar types of tools, and so in essence their technology was remarkably uniform. Since intertribal marriages average 15 percent per generation, the rate of gene flow between isolates is known and is relatively constant in rate. No other continent on Earth presented such precontact simplicities and uniformities. Australian populations might be expected to show interesting aspects of human evolution, and they do.

Since the arrival of White colonists in Australia beginning in the late eighteenth century, the Aborigines have been buffeted by new diseases, dispossessed of their land, and occasionally shot like animals. Today they number no more than 25,000 unmixed survivors. During the two years of field work beginning in 1953, I intensively surveyed the remaining Aborigines who had come into White settlements, fringing the perimeter of the Great Western Desert. In addition to a great many measurements and observations, 1,850 natives gave samples of blood for analysis, preliminary findings on which are given below. The data have been grouped into 28 tribal populations, or minimal-sized genetic isolates. They range from the southern tip of the desert in the state of Western Australia to its most northerly regions. The blood was tested by R. T. Simmons, and the gene frequencies calculated by J. J. Graydon, both experts at the Commonwealth Serum Laboratories in Victoria. Their identifications enabled the plotting of the variations in genes at no less than five loci.

While the distribution of the 28 tribes for which good genetic data are available covers a considerable area in Australia, they only represented about four percent of the total number of genetic isolates which existed before their destruction by White settlement. Thus they represent a very small minority of what the continent once had to reveal. On the other hand, there is no other place in the world where data this extensive can be collected on so many isolates among people at the hunting and gathering level. If these data do not reveal everything, at least the patterns they show will give us a part of the evolutionary story. Surrounding the 28 isolates are areas in which natives were also examined but so few per tribe that they have been thrown into regional series, by pooling them together. These data are less sensitive than the tribal data, but they help extend the boundaries of the clinal boundaries.

In this portion of Australia there are only two alleles at the A-B-O locus. The gene _b_ is absent among the Aborigines here. The frequency of the gene _a_ is shown in Figure 16–10. At first sight the clines, defined by the isogenes, appear to be both complex and confusing. But the observer will note that there is a very steady slope to the clinal surface which shows high values in the south generally and much lower values in the northern portion of the continent. This feature is regular enough to demonstrate convincingly that the gene _a_ is under selective control here, as it appears to be in the rest of the world. The picture is disturbed by the fact that in its center there is a very high peak value, actually .53, for the Mandjiljara tribe, while adjacent to it on the north, the Wanman tribe only shows a value of .13. This is an enormous difference which is reflected by the extraordinarily steep slope of the clinal surface passing from one tribe to the other. No less than eight isogenes, each representing a difference of .05 in gene frequencies, separate these tribes that live side by side and exchange daughters in marriage. The significance of this unexpected feature will be discussed shortly. The only other marked aspect of the cline involves a trough in which the value of the gene falls to .07, which is of course a very low value, and below that usually met in the rest of northern Australia. This might also be considered a feature deviating from a general sense of the clinal surface. In summary, the gene _a_ is distributed in this region in such a way that values decrease rather uniformly in going from south to north, but 3 of the 28 tribes fall out of the pattern in that their values somehow have been disturbed and altered compared to those of their neighbors.

At the M-N locus the Australians are less complicated than other people, and there the system contains but two genes, _m_ and _n._ Australia, together with New Guinea, represents the portion of the world in which the value of the gene _m_ is lowest. The distribution of the gene _n_ is shown in Figure 16–11. In the southern portion of our series is a very large area in which the value of the gene _n_ oscillates around .95. In two tribal groups, _n_ rises to .97 and .98, respectively, and so these represent maximal values for the known world populations. But these tribes are surrounded by other comparable values, so that the whole area is a kind of flattened peak in which _n_ shows very high frequencies. Proceeding to the north, the values of _n_ decrease rather steadily until values below .65 are reached in a few places in the north. In fact, a belt of these lower values extends from the coast down toward the desert. While the isogenic lines look rather complicated in Figure 16–11, the real clinal surface slopes quite gradually. In a general way, the clinal surface for the gene _n_ shows a rather regular slope from the very high values in the south to the very much lower ones in the northern coastal area. It can be concluded that the gene _n_ is under effective selective control and so is adaptive in some unknown way. In the Worora tribe on the northwest coast there is a relatively sharp increase in the frequency of _n_ which rises to .83, and so creates a sharply bounded peaklet in the generally low value surfaces found to the north. With this exception we may again conclude that the gene _n_ is under effective selective control and is adaptive in some unknown way.

At the Rhesus locus four different chromosomal arrangements are found among the Aborigines of Australia. These are R^1, which is the commonest arrangement, R^0, which occurs in moderate values, R^2, which has about an equal frequency, and R^z, which is a very rare arrangement in the rest of the world but is somewhat commoner in Australia. Let us look at the clinal surface of the chromosome R^0 shown in Figure 16–12. This cline shows an extended central peak area surrounded by diminishing values on all sides. The contours of

Figure 16-10

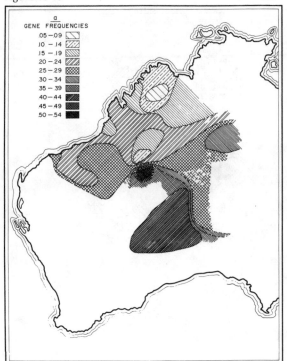

Figure 16-11

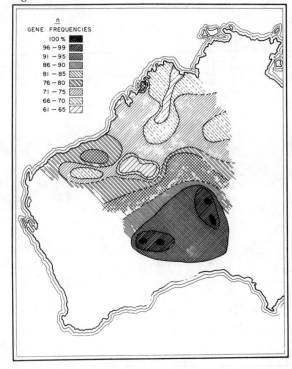

the isogenes in this distribution are relatively smooth and only at the central peak show any very abrupt slopes. Its pattern is unlike those previously discussed but still is very evidently under selective control in that its form is determined by adaptive differences in the tribal isolates. Aside from the culminating central peak, the other extreme feature of the cline consists of a single tribe, the Indjibandi, at the extreme left of the distribution, where depressed values of the chromosome approach .00. Two other areas in the extreme south do not deviate so greatly from surrounding values. So this clinal surface, like the others, is characterized by isolated peaks and sinks. But in this case, the peak involves three adjacent tribes, and is part of the general clinal structure. Overall, this configuration suggests that an important evolutionary event has been caught in the act. It may well stand as evidence for the ongoing process of genetic coadaptation in human populations.

The chromosome R^z is one of the rarest in most populations, but in Australia it rises to an average frequency of perhaps .04. In the 28 tribes from which these data are derived, this chromosome rises to near world record frequencies. In general, its clinal surface in Figure 16–13 reveals a low, undulating topography with the exception of three strongly contrasting and isolated peaks. Along the center of

Figure 16-12

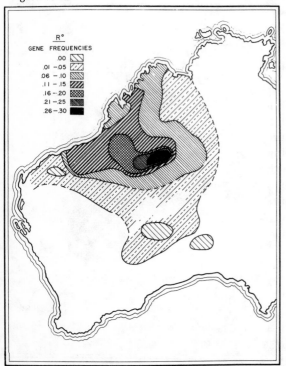

Figure 16-13

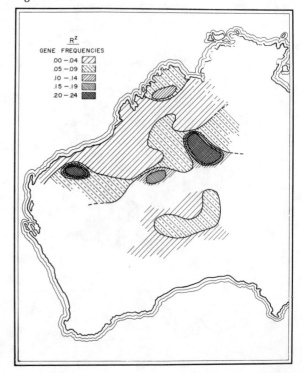

the diagram there occur three tribal groups, each well separated from the others in space, in which the value of this rare chromosome reaches extraordinary heights. Again, the Mandjiljara are out of line with a value of .18, which contrasts abruptly with the bordering tribes to the north. Even more remarkable are the Indjibandi, whose frequency of .20 raises a peak separated from the low values in the north by a full four isogenes. Most remarkable of all is the Ngardi value of .24, which is near the world's maximum recorded frequency for this chromosome. They are separated from surrounding tribes by two to three isogenes, while to the north the small Wandjira tribe shows a value of .00 for the R^z chromosome. This is a remarkable situation, for in two nearby Australian tribes with ordinary social relationships involving the common exchange of daughters, the two polar values for the R^z chromosome are found. It, of course, raises the question of how such differences are caused, and whether they can persist for any length of time.

What the Australian Aboriginal Blood Group Clines Reveal

In any new field, such as the analysis of clines, the first results are always obtained at a relatively simple level of scientific inquiry. As analysis becomes refined, more subtle questions can be asked and, with some luck, answered. The blood group data detailed earlier for 28 Australian tribes at this time only allow some answers to relatively simple questions, such as what evolutionary forces have been acting upon these populations, and what is estimated to be the relative contribution of each.

Before asking the question, let us again recall what the clinal surfaces revealed. In each case, for the gene a, the gene n, and the two Rh chromosomal arrangements, R^0 and R^z, the clinal surface was relatively simple over the greater portion surveyed, but in each instance isolated peaks and depressions broke general trends. For the gene a there was the face-to-face positioning of the gene's maximum value adjacent to a minimum value. The clinal topography for the gene n showed a generally uniform sequence from the very high in the south to lower ones in the north. The cline for R^0 showed a central ridge of high values culminating in a peak, with descending values distributed in a more or less concentric pattern. But the peak sloped sharply down through four isogenes to the east. The chromosomal arrangement R^z showed a relatively uniform topography out of which emerged three isolated tribal peaks, all of which attained very high values. The general pattern was one of prevailing uniformity interrupted by localized peaks and troughs.

The most important microevolutionary aspects of the Australian blood group clines involve the partitioning of the evolutionary forces which produce them. Mutational pressure may be immediately discounted as operating so slowly as to have in effect no influence in producing the kinds of clines which have been displayed. If a generalized rate of mutation is perhaps one per 100,000 gametes per locus per generation, then this process can hardly be conceived of as ordering the gene frequencies so as to produce the observed gradients. It may be ruled out as contributing to these results.

The possible role of gene flow should be examined next. The tendency of gene flow proceeding in all directions through intertribal marriage operates to level the values of gene frequency to a stabilized common value if no other forces are acting. Its operation is analogous to water seeking its own level. This has not occurred in Australia, so that it may be concluded that while gene flow does operate, its effects are overridden by more important evolutionary forces.

Selection is the evolutionary force which is present everywhere, acting on all populations at all times. The sum total of its effect, as indicated earlier, is to produce clinal surfaces with relatively uniform gradients. There is no way of predicting how these surfaces will slope for a given gene, but the gradients should be slight in character, rather than showing abrupt genetic escarpments. All four of the clines presented for Australian Aboriginal blood group data show this basic anticipated characteristic, and so we may conclude that selection is the major evolutionary process involved in their construction. We are still in ignorance as to what aspects of the environment cause differences in fitness for our various blood group genes, but there is no doubt some do. The only surprise in this finding is that blood groups should reveal the action of selection so clearly in a limited series of minimal breeding populations.

Before deciding whether random genetic drift may have acted in conjunction with selection on these Australian tribal populations, let us review how it might manifest itself. The direction of the change it produces is unpredictable, so it might cause an increase in gene frequencies as often as a decrease. The effective size of Australian breeding populations is small enough so that drift may be anticipated to operate on an intergenerational plane. But certainly the way in which its action would proceed the fastest would come at times when the total tribal population suffered reductions to much smaller than usual numbers, as in the case of severe droughts. The surviving members would become the founder stock of the next generations within the tribal territory. The founder type of drift produces greater changes faster than any other known evolutionary process. The majority of tribes investigated live in a desert in which less than ten inches of rainfall fell annually on an average and in which droughts would be frequent. One further manifestation of drift is that its effects should appear more or less at random over the landscape and act so as to disrupt the smooth clinal gradients produced by selection.

All of these attributes are found in the four clines for the blood groups presented earlier. The steep-walled peaks and sinks are best explained as the consequence of drift acting in the not too distant past. For once the initial disturbance caused by drift has occurred, then both gene flow and selection will tend to restore the deviating values to those which conform to the general gradient of the clinal surface. Accordingly, it might be expected that only recent consequences of drift would show themselves at a given time, such as when this survey was conducted. We have already eliminated mutational pressure and gene flow as being important in determining the clinal slope, so the only question that remains is whether selection could produce the observed peaks and troughs. With one exception, these disturbances in the clinal surface are all limited to single tribes, and it is impossible to imagine that selection would act upon a single breeding population and not affect those around it. This is particularly true since most of the country is exceedingly uniform in its desert characteristics. All of the evidence points to the fact that drift has been operating strongly upon these tribes as an erratic force, while selection has been operative as a regularizing evolutionary force.

Evidence for the Operation of Genetic Coadaptation

In the Dobzhansky experiment discussed earlier, drift of the founder type served as a trigger, while selection subsequently operated to find a new genetic plane of equilibrium for each of the small test populations. The result of the two processes may be considered as genetic coadaptation. It is significant that the genetic changes brought about by drift can become important in subsequent evolution. This is a direct answer to those evolutionists who have always considered drift to be a trivial affair of no lasting importance.

It is tempting to regard the elevated ridge shown in the central distribution of R^0 as an example of this kind of genetic coadaptation. The total known clinal surface is organized around this ridge of high values. The slopes descending from it are sufficiently uniform to indicate that its elevation has influenced the surrounding clinal surface. The steep descent of the surface at the eastern end of the highest peak suggests that the disturbing factor has operated recently enough so that the influence of subsequent gene flow and selection have not smoothed it out entirely. This example is a plausible clinal example illustrating how initial drift assisted by subsequent selective processes changes the total genetic gradient in ways which are designed as genetic coadaptation.

The Broader Evolutionary Lessons from These Blood Group Examples

A number of years ago Sewall Wright indicated that his studies of the mathematics of evolution showed the conditions under which the process would proceed at the fastest rate. He specified that the species should consist of a large number of small, isolated breeding populations. In them, from time to time, the chance effects of drift would produce changes in gene frequency in an erratic pattern. Most of these populations would become less adapted and perhaps be lost. But in a few breeding populations the chance shift in genetic values would be an improvement and that population would prosper. This landscape of small populations has partially isolated breeding units and the genetic exchange between them is slow but constant. Under these unique circumstances selection would override drift, and the favorable genetic combinations would be spread throughout the entire population. What we have just demonstrated is that populations structured like the Australian Aborigines, continuously distributed in space and constantly exchanging women, conform to Wright's mathematical requirements. For exactly what Sewall Wright considers the most favorable combinations for rapid evolutionary change go on in Australians who at all times remain structured into small breeding populations. In effect, this example shows that *genetic coadaptation in these populations occurs as predicted by Wright's mathematical hypothesis of shifting balances.* Nature confirms his hypothesis. This is an important conclusion.

The same Australian blood group data provide new perspectives on race formation and race classification. Elsewhere I have published the differences for all genes at all five loci for the 28 tribes used here. The differences between these adjacent and interbreeding tribes average .08 for all loci. This is an enormous genetic difference and must be seen in its proper perspective. The best comparison is provided by taking suitable populations for three widely separated areas of Eurasia, namely Europe, peninsular India, and Eastern Asia. When comparisons are made between each of these three polar populations, which are reasonably representative of three major population groups, the Caucasoid, the Veddoid, representing aboriginal India, and the Mongoloid, at the same five loci they, too, have an average difference that approximates .08. This comparison points to the startling fact that adjacent and interbreeding Australian tribes differ as much genetically from each other as the three above major populations do.

Yet another comparison is possible, for if the range of gene frequencies is taken for all genes at five loci for our 28 Australian tribes and compared for all known gene frequencies for the whole of Eurasia, then the Australians do not suffer in their overall range of variability. In the majority of cases the Australians show a greater variability than the polar Eurasiatic populations. In one instance they are equal, and in a minority of cases they show somewhat less variability. The thrust of this comparison is that where genetic surveys have been done, as in this case, on the real breeding units, a totally new picture of variability emerges. And it is this picture which must be projected back into Pleistocene times, when all people lived in equally small breeding populations.

It seems clear that the Australian data stand directly opposed to those professional investigators who believe they still can use genetic similarity as a basis for making racial classifications. Certainly at the level of the serologically simple gene this is not true. Remember that the extreme values for the chromosomes R^z, .00 and .24, respectively, were found in the nearly adjacent Ngardi and Wandjira tribes. Even though such extraordinary differences are the consequences of founder effect drift, they show that man's evolution in many ways has followed occasional paths of irregularity and even indeterminacy. Those who would still play the classifier's game to advance our knowledge of human evolution could spend their efforts more advantageously.

Bibliography

ALLISON, A. C.

1954a The Distribution of the Sickle-Cell Trait in East Africa and Elsewhere, and Its Apparent Relationship to the Incidence of Subtertian Malaria. Transactions of the Royal Society of Tropical Medicine and Hygiene. Vol. 48, pp. 312ff.

1954b Protection Afforded by Sickle-Cell Trait Against Subtertian Malarial Infection. British Medical Journal. Vol. 1, pp. 290–294.

1954c Notes on Sickle-Cell Polymorphism. Annals of Human Genetics. Vol. 19, p. 39.

BIRDSELL, JOSEPH B.

1953 Some Environmental and Cultural Factors Influencing the Structuring of Australian Aboriginal Populations. The American Naturalist. Vol. 87, No. 834, pp. 171–207.

1975 Preliminary Notes on Further Research on Man-Land Relationships in Aboriginal Australia. Memoirs of the Society of American Archeology. In press.

BUETTNER-JANUSCH, J.

1966 The Origins of Man. New York: John Wiley and Sons, Inc.

GUILER, ERIC R. and DORIS M. BANKS.

1958 A Further Examination of the Distribution of Brush Opossum, Trichosurus Vulpecula, in Tasmania. Ecology. Vol. 39, No. 1, pp. 89–97.

LIVINGSTON, FRANK B.

1958 Anthropological Implications of the Sickle-Cell Gene Distribution in West Africa. American Anthropologist. Vol. 60, pp. 553ff.

1962 On the Non-Existence of Human Races. Current Anthropology. Vol. 3, pp. 279–283.

1967 Abnormal Hemoglobins in Human Populations. Chicago: Aldine.

PEARSON, J.

1938 The Distribution of the Tasmanian Brush Opossum. Papers of the Royal Society of Tasmania for 1937, p. 21.

CHAPTER 17

GEOGRAPHICAL
VARIATION
IN VISIBLE
CHARACTERISTICS

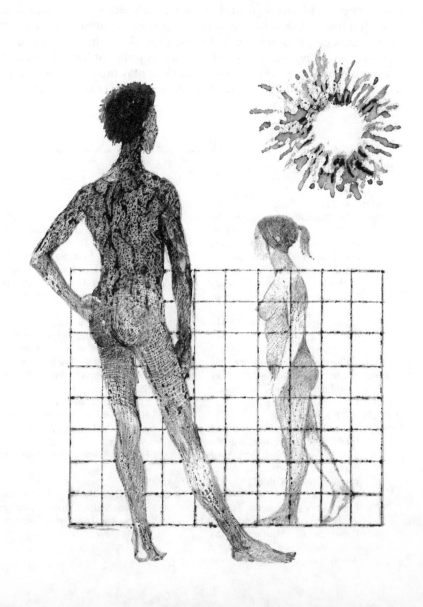

The genetically simple blood groups provide clear-cut answers to some of the preliminary questions as to how evolution has proceeded. They are hidden traits and so are not exposed to selection in an immediately visible form. We have seen that these blood groups manifest evidence of genetic coadaptation in man, and that in all cases examined, the variable loci indicate that selective processes mold the blood group frequencies geographically. Thus, they provide excellent markers by which the details of evolutionary process can be studied. On the other hand, the surface phenotypic characteristics of man belong to a genetically different category. All important ones are genetically complex and subject to some modification of expression under differing environmental conditions. This contrasts notably with the blood group genotypes which are fixed at birth and do not vary under any known changing environmental conditions. It should not be surprising if these two categories of traits reveal different patterns geographically.

The Nature of Polygenic Inheritance

Genetically complicated characteristics are the expression of numbers of genes located at more than one locus. They have been called *multifactorial characters* or *polygenic characters*. The nature of this complex inheritance has been studied primarily mathematically, because analysis is difficult in real situations. At this time, no multifactorial characters have been analyzed in man.

Polygenic characteristics are determined by allelic series of genes located at two or more loci, usually on different pairs of chromosomes. That is to say, genes at one locus are not *linked* with those at others, so each set can and does segregate separately in the production of gametes. This property produces a very great range of genotypes, which is one of the complicating factors in this form of inheritance. In order to make their theoretical approach easier, geneticists have made some simplifying mathematical assumptions. They have assumed, for example, that the genes at the various loci are equal and additive in their effect. This assumption is necessary to simplify calculations, but it is certainly incorrect.

The nature of polygenic inheritance is best understood by the series of examples given in Table 17–1. We shall examine the pattern of inheritance in cases of a pair of genes, or alleles, existing at first one locus, then a pair at each of two loci, and, finally, three biallelic loci. In our example, let us further assume that capital letters contribute equally to the expression of the phenotype, while small letter genes add nothing to its manifestation. In the initial example, the single locus with the allelic genes A–a, we have but three possible genotypes AA, an intermediate heterozygous genotype Aa, and the homozygous aa. Under our assumptions, such a system will produce three different categories of phenotypes, and these should

TABLE 17-1

The Way in Which Polygenic Inheritance Increases Complexity

Number of Loci		1		2		3
Phenotypes by classes (1)		AA	(1)	AABB	(1)	AABBCC
	(2)	Aa	(2)	AABb AaBB	(2)	AABBCc AABbCC AaBBCC
	(3)	aa	(3)	AAbb AaBb aaBB	(3)	AABBcc AAbbCC aaBBCC AABbCc AaBbCC AaBBCc
			(4)	Aabb aaBb	(4)	AABbcc AAbbCc AaBBcc aaBBCc AabbCC aaBbCC AaBbCc
			(5)	aabb	(5)	AAbbcc aaBBcc aabbCC AaBbcc AabbCc aaBbCc
					(6)	Aabbcc aaBbcc aabbCc
					(7)	aabbcc

be individually recognizable. This is the type of simple Mendelian inheritance upon which most examples in genetics have been built.

Now let us take a more complex case, although the simplest possible in a polygenic situation. We will consider two loci, each with a pair of genes. The first will have A–a as the alleles, and the second will have B–b. With our previous assumption, these genes can combine independently through random assortment in meiosis, and the resultant zygotes will fall into five different genotypic categories. If we further assume that both polar homozygous genotypes cover the same visible range as in our preceding example, then it follows that the three intermediate categories are more difficult to distinguish one from another. Increasing the number of genetic factors contributing to a phenotype blurs the differences between the various classes of phenotypes. Furthermore, the situation is complicated by the fact that in the intermediate category in this example three different genotypes produce exactly the same phenotypic expression.

Let us proceed to a yet more complex situation in which a character is determined by pairs of alleles at three loci, and each capital letter gene contributes equally and additively. Add the genes C–c at the third loci. When all possible genetic combinations are listed, we now have seven classes of phenotypes and twenty-seven genotypes. Let us look at the three central groups. The third class consists of a phenotype which is produced by any combination of four capital letter genes. There are six such possible combinations, and each would yield identical results in the model we are using. In the fourth and middle class, consisting of the phenotypes produced by three capital genes acting together, we have seven genotypes. Finally, in the fifth class, in which the two capital letter genes are present, there are once more six genotypes which produce identical phenotypes. It is fair to say that in this type of situation, the 19 different genotypes contained in these three classes would each produce a phenotype which would be extremely difficult to distinguish from the others. In short, genetic complexity blurs the distinction between the phenotypes produced.

In real life many of the visible or phenotypic characteristics of man are produced by genes at more than three loci. The pattern set in the preceding model indicates how rapidly any increase in the number of loci complicates the situation. With the polygenic characteristic involving three or more loci, we reach a new conceptual category, which really involves continuous variation in phenotypes. With 10 biallelic loci the number of possible genotypes becomes 3^{10} or 59,049. Essentially in these circumstances, differences can only be expressed in finely graded measurements. With such complexity there is no real point in trying to analyze polygenic characteristics in terms of the actual loci and the number of alleles at each locus that contribute to the phenotypes in question. Nature further departs from our model in that the contributions of all genes are not in fact equal. My own work in skin color on Australian Aborigines and their hybrid descendants clearly indicates that some of the genes involved in producing a skin pigment have much greater effect than do others. This in a sense simplifies a situation from one point of view, but complicates it from another. Polygenic characteristics do not provide the best data on evolutionary processes for reasons that now should be clear.

The Effect of the Environment upon Gene Expression

The blood group genes considered in the preceding chapter are especially useful since no environmental variables can alter their expression. But most other characteristics can be modified by changes in environment, and this is particularly true for polygenic traits. Let us consider human stature as an example. It is determined by an unknown number of genes at various loci, so that it is thoroughly polygenic in character. In addition, it depends upon growth processes which themselves are easily modified by changing environments. One of the most visible variations in the individual's environment involves the amount, quality, and variety of food he eats. No one can grow taller than his genetic code provides, but a great many people, probably indeed most, fall below the genetically determined standard to some unknown degree. Starvation stunts growth notably. The impact of a number of diseases also leaves its mark by interfering with growth processes. Diets well above the starvation level, but deficient in some of the vital elements necessary for growth, including the trace minerals, will produce adults who have not grown to their full genetic potential.

Diet. The effects of diet are more marked upon human growth than would be suspected from records of nutrition and health. For the last several generations in this country, children have averaged almost a full inch taller at maturity than their parents did. This trend, which is called the *modernization of growth,* has resulted in producing a very much taller general population. It will not go on indefinitely. The American diet fifty years ago certainly seemed both ample and varied for most people, but the increase in stature which has resulted since indicates that it had serious deficiencies. Since our population has not changed much genetically in this time interval, the differences must be due to nongenetic changes affecting growth patterns. Interestingly, this same modernization of growth has occurred throughout the countries of Western Europe and the most modernized of Asiatic countries, Japan. More varied foods, provided by better distribution systems, hold most of the answer.

The effects of changed nutrition and environment are perhaps most strikingly seen in the children of people who immigrated to this country as adults. They frequently have children here who, when grown, may tower four, five, or even six inches above their parents of like sex. Better nutrition is involved in such cases, and greater bodily activity in the form of sports very likely contributes to the dramatic difference. Professional sports manifest the same kind of change in even more sensational fashion. Forty years ago, the college middle lineman who exceeded 200 pounds in weight was something of a giant. Frequently, he was so poorly constructed that he couldn't move very rapidly. But today, in the same position, men weighing 270 pounds, and as fast as cats, are relatively common. Indeed, the entire inside line of most first-rate professional football teams comes in such sizes.

Let us go back to Table 17–1 and view it again. In addition to being appalled by its genetic complexity, let us further consider that each of the genotypes represented may be considerably influenced by differences in environment of one sort or another and so further blur the differences in environment between the various classes of phenotypes. The range of environmental variation is so great that the analysis of polygenic characters now literally becomes impossible in natural populations. They do serve a useful evolutionary purpose, for when used with discretion and understanding, they have some evolutionary morals of their own.

The Method of Anthropometry

Biological anthropology, in common with all of the other emerging sciences, passed through a series of descriptive phases of development. Verbal descriptions of strange peoples were brought back to 17th- and 18th-century Europe by explorers and colonial voyagers. Such descriptions were good enough to enable Linnaeus to make some preliminary sense out of the distribution of the various populations of mankind. In the 19th century, researchers began to seek finer methods of description, and so they devised a series of measurements and the instruments with which to make them. Today the standard instruments are three in number. One is a jointed rod with sliding arms with which to measure stature, and, when broken down, various gross body breadths and thicknesses. The other two are calipers, which were modified from the measuring tools long used in carpentry and other trades. A spreading caliper was devised to allow the measurement of the major dimensions of the human head, both upon the dry skull and the living person. A sliding caliper with blunt points on the side for measurements on the living and sharp points on the other for cranial measurements provided the means of taking such smaller measurements as are necessary on the face as well as some of the detailed regions of the skull.

These three anthropometric instruments served to measure dozens, and, in the case of some very thorough German anthropologists, hundreds of different portions of the human anatomy. The linear measurements they provided were compared to each other, and so a very large number of ratios or indices were constructed upon the measurements. In a very crude sense, the measurements and indices allowed a rather detailed description of the major and minor features of the human body which could be incorporated within their domain. Since measurements were frequently taken to the nearest millimeter (approximately $1/25$ of an inch), this measuring of men dead and alive appeared to be very scientific. As a matter of fact, at the level of sheer description, it was done with sufficient accuracy to serve its general purpose.

Unfortunately, these anthropometric measurements have served primarily to describe the populations examined. They have not been lifted to the level which would allow investigators to explore evolutionary processes. They have remained static tools which have impeded progress in the study of human evolution. Many researchers, given the bewildering array of measurements and indices which are possible, have become bogged down with the problems of classification rather than gone on to investigate population dynamics. Classification itself is a descriptive exercise and as was shown in a preceding chapter cannot possibly be done within a species where clines demonstrate both continuous variation and independence between characters. In my own field research in Australia, I have taken more measurements, and calculated more indices than most investigators, but I do recognize that their proper use is to serve in the investigation of processes and not in the establishment of rigid and indefensible categories. One can only hope that some of the millions of measurements taken in the past can be salvaged by incorporating them in new problems on process rather than obsolete matters of classification.

The World Distribution of Stature in Man

Man varies more in body size than most other mammals. One of the most direct expressions of body size is total height or stature. As a species we are polytypic in that regional populations differ greatly in their average stature, the shortest being below five feet for average males, and the tallest approaching six feet. At the same time, each regional population is highly polymorphic, that is, shows great range among individual members. In this country, members of our population exceed the world range in terms of average values, for there are perfectly normal men standing no more than five feet tall, while the tallest among us exceed seven feet.

The distribution of stature, which is polygenic, is very susceptible to changes in environmental factors, such as nutrition. For these reasons, the worldwide distribution of stature has its evolutionary meanings blurred. Even so, the differences in the extremes of human height must have a considerable genetic component determining their expression. It is very doubtful that rain forest Negritos, given the best of diets, would grow very much larger than they do now. By the same token, among the world's tallest peoples it is unlikely that a bad environment would depress them to much below medium stature levels. Human males on a worldwide basis average about five feet, six inches in height. Very tall populations exceed five feet, eight inches, while some groups are so-called pygmy stature, that is, below five feet. Let us begin, in Figure 17–1, with an examination of the world's tallest populations. In Europe, the five feet, eight-inch level is exceeded in Ireland and Scotland. These are the tallest populations in the Caucasoid race. We find among the natives of the New World some really tall populations. The buffalo

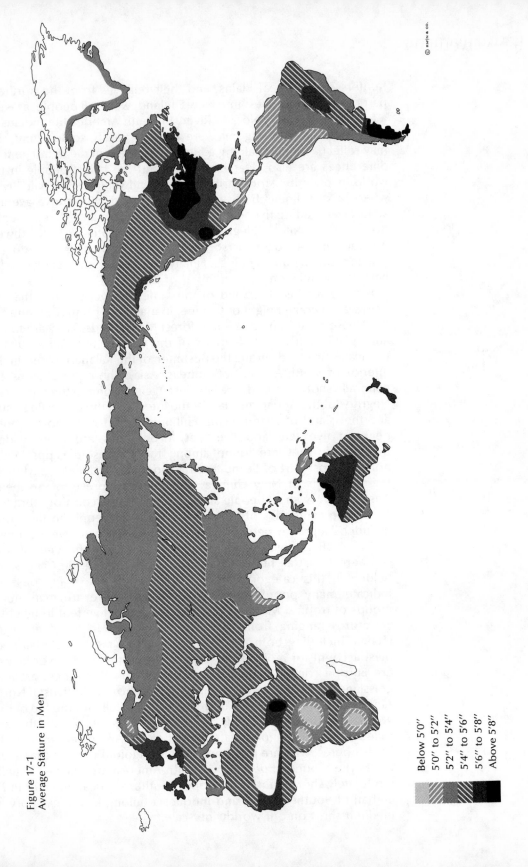

Figure 17-1
Average Stature in Men

Below 5'0"
5'0" to 5'2"
5'2" to 5'4"
5'4" to 5'6"
5'6" to 5'8"
Above 5'8"

hunters of the Great Plains, and their relatives from the northern part of the Middle West into New England, were tall people, as were a few tribes on the Colorado River. In South America the Indians of the great plains of Patagonia were also very tall. To some extent, this may reflect a diet containing a high component of meat, but genetic differences are also clearly involved. But the tallest people in the world involve the African tribes at the headwaters of the Nile River, where a very linear body build is combined with above average statures exceeding five feet, ten inches among males. A representative of these Nilotic Negroids is shown in Plate 17–1. Northwest Australia has a dozen or so tribal groups who also fall into the category of tall human beings. Finally, the racially mixed Polynesians have tall populations on a number of islands.

Turning to the other end of the scale, the populations that fall below an average height of five feet in males are primarily limited to the Negroid race and a few rain forest Mongoloids. They are found among the rain forest Negrillos of the Congo, the Camerouns in tropical Africa, and among the Bushmen of the Kalahari Desert in the interior of South Africa. In Southeast Asia, where populations are generally short-statured, the smallest again belong to the rain forest Negritos of the Andaman Islands, the interior of the Malay Peninsula and the Island of Luzon in the Philippines. But very short people, between five feet and five feet, two inches are more widely distributed. They are found among the Caucasoid Lapps in the northernmost part of Scandinavia, in southern India and Ceylon, in the highlands of New Guinea, and the rain forests of northeast Australia. Very short people also occur along the coast of Labrador. These two categories of the shortest-statured people in the world included Negritoid people in Africa, Southeast Asia, New Guinea, and Australia. Large blocks of population among the Veddoids of southern India fall here as well. The Lapps among the Caucasoids belong in this category. Among Mongoloids the map does not indicate many populations this short, but there are many local groups of tropical Mongoloids little more than five feet in height.

Statures ranging from five feet, two inches to five feet, eight inches, including both moderately short and moderately tall peoples, account for most of the populations of the world. Northwestern Europe contains taller populations within this range, as does grassland Africa and northern Australia. Most of the Indians of North America and a wide belt in South America also fall into the top of this group. On the other hand, the shorter members within this so-called medium category include most of the Arctic peoples in both hemispheres, who are predominantly Mongoloid in their characteristics. The mountain peoples of South America are likewise usually moderately short. Southern Europe, Southern Africa, much of India and all of Southeast Asia and Indonesia belong in the category of medium short on our worldwide scale.

Plate 17-1
This Nilotic man from East Africa
illustrates the linear body build
characterizing people generally
in hot, dry climates.

The reading of Figure 17–1 indicates that differences of stature fall into clinal patterns. To this extent it must be concluded that it has some genetic basis which is subject to adaptive changes under evolutionary stress. But it is equally certain that the harshness of the environment depresses stature, although the amount involved is difficult to measure. Arctic peoples everywhere are short. Tropical peoples tend in this direction, as do dwellers in the very high mountains. Perhaps the most important conclusion from this survey is that great variations in stature are found among the major racial populations. The Negroids show the most extreme differences, containing both the shortest and the tallest people on our planet. Caucasoids are not far behind in their range. Mongoloids in both the Old World and the New World also demonstrate the same variability. Only among the people of India, a subcontinent in its own right, are the full extremes shown by other groups not realized. If our data were more detailed, it is probable that small regional populations there would show the full variability. We must conclude that stature is only in part a matter of population group, and that it is subject both to adaptive evolutionary forces, variations in nutrition, and other environmental factors. Like all genetically complex characters, the full meaning of stature is difficult to unravel.

Regional Variations in Human Body Shape

Man is a warm-blooded animal, and it is realistic to expect him to adapt to the climates of different regions much as have the other mammals. The climatic extremes include two kinds of problems, stress caused by heat and stress caused by cold. Human populations have found three types of solution: changes in body size, changes in body shape, and adaptive change of a physiological nature.

Changes in Size.   *Bergmann's Rule* was established by that author as early as 1847, when he pointed out that in polytypic warm-blooded species the body size of regional populations usually increases as the temperature of its habitat decreases. This generalization that the colder the climate the larger the animal holds for a majority of warm-blooded species, but there are some exceptions. The rule is based upon the idea that in cold climates heat loss from the animal's body must be reduced. One of the two ways of doing this is to increase body bulk.

Some simple geometrical considerations show how Bergmann's Rule operates. In Figure 17–2 three spheres are represented which have radii from left to right of 1, 2, and 3 units. In each the volume is proportional to the radius cubed, while the surface of the sphere is directly proportional to the radius squared. Therefore, the ratio of skin area to volume is $K\left(\frac{1^2}{1^3}\right) = 1$. If an animal doubles in its linear measurements as shown in the second instance, this critical ratio becomes $K\left(\frac{2^2}{2^3}\right) = \frac{4}{8} = \frac{1}{2}$. Since heat loss is directly proportionate to the amount of skin exposed to the air, and heat production is proportionate to the mass of the animal, it follows that the animal which is twice as large as the initial member of the series will lose heat by radiation only one-half as rapidly. The final sphere with a radius of 3 units gives a skin-to-volume ratio of $K\left(\frac{3^2}{3^3}\right) = \frac{9}{27} = \frac{1}{3}$. This geometric example makes obvious the advantages of increasing body size in cold climates to minimize heat loss, hence utilizing energies more efficiently. In this instance, we have considered changes in size alone. It is significant that Arctic mammals are consistently larger than their relatives in the temperate and tropical zones.

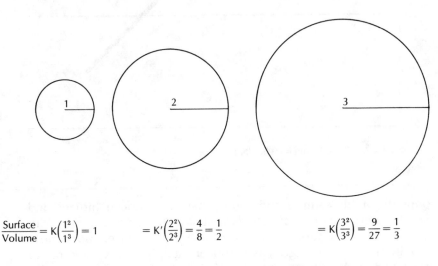

Figure 17-2
Effect of Size Changes in Altering the Critical Ratio of Surface to Volume

$\frac{\text{Surface}}{\text{Volume}} = K\left(\frac{1^2}{1^3}\right) = 1 \qquad = K'\left(\frac{2^2}{2^3}\right) = \frac{4}{8} = \frac{1}{2} \qquad = K\left(\frac{3^2}{3^3}\right) = \frac{9}{27} = \frac{1}{3}$

For those who believe that men living under economically simple conditions should not behave as other animals, the data provided by D. F. Roberts (1953) will come as something of a shock. While thousands of different populations have been measured by anthropologists, few bother to weigh their subjects, since scales are awkward and bulky to carry into out-of-the-way places. Roberts's data were therefore limited to 116 series of males drawn from all over the world. Series from England and Africa were more numerous than others. In correlating population weight with mean annual temperature regionally, he obtained a scattergram which clearly demonstrated that weight increases in man as temperatures fall. The correlation between the two characters was $-.60$, a comfortably high value. The slope of equation which summarizes this relationship is shown in Figure 17–3. Roberts analyzed the regional com-

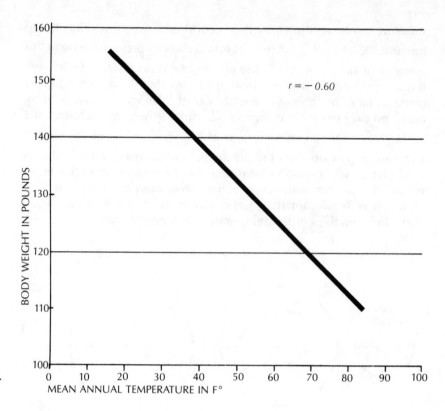

Figure 17-3
Relation of Human Weight to
Climatic Temperature (after D. F.
Roberts, 1953)

ponents of his sample and found that in American Indians and
Southeastern Asiatic Mongoloids weight increased less rapidly with
a drop in temperature, while in Africa weight increased much more
rapidly than the average. His European sample had a slope rather
close to that of the total series. A series of 33 female samples showed
an even higher coefficient of correlation.

Roberts's evidence demonstrates a strong inverse relationship
between weight in human population and the average temperature
at which people live. Bergmann's Rule is operating upon man in
spite of his culture. That this should be so is better understood if it is
realized that in a mild climate, such as Southern California, somewhere between 80 and 90 percent of the food we consume is
devoted to simply maintaining a normal body temperature of 98.6
degrees. Obviously, the situation would be very much worse in an
Arctic environment. For even with the protection provided by
tailored fur Arctic clothing, heat loss remains a critical problem.
Human populations living in the Arctic have not doubled their
stature in order to reduce heat loss through radiation, for this
solution would be disastrous. It would require eight times as much
food to stoke their inner fires. This would, of course, lead to

extinction by famine. Arctic peoples, in fact, are short in stature, but heavy in weight. That their evolutionary solution is a good one is demonstrated by the fact that hunters in both North America and Asia survive under the most rigorous of Arctic conditions. Physiologists have sometimes pointed out that the interior of their huts and igloos usually is heated to a comfortable 75° F. This is not the point, for they must survive to do their hunting at exterior temperatures as cold as 60° F. below zero. Cold stress is real in the Arctic, and men die from it.

The situation in the tropics is the opposite, for there men suffer under heat stress. Bergmann's Rule works in the hot climate also, for the three figures in Figure 17–2 can be read in the opposite direction. If we start with the middle figure and compare its surface volume ratio with that of the smallest figure, we shall see that by reducing the radius to one-half, the ratio of skin to body mass has doubled. Therefore, men in tropical regions should be small and usually they are. This allows a greater relative skin area through which heat can be lost than if they were of normal bulk.

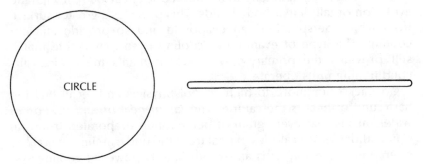

Figure 17-4
Effect of Changes in Shape upon the Surface/Volume Ratio

Changes in Shape.   The other way in which mammals, including man, have adapted to climatic stress is by a change in shape. This again involves optimizing the relation between skin area and body volume. The principle can best be understood by a two-dimensional shape shown in Figure 17–4. On the left-hand side a string has been positioned in the form of a circle. It encloses the maximum possible area for its given length. If it is now stretched between the fingers of two hands, the elongated shape enclosed within its perimeter is very much smaller in area than the original circle. The reader may remember this principle from plane geometry. Translating the idea into three-dimensional solid shape, it is immediately obvious that a sphere is the shape which provides the smallest surface enclosing a given mass. Any deviation from a spherical shape will tend to enclose a reduced volume within the same surface area. In nature, the elongation of shape operates so as to increase the relative amount of skin covering a given mass. Therefore, it could be predicted that both man as well as other animals would show a compact shape under Arctic cold stress, and a long or linear shape under tropical heat stress. Generally, this is true.

Physiological Changes.   *Allen's Rule,* also of interest at this point, states that extremities become reduced in size in warm-blooded animals in cold climates in order to reduce heat loss through radiation. In man this doesn't apply importantly to such extremities as nose and ears, but it is operative in arms and legs.

Arctic peoples everywhere tend to be short and heavy. The reduction in stature, together with the relatively high weight, thus combines both the principles of size change and shape change to minimize heat loss through radiation under conditions of Arctic stress. Eskimos are very short by our standards, but the average male weight of more than 140 pounds is relatively high.

Men in the tropics show two general patterns of adaptive body build. All are relatively light in weight. In open grassland and woodland country men tend to further reduce heat load through the evolution of tall, linear body builds. This provides a greater surface from which perspiration can evaporate and so provide efficient cooling. This type of evaporative cooling is so remarkable that it still provides the primary way in which man's mechanical air-conditioning units operate.

Go back once more to high school physics, and recall that the basic unit of heat is the calorie. The *latent heat of vaporization of water* means that every gram of perspiration evaporated from the skin withdraws 540 calories of heat from the body, assuming that the operation is ideal. This is an extraordinarily powerful cooling system, so that it is not surprising that linear body builds characterize tropical men living in open or nearly open country. When combined with body weight, they can rapidly dissipate heat, as long as they have the internal water with which to do it.

Some mammals have adjusted to the dry tropics in various physiological ways which minimize the need for drinking water. Some rodents and a few large desert antelopes, like the oryx, literally live on the moisture obtained from the food they eat. Man has not evolved in this direction, for he is prodigal with his water usage. An active man in the desert may require two gallons of water for sheer survival, and much of this is used in his keeping his body temperature constant through sweating. Behaviorally this means that human survival in arid regions in the tropics depends upon ample surface water of known locality and reliability. Desert dwellers know this and guide their movements accordingly. If the body should not be able to perspire under conditions of heat stress, the condition we call heat stroke follows, and body temperature rises to lethal levels. The victim falls into a coma, brain tissues are irreversibly damaged, and death is the conclusion.

The efficiency of sweating as a form of evaporative cooling, of course, depends upon sweat glands, both in terms of their number per unit of area, and possibly upon their type. Years ago anatomists thought they had detected evidence that people in the tropics had evolved a higher number of sweat glands for a given area than those living in temperate and Arctic zones. This seems to be a reasonable expectation, but today the evidence is ambiguous.

The other type of adaptive solution to buffer heat stress has been evolved in tropical rain forest dwellers. These people face the daily task of doing their hunting on animal game trails, for this is the only way to move through the nearly impenetrable tangle on the understory of the rain forest. Normal-sized Europeans who have attempted to go with the Negrillos of the Congo through their country have marveled at the agility of these little men and been chagrined by their own ineptitude. It is reasonable to believe that in this type of environment linear body builds would not be a practical solution to relieving heat load, and that the best solution involves the alternative option, reduction to very small stature and light weight. It is significant that in the rain forests of the tropics throughout the world people are short and usually light in weight. The Mongoloids of both the Amazon Basin and Southeast Asia show these characteristics. The Negrillos of the Congo Basin and some areas of Southeast Asia carry the tendency even further. Rain forests do not have temperatures reaching great heights, as do the arid deserts, but the humidity is so high that comfort and quite possibly life depend upon having a high ratio of skin area to body weight so that what evaporative cooling may take place does so efficiently.

Most of the ideas sketched above were developed by Coon, Garn, and Birdsell (1950) in a little book on regional populations. The principles of weight and shape changes advanced then seem just as sound today. But it must be admitted that people living in more or less neutral climates, as in the temperate zone, do not follow these trends as strongly as those living in extreme climates. The intermediate climates allow considerably more latitude in body build for successful survival, and even within a given breeding population, considerable variation in body build is evident. When I examined more than 100 men of the Nangamarda tribe living in the desert of northwest Australia, the variation in physique was remarkable. As a group, they, of course, showed linearity as might be expected, but among them was a minority of individuals with robust muscular builds, involving barrel-like chests, heavy thighs, and thick calves. There were enough men of this type to raise the body weight of the group well above what might have been expected for their hot, dry environment. Selection shows in the average attributes of these people, but the fact that so wide a range of individual variation can be tolerated also demonstrates that individuals may have considerable latitude in terms of survival. It should always be remembered that evolutionary processes work upon a statistical basis affecting gene pools, and this allows many individual exceptions. Marshall T. Newman (1953) has tested ecological rules against the Indians of the Americas with positive results.

The Biological Meaning of Skin Color

Man's recognition of his own visible differences primarily depends upon the label provided by variable skin color. Europeans have long referred to other populations as yellow, brown, red, or black. Imbedded in the European value system is the idea that the color white stands for goodness or purity, and any deviations from it are bad. This culminates in the belief that black is particularly bad. The phrase, "as black as Satan," is, of course, derived from ancient religious dogma. But it is safe to say that in socially dominant white populations, virtue is measured directly by the degree of depigmentation. Hitler made much of the blue-eyed, white-skinned, blonde-haired "Nordic" type as the ideal for his Reich, which was to last 1,000 years. Small wonder that in self-defense, the American black minority has focused upon its own value with the phrase, "Black is Beautiful." Since differences in skin color are so involved with racism, it is useful to see what they mean biologically.

The general distribution of the intensity of pigment in the human skin leaves no doubt that it is basically adaptive in nature. Before examining the world distribution in detail in Figure 17–5, a few points must be made. Not all native populations have long resided in the areas in which they are found today. For selective processes to be evaluated, we must discount the results of relatively recent migrations. The first men to enter the New World seem to have done so approximately 50,000 years ago, so men are relatively recent everywhere in this hemisphere. There is, however, no way of judging how long any of the American aborigines have lived in the places they were found at the time of contact. There undoubtedly has been a steady drift of peoples from the point of entry of the Bering Strait toward the south. Men reached the tip of South America at least 10,000 years ago. But the various niches in the environment have been occupied at different times. The great wet rain forests of the Amazon Basin are unattractive for hunting peoples. They may not have been penetrated earlier than 5,000 years ago and probably did not support a population approaching recent numbers until local food plants, particularly root crops, were domesticated and became the basis of their economy. The Indians living there today may have been under the selective pressures of this environment for only a few thousand years.

There are other places in the world where the recency of occupation must be discounted. Obviously the outer Pacific Islands of Micronesia and Polynesia have been available to men for perhaps only five or six thousand years, that is, since the time that technology made it possible to construct ocean-going craft. On the mainland of Asia there are other population displacements of relatively recent nature. Today Southeast Asia is peopled by a diminutive variety of Mongoloids whose arrival in that area is certainly connected with the spread of agriculture in the region. Pockets of earlier peoples, represented by still-surviving hunters, are not Mongoloid in regional type at all, but diminutive Negroids, whom we call the Negritos. They may be taken to represent the long-time inhabitants, and probably the original form of modern man in the area. India has been inundated by Middle Eastern Whites coming over the Khyber Pass, among others, and dominating the river valleys and plains in the northern part of the peninsula. The original inhabitants of India may best be judged as being like the jungle tribes who still survive in forested South India. They differ as a regional group and in skin color from the so-called Aryan-speaking invaders of the North. These late replacements of people must be taken into account in evaluating the evolution of human skin color.

The distribution of human skin color is shown in a very generalized way in Figure 17–5, where five categories of color differences are recognized. They are not all of equal value and have been chosen primarily to recognize differences rather than to measure them accurately. All Europeans are white-skinned, but those in northern Europe are much more depigmented than those in central and southern Europe. The former we would call fair, the latter brunette Whites. Many Mongoloid populations have very little more skin pigment, or *melanin,* than do southern Europeans. But at the same time their skin tones are somewhat yellowish and so are usually distinguished from European brunettes. The actual difference in the amount of pigment present is not great, and these Asiatics whose skin color has a yellow overtone presumably are that way because of another pigment, *carotene,* which they have in somewhat greater quantity than do Europeans. At the darkest end of the skin color scale there are some populations with so much melanin that to our eyes they appear bluish-black. They are concentrated in equatorial regions. But at the same latitude are larger populations whose skin color is best described as very dark brown.

Up to this point, we have been talking about the exposed skin color such as one sees on the face or the hand, which is constantly under solar stress. If we were to revise our scale so as to measure skin pigment protected from sunlight, the results would differ

Figure 17-5
Unexposed Skin Color

Fair-White
Light Brown
Medium Brown
Dark Brown
Very Dark Brown ("Black")

© RAND M NALLY & CO.

considerably. Many peoples with apparently dark skin suffer from sunburn each year as spring progresses into summer. Obviously even with a great deal of melanin in the skin, further amounts are needed as solar radiation intensifies. The map is plotted in terms of unexposed skin color.

Let us limit our preliminary examination to the peoples of the Old World. The two darkest categories of skin color are concentrated around the equator, as might be expected if the intensity of melanin is a function of the intensity of solar radiation. In West Africa, nearly the whole of the equatorial region of that continent shows very dark skin color. Dark-skinned Bantu people were overrunning a yellow-skinned group of Negroids, known as the Bushmen and the Hottentots, as they moved southward toward the edge of the tropics. The Arabian peninsula extends into the tropics, but it has always been controlled by people of essentially Caucasoid origin. Nonetheless, in this environment they became much darker skinned than other Caucasoids; enough so that as refugees in Israel, the Yemenites attract considerable attention. Further to the east, tropical India preserves in its jungle tribes and, to some degree, in the lower castes of its social system very dark-skinned people who must be considered representative of the preinvasion population of that subcontinent. Southeast Asia, as indicated earlier, has been overrun by agricultural Mongoloids, but the original scene is preserved only in a few isolated hunting Negritic tribes. Even so, it is interesting that the equatorial Malays are much darker in pigment than other Asiatic Mongoloids. Recent Mongoloid waves of people have so overrun Indonesia that the original inhabitants are indicated only by the remnant interior tribes of a few of the more easterly islands, where darker skins and frizzy hair are still apparent. In New Guinea and its outlying continental islands, and in Melanesia, skin color is everywhere dark brown, except for a few enclaves which have been influenced by a Polynesian or Micronesian backwash of peoples. Indeed, in some of the Solomon Islands lying just to the east are peoples with skin colors as dark as any in the world.

Continental Australia shows its own gradation of skin color values. By European standards all of its inhabitants are classed as dark-skinned. But the unexposed skin color of the Aborigines in the southern portion of the continent are closer to medium brown in the values of our scale, and only as one proceeds north into the tropics are populations to be classed as dark brown in skin color. Among them only a few individuals approximate the blue-black tones that are found in some tropical Africans and in the Solomon Islanders. Man has populated the Australian continent for more than 40,000 years, but this is enough time so that we may consider that this skin color gradient to some small degree reflects adaptation to local solar stress. Yet there is good reason to believe that peoples with different ranges of skin color contributed to the populating of the continent.

Turning to the other end of the scale in the Old World, the most depigmented populations occur in the extreme northwest of Europe, in regions that were glaciated no more than 12,000 to 15,000 years ago. Man moved into the area after the retreat of the last glacier. Today the whole of the region suffers from an oceanic climate in which the summers are both cloudy and rainy. Coon, Garn, and Birdsell (1950) suggested that under this type of climatic stress, with its minimal sunlight, selection may have favored a reduction of melanin to the point where it amounts to virtually partial albinism. Vitamin D is primarily synthesized within the human body, and it requires sunlight to fall upon the skin for the process to proceed. The hypothesis that northwest Europeans have been especially selected to maximize the use of the little sunlight that they receive is both reasonable and very likely demonstrable.

Depigmentation seems to have affected two northern European populations somewhat differently. Fair skin occurs both on the Scandinavian side of the Baltic and on the Finnish side. But both skin and hair color tones differ somewhat between these two populations, and it looks very much as though they had proceeded independently toward depigmentation.

Between the extreme populations with the heaviest pigmentation and those with the least pigmentation occur vast blocks of humanity whose skin color ranges from the brunette tones of southern Europeans through the variety of light to medium yellowish-browns which characterize most Mongoloid peoples. They essentially occur in regions outside of the tropics. Tests show that skins of this type do provide considerable protection against solar radiation. Their distribution suggests that it is adequate for human needs in the regions where it is found. It is, of course, extremely interesting that the aboriginal inhabitants below the tropics of Africa also have yellow-brown skin tones very like some Mongoloids. That solar radiation is the primary determinant is also demonstrated by the skin color of the Negrillos who live in the climax rain forests of the Congo Basin. They, like the Bushmen and the Hottentots of South Africa, are merely specialized groups of the Negroid population. But in the rain forest environment with the top canopy 200 feet above the ground, and multilayered stories of intervening foliage, very little, if any, sunlight reaches the ground. Living in a kind of perpetual shade, these diminutive Negroids in the tropics also show yellowish to reddish-brown skin color, with a great deal less melanin than is found in other tropical Negroids. The distribution of man in the Old World is clinal and testifies to the adaptive nature of melanin in the skin.

In North and South America skin colors are everywhere intermediate in range, and not unlike the Mongoloid populations of Asia from which these people sprang. Their skin colors sometimes have a reddish or ruddy component; hence the term "red men" for the Indians of the New World. But considering the recency of their arrival, these populations also show some selective gradients. The darkest pigments seem to have developed among the Indians living along the Colorado River where it flows through its desert valley on the way to the Gulf of Mexico. The exposed skins of these Indians are darker brown than those of most others. Nonetheless, they would remain within the scale of moderate pigmentation, and it is likely that their unexposed skin is considerably lighter. Darker pigmentation is tropical in its American distribution, except for the dense rain forest environment of the Amazon Basin. There, skin color turns to yellowish-brown as it does along the cloudy stretches of the Pacific Coast of Canada. The Eskimos of the Arctic and Alaska are lighter in skin tone than people exposed to greater solar radiation. The distribution of melanin in the New World does not show the dramatic differences found in the other hemisphere, but this indicates no more than the fact that some tens of thousands of years is insufficient time for selection to produce total regional adaptation.

It is easier to conclude that skin color darkens as a response to solar stress than it is to explain the exact mechanisms. There is general agreement that melanin particles of pigment in the outer layers of skin serve to protect the deeper layers from damage by ultraviolet light. Negroes and other dark-skinned tropical people not only have more heavily pigmented skin, but it is thicker than the skin of northern Europeans. Scientists have investigated the problem in terms of frequencies of skin cancer under various conditions. The results show that there is no doubt that skin cancers are most frequent in areas exposed to sunlight, such as the face. In the United States, Whites have seven to eight times the frequency of skin cancer as Negroes living in the same cities at various latitudes. This is direct evidence as to the damaging effect of the sun, but skin cancer in itself is not very likely to be the agency that makes for differential fitness, for it is a relatively mild and slow-acting form of cancer that seldom causes death.

On the other hand, rare melanoma, the dread black cancer, kills quickly and affects the young. Its frequency increases as one approaches the tropics, but it is too infrequent everywhere to be of evolutionary significance. Perhaps a better guide to the problem is to be found in the fact that Whites living in tropical countries soon learn to protect themselves against midday sun either by pith helmets and the like, or a swaddling of heavy clothing all over the body as among desert-dwelling Whites, or simply by not going out of doors at that time. The Noel Coward line that "only mad dogs and Englishmen go out in the noonday sun" is very expressive. Dogs sleep in the shade at noon, unless rabid, and native peoples everywhere in the tropics follow their example. It is noteworthy that albinos in the tropics do not venture out of their huts at any time during the day, for their unpigmented skin burns seriously enough to threaten life.

But the selection for heavier pigmentation in the tropics may be more complex than it appears. The melanin in the skin is accompanied by about the same degree of melanin in the iris of the eye and in the color of the hair. In general, the pigment-making processes in the body seem to affect all three regions to a corresponding degree. This opens up the possibility that pigment in the eye may also contribute importantly toward the differences of fitness in various climatic regions. The iris of the eye comes in various tones of brown among all the peoples of the world, irrespective of skin color, aside from the partially depigmented populations of northwestern Europe. When melanin is lacking in the outer layers of the iris, but present in the posterior layers, the color of the eye may appear to be blue, gray, or even faintly green. Such an iris less effectively buffers the inner eye from direct sunlight. Perhaps it is an acceptable eye in regions only where solar stress is no problem, as in cloudy northwest Europe. But there is another aspect of the eye that deserves consideration here. The rear surface of the anterior of the eyeball contains the retina, with its rods and cones which receive light and send their stimuli to the brain. The retina, in addition to these light receptors, contains a variable amount of melanin. This is slight in blonde Europeans, increases in brunette ones, and becomes very intense in Africans. The retinal melanin serves to cut down the glare of reflections caused by incoming light, and so to allow the perception of clearer images. Perhaps a part of the difference in fitness which has produced greater degrees of general pigmentation in the tropics is contributed by the need for more melanin in the retina.

This retinal pigment even provides a possible explanation for the brunette characteristics of Arctic Mongoloids such as the Eskimo. They are only a light yellow-brown in skin color, but the question has been raised as to why they should not be depigmented as are northern Europeans. The Arctic is a curious place in which one-half of the year is continuous night and the other half is never-ending daytime. When the sun is above the southern rim during their summer, reflections off the snow are so intense that Eskimos everywhere must wear snow goggles to prevent snow blindness. These consist of protectors made of either wood or bone and so slitted that only a narrow band of light can pass through the pupil and fall on the retina. It is obviously a device to cut down on the intensity of reflected sunlight. These are very severe conditions, and it is likely that retinal pigment is more important to Eskimos than pigment in the skin itself. Retinal pigment should lend itself to a variety of experimental investigations, and in time there may be clear-cut answers about how it contributes to human adaptiveness.

In recent years, physiologists have experimented with people and attempted to formulate some evolutionary conclusions. In one instance, the reflectivity of skin was measured upon White and Black American soldiers marching in the desert. As might have been predicted, the white skin reflected sunlight better than the black. The author concluded that black skin did not confer protection against sunlight, since it absorbed more. But against this stands the fact that the sunlight did not penetrate as deeply, for the skin of the Negro is thicker. Furthermore, the fact that dark skin heated faster meant that individuals having it began to sweat earlier. The turning on of this evaporative cooling mechanism quicker is certainly no evolutionary disadvantage. While we do not know many of the details involved in the darkening of skin color with increasing solar stress, it nonetheless stands as another instance of adaptive processes at work in man.

Variations in Pigment Patterns

Pigment in man presents a few other very interesting problems. Skin, hair, and eye color in general vary together. But in northern European populations, one will occasionally find a valid blonde with brown eyes, a disharmony in pigment. Such individuals usually tan better than ordinary blondes. In Ireland there is the curious circumstance that while much of the population of the islands is brunette in hair color, the same individuals frequently have very white skin, and a disproportionately high frequency of blue eyes. The blue-eyed, black-haired Irishman is common enough—the combination has become a kind of stereotype. There is no known reason for this attractive disharmony in pigment among the Irish, but it may go back to the fact that as Ireland was repopulated at the end of the glacial period, the founding stock may have been small, with some interesting combinations in its own gene pool, and possibly was affected by genetic coadaptation.

One of the most interesting and unexpected aberrations in pigment involves blonde hair among the dark-skinned Australian Aborigines. I have spent a good deal of time investigating this situation firsthand. Figure 17–6 shows the situation as it appears from my first two years of fieldwork. Subsequent work in the years 1953–54 showed the outer boundary of blondeness should be extended to include all of northern Australia and go further to the east, with probably only the southeastern corner of Australia lacking the characteristic. Of course, in these outer reaches of the distribu-

Plate 17-2
Blonde Australian Aboriginal Boy

tion, the frequency is very low and the appearance less striking. This characteristic is manifest as striking blondeness in infancy and childhood among the desert Aborigines. The appearance of hair as blonde, as in most northern Europeans, among a dark-skinned people is stunning. True, the hair darkens with age, but in some individuals real blondeness persists well into adulthood. The hair is light enough in color so that the natives themselves call it "white" hair. A young blonde Australian boy is shown in Plate 17-2.

The distribution shown is clinal in terms of the distribution of phenotypes. The region in which more than 90 percent of the population was blonde in infancy covers a half dozen or more western desert tribes. In the Pitjantjara, the frequency reaches 99 percent. From this peak, values generally descend in a fairly uniform rate except that a sharp escarpment in phenotypic frequency occurs on the eastern boundary of the peak values. This has been produced by a migration of people from the north into the Central Desert region. The genetic discontinuity is real. This blonde or tawny hair is produced by a single gene substitution at an unknown locus and while visible, it is not polygenic like the other traits in this chapter. It is dominant over the normal allele for dark brown hair, as is shown by the uniform occurrence of blonde hair in first generation aboriginal hybrids in that region. This suggests that it is genetically very different from blondeness in northwestern Europe. Two phenotypic variants of blondeness can be recognized, a rare type, presumably homozygous for the tawny gene in which the hair is lighter in infancy and remains blonde into adulthood. The other more common phenotype is presumably caused by the gene in a heterozygous condition, and this type of blondeness disappears prior to puberty in boys and lasts but a few years longer in girls. Individuals with blonde head hair also have blonde body hair. When blondeness is lost with maturity, the hair of the head still remains lighter in color than in normal Aborigines, and it appears ash brown rather than golden brown in transmitted light. This is fortunate, for it makes it possible to identify adults in any age who once had "white" hair as children.

Blondeness in aboriginal Australia was the first case of transient polymorphism reported for man (Birdsell, 1950). The pattern of distribution indicates that it was adaptive in nature, and time factors indicate that it spread relatively rapidly. Here again we have no way of knowing how the tawny gene conferred any added fitness to the individuals born with it. It is ridiculous to think that, in itself, having blonde hair would be of any advantage for people living on a dry continent. The probable solution will ultimately be found to involve some biochemical advantage which the gene conferred, while the outward blondeness is an inconsequential phenotypic manifestation of its action. Indeed, the pattern of spread in Australia suggests that this gene may have proved coadaptively advantageous in the gene pool of the Aborigines, since it may be a better mixer than the gene which produces normally dark brown hair. Blondes of this type occur sporadically in New Guinea and in western Polynesia, but there it has not spread. If Australia had been untouched by White colonists for another few millennia, on arrival they would have found an entire continent of dark-skinned people crowned with blonde hair in childhood. Given yet more time, the majority of adults might have evolved into blondes, too.

The Variety and Distribution of Hair Form

Variations in the form of hair on the human head are included in all attempts at racial classification. Differences in hair form result from two variables, the flatness of the hair shaft and its thickness. Hair which is round in section and thick in diameter is straight in form. Such hair is characteristic in both Old World and New World Mongoloids. These people are remarkably uniform in this characteristic, with a tendency toward waviness found in only a few individuals in some scattered regions. Face and body hair tend to be scant among these peoples. It has been suggested (Coon, Garn, and

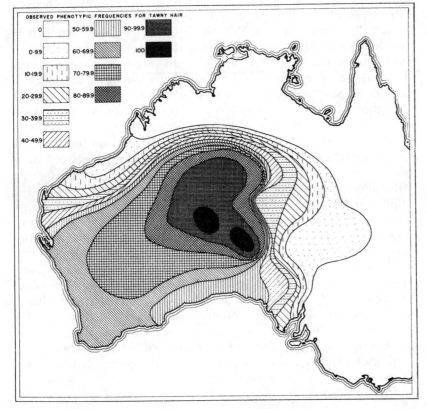

Figure 17-6
Observed Phenotypic Frequencies for Tawny Hair

Birdsell, 1950) that, in the Late Pleistocene, Arctic selection tended to favor hairless faces, to minimize the danger of frostbite and consequent gangrene, and that the seed stock of the Mongoloid race evolved under these conditions. This hypothesis would be difficult to test, and it has not been tested.

Many of the peoples of the world are characterized by medium to fine diameter head hair, with small to moderate flattening. The result is hair which is straight to wavy in form. As indicated on Figure 17–7, such hair characterizes all Caucasoids, the people of aboriginal India, and the natives of Australia. It is likewise found among the primitive Caucasoids of northern Japan, the so-called hairy Ainus. Strangely enough, these attributes often go with early balding in males. The genetic relationships between these characteristics is not known, but male hormones do trigger the effect of the gene which produces baldness. The complex of straight to wavy hair, heavy beards and body hair, accompanied by baldness, occurs as a constellation in northwestern Europe, the Ainus of Japan, and the Aborigines of southern Australia. The combination seems to be associated in marginal peoples and hence presumably early types of Caucasoids.

Tight, spirally-curled hair is a consequence of a very flattened hair shaft of moderate diameter. This type of hair is so firmly associated with the Negroid race that it is tempting to think of it as a race marker. Its inheritance in Negroid hybrids indicates that it is a complex or polygenic trait. The distribution of spirally coiled hair is limited to the Old World tropics. It occurs throughout sub-Saharan Africa, is not present in India, but crops up again among the Negritos of Southeast Asia. In a somewhat diluted form it is found throughout New Guinea and Melanesia, the rain forests of northeastern Australia, and characterized the now extinct Tasmanian Aboriginals. In Australasia, the intensity of coiling is always less than in African populations, a fact which is taken to indicate that some hybridization with wavy hair groups has occurred in the past.

The most intensely coiled hair, known as peppercorn hair, is found primarily among the Bushmen and Hottentots of South Africa and as an occasional variant in the Negrillos of the Congo and the Negritos of Southeast Asia. The hair tends to coil together centrally, leaving apparent bare spaces of scalp between the clumps. It is certainly the most specialized of all human hair forms and nothing quite like it occurs among any of the other primates or mammals. Among most populations with spirally coiled hair, beard and body hair are weakly developed. Where it is present, it, too, will be spirally coiled. One of the puzzling things about the distribution of this hair form in the Old World is its absence in India. This raises a number of problems from the point of view of human evolution that will be discussed in the next chapter.

Figure 17-7
Head Hair Form

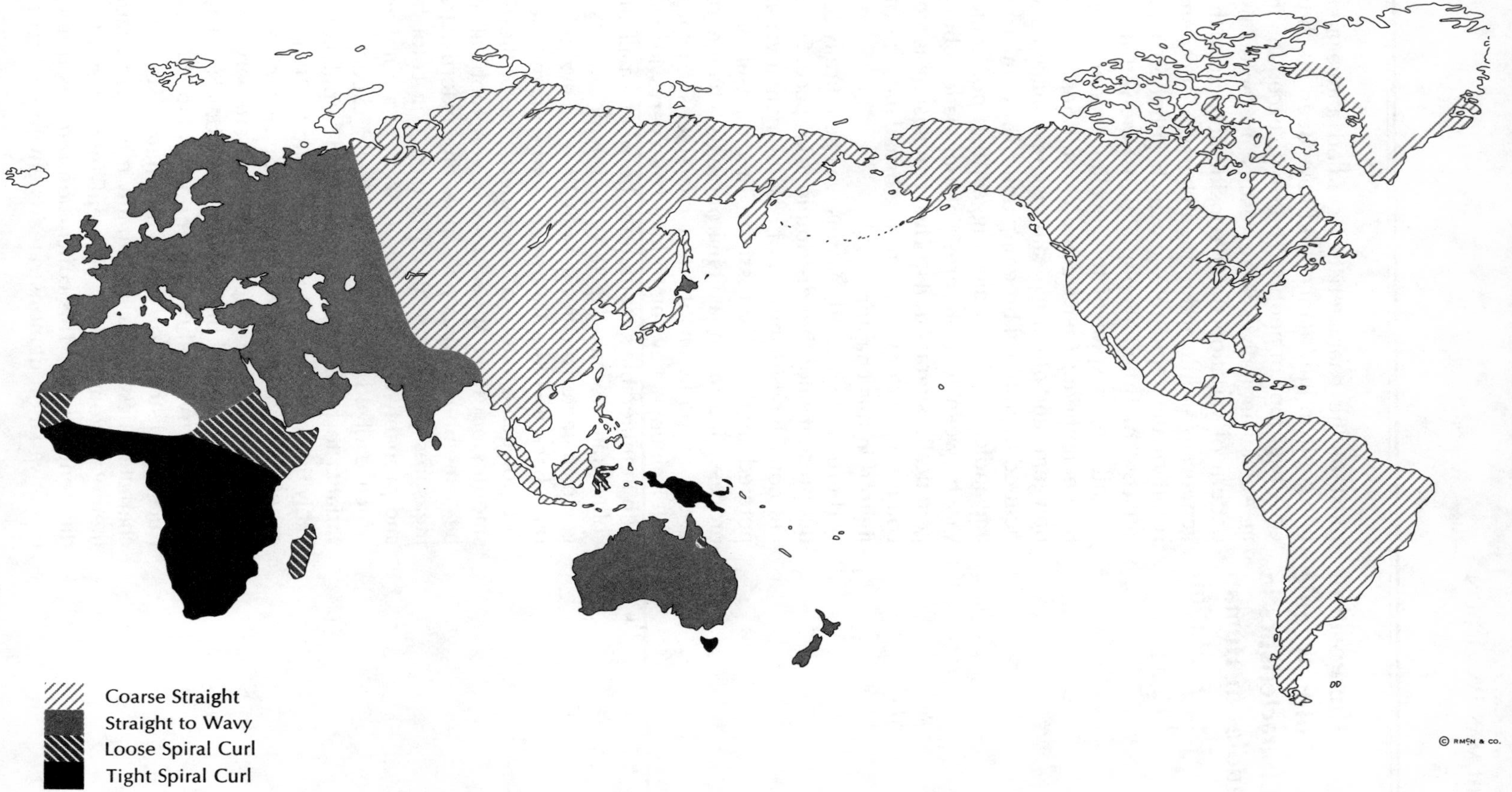

Coarse Straight
Straight to Wavy
Loose Spiral Curl
Tight Spiral Curl

The Position of Polygenic Characteristics in Studies of Human Evolution

Most of the characteristics studied above are externally visible, polygenic in inheritance, and have been much used in attempts to classify groups of people. The single exception involves blonde hair in Australian natives. The results presented by the geographical pattern of variation differ considerably from those evident in the genetically simple and hidden characteristics. Among the latter, as a result of the rapid rate at which genetic coadaptation proceeds, adjacent or breeding populations showed differences of a major magnitude, in spite of the leveling tendency of genetic exchange between them. There are no published results to indicate how polygenic and phenotypically visible characteristics vary on the scale defined by minimal breeding units. My own array of measurements and indices for Australia suggest that polygenic characteristics do vary between adjacent breeding populations but much less than the genetically simple traits. This conclusion is consistent with, but could not be derived from, the broad patterns of regional variation figured in this chapter.

There are two good reasons why genetically complex characteristics present smoother distributions in space. First of all, an unknown but considerable amount of environmental variation may be incorporated in their expression. This in itself acts as a smoothing process, inasmuch as environments change relatively slowly in space. Secondly, insofar as a characteristic expressed is genetically determined, it may be produced by a very wide variety of genotypes, with innumerable substitutions between small additive genes which collaborate to make up the phenotypic expression. Since no polygenic characteristic in man has been analyzed in detail, it is best to refer back to Table 17–1 to sense the magnitude of this possibility. The polygenic nature of the inheritance of these exterior traits also smooths out their appearance in geographical space, but it should be remembered that like-appearing external traits are not direct measures of relationship. The dark skin color found in Africa can and probably does differ genetically in a number of ways from that which evolved in India. One would expect certain basic genetic variants to be incorporated in both populations, but it is equally likely that each has a few unique genetic intensifiers of melanin. The genetic buffers of malaria should be remembered.

The distribution maps presented are very generalized since information on local breeding populations is not available. Nevertheless they serve to show the important fact that in all these characteristics the distribution is clinal in nature. Each of these phenotypic characteristics, and a great many more, are adaptive in nature, and their variations reflect different intensities of selection acting in different regions. This conclusion remains true even though a part of the variation demonstrated is environmental rather than genetic in origin. As it now stands, it can be maintained that both the invisible simple genetic characteristics and the exterior polygenic ones are generally adaptive in nature. It follows that *all peoples originate as adaptive regional types.*

The clines of polygenic characteristics are generally independent of each other, again confirming the same conclusions found for genetically simple traits. Therefore, the phenotypically visible characteristics offer the same type of problem. They refute the assumption that a combination of characteristics can serve to tightly define regional populations. This is an important point, for while many anthropologists who attempt to classify people are aware that it cannot be done in terms of a single characteristic, a number of them persist in the effort, reinforced by more complex statistical techniques which take into account simultaneously a greater number of traits. These numbers jugglers do not recognize the basic biology of the processes which produce regional patterns of variation.

The fact that polygenic characteristics vary continuously in space reinforces the notion based upon the genetically simple characteristics that there can be no boundaries or fences erected around human populations, except where isolating factors such as water gaps provide discontinuities in space. This is an important conclusion, for it strikes at the very heart of attempts to pigeonhole human populations, with the further implication that analyses of regional variations in man should be done in terms of clines rather than as a mail-sorting exercise. There are no published data which make the *analysis of clines* very profitable at this point, but when they do become available they *will provide preliminary answers about the processes of evolution, rather than pigeonhole populations.*

Bibliography

BIRDSELL, JOSEPH B.

1950 Some Implications of the Genetical Concept of Race in Terms of Spatial Analysis. Cold Spring Harbor Symposia on Quantitative Biology, Vol. 15, pp. 259–314.

1951 The Problem of the Early Peopling of the Americas as Viewed from Asia. Papers on the Physical Anthropology of the American Indian. New York: The Viking Fund, Inc.

COON, CARLETON S., STANLEY N. GARN and JOSEPH B. BIRDSELL.

1950 Races: A Study of the Problems of Race Formation in Man. Springfield: Charles C. Thomas.

HOOTON, EARNEST A.

1946 Up from the Ape. Revised Edition. New York: The Macmillan Co.

NEWMAN, MARSHALL T.

1953 The Application of Ecological Rules to the Racial Anthropology of the Aboriginal New World. American Anthropologist. Vol. 55 No. 3, pp. 311–327.

ROBERTS, DEREK F.

1953 Body Weight, Race and Climate. American Journal of Physical Anthropology, N. S., Vol. 11, pp. 533–558.

CHAPTER 18

SOME
PROBLEMS IN
REGIONAL VARIATIONS
IN MAN

The use of the term "race" has been discontinued because it is scientifically undefinable and carries social implications that are harmful and disruptive. Still it is necessary to deal with the reality of regional variations in human patterns. These are of interest in terms of the origins of population variability, the movement of peoples, and in fact both the long- and short-term evolution of our species. It is worth looking briefly at some examples of this kind of population research.

How the Other Natural Sciences Have Handled the Problem of Race

Subspecies are as difficult to analyze in frogs or fleas, in herbs or trees, as in man. In the natural sciences we find a descending order of interest. Paleontologists working with the fossil record are happy if they can assign a new find to its proper genus. While species names are assigned taxonomically to identify fossil finds in more detail, few are numerous enough to allow population studies at the species level. The analysis of subspecies becomes possible only where large numbers of fossil specimens are available.

Scientists who work with living animals and plants at one time tumbled over themselves to invent new names by splitting up many natural populations into subgroups to which they gave subspecific labels. Today it is recognized that these *splitters* in classification worked with inadequate data and concepts. Their primary fault lay in the fact that they viewed both the species and the subspecies as unchangeable types and ignored the reality of population variation. In recent decades, among systematists a reaction has set in, for it is now realized that many species are both *polytypic* and *polymorphic*, that is, species not only consist of regional populations which vary in type among themselves, but each population contains many different forms owing to the variation among its individuals. As a consequence, most modern classifiers tend to be *lumpers* in that they put together categories which were once improperly separated. In zoology, the study of subspecies and the evolutionary processes which produced them is generally considerably less advanced than it is in anthropology.

The Distribution of the Four Visible Major Regional Populations

In the early days of evolutionary anthropology, and before the invention of a series of complicating techniques, all observers agreed that there were at least three major groups visible among the populations of mankind. These were essentially populations covering most of a continent and differing in appearance from all other major populations. The early anthropologists, being Europeans, recognized that the populations of that subcontinent, the Middle

East and North Africa belonged together in spite of the differences in pigmentation which resulted from considerable blondeness in the northwest of the region. Today no one would seriously quarrel with the essential uniformity of the Europeans, or *Caucasoids*, as one of the important visible population clusters.

Equally apparent were the people south of the Sahara Desert in Africa, who, in spite of their great diversity in stature and even skin color, are properly grouped together as one major block of mankind. They are called *Negroid* because of their generally dark skin pigmentation. Visible to all are the populations of Eastern Asia, with their characteristic straight, coarse hair, eyefolds, and moderately pigmented skins usually showing yellowish tones. The *Mongoloids* are a numerous and successful subspecies of modern man. No one disagrees with the choice of these three great visible blocks of humanity. But some argument arises as to the question of whether there are any fourth or fifth equally significant groups of populations which differ from the first three. Sometimes early investigators broke up the Mongoloids into subdivisions, and so designated Malays and American Indians as equivalent in value with the three primary apparent divisions. No one holds this position today.

Two other populations have proved troublesome to classifiers. First and foremost are the original inhabitants of India who today are best represented among the jungle tribesmen in the south of that peninsula. They are dark-skinned, linear in body build, but much smaller than most Africans, and they differ from the latter in having hair which is fine in texture and straight to wavy in form. Amusingly, one early anthropologist decided they were dark-skinned Caucasoids because of their hair form. Another pioneer claimed that they were wavy-haired Negroes because of their skin color. Neither was correct. India has been inhabited from at least the beginning of the Middle Pleistocene, and so man there has been subjected to a long period of evolution under local conditions. It is more reasonable in light of today's knowledge to consider that the natives of India are their own kind of humanity and constitute a fourth major visible segment of humanity. Those who hold this view usually use the term *Veddoid* as a descriptive title for them, because the Vedda, a small hunting group in Ceylon, the island to the south of India, represent the type well. Since these Indians are both numerous and ancient in type, they may reasonably be admitted as a fourth major population. Plate 18–1 shows a typical young Veddoid male.

The continent of Australia originally was inhabited by about 300,000 people who had a certain superficial resemblance to each other and whose characteristics set them apart from most other populations. A few writers have elevated them to the status of a fifth major group. I cannot accept this position, for, having studied them

Plate 18-1
Young Vedda tribesman from
Ceylon represents the basic early
population of India.

intensively in all regions of Australia, I find that they show patterns
of difference within the continent which equate them with other
Asiatic populations. The details of this position are given a little later
in this chapter.

It is of importance to consider briefly the consequences of
differences in technology on the fate of regional populations.
European peoples, because of their participation in the industrial
revolution and colonial expansion, are much more numerous today
than they were four centuries ago. The British in particular have
increased relative to the other peoples on this planet. They prospered in numbers at home with the change from handcraft to
industrial technology. As settlers with a higher level economy, their
numbers increased particularly in the colonial areas where the
original inhabitants were displaced or for a variety of reasons driven
to near extinction. The aboriginal Tasmanians are gone. The original
inhabitants of the continent of Australia have been reduced from
about 300,000 people to perhaps 25,000 unmixed persons. Today
these countries hold about 13 million persons largely of British Isles
origins. Similar types of population changes have occurred in both
North and South America where the Portuguese and Spanish shared

with the British and French in the replacement process. From these and similar examples it is clear that regional populations wax and wane in time, and in recent centuries many have been driven close to extinction while others more fortunate in their technologies have increased inordinately. As for the future, it will be worth watching the populations of mainland China, who today comprise between one-quarter to one-third of the total world's population.

It is interesting that today the designation of the major subspecies of man is somewhat less clear-cut as the result of advances in our scientific knowledge than it was a century or two ago. The clinal distribution of simple genetic characteristics is one of the disturbing factors. The great diversity of both visible and invisible characteristics within each of the major divisions is another complication. Therefore, it is reassuring to turn back the clock to 1775, the year in which the philosopher Immanuel Kant proposed the first scientific classification of the subspecies of man. It is striking that the gifted philosopher should have done better in his analysis than did the pioneer classifier Linnaeus, or the various other natural historians such as Buffon, Cuvier, and the early anthropologist, Blumenbach. The great virtue of the scheme proposed by Kant was that he was the first to suggest that the major populations of man had evolved as a response to the selective pressures of the regions in which they lived. The philosopher suggested that Caucasoids had evolved in a damp, cold climate, a point which seems self-evident today. His second major group equates with Mongoloids in the modern sense, and he conceived of them as having been produced by the selective pressures of a cold, dry climate. The Negroids of Africa evolved under hot, dry conditions, and so did his fourth group, the inhabitants of peninsular India. While some of Kant's reasoning can be faulted in the light of our modern knowledge, nonetheless his scheme stands as the first classification based upon evolutionary processes, and there is very little reason to change it today, except to provide better evidence for it. It is ironic in the history of the natural sciences that the ideas of Kant were totally forgotten, influenced no later generations of anthropologists, and finally were recreated as an independent invention by Carleton S. Coon (Coon, Garn, and Birdsell, 1950).

The Boundaries Between the Four Major Human Groups

The visibility of the major groups of mankind arises in part because of isolating factors. Let us first examine the position of the Negroids of sub-Saharan Africa. Their position is truly an isolated one. While a few highlands and oases in the Sahara Desert are sparsely inhabited, most of its great expanse, which stretches virtually from the Atlantic on the west to the Indian Ocean on the east, is uninhabited. It stretches as a great inhospitable barrier of aridity between the Negroids of the continent and the Caucasoids living on its northern Mediterranean shores. There are few places where population continuity connects the two great groups. Along the Atlantic stretch of the Sahara there are scattered populations which intergrade between Negroid and Caucasoid over a distance of almost 1,000 miles. The River Nile in the east cuts its narrow valley from the unmixed Negroids of East Africa to the Mediterranean Caucasoids of its lower reaches. The populations along the river again show intergrading traits between the two great groups occupying its ends. This kind of isolation has certainly contributed to setting apart the Negroids from the Caucasoids.

The Caucasoids in Europe, the circum-Mediterranean area of the Near East and in the Middle East, are well isolated from the Negroids to their south by the desert belt, but no such topographic feature buffers them from the Mongoloids of the Far East. The populations ranging from southern Russia well into central Asia show a visible blend of Caucasoid and Mongoloid features. The transition from one great group to the other is gradual and knows no boundaries. Even though central Asia has a northeasterly trending line of great mountains, this has not effectively isolated the two groups from each other. Consequently, there is a belt of populations, several thousand miles in width, which shows transitional features between Caucasoids and Mongoloids. There is no possible way of drawing a boundary between these two great regional groups.

Peninsular India is bounded on its northern side by a great arc, the mighty Himalayan Mountains. Thus it is virtually protected from the Caucasoids on its western boundary and the Mongoloids inhabiting the plateaus of Tibet and the deep river valleys and gorges to the east. True, in recent millennia the Middle Eastern type of Caucasoid spilled into its northern plains, and Mongoloids came filtering down toward the lower levels, both from Tibet and Southeast Asia. But, in general, these are relatively recent changes involving the advent of agriculture, the use of horses, and aggressive military behavior. If we roll back the years to the end of the Pleistocene, India must be considered a kind of human sanctuary. As of that date doubtless a clear-cut boundary could have been drawn between the Veddoid series of populations in the peninsula, separating them from both Caucasoids and Mongoloids. General evolutionary theory suggests that some degree of isolation is necessary for subspeciation, and this also applies to man. Only the great belt of gradual mixture in the central reaches of Eurasia stands as an exception to this rule, and there distance has served to isolate the extreme populations of the western and eastern portions of that great land mass.

Plate 18-2
Variability in North American
Indians is shown by these four
types coming from the
Southwest of the United States.

(a) This Navajo man shows
generalized American Indian
features.

(b) This Kiowa, Big Bow, shows
the features characteristic of
many Plains Indians.

(d) Old Cahuilla man from Palm
Springs, showing extreme facial
hair rare in American Indians.

(c) Two Shoshoni-speaking
Paiutes from southern Nevada
illustrating generally
non-Mongoloid features.

All of the great regional groups of mankind contain much variability within their own domain. It is both visible and convincing. Consequently, from earliest times anthropologists have been tempted to create categories representing minor groups within the great ones. Where local isolating factors exist, the subdivisions may be intellectually defensible. But more often there are no real topographic boundaries between the populations abstracted by the classifier. Worst of all, such subdivisions often rest upon the concept of type instead of the reality of populations.

Continuous and intergrading population variation is the rule wherever topographic barriers do not completely isolate peoples. Such variation is clinal in character and is largely produced among precolonial peoples by local adaptive processes. Ultimately, this kind of regional variation will yield to analysis, but at this time only initial observations have been made. It is possible (Birdsell, 1972) to quantify clinal variation under the special circumstances in which data are derived from separate breeding populations more or less continuously distributed in space. This has been done for 28 tribal isolates in Western Australia, but the method, while promising, has not yet reached the point where readers of this book need be concerned with it.

It is allowable to use regional names where topography or prehistory have created isolating factors. There is no doubt that both North and South America were ultimately peopled by small human groups straggling across either a land bridge, or the winter ice, in the region of the Bering Strait. Since the natives of the New World have been effectively cut off from Asia for 10,000 years or more, they have undergone evolutionary changes on their own. Consequently we can realistically speak of the American Indians as consisting of a regional series of populations. Among them are very tall men, as well as very short populations (see Figure 17–1). Indeed, nearly the world extremes for stature are found in these two continents. But the general view today is that much of the regional variation which is found in the Americas has largely been created by local adaptive processes. This position has been well documented by Marshall T. Newman (1953). A variety of Indians from the Americas is featured in Plate 18–2. There is no doubt that they are in part Mongoloid in origin, for their features suggest this, as well as the relationships of geography. But except for the Eskimo of the Arctic coast, no American Indians are extremely Mongoloid in their facial features. Many anthropologists, therefore, have concluded that they may contain a submerged element somehow related to the Caucasoids of Europe. Again, in terms of geographical factors, the archaic Caucasoids who are now reduced to a marginal position in northern

The Subdivisions Within Major Regional Populations

Japan probably are the contributors of the White substratum. I put this proposal forth many years ago (Birdsell, 1951). In the time that has elapsed, no very convincing proof has come forth as to the reality of this proposal, except that the crania of the earliest immigrants to the New World uniformly are less Mongoloid than living Indians in the same regions.

Let us see what variation sub-Saharan Africa holds. In many ways, the Negroids are the most specialized of living major groups; that is, they have evolved furthest away from generalized anthropoid characteristics. Their skin color may not fall in this category, for if man evolved in the tropics, as the fossil record would now suggest, dark skins must be very ancient and, indeed, the ancestral type for all modern men. But in their spiral hair form and frequently linearity of body build, the Negroids can only be classed as specialized. The thick everted lips are very advanced features, since all of the great apes have thin lips. The Negroid people contains both the shortest and tallest populations of mankind (Plate 18-3). Among the slender Nilotics of East Africa, average statures are the world's maximum at five feet, ten inches for males. Among the rain forest Negrillos of the Congo Basin, the average man is not much more than five feet in height. Other regional tendencies can be detected, for the populations below the tropics, such as the Hottentots and Bushmen, have a yellow-brown skin that is quite light in tone. They show the most advanced Negroid characteristics in their peppercorn hair, and in a curious condition known as *steatopygia*. (See Plate 18–4 for a so-called Hottentot Venus.) This consists of enormously protruding buttocks, which involve masses of fat, and some additional slant in the sacrum. It is a highly specialized characteristic and is common in only one other group. Its functional meaning is unknown. The old idea that it was a food storage deposit, like the fat hump of a camel, is, of course, nonsense. It may simply be one of those curious traits in evolution which come along more or less opportunistically through genetic coadaptation rather than through selection.

The Bushmen and the Hottentots serve one further purpose for us, for it once was claimed, and by some respectable authorities, that they were somehow related to the Mongoloids. This belief rested upon two observable features. First of all, their skin tone is yellow-brown as among Mongoloids. Second, they often showed, particularly among the women and infants, a form of eyefold known as the internal *epicanthic* fold which is also highly characteristic of Mongoloid peoples. Given these two resemblances, some anthropologists ignored the fact that they had the most specialized Negroid hair form in the world and are essentially unique in their steatopygia. Today it is generally recognized that any people with intermediate brown skin usually show yellow tones and that in-

Plate 18-3
These African pygmies, or
Negritos, show a tendency
toward infantile facial features
and body proportions.

Plate 18-4
A Hottentot "Venus," showing
an extreme degree of steatopygia
common among the women of
this population. Note the
peppercorn hair form.

Plate 18-5
(a) The chiefly families of Polynesia were of great size, as demonstrated by Queen Salote of Tonga, her son, and daughter-in-law. The late queen was six feet, three inches tall and of proportionate weight.

(b) Two Polynesian girls from Rapa Island demonstrating that these people are usually good-looking, but not always the beauties reported by early sailors.

dividuals and races with low nasal bridges frequently show the internal eyefold. Both Bushmen and Mongoloids have very low nasal bridges, and it is more extreme in infants and females than in adult males. This example indicates how a naive belief in the hypothesis that in a few traits similarities in form will reveal relationships in descent can be highly erroneous.

Other regional populations of man can be set apart geographically and to some degree biologically. The outer islands of the Pacific were late in receiving their human migrants who sailed across great expanses of open Pacific waters to reach them. Once there, they lived in effective isolation and in essence continued their own evolutionary pathways. The great outer triangle known as Polynesia, or World of Many Islands, has its own unique population. Native Polynesians show in their language and customs that they reached their Pacific home from a point of departure somewhere in Southeast Asia. Polynesians are both tall and robust, the latter attribute involving both muscles and fat. They certainly would rank among the world's heaviest people because of these combined characteristics. Their facial features are an interesting and attractive blend of Mongoloid features, presumably diluted by combination with those of primitive Whites. They are brown-skinned, wavy-haired, and an attractive people.

The Polynesian mystery is heightened by the fact that the group of Pacific islands which provides a link connecting them with Southeast Asia, and known as Micronesia, is peopled by a population of moderate height which shows little tendency toward obesity. Aside from differences in size, the Micronesians really look very like Polynesians except that in a few islands the suggestion of some sort of Negritoid contribution is evident in hair form. The mystery remains as to why Polynesians should have become so relatively enormous, while Micronesians are more like Southeast Asiatics in general build. But the lifeway of both groups of islanders is very similar and what has produced these visible differences is not known. There seems to be a suggestion of preferential mating in terms of great body size among the ruling houses in Polynesia, but it is not certain how this would affect an entire population. The Polynesians shown in Plate 18–5a are characteristic.

The Problem of the Oceanic Negritoids

With Africa clearly the evolutionary homeland of the great Negroid group, a problem arises from the fact that some Negritoid peoples inhabit Southeast Asia and Australasia. Their distribution indicates that they are in a marginal position in that part of the world, having been pushed into pockets and corners by later migrants. Undoubted Negritos occur in several places in Southeast Asia. The

Semang inhabit the mountainous interior of the Malay peninsula, where they are surrounded by small agricultural Mongoloids. On an archipelago of islands in the Bay of Bengal between Malaysia and India are the least mixed of the so-called Oceanic Negritos, the Andaman Islanders. The interior of the Island of Luzon in the northern Philippines has a marginal group of Negritos, the Aetas. Indonesia preserves some frizzly-haired peoples in the matrix of small Mongoloid horticulturalists, but the whole island of New Guinea essentially belongs to a Negritoid range of peoples. Farther on, the arc of islands known as Melanesia, or the Black Islands, all contain Negritoid people of one type or another. Mixed with Polynesians, and so much larger in size, finally come the inhabitants of the Fiji Islands.

In the climax rain forests of northeast Queensland is a group of 12 tribes diminutive in stature and Negritoid in character. Finally, in Tasmania, now separated from continental Australia by Bass Strait, were tribes of Aborigines with spirally-curled hair, dark skins, and some Negritoid casts of features. The Tasmanian Aborigines are now extinct, but reports from the living, as well as skeletal remains, indicate that they were not as reduced in stature as most people of this type. In short, they were much like the mainland Aborigines of Victoria except for hair form and skin color and consequently appear to be a blend of the two racial types. This is a most likely solution, for less than 10,000 years ago Tasmania was a peninsular extension of mainland Australia.

The problem can be stated in simple terms. Since the Oceanic Negritos so visibly resemble those of Africa, they must be explained by one of two polar choices. Either Negritoids evolved twice in different parts of the Old World quite independently, or the Oceanic Negritoids reached that area as a consequence of an early migration from Africa. Both of these alternatives pose serious problems. Let us first examine the proposition that the Oceanic Negritos are the products of local evolution. A number of anthropologists agree with this point of view, pointing out that rain forest peoples are generally small, and that any group who stayed in a rain forest environment for a long time would become Negrito-like in character. The position of these "easy evolutionists" is deceptively simple and demonstrably incorrect. Granting that selection might produce reduction of stature in the depth of rain forests, as indeed it has among Mongoloids both in Southeast Asia and South America, there still remains the difficulty that these peoples did not evolve dark skin and tight spirally coiled hair. A Negrito is more than a dwarf human being. This difficulty seems to eliminate the proposition that Negritos evolved independently in both Southeast Asia and Africa.

The other proposition, namely, that Negritos migrated from Africa to Southeast Asia, has its own problems. There is not one bit of fossil evidence to indicate that they made the trip. But the region traversed contains no fossil men in the great arc from the headwaters of the Nile through India into Southeast Asia. Negative evidence of this type does not count. On the other hand, almost all of the Negritic peoples living today exist in rain forest environments. There is nothing to suggest that rain forests ever extended across Ethiopia and the Arabian peninsula to reach those in the foothills of the Himalayas. Perhaps Negritos have lived outside this special environment in the past, as the somewhat Negritoid Tasmanians did until exterminated by colonists.

The best evidence that the Oceanic Negritos are, in fact, derived from the African rain forest people lies in some of their physical characteristics. Their dwarf stature is not to the point, since rain forest peoples usually do have this characteristic. But among the Negritos of the Andaman Islands, who are the least altered by surrounding peoples, the spirally-curled hair in some individuals takes on the appearance of peppercorn hair, which is found in the Congo as well, being characteristic of the Bushmen and the Hottentots. This is one link with the specialized African peoples. A second trait in common is that both the Negrillos of the Congo and the Andaman archipelago Negritos have a great deal of red pigment in their hair which manifests itself when light passes through it. This characteristic is found in some Mongoloids, and is not unique to the diminutive Negroids, but it is lacking in most populations, so it does count in favor of the hypothesis of migration. Perhaps the most telling point is that on Little Andaman Island, the most southerly of the archipelago, some adult women show a marked degree of steatopygia (Plate 18–6). This also occurs occasionally in rain forest Negrillos in Africa, and, of course, is characteristic of their specialized relatives in the southern part of that continent. The fact that the Oceanic Negritos carry two specialized traits which are only found in African Negrillos, that is, peppercorn hair and traces of steatopygia, weighs heavily in favor of the migratory hypothesis. The further concordance in the intensity of red pigment in their dark brown hair is one more added factor. The gene frequencies of the blood groups for the two populations show no particular agreement, but as we have already seen in the much more rigorous circumstances examined in Australia, this would not be expected.

Who Are the Ainu?

The problem of correctly identifying origins among widely separated populations is again encountered in East Asia. Ringed by Mongoloids on all sides, a few thousand Ainus lived on the northernmost Japanese island, Hokkaido, and the neighboring Russian island to the north, Sakhalin. Like all northern Asiatics, they are short in

Plate 18-6
Negrito Woman from Little
Andaman Island, with highly
developed steatopygia.

stature, but here resemblances cease. The Ainu have a light skin, with ruddy tones, as do Europeans. They have a great profusion of beard and body hair (Plate 18–7), a totally non-Mongoloid characteristic. Their eye folds are shaped as in Europeans, rather than with the epicanthic fold characterizing Mongoloid people. They have a nose of rather high relief and greater breadth than their neighbors. Finally, baldness is relatively common among aging men. All of the ways in which the Ainu differ from their Mongoloid neighbors point in the direct of the populations of northwest Europe. Their skulls, furthermore, resemble the rough-hewn Caucasoids of the Upper Paleolithic of Europe. What do these divergences from the Mongoloid matrix of East Asia mean?

The problem of identifying the affinities of the Ainu is complicated by the fact that they are separated by thousands of miles from the nearest Caucasoids. Furthermore, their blood group genes are not particularly like those of the peoples of Europe, although they differ in some ways from the neighboring Japanese. This fact does not present great difficulty, since it would be expected that an isolated White population, presumably cut off from gene flow with Europeans for many thousands of years, would develop its own constellation of gene frequencies through regional selection and the process of coadaptation in its pool.

Archeology provides no clear solution to the problem. Admittedly, the skull of an adult man from a cave outside Peking can hardly be distinguished from those of the Ainu except for its greater size. An older cranium from Liu-Kiang in Kwangsi Province in southern China also resembles an Ainu in type. The datings on both these crania are uncertain, but they do at least serve to indicate that at one time an Ainu-like population was present on the eastern Asiatic mainland in Late Pleistocene times. The problem might be solved by the "easy evolutionists" by asserting that the Ainu merely are a Mongoloid population who have evolved in a strange direction and more or less by accident have come to look like Europeans. But most authorities would agree that the resemblances between the Ainu and the Europeans of the Upper Paleolithic are too numerous to mean anything but some sort of common ancestry, and, thus, close relationship. As unsatisfactory as the latter view is, for it does not rest on continuous archeological evidence proving the point, it seems necessary to assume that in the Late Pleistocene, Caucasoid populations ranged not only through Russia but eastward through Siberia to reach the Pacific shore, and that the present-day Ainu are marginal descendants of such peoples. Genetic exchange has been going on between the marginal Ainu and the conquering Japanese for thousands of years. Many Japanese in north Honshu show diluted Ainu-like characteristics. At the same time the surviving Ainu reveal some Mongoloid admixture. In addition to adapting the costumes and customs of the dominant Japanese, they have also taken on some of their characteristics genetically. A trace of Mongoloid eyefold is present in some individuals, particularly the women. A typical Ainu is illustrated in Plate 18–7b.

Ideas About the Origin of the Mongoloids

The origin of the adaptive Mongoloids remains in question. They have expanded their home range extensively since the end of the Pleistocene. All of North and South America have been populated by peoples from eastern Asia, presumably coming in small groups over long periods of time. The marginal position of Negritoid people in mainland Southeast Asia, in the Philippines, and in Indonesia serve as testimony to the expansions of waves of horticultural Mongoloids into the Asiatic tropics. The Mongoloids are a highly successful group, surviving apparently as well in the steaming tropics as in the arid Arctic. A plausible case has been made (Coon, Garn, and Birdsell, 1950) for the hypothesis that the Mongoloids

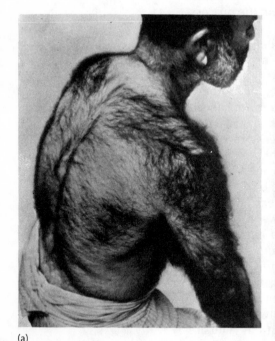

(a)

(b)

Plate 18-7
(a) Ainu Back, showing body hair

(b) Ainu Man, with
well-developed facial hair

evolved under extreme Arctic stress into a new human type. Northern Mongoloids still retain the compact, large-trunked bodies which suit them to cold stress through reducing the loss of body heat by radiation. But the primary point of Coon's hypothesis is that Mongoloid facial features have also been modeled for survival under extreme cold.* The general idea is that rounded surfaces are less subject to frostbite than projecting features. The forward positioning of the cheekbones in Mongoloids, their covering with fat, and the draping of the eyes with fat-filled eyelids all could be explained as an adaptive response to extreme cold. The lack of beard and body hair could arise from selection against facial hair on which frost formed by the exhaled breath could freeze and produce frostbitten features. Insofar as facial hair and body hair are expressions of the same biological mechanism, the reduction in one through adverse selection should result in the reduction in the other. Features acquired by evolution under cold stress need not prove disadvantageous in hotter climates. Tropical Mongoloids have reverted to smaller size and lighter body build, but their characteristic facial features and lack of face and body hair remain. An extreme type is shown in Plate 18–8.

If this hypothesis cannot be demonstrated either by the fossil record or the experiments of physiologists, nonetheless it will survive as a reasonable idea until better ones are put forth. In any case, the Mongoloid population shows great fitness in the modern world and comprises a great and increasing proportion of all mankind. The recent expansion of Mongoloid people is especially difficult to unravel since the phenotypes of the race tend to predominate in character in hybrids produced by intermixture. The individual shown in Plate 18–9 was born to an Australian aboriginal mother and fathered by a Chinese immigrant. Such first-generation hybrid offspring are largely Mongoloid in the appearance of their features as well as in their skin color. They point to the principle that regions which appear predominantly Mongoloid today may contain in the gene pool of the population an expectedly high proportion of genes derived from earlier hunting and foraging peoples. This is certainly true in Southeast Asia.

The analysis of human groups is primarily an evolutionary problem. If the movements and relationships of various human populations can be identified correctly, then a reconstruction of prehistoric population changes may become possible. The detailing of past

The Population Components in the Australian Aborigines

*R. G. Steegman, in a series of well-designed laboratory experiments, seems to demonstrate that Coon's hypothesis about the Arctic engineering of the Mongoloid face is not confirmed in his data. His results are complex, and do not as yet point in any particular direction.

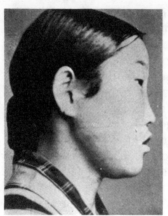

Plate 18-8
Northeastern Siberian Woman,
showing extreme development of
Mongoloid features

Plate 18-9
This Mongoloid-appearing man
had an Australian aboriginal
mother and a Chinese father.

human evolution is not technically feasible today, but it is interesting to examine another of its problems. The continent of Australia is the writer's research area, and it presents evolutionary problems of a challenging character. All aboriginal Australians have a kind of appearance of their own and would be difficult to confuse with any other population in the world. True, some individuals in New Guinea or South India might disappear in an Australian group, but, in general, differences would be evident. Furthermore, the variation which occurs on the continent is continuous in nature, and abrupt changes are seldom found. As shown in earlier chapters, the regional variation in such hidden properties as the blood group gene are enormous in magnitude, indicating that the observed differences have largely arisen after the peopling of Australia. But beneath the apparent visible uniformity and the identified hidden variability, there are certain patterns which are worth examination.

A decade ago, the depth of Australian prehistory was to be measured in only several thousand years of time. Our knowledge of the prehistory of Australia has advanced enormously in the last few years, however. At a number of places human tools have been dated as going back into the last part of the Pleistocene. The oldest occupancy to date has been found at Mungo Lake in the arid reaches of western New South Wales. There, in a series of fossil sand dunes, aboriginal tools have been found going back to almost 40,000 years ago together with a shell midden and a hearth which has been dated at 44,000 years old. The tools represent a large flake type of industry with rather simple chipping and obviously too large to be hafted. These same types of tools appear in all the other early sites on the continent, terminating only about 6,000 years ago. They were still in use by the Tasmanians at the time of their discovery.

Mungo Lake has also provided four skeletons to date. They are naturally in very fragmentary condition, but the best reconstructed of these, known as Mungo I, reveals a long-headed, very delicate cranial vault, probably from a young adult woman (A. G. Thorne, 1971). While this individual falls within the extremely gracile range of living southern Australians, she would fit even better into the Barrinean rain forest populations. The other two cremation burials have not yet been reconstructed, and a recently-found extended burial has not yet been published. The site is of considerable importance, for it promises in time to provide at least a modest sample of populations living there some 20 to 40 thousand years ago. Whether they will reveal consistent Negritoid traits, or will be more representative of the type of Aborigines who lived in south-eastern Australia into historic times, remains to be seen.

Another very important, more recent site has been systematically excavated at Kow Swamp in northern Victoria. There the burials are not cremated and so provide remains in better condition. Of the restorable skulls, there is a common pattern which, as Thorne (1971) points out, deviates from normal aboriginal standards for the area in having very sloping foreheads and much enlarged mandibles in their anterior portion. These characteristics which are uniform in the population are claimed to suggest ultimate derivation from Solo man, who lived some 300,000 years ago in Java. Since these somewhat brutalized types lived from 10 to 15 thousand years ago, that is, much later than the Mungo Lake group, there remains the interesting problem of how the archaic population could have survived, even very locally, in country rich in foodstuffs so late into time.

Early settlers of Australia must have come out of Southeast Asia, making their way between visible islands by raft. It is reasonable to presume that the journey was aided by the low sea level which resulted from the tying up of great masses of water in the continental glaciers of the Würm ice advance. At that time, with sea level 270 or more feet lower than it is today, Australia and New Guinea were united by a sunken land bridge, the Sahul Shelf, approximately 1,000 miles in width. Furthermore, the great Indonesian islands of Sumatra, Java, and Borneo were connected to Southeast Asia as a dry land extension forming what is called the Sunda Shelf. Between these two emergent land masses there extended two chains of lesser Indonesian islands. In both routes water gaps were reduced but not bridged. It has been estimated that at every point in proceeding from the Sunda to the Sahul Shelf, the next land mass lying to the east was visible and separated by no more, frequently by less, than 25 miles of open water. Men did make the crossing in small rafting groups over many thousands of years of time. The question arises as to what kind of people they were.

The types of population available to raft out into Australasia can be examined from two points of view: (1) what kinds of people were available in Southeast Asia; and (2) what kind of variation in the populations of Australia and the adjacent region can be related to the mainland peoples? Four different kinds of populations can be distinguished today in Asia. They are the dominant Mongoloids, the numerous peoples represented by the jungle tribes of South India, the marginal Ainu of northern Japan, and the isolated remnant populations of Negritos on the islands and mainland of Southeast Asia. Since there are no evidences of Mongoloid contributions to the Australian population, aside from a little recent gene flow originating from Malay fishermen along the north coast, that group can be discarded as contributing to the Australian Aborigines. The Asiatic Negritos are a very distinctive population, and if they had reached Australia, should have left visible markers in the living populations. Traits to be looked for would include spirally-curled hair, or its derived forms, dark skins, and reduced stature. Since steatopygia was only slightly developed in most Oceanic Negritic populations, it is not to be expected that any traces would remain where the Negritic element is much diluted.

Tasmanians. There are two areas of Australia where Negritic contributions are undoubted. The extinct Aborigines of Tasmania have always been recognized as differing from their mainland neighbors in their spirally curled hair. Their skin color seems to have been dark, but their stature was not much reduced. In facial features and cranial characteristics they look much like the Aborigines of Southern Australia. The best solution for their origin is to consider that they represent a well-blended mixture in which Negritic characteristics were still expressed but much diluted. Billy Lanné, the last surviving Tasmanian man who died more than a century ago, is shown in Plate 18–10a. Tasmania is a marginal area in the sense that it is distant from the northern coastal regions by which men must have reached the continent. The fact that it was cut off by water some 9,000 years ago has helped to preserve its human population in isolation.

Barrineans. There is another marginal area in Australia whose isolation depends upon a dramatic shift in its flora, rather than to a water gap. In northeast Queensland, in the country back of the pleasant coastal town of Cairns, rise two plateaus covered with dense climax Indo-Malayan type rain forest. Rainfall reaches values up to 160 inches annually in this area, so that it represents a very special environment. Rain forests in general are poor places for hunters, for there is very little meat on the ground. They are unattractive places in which to live and so are marginal in terms of hunters' preferences. The Cairns rain forests contain a group of 12 tribes of Aborigines first scientifically investigated by N. B. Tindale and myself (1939) which indisputably show evidence of a partial Negritic ancestry. They have been called *Barrineans,* after a rain forest lake. Almost three-quarters of them to some degree or other show forms of hair which are derived from the spiral curls which characterize Negritos (Plate 18–11a). Their stature is much reduced so that men average five feet two inches in height. Many of their faces are infantile in features in contrast to the normal Aborigines. Like Negritoids elsewhere their dark brown hair shows an excess of red tone when viewed through transmitted sunlight. People from the rain forests of Queensland have a Negritic cast to their features, but, of course, they are not to be compared with the Andamanese, who are the least mixed of all the Oceanic Negritos. The evidence of these two marginal populations strongly suggests that some Negritic

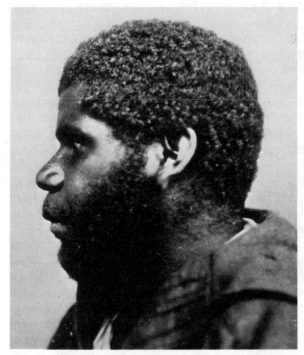

Plate 18-10
(a) Billy Lanné, the last Tasmanian man to survive. He died by drowning in mid-nineteenth century.

(b) Highland New Guinea man with genetic composition similar to Tasmanians. His woolly hair has been extended by braiding with rope.

peoples did reach Australia early, very likely as the first wave of colonists, since their traces remain today only in isolated and marginal areas. The fact that the island of New Guinea to the north is predominantly Negritoid in racial type stands as strong confirmation of this idea.

Murrayians.   Most of Australia was peopled with natives of average size. Everywhere they had large brow ridges, big teeth, and broad noses. But beyond these features in common, marked regional differences are evident. The Murray River, and its major tributary, the Darling, comprise the only permanent water drainage system in Australia. Within its boundary, the Aborigines deviate markedly from those living in the northern stretches of the continent. These *Murrayians* (Plate 18–11b) are of moderate stature but heavily built, with big-chested torsos and body proportions approaching those of Europeans. Their unexposed skin color is a medium brown, which tans to a fairly dark brown in exposed areas. Adult men have strikingly heavy beards and body hair, and are given to baldness. With their relatively high-bridged noses, these people look like rough-hewn Europeans, or like the Ainu of northern Japan. They deviate from the general aboriginal standards in the direction of primitive Whites. Indeed, in all their measurements, except stature, they are closer to the Ainu than they are to northern Australian Aborigines.

These people have been on the continent for a long period of time. The Keilor skull discovered in southern Victoria is a classic Murrayian and dates from about 10,000 years ago. Their distribution in Australia covers most of the southern one-third of the continent and extends up the east coast to the rain forest area in northeast Queensland. This puts the Murrayian in a marginal position relative to the northern coastal populations. On the other hand, their position is not as extreme as that of the Negritoid Barrineans, so it may be presumed that they entered the continent after these diminutive people. The Murrayians represent the second wave of people coming out of Southeast Asia and into Australia and New Guinea. In their least mixed form they are best represented in southeastern Australia.

Carpentarians.   In all of coastal north Australia, around the Gulf of Carpentaria, the portal by which it was populated, are Aborigines of good stature, linear body build, very dark skins by Australian standards, and a notable lack of body hair. They clearly have body proportions suited for life in the hot, dry tropics, but the question arises as to whether it was achieved in Australia or molded somewhere in southern Asia. These people have low-bridged broad noses, large brow ridges, faces very wide relative to their very narrow skulls, and so do not look in any way like primitive

(a)

Plate 18-11
(a) This Barrinean mother shows the infantile features and curly hair that denotes a Negritoid component in the rain forest tribes of northeast Australia.

(b) This heavily-bearded Murrayian from southeast Australia shows in extreme form the muscular, stocky body build of the Aborigines from this area. The possum skin robe is standard dress.

(c) This magnificent Carpentarian from North Australia displays his beautifully proportioned linear body. These people are tall but lightly built.

(b)

(c)

Europeans. In their general features they recall the jungle tribes of southeast India, except that these *Carpentarian* Australians (Plate 18–11b) are very much larger in size and more heavily boned. It is tempting to relate the two groups, although the reduction in size among the Indians is difficult to explain. In any case the Carpentarians reached Australia relatively late, for they introduced a domesticated wolf, which has gone wild as the dingo, and which has radiated explosively throughout the continent. The dingo came too late in time to cross over to Tasmania. This sequence of events indicates that the dingo must have been introduced to the continent sometime after rising sea level had separated the island of Tasmania from the mainland, that is, later than 9,000 years ago. From their position along the northern coastal stretch, the Carpentarians were obviously the last major group of immigrants to reach Australia, and, like the others, must have come by raft across the series of small water gaps in Indonesia. Some of them must have brought their dogs as well as their wives and children on these flimsy crafts.

When a continent reveals the regional variation that Australia does, it can be explained in one of two major ways. Evolution does proceed, and given enough time it can produce considerable differences. It is possible to maintain that the differences described above all evolved in Australia as an adaptive response to different kinds of selective pressure. On the other hand, without denying the ever-present effect of local evolution, differences of the magnitude which occur in Australia can also be explained as the residual differences which remain after the blending of the very different groups of immigrants who reached the continent. My own position has always favored the latter of the alternatives, although certainly some evolutionary changes can be recognized as occurring within the continent. For example, the increased linearity of the bodies of the desert dwellers of the interior must involve adaptive changes in body build. But the final demonstration as to which of these alternative hypotheses is more nearly correct will turn upon the analysis of clines for both genetically simple characters and for the polygenic traits. Much of the clinal variation will be expected to reflect local selective pressures, particularly in the blood group genes. But if the polygenic characters consistently show a clustered pattern of unexplained increases in frequencies of the extreme forms of traits in areas which are marginal either in terms of distance, isolation by water, or isolation in terms of a changing ecology as in the rain forest, then the evidence will strongly favor the idea that the differences primarily originated when different peoples entered the continent. This future analysis will be a necessarily complex one, but a preview indicates that its results will favor the idea that three separate waves of people entered Australia, and that each can be related to populations still surviving on the mainland of Asia.

The unraveling of the consequences of ancient gene flow and hybridization is complicated and difficult. But in many regions of the world today new population divisions are in the process of being formed. Our fiftieth state, Hawaii, is in the act of producing a new, interesting, and attractive group such as never has existed in the world before. The original Polynesian inhabitants have contributed some genes to the new population. Probably more important are those provided by the very sizable Japanese and Chinese immigrant populations which have been settled in the Islands for some generations. Early Portuguese arrivals, from the Cape Verde Islands, brought in a very minor complement of African Negroid, as well as southern European genes. And, finally, Whites of continental American ancestry are contributing steadily. It should be noted that two of the contributors to the neo-Hawaiian population were themselves markedly hybrid in origin, namely, the original Polynesians and the Portuguese.

The social conditions which allow this great mixture are of some interest. The original Polynesians were an unprejudiced group who accepted all immigrants without bias. Hybrid descendants sired by early whalers introduced a new element into the population. With the arrival of later immigrants the Islands essentially remained an open society save for the very rich White planters, who came to dominate it financially. While matings between various groups were not completely at random, there was a substantial amount of gene flow in all directions. Hawaii is not quite a paradise, for minor levels of discrimination are present. But it is safe to predict that its population will approach a condition of panmixia because of the increasingly random mating pattern practiced. Today the Islands have not reached the stage of complete assimilation, but the process of mixture has proceeded far enough so that it can be noted that the population is physically vigorous, handsome in feature and body, and obviously good to look at. Many of the women of Hawaii of mixed origin would be classed as beauties anywhere and by anyone.

The results of hybridization between different human populations is not predictable in any simple fashion. If the mixture is long standing, involving a good many generations, then the resultant populations may be to some degree intermediate between the parental types. Recent hybrids do not necessarily fall into this position of intermediacy. The F–2 European-Tasmanian pictured earlier in Plate 14–2 could be a European from almost any country, although his skin color is that of a southern European. Here again, in the second generation of hybridity, we have an individual who reveals no real traces of the aboriginal genes which constitute about one-half of his genetic makeup. In fact, the rest of this hybrid

The Process of the Evolution of Human Groups in the World Today

European-Tasmanian population, of which I have examined well over 100 individuals, mostly could pass as southern Europeans in appearance. Quite clearly the Tasmanian genes are expressing themselves very little even down into the fifth, sixth, and seventh generations of mixture.

There are other regions of the world in which new populations are being created at a notable rate. In much of South America a European aristocracy, economically speaking, rests upon a very broad population base in which three major elements are now blending. The aboriginal Indians of the continent have contributed to the new group. Negro slaves brought in in early times are also effective providers to the gene pool. And, of course, the Spanish and the Portuguese have passed their genes on into the lower-class elements from which they like to stand aloof. Mating is not so random here as in Hawaii, so that it will take a longer time for the blending of the population to proceed to completion. But it can be assumed safely that a new type of population, sometimes called Ladino, is in the process of formation and that the people of South America and much of Middle America will in time fall into a well-blended whole.

Wherever different kinds of people have met in the past, or do now in the present, gene flow has proceeded between them, and new types of populations have come into being. Although the White supremacy government of South Africa does not like to recognize its responsibility for the act, the so-called "Cape Coloured" of Cape Town, numbering several hundred thousand individuals, are the descendants of the original colonial Boers and native Hottentot and Bushmen women. For their present attitude of righteousness to arise from so much bedding of native girls is a remarkable act of self-justificaton.

Bibliography

BIRDSELL, JOSEPH B.

1951 The Problem of the Early Peopling of the Americas as Viewed from Asia. Papers on the Physical Anthropology of the American Indian. New York: The Viking Fund, Inc.

1967 Preliminary Data on the Trihybrid Origin of the Australian Aborigines. Archaeology & Physical Anthropology in Oceania. Vol. 2, No. 2, 100–155.

1972 The Problem of the Evolution of Human Races: Classification or Clines. Social Biology. Vol. 19, pp. 136–162.

COON, CARLETON S., STANLEY M. GARN, and JOSEPH B. BIRDSELL.

1950 Races: A Study of the Problems of Race Formation in Man. Springfield: Charles C Thomas.

NEWMAN, MARSHALL T.

1953 The Application of Ecological Rules to the Racial Anthropology of the Aboriginal New World. American Anthropologist. Vol. 55, No. 3, pp. 311–327.

THORNE, A. G.

1971 Mungo and Kow Swamp: Morphological Variation in Pleistocene Australians. Mankind. Vol. 8, No. 2, pp. 85–89.

TINDALE, N. B. and J. B. BIRDSELL.

1941 Tasmanoid Tribes in North Queensland. Records of the South Australian Museum. Vol. 7, No. 1, pp. 1–9.

CHAPTER 19

THE
RELEVANCE
OF RACE
IN THE WORLD
TODAY

The problem of "race" falls into two totally separate categories. First of all, the evolutionary biologist looks at regional variations among populations to see what light they may throw upon evolutionary processes including the short-term ones. Second, "race" represents a kind of boogeyman in the popular mind of many. It is combined with emotionalism, ignorance, and prejudice frequently originating in parental attitudes. If "race" is not real, certainly racism is, and in certain quarters it is guided by propaganda to inflame public feelings. Recently there has been a surge of types of scientific racism among psychometricians. They have presented strongly worded and detailed cases for the intellectual inferiority of several groups of people. Their arguments fall into two assumptions. First, they have assumed that what is called "intelligence" is inheritable at the 80 percent level. This is considerably higher than properly chosen data would indicate. In the second place, they have assumed that their test instruments are largely culture free, whereas in truth even in a single society there are many subcultures, some of them grossly underprivileged. In the modern world it is not enough for an educated person to know that modern human variation is a product of microevolution; it is necessary to be able to grapple with and disarm racist dogma and argument.

In a wretched portion of the Central Desert of Australia lived a tribe known as the Walpiri. They were distinguished from the tribes that surrounded them only by slight dialectal variations in language and by no aspects of culture and physical appearance. Yet they considered themselves to be a superior people and cast their beliefs in the following terms (Meggitt, 1962, p. 34):

An Attitude of Self-Esteem Is Universal

There are two kinds of black fellows, we who are the Walpiri and those unfortunate people who are not. Our laws are the true laws; other black fellows have inferior laws which they continually break. Consequently, anything may be expected of these outsiders.

We have here the age-old manifestation of a people's pride in self, a system of beliefs which elevates their own imagined position in the world by lowering the position of others. This *ethnocentrism* has characterized most small societies in the past and many that still exist today. Such attitudes serve the social purposes of increasing group solidarity and instilling local pride in the members of each society. Ethnocentrism is akin to, but not identical with, race prejudice, and like the latter it inflates the ego of the members of the ingroup. To a considerable degree it is based upon a lack of, or breakdown in, communication, for folklore says that if you cannot speak with strangers there is something bad about them. In Australia, where each tribal group was set apart by great or small language differences, if you could not understand the language of strangers, you feared them and speared them. The inability to communicate is a large factor in both ethnocentrism and racism. While such attitudes can be understood in an ancient world of small societies, they are hardly acceptable in a world where man has walked upon the moon and where he monitors the radio emissions of distant galaxies.

Man's Capacity For Hate

Racism is the culmination of a complex of attitudes which involves hatred of members of outgroups. One of the most powerful forces propelling these enmities lies imbedded in man's system of beliefs in the area known as religion. The conflict between Protestants and Catholics in Northern Ireland since the early 1970s is primarily a consequence of differences in religion, although certainly political and economic overtones are involved, too. The mutual massacre by Moslems and Hindus in the exchange of populations in India a quarter of a century ago was essentially based upon differences in religious beliefs. Unnumbered victims of torture in the Spanish Inquisition were murdered in religious zeal.

But learned behavior provides other reasons to hate. In the emergent nations of Africa today tribalism stands for potential disasters in a variety of forms. In Nigeria, the secession of Biafra primarily arose from the fact that the Ibo, the dominant tribe inhabiting an oil-rich section of the country, placed its own tribal welfare above that of other tribes in the nation. In East Africa, hostilities between the dominant Kikuyu and the less influential Luo tribesmen culminated in the assassination of Mboya, one of the most promising poltical leaders in a new nation which desperately needed them. Tribalism, of course, reflects much of the same spirit expressed by the Walpiri, even though in Africa tribes may number well over a million people. Ancient hostilities are remembered, language differences remain, and each tribe believes in its own superiority and hence the inferiority of others.

Social self-esteem can produce enough human conflict, but when it is tied into a system of economic exploitation, its consequences are magnified. Except for the few vanishing societies in which hunting and collecting is the basic economy and wealth is unknown, men everywhere value economic goods and try to improve their own lot at the expense of others. The reader should note that this is not an exclusive fault of capitalism in the United States. It operates almost universally. In the whole of Latin America, the conflict between the landed and the landless is primarily economic in nature, even though the tenants tend to be more Indian and the landlords more Spanish in their physical composition. But if there were no racial differences, the conflict would still be as disastrous and as prolonged. In South Vietnam, the same contest between those with no land and those with too much goes on in spite of the visible necessity of economic reform. In Southeast Asia, there is no reason to believe that the privileged are any different in racial makeup from the exploited. It would seem that people everywhere who have wealth try to increase it, whatever the consequences to others in terms of the hardships that oppress the poor.

In earlier sections, it was pointed out that population differences are a matter of degree and that any group whose genes proved different in frequency from others is biologically different. But in the practical affairs of men, such differences become important when they become visible. Distinctive costumes and behavior set peoples apart, but the big visible differences are in skin color, hair form, and details of the facial features. Where these occur, then the problems of social difference or economic difference reach a new level, that of racial conflict. Even in those societies which pride themselves on being without prejudice, internal evidence denies the claim. In the West Indian Islands, where there is usually a predominantly Black substratum to the population and a lighter skinned to even White elite at the top, the reality of prejudice is reflected in the large number of labels used to describe differences in skin color. In such societies a man with a genuinely black skin can become socially White when he becomes rich enough. In the larger society of Brazil, some of the same values prevail. In Brazil's population, derived from Portuguese, Indians, and Negro slaves, differences of origin are not officially recognized. But, at the same time, social customs are such that people are ranked very carefully in terms of skin color and hair form.

The problem of minorities in the United States today is so massive and so far from proper solution that it will be helpful to see how such problems came into being elsewhere. In the emergent nations of East Africa, citizens of Indian origin who migrated there years ago as coolie labor, but soon constituted a middle-class trading element, have been made unwelcome since these countries attained their independence. There the African clearly showed himself in a prejudicial posture, even though a fair amount of conflict arose from economic competition between Africans and Indians. The Indians proved difficult to assimilate, and so they were told to leave. The two peoples are separate in language, in behavior, including customs, and also in appearance.

The British, who have always prided themselves on a broad form of tolerance, are now showing prejudice in a well-developed and virulent form. It has been occasioned by the open-door policy of admitting empire citizens to the home country. Indians from East Africa have chosen to migrate to England rather than return to India, presumably owing to the very much higher standard of living in England. The same policy has encouraged an inflow of British citizens from the West Indies. These two groups, Blacks and Browns, now number more than a million persons in a total population of 45 million. As recent migrants they settle in their own ghetto groups, they speak their own dialectal versions of English, or even their own Indian languages; they may dress a little differently, and, of course, they are phenotypically visible in the light-skinned British population. They are easly visible, and they threaten economic competition with the lowest stratum of British society, which reacts in a predictably hostile fashion. The problem in England is serious enough that at least one high-ranking politician has openly appealed to the public to support "racial" segregation. In short, he proposed to save England for the White man.

Even in Russia, where the Marxist dialectic claims a patent on the idea of the brotherhood of men, "racial" problems have reared their heads. A few years ago, in Moscow itself, rioting occurred between the White populace and the body of African students being schooled in Russian universities. These African Blacks were invited to and subsidized in their education in Russia, and presumably they were to serve as a seed stock for spreading Marxian ideas in their native homelands. But their differences in speech, costume, and skin color defined them as a minority in the metropolitan population, and they were subjected to physical violence. It is quite clear that in the present state of man's ignorance and intolerance, prejudice and violence inevitably spring up wherever a substantially different minority is injected into a matrix population. These are not instinctive reactions, in the sense that they lie in the code of inheritance of the individuals, but directly reflect the learned values of life. Research has shown that prejudice can be instilled in American White children as early as the age of two or three by the transmission of parental values to offspring. This comes from a reflection of racist attitudes in unconscious but consistent ways. Prejudice is primarily a product of ignorance, fear, and intolerance. It can be overcome, but success requires persistent, directed effort on a broad social plane.

Neoracist Claims and Answers

"Racial" hatreds at the gut level characterize the common man in his fear of those who are different and who may compete economically. The lower the economic status, the greater the prejudice. It is fair to say that these common antagonisms would arise without direction from above, wherever circumstances are right to create them. But they are fanned by the writings and propaganda of a small number of well-educated and intelligent men (I shall call them neoracists) who use their very real talents to heighten racist hatreds throughout the English-speaking world. Unfortunately, they include some academic professionals who practice in fields in which they should have learned better. The roster of their contributors includes psychologists, sociologists, and a few individuals from odd corners in the natural sciences. Others among their membership involve men high in managerial capacities in industry. Since as a group they show a flair for persuasive writing, much of their propaganda is convincing to the lay reader. Because the writings of the new racists sound plausible and are built upon truths and half-truths, as suits their purpose, a series of the major statements upon which their claims are built is listed. Answers, based upon modern information derived from both the natural and social sciences, are given in order to enable the reader to answer these types of assertions.

(1) Race Is Real

The racist is not concerned with the scientific definition of race and the impossibility of applying it among undisturbed populations. The racist operates as a propagandist and not as scientist. The man on the street agrees with him for he can distinguish Whites from Blacks with ease. This visibility is the result of the displacement of Africans from their continent of origin to one dominated by a White society. In this context the biological differences are great and of course can be seen by everyone.

(2) Races Are Different

When the neoracist makes this assertion he is preparing for a later twist in logic. But here we can freely grant and indeed emphasize that groups are biologically different, for their pools contain genes in different frequencies, although they are largely genes that are found widespread in the human species. The point is that it is primarily not the kinds of genes which are responsible for the differences, but rather variation in their frequencies.

To approach this topic from another point of view, we have demonstrated much earlier that every human being is genetically different, that is, born with a genetic code which is unique to him. Obviously, it is also true that clusters of individuals must differ in their genetic properties. Our Australian data show that the differences between even adjacent and interbreeding tribal isolates are surprisingly great in magnitude, and so again we can readily grant that in terms of inheritable properties populations are indeed different.

(3) Some Races Are Superior to Others

This is the payoff of the careful maneuvering in the preliminary rounds. If "races" are real and different, the racists say, it follows as a logical conclusion that some must be better than others. There are very few Whites, and this is the group toward whom propaganda is directed, who do not have an unconscious feeling that they do belong to a superior group, and so they willingly go along with the racists' logical structure.

With false generosity some of the racists allow certain kinds of physical superiority to Blacks, such as in the shorter sprints, basketball, and football. Then, having given them the benefit of a kind of animal superiority, they point out that in mind and personality Blacks are "feckless and irresponsible," and that it really takes the Whites to show intellectual superiority and leadership. This is a game of give and then take away, designed to foster a belief in White superiority where it counts—the mind.

But difference does not mean higher and lower levels of innate ability, and this is the point at which racists' blandishments can be rejected. In the first place, populations evolve initially in terms of the forces of local adaptation. We may conclude, then, that in the original Pleistocene context the human groups, even down to the minimum breeding isolates, were better adapted to live in their own environment than other groups would have been. This involves the

Plate 19-1
An Australian Cartoonist's View
of Apartheid

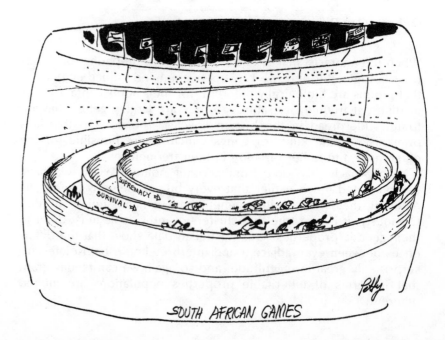

principle of maximizing population fitness to the local living scene. This is the origin of human genetic differences, and it involves no such thing as higher or lower types of inheritance. On the contrary, it stresses that differences arose to produce the best adjustments of population to their living space and climate. This is one of the major conclusions of modern evolutionary biology, and when modern man realizes its importance, some of his hatred for his fellowmen may vanish. *Differences arose among regional populations through adaptive processes.*

It has been argued by some of the neoracists that if the Blacks of sub-Saharan Africa evolved in the equatorial stretches of that continent, they may be more fit to live there than are Whites, and, conversely, Whites have biological advantages in the temperate zones. That is to say, populations should remain in the places of their origin. This again rings of plausibility until one realizes that human populations in adapting in the Pleistocene were doing so at a simple hunter and gatherer level of economy. Since that time, world technology, fueled by the agricultural revolution, has gone upward at such a dizzying rate that today most of the world's three billion people are directly influenced by modern technology, and soon all of them will be. In the cultural revolution of the last 10,000 years, man has primarily progressed by changing his behavior adaptively rather than by organically changing the content of his gene pool. True, there has been some biological evolution, but most of man's progressive steps upward have been through adapting his behavior to new scenes.

For example, it is quite likely that a Black child born and reared in a New York City ghetto, and endowed with a heavily pigmented skin, may not receive enough sunlight to synthesize an adequate amount of vitamin D. With a dark skin evolved in an equatorial climate, such a child may suffer rickets in infancy and thus become biologically deprived in a number of ways. Racists argue that this is a good case to demonstrate that Blacks should have remained in Africa. To the contrary, Harlem children can and do receive therapy in the form of an occasional teaspoon of cod liver oil and, with this small addition to their diets, do as well in the climate of New York as do children of northern European ancestry. In this and many other ways, simple behavioral flexibility adapts people of all kinds to survive in a wide variety of environments without biological handicap. The genetic differences that evolved in Pleistocene men are of relatively little importance in these days of modern medical and biological therapy. There is greater need for cultural therapy, but this will be discussed later.

Biological adaptation is always in some sense a statistical phenomenon, and not all descendants of northern Europeans function in a superior fashion in northern climates, and not all men originating in the tropics are deficient there.

(4) Intelligence Tests Demonstrate That Whites Are Superior to Blacks

This is a complex issue, but the statement is wrong on a number of counts. The very fact that psychometric tests are called intelligence tests, not just by the lay public, but by the very psychologists who administer them, is misleading. What is actually tested is individual knowledge and not inborn intellectual capabilities. It is very important to distinguish between the genetically coded capability of the brain and central nervous system and what the individual can do in a given culture as a result of repeated conditioning experiences and education. Intelligence tests reflect to one degree or another the learning success of an individual in a given culture, and only to some unknown degree his inborn intellectual ability.

The problem is complicated by the fact that individuals do vary greatly in their genetically determined intellectual capacities. We recognize that in our own society some unfortunate individuals are born and live out their lives at a level which has been labeled as that of a low-grade moron. At the other end of the scale are fortunate individuals whose intellectual abilities are so great that we call them geniuses. This range of individual variability is real to the extent that it is not masked by *different learning experiences* or the *conditioning of culture.* It is true that morons perform better under certain special learning circumstances, and it is equally likely that the mind of a genius may be undiscovered and undeveloped if his cultural environment is improper. On the other hand, the basic range is real and corresponds closely enough to the variables of genetically determined brain function.

There is every reason to believe that the same individual range of innate intellectual competence exists in all populations and in all societies. Where the level of technology is high enough, the performance of a genius becomes manifest, and he is recognizable as an individual. In technologically very simple societies, such as among hunters and gatherers, or primitive gardeners, the technological tools by which genius can demonstrate its qualities are not present. Where the need for intellectual abstraction is rare, where time is unmeasured and numbers are virtually meaningless, one cannot expect a Newton or an Einstein to appear. But in societies such as those found in aboriginal Australia, the potential of intellectual genius will manifest itself in other ways. For example, the Aborigine developer of the four-section system for regulating marriage—an involved set of rules which determines the choice of a proper potential marriageable partner and further places all individuals in the small society in a specified relationship with each other—must have been a person who would have operated in our culture at what we label a genius level. The general dogma in the biological sciences today is that human beings everywhere have the same potential intellectual range, from the heights of genius to the depths of imbecility, and that some cultures mask easy recognition of this potential.

Psychologists have never invented a test which does not carry buried within its content biases based upon cultural differences. There is no psychometric test which is culture-free. When psychologists claim that their tests are valid within a given culture, they are making the erroneous assumption that a society is characterized by a single culture. It is well known to anthropologists and sociologists that a complex society such as ours is made up of many subsocieties, each manifesting a different subculture. To argue that a psychological test levels all cultural differences because it does not depend upon language for its performance is to ignore the fact that cultural differences are not imbedded in language alone. Some years ago when psychologists were more naive than they are now, tests of this type included such things as pictures of two men playing tennis across a stretched net, but with no visible ball in the air. Another illustration showed an electric light bulb with no filament. These may have been fair enough tests for children of the tennis-playing set who lived in houses lighted by electricity, but children who had never seen a tennis game and to whom electricity was unknown could hardly point out the error. Predictably, Australian and Eskimo children got low scores. These rather extreme examples illustrate the types of difficulties which exist in trying to test intellectual capability through tests which primarily focus on learned experiences. The whole problem of measuring White and Black intelligence is compounded by the fact that psychometricians, imbedded in their own technique and looking at the world with a very limited focus, usually argue that within a given culture they are testing effectively individual differences in intelligence. The fact remains that they do not, and they cannot.

Over a good many years, testing by a great many psychologists, using a number of different instruments, has indicated that in the United States Black children achieve lower scores than do Whites. The differences average twelve to fifteen points depending upon the types of tests. Some psychologists, without themselves being necessarily racist, have come to believe that this indicates the magnitude of difference in the overall average inborn intelligence between Whites and Blacks in American society. Geneticists and anthropologists have never conceded the point. Both argue quite correctly that the test instruments do not directly measure the innate capacities of the mind and are much blurred by the individual's own learning experience. The former point out that the inborn capability of the brain is primarily determined in terms of biochemical functioning, an area not tested by the psychological instruments, while the latter emphasizes that the subculture in which Blacks live is totally different from that in which Whites live at any economic level. The fact that psychologists may try to even out differences arising from variations of economic status, achieved school position, and the geographical factors of residence in the North versus the South does not alter the fact that they cannot cancel out *the impact of being born a Black child in a White society.* To be born Black in the United States involves more than a depressed economic status, deficient diet in childhood, life as a rule in a fatherless family, and a whole series of other socioeconomic variables. A Black bears the crushing label of being inferior by birth.

Even in good educational systems, research has shown that scholastic achievement rather directly follows the level of the teacher's expectation. Without claiming that prejudice is directly involved here, it nonetheless seems apparent that the average White teacher expects a lower level of intellectual attainment from Black students, and their grades indicate that the teacher's beliefs win in this conflict. There is no way in which the handicap of being a Black in our society can be eliminated from psychological test procedures.

It will be of interest to those readers who have optimism that there is evidence that White identical twins raised separately in different environments may show differences in test scores amounting to twelve or more points. This is almost exactly the average difference between Whites and Blacks. Since none of the identical twins is ever raised in an environment equivalent to that which results from being a Black in a White society, it is reasonable to suspect that all of the difference between Blacks and Whites may be due to environmental variables, and if they could be equalized, the difference would disappear. In some future, happier society it may become possible to demonstrate that under equal environments Blacks show slightly better test scores than Whites. But the major point to be remembered is that no single psychological test, nor any group of them, really measures inborn intelligence, for none is culture-free.

(5) The Brain of the Negro Is Anatomically Inferior

Racist literature asserts again and again that the brain of the Negro is inferior to that of the White. The motivation is clear enough, since from the earliest days of slavery it has been important to the conscience of the "master" to believe that the Black was not quite human, or at the very best was an inferior human being. Since mind separates men from other beasts, it was but natural that the campaign of debasement of the Black should involve claims of organic deficiences in the material basis for intellect, the brain.

Since the brain is the seat of the human intellect, it is worth giving a resumé of some of the things that are really known about it, as well as a brief list of some that are unknown. In general evolution, as pointed out in earlier sections, the ascent of all of the progressive mammals has been characterized by substantial increases in brain size and complexity, as measured anatomically, during the whole of the Tertiary period. Coming out of the Mesozoic with a brain little bigger than a reptile's, the early mammals evolved increasingly larger and better-organized brains. With the reproductive system requiring prolonged contact between mother and nursing young, the emphasis shifted from the development of instinctive behavior to a new stress on learned and flexible behavior. This itself justified the evolution of better brains. This progressive evolution occurred in all but a few mammalian lineages, and so the ancestors of elephants and horses, cows and whales, as well as the primates, show consistent organic improvement in brain structure. It culminated in the primate in the rapid expansion of brain in the hominids in the three million years of Pleistocene time.

It is worth plotting the form of this progressive evolution among human ancestors to stress that while average size of the brain did increase through time, at any given instant the human species was characterized by an enormous range of variation in capacity. This is shown in Figure 19–1. Starting with the australopithecine grade of human organization in the Middle Pliocene, the average cranial capacity can be estimated at about 500 cubic centimeters, but it rises to a theoretical upper limit of perhaps 670 cubic centimeters and falls to a lower limit of perhaps 330 cubic centimeters for a small female. Thus, among these early human ancestors living at the same instant of time, some would have had brains twice the absolute size of others. In part, this represents difference in sex, which is associated with differences in body size, and this, of course, is reflected in brain size. But there are other individual size variations involved about which we know very little even at the present time.

In the Middle Pleistocene, in the stage of *Homo erectus,* we can estimate cranial capacity as now doubled to average about 1,000 cubic centimeters, but the full range within this population would probably go from lower limits of 670 cubic centimeters to perhaps 1,335 cubic centimeters as an upper limit. Thus, among the men of the Middle Pleistocene, some would be as small-brained as australopithecines, and others as large-brained as modern man. Turning now to *Homo sapiens,* we find among living people a perfectly normal range of variation from a little below 900 cubic centimeters for an occasional very short, slender woman to more than 1,800 cubic centimeters for a rare, large-bodied male. The average of both sexes is about 1,350 cubic centimeters.

These figures illustrate a seeming paradox. It is perfectly true for mammals in general, and for man in particular, that average brain

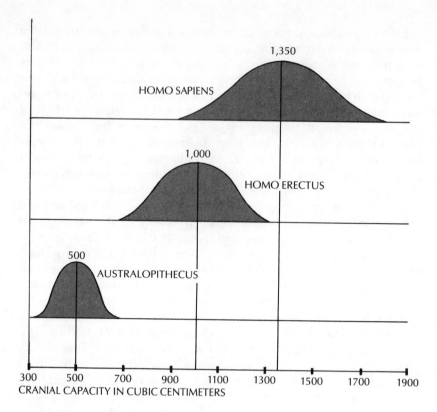

Figure 19-1
Range of Cranial Size in Three
Human Grades

size has increased over time, and this is a part of the story of progressive evolution. On the other hand, at any given instant of time, there is an enormous range of variation within a single population, and these size differences are not meaningfully connected to intellectual capacity of a genetically determined type. Anders Retzius, often called the founder of physical anthropology, was a gifted Swedish anatomist who for many years investigated the relationship between the anatomy of the human brain and its possessor's intellectual capacity in life. His materials consisted of the brains of eminent people who willed them to him for his study after their death, as well as brains which came, via the dissecting rooms, from the undistinguished segments of Swedish society. He apparently engaged in this research with the optimistic hope that some sort of correlation would be found between brain size, or the anatomical details of the brain, and a person's intellectual competence as demonstrated in life. After many years of diligent effort, he was forced to conclude that no such relationships could be detected. Some big-brained individuals were dunderheads in life. Some small-brained people proved themselves brilliant intellectually. An internationally known Russian mathematician had a brain as small as 1,100 cubic centimeters, well below the female average, and so simple in its cortical convolution as to remind one of the brain of a chimpanzee. Yet in life she was an extremely bright woman. When Anatole France, one of the world's distinguished writers, donated his brain to science, it was found to have a capacity of barely 1,100 cubic centimeters, which, of course, is small for a man. Yet his literary contributions were great. It is evident that gross brain size does not reflect innate intellectual capacity, nor does brain size when measured relative to body size. All of the evidence indicates that qualities other than the volume of the brain are more important in determining the level of its functioning. But there are many traps in this area, and it is with some amusement that one can recall that a very eminent physical anthropologist tabulated the hat sizes of American men of science and concluded that they had larger heads, and hence presumably larger brains, than the average American. This same pioneer investigator, without apparently connecting the two facts, also noted that American scientists were taller and heavier than the average citizen. Seemingly, he did not understand that there is a positive association between body size and brain size, all other things being equal.

(6) Whites Have the Highest Capacity for Civilization

This is not only another example of the technique of the big lie, used so impressively in his day by Adolf Hitler, but it reveals a total ignorance of the facts recorded in human cultural evolution. As the racists tell their story, all of man's technical progress results from the early development of civilization in the Middle East by Whites, and all subsequent human progress can be traced back to these gifted people. The other Old World civilizations, such as those which arose in Egypt, India, and China, were derived from the Near East, a fact which demonstrates that their people had somewhat less "capacity for civilization." And of course it is repeatedly pointed out that no great civilization arose in Africa south of the Sahara, proof that the African Black also has no "capacity for civilization."

These assertions demonstrate that the neoracists either don't know the facts of human cultural evolution or, as is more likely, distort them to serve their purposes. But their statements contain a curious mixture of truth and falsehood, and it is important that the reader understand what the record of prehistory does show. Civilization as we know it did arrive in early post-Pleistocene time in that area of the Middle East known as the Fertile Crescent. The exact place and time of its origin changes as the archeological record unfolds. Once thought to originate in the lowland valleys of the Tigris or Euphrates Rivers, the seeds of civilization can now be traced back to the hilly highlands to the north. There, about 9,000 years ago, the inhabitants began to settle into relatively permanent villages of modest size. The record seems to reveal that their populations were supported by the continuation of hunting, particularly of the wild cattle, goats, and antelope of the region, but settlement sizes strongly suggest that the domestication of grains was also beginning, although possibly on a very informal basis. The mainspring of Old World civilization lies in man's learning to control his two primary food sources, plants and animals. As husbandry advanced, surplus foodstuffs could be and were produced, larger populations were supported, and in time the social controls governing these small societies had to be elaborated.

While the racists claim that it was White genius which resulted in the original domestication of plants and animals, the record is not as simple and straightforward as this. Certainly the impetus of Middle Eastern civilization spread both to Egypt with some delay and to India and the Far East. But in both India and the Far East there were separate centers of domestication, and new animals and plants were brought under man's control and provided the basis for his growing numbers. The Dravidian speakers of the Indus Valley and the early Chinese deserve much credit, for when the *idea* of the agricultural revolution reached them, they remolded it in their own terms, and achieved their own civilizations independently. Recent evidence from Southeast Asia suggests domestication of animals there earlier than in the Near East. Thus in the Old World, early civilizations must be credited to three major racial groups—Caucasoids, Veddoids, and Mongoloids.

At about the same time as the agricultural revolution began in the Middle East, similar events occurred in Central America in the New World. Here, totally independent of ideas from the Old World, people began the domestication of plants and the few animals available for this process. Even though the resources of their natural environments contained fewer promising plants and animals for conversions to man's domestic uses, these proceeded so successfully that, by the time the Spanish reached the Western Hemisphere, the civilizations they found there in some ways excelled those they had left behind in Europe. While gunpowder was unknown, and no American horses survived to domesticate, the little Mongoloid peoples of Central America proved themselves to be superior astronomers, as witnessed by the fact that they had devised a more accurate yearly calendar than was known in Europe at the time. Further, they had invented the mathematical concept of zero, a very important intellectual contribution, before it was conceived of in the Old World. The Central American Indians independently developed urban life, elaborated a complex religious theology and theocracy, divided society into many classes and offices to facilitate social management and exploitation, and, unfortunately, used war both to gain control and to gain wealth. Even worse, the Aztecs used their military power to capture hundreds of thousands of victims for sacrifice to their gods. All of these attributes are considered to be marks of higher civilization and were achieved with no stimulus from Whites. But there is one point of considerable interest. The Mexican plateau, where archeological evidence now suggests plants were first domesticated, lies at a kind of crossroads between the great landmass of North America above and the whole continent of South America below. Ideas travel both further and faster than men's feet, and there is ample evidence of the exchange of both objects and ideas in the ancient Americas. It seems plausible to suggest that Mexico was the favored region for the rise of New World civilizations because more ideas funneled through it.

When we return to the question of why the ancient Middle East should have been the site of the rise of civilization in the Old World, its geography offers some clues. The Fertile Crescent lies astride the gates of communication between the Nile and North Africa on one hand and Asia and Europe on the other. Ideas flowing from Europe to Africa or Europe to Asia or Asia to Africa all would pass through the constricted region we are considering. It might even be said that in the ancient world it represented a kind of Broadway and 42nd Street corner where it is said everyone passes at some time or another. The ancient Middle Easterner was the middleman for more idea diffusion in the Old World than was any other population. Coupled with this geographical good fortune was a native abundance of wild grasses which, through domestication, became transformed into some of the most important of our modern cereals. Likewise, large numbers of cattle, sheep, and goats provided an early basis for domestication, while donkeys, pigs, and horses were later added to the list. In view of these regional advantages, it is safe to conclude that any human population which had lived in the Middle East in terminal Pleistocene times would have gradually gone through that sequence of events which led to the agricultural revolution and so laid the groundwork for the later development of what we call civilization. If this region had been inhabited by Mongoloids, they would have been the carriers of civilization. Had it been the dark-skinned people of India, they would be the present leaders in technology. Had the region been inhabited by African Blacks, the same would have been true. It is ironic that accidents of geography gave Whites their so-called superiority and "capacity for civilization."

In insistent assaults upon the African Black, racists always raise the question of why no great civilizations arose south of the Sahara. There are a number of perfectly appropriate answers to that question. First of all, let us consider the geography of Africa. That portion which lies above the Sahara belongs to the Mediterranean world of Europe and the Near East. Its cultural relations lie there and, like those areas, it is peopled by brunette Whites. The influences that quickened the Middle East also in time reached the Nile and North African shores. Below the Sahara, and what a formidable barrier this desert is, we come to equatorial Africa, the land ranging from dry grasslands to perpetually wet rain forests. It lies in isolation, connected with countries to the north only by very narrow channels of communication. The most important of these is the Nile River, which arises in East Africa and flows north into the Mediterranean. But as fertile as its flooded inner valley may be, the harsh desert encroaches on each side, and the Nile River and its valley provide a very narrow linear corridor extending from north to south. On the other side of the continent a narrow strip of Atlantic coastline no doubt allowed some trickle of ideas to reach south of the Sahara. It is not overstating the case to point out that insofar as the diffusion of ideas is concerned, Africa south of the Sahara could receive them only through two very narrow conduits, and this, of course, cut down not only the speed with which they traveled but the number which came through. In terms of the all-important exchange of ideas, Black Africa might as well have been situated in the mid-Atlantic Ocean. And yet, in spite of this inherent geographical disadvantage, civilizations of considerable substance did arise there. Some of the great kingdoms of West Africa did reach substantial levels of achievement in terms of social organization, urbanism, and particularly in some of their arts. At the same time, on the coast of East Africa, there are the interesting archeological remains at Zimbabwe, where a great medieval stone-walled city, with its tower and keeps, gave rise to a local peak of civilization. While much of its art has perished, it shows that goldworking had reached a high level and that other arts flourished. These ruins show evidence of trade with Arabs, so while they are not of Arabic inspiration they do owe their existence to trade with the outer world. Here again we have an illustration that commerce, even if carried on only in material goods, and possibly in terms of slaves, does exert its influence in the development of what we'd like to call civilization. So the exchange of goods, as well as of ideas, provides opportunity and enrichment. There are many aspects of culture in Black Africa which are worthy of comment, but this can be left to Black Studies courses, for we are headed elsewhere.

It is one of the ironies of the neoracist position that it associates a superior "capacity for civilization" (on the basis of early Middle Eastern civilization) with superior intellectual endowment. Neoracists make this claim because they, and the public which they seek to influence, do not understand the timing of the organic evolution of the human brain. As much as can be told from bony crania, and this is all we shall ever have to go by, the modern brain of man was fully developed before the end of the Pleistocene, 40,000 or more years ago. All people at this time lived by simple hunting and gathering, a life which in almost all regions required a daily search for food, for means of preservation were mostly wanting. There was no surplus of goods to provide a basis for wealth, there was no stratification of society, and life was simple and, at times, hard. It is interesting that wherever the evidence is available, be it Australia or sub-Saharan Africa, the Far East or Europe itself, these Late Pleistocene people were bigger brained than their modern descendants. So that the human mind, with its enormous capabilities and its billions of cortical nerve cells, evolved long before the agricultral revolution set the foundations for civilization. There was no "capacity for civilization" in those days in the sense meant by racists, and man's brain already had reached the pinnacle of its evolution in size and, presumably, in innate capabilities. We do not know why brain size has declined by approximately five percent since the end of the Pleistocene, that is, in the last 10,000 years, and during the very period when civilizations were beginning to emerge. Men are not smaller-bodied now than formerly; indeed, the contrary is true. Some climatic change has occurred, but it could only affect brain size through changes in body size, and the evidence contravenes this suggestion. Perhaps the only clue is the fact that domestic animals, when compared with their wild ancestors of comparable size, also show a substantial decline in brain size, approximating 10 percent. It is possible that with the advent of agriculture and larger settled groups, life in some ways became easier, and man acquired some of the characteristics of tamed animals as he adapted to the new conditions. This does not mean that man has become a domestic animal, but merely that the demands of growing society may instill a kind of tameness in him as a price for his survival. In any case, modern men everywhere are smaller-brained than were their Pleistocene ancestors, a fact that negates the assertion that certain races have a greater "capacity for civilization" because of brain size.

(7) The Purity of the White Race Must be Preserved by Preventing Its Mongrelization

The racists who stoutly maintain this position are clearly in error in terms of the basic concepts of purity, mongrelization, and the consequences of gene flow between races. Let us take these in appropriate order. Racial purity, genetically speaking, is a meaningless term. During the whole course of human evolution, genes have been exchanged between populations on a face-to-face basis and indirectly through intermediaries over a long distance. The fact that man has been able to maintain himself as a single evolving species bears witness to this constant and continuous exchange of genes throughout the species range. Furthermore, any evolutionist would have to stress that the process on the whole has been beneficial. Purity to the geneticist is the equivalent of homozygosity or the idea that the genes at each locus are all identical. This kind of "purity" could only result from many generations of the most intensive inbreeding and it would lead to drastically lowered fitness physically and reproductively. Even in the cases of domestic plants and animals where the breeders are intentionally trying to fix a given type of appearance or performance in the strain, care must be taken to preserve genetic variability, that is, heterozygosity, so that overall fitness is not lowered. Racial "purity" genetically would be a disaster.

The term mongrel is used as an insulting epithet by angry men whether they are racists or not. It derives from the fact that man has purposely bred his domestic animals, and especially his dogs, in such a way so as to concentrate different desirable behavioral characteristics in various breeds. Thus the German shepherd dog has been highly bred, its instinctive behavior selected from many generations, so that it can by proper training be turned into an intelligent man hunter, a guide dog, or a war dog. The American pointer is a specialized bird dog whose breeding has been particularly directed for producing a wide-ranging quail hunter, particularly in field trials where judging is done from horseback. Both breeds are highly specialized behaviorally. Mongrel dogs have no particular breeding since no man interferes with their nearly random mating patterns. In them wild-type genes reassert themselves, and there results a more generalized type of behavior which allows for survival under more varied and depressed conditions. Ownerless mongrels can and do survive either as wild country dogs or outcast city dogs under conditions in which well-bred dogs with their specialized behavior could not. Mongrels are better adapted for this kind of survival test than are specialized breeds. Humans cannot be mongrels in any sense, since at no time in their evolution have they been selectively bred for genetically determined forms of behavior.

Another point to consider involves the known genetic effects of increasing genetic variability in individuals through cross-breeding populations. The subject is complicated, but all evidence indicates that an increase in genetic variability heightens the biological and physical fitness of the individual. Among man's domestic plants in particular, and in a few cases his animals, it is possible to produce hybrids which exceed both parental lines in some of the attributes valued by man. Thus hybrid corn is of great economic importance in this country today, for it yields more bushels per acre than any pure-line strain of corn. But it should be noted that not all crosses produce these desirable results, but only a few, and the search for them is difficult. However, many of the world's populations are the consequence of relatively recent racial crossing and are notably vigorous. The Polynesians, who reached their outer-Pacific empire islands only in recent millennia, are quite clearly a hybrid population. This mixed group is among the world's tallest and perhaps the world's heaviest peoples, and there is no doubt about their physical vigor.

It is genetic dogma, based upon controlled crosses in such experimental animals as *Drosophila,* the fruit fly, and other creatures which are easy to breed and manipulate, that increases in genetic variability tend to produce greater fitness in both individuals and population. This can be measured in terms of longevity, fertility, or a number of other ways. More impressive, the very existence of balanced genetic polymorphisms in natural populations is strong evidence for the fact that the heterozygous genotype confers greater fitness than do the homozygous genotypes. This has been detailed in earlier sections, with the particularly useful example of the sickle cell trait. But even in normal characteristics the heterozygous genotypes generally have greater vitality than those which are based upon homozygous genotypes. Since all population studies show a great deal of genetic polymorphism, and in some cases this can be demonstrated to be of long-term evolutionary standing, the geneticist naturally concludes that a certain hybrid vitality goes with genetic heterozygosity. Genetic inbreeding, that is, mating between close relatives, in man as well as in other creatures, results in a demonstrated lowering of fitness due to the lessening of heterozygosity. For these reasons, geneticists are in general agreement that the increasing of heterozygosity produces a consequent increase in vigor and fitness.

There is one technical area in which a few geneticists have voiced reservation about the benefits of hybridizing. It involves a concept known as the *supergene,* which consists of a series of genes linked in sequence, which function physiologically more or less as a unit. It is considered that the supergenes arise in evolution through a series of chromosomal changes which create a long section of code in which genes operate best as neighbors of each other. This is considered to be an adaptive unit, and its presence has been particularly noted in the fruit fly. There the integrity of supergenes is routinely maintained by the existence of *inversions* which involve these blocks of genes. When an individual fly is homozygous for the inversion, crossing over between the two chromosomal strains does not change its content. When it is heterozygous for the same inversion, crossing over is prevented, and so the unity of the supergenes is maintained. A few of the geneticists who have worked intensively with the fruit fly have wondered whether racial crossing in mammals, including man, might not destroy such adaptive groups of genes. This is a valid scientific inquiry, and it can only be answered obliquely at the moment. It is not now possible to identify inversions in human chromosomes, and so the question cannot be answered directly. On the other hand, data from aboriginal Australia, some of which have been discussed in earlier sections, strongly suggest that gene pool adaptation there, and presumably in other human populations, is attained primarily by genetic coadaptation between individual genes, since their contents fluctuate rapidly in space and in time. The evidence suggests that supergenes may characterize fruit flies, but other species such as man attain their maximum adaptation by other means which primarily involve the relations between single genes, irrespective of their location in the linear code. The concept of supergenes does not yet seem to apply to man.

In discussing the consequences of hybridizing in human populations, it is important at the outset to recognize that social judgments and genetic judgments are based upon totally different qualities. There is probably no society in the world which does not frown upon matings and marriages between its own members and those belonging to other racial groups. In part this goes back to old primitive feelings about preserving the integrity of one's own ingroup, and in part it rests upon real social ambiguities as to how to treat the individuals in mixed marriages and, of course, their children. Society is notoriously conservative and in a hidden sense its racist attitudes are ingrained. Thus, even in aboriginal Australian society, the first generation of hybrids produced by White fathers and aboriginal mothers was routinely killed to solve this problem. In more complex societies, where different races occupy different social statuses, the problem is usually solved by depressing the social position of the mixed married couples and insisting that their children belong to the lower social group. This is manifest in so many ways that examples occur everywhere. In the United States, this attitude is reflected in the fact that an individual with any known Negro ancestry is classed as a Negro. And this in spite of the fact that many individuals with only one Black grandparent can physically pass as members of the dominant White stratum of society. In short, social penalties have been applied everywhere as sanctions against those who dare mixed "racial" marriages.

But the genetic picture differs considerably from the social one. The principle that increasing genetic variability in the individual maximizes his fitness seems scientifically proved. Obviously, the offspring of racially mixed marriages are going to show a greater degree of this heterozygosity and presumably greater intrinsic biological fitness. This of course may be depressed if the offspring of mixed marriages are raised in an impoverished environment and on an inadequate diet. In a world where our physical mobility as individuals accelerates each year, there are increasing opportunities for the members of differing biological groups to meet, to court, and possibly to marry each other. When this is combined with a general diminution in "racial" intolerance, a clear-cut trend at the moment, then it seems certain that the future will see an increased flow of genes between different peoples throughout much of the world. As a consequence, if the trend continues over sufficient time, population differences of a genetic sort will tend to be diminished everywhere. The world inevitably will tend toward a light-brown population, and become both biologically and socially better.

The Invisible Race: A Case in Which Environment Is Everything

One of the continuing difficulties in unraveling the contribution of inheritance and environment involves situations in which the two factors are almost indistinguishably mingled together, and neither can be accurately evaluated. Fortunately in Japan there is a situation in which biology is constant and the environment varies.

In that country there exists a people, the Eta, or filthy ones, or as they are now more usually known the *Buraku-min*. These are a depressed group of Japanese who even today remain socially untouchable. They generally live in one of the country's five or six thousand Buraku, or ghetto slums, and they number somewhere between 1 and 3 million persons. They have the reputation of being violent, untrustworthy, ignorant, and emotionally unstable. They are of low moral character, and this is reflected in their behavior. They lack any notion of either sanitation or manners. Buraku families are, in fact, on relief twice as often as other Japanese families. Their rate of juvenile delinquency is three and one-half times the national rate. Some of their high schools have been turned into blackboard jungles. They are said to be "the last to be hired, the first to be fired." Their overall status is the lowest in the country.

The origin of the Buraku goes back to the seventeenth century when religious pressures resulted in the social segregation of a group of Japanese workers who lived by leather-working, butchering, and other despised occupations such as burying the dead, executing criminals, or telling fortunes. They remained a formal outcast group until 1871, when the state proclaimed the caste system illegal in Japan. Even so, today they still remain at a great social disadvantage. From our point of view the most interesting thing about the unfortunate Buraku is that they are racially indistinguishable from normal Japanese. They are an *invisible race*, identifiable only by such social criteria as place of birth or residence. Since the Buraku children are distinguished only by their depressed social background, it is important to inquire how they rate on educational scales, for if their scores are depressed it will be solely for environmental reasons. In a series of so-called cognitive tests they scored 16 points lower in IQ than other Japanese children coming from normal families and backgrounds. This spectacular deficiency is the same, or perhaps slightly greater, than that scored by American Blacks as compared to Whites.

Since the major contention of the neoracist school is that mental inferiority is highly heritable, this example goes a long way toward destroying their contention. For here we have a case in which large numbers of people are genetically indistinguishable from the rest of the Japanese, but who for a half a dozen generations have lived under depressed material and social conditions. The consequence is that they show the same disadvantage in so-called intelligence tests as do American Negroes. Hence, a great part, and perhaps all, of the poor showing by Blacks in America is likely due to environmental causes and does not involve biological or intellectual inferiority.

The Problem of "Race" Today Will Become the Problem of the Poor Tomorrow

The problem of "race" today is real enough, but in the near future it will be transformed into the problem of the economically underprivileged. Enough legal momentum has been generated so that progress for the poor is inevitable, whether they are Black or White or Brown. The pace of improvement will seem frustratingly slow to those most involved, but it will be quickened with the death of the unchangeable old and the ascent of the idealistically oriented young. Let us examine some of the dimensions of change.

The study of the gross anatomy of the brain, Black or White, is of diminishing interest, for it has yielded relatively little, except for the assignment of certain motor functions which lie on either side of the great central fissure. The rest of the cortex, which involves far and away the greatest amount of its surface, is functionally little known. But it is now becoming apparent that chemistry rather than form is the open door for future knowledge about the way the brain functions. For example, memory storage remains an unknown, anatomically. But the biochemists have some suggestive information to indicate that the mechanism involved somehow is associated with RNA, or ribonucleic acid, which in some of the simplest organisms is the material of their genetic code and inheritance. Other evidence suggests that the natural functioning of the brain, and the quality of its performance, is directly related to specific chemical reactions. As biochemists explore these new avenues of inquiry, it may in time be possible to quantitatively define brain functioning, and even objectively to measure its variations in time, and between individuals. This is one of the directions of future research.

Even more exciting glimmerings have been obtained from studies of brain growth and performance in terms of differences in the environment surrounding the individual. This is a new dimension added to the old question of how much heredity, as opposed to environment, is responsible for individual behavior. Experiments with that most useful of mammals, the laboratory rat, have shown that changes in the environment affect the behavior of individuals, and the actual size, and presumably the functioning, of their brains. One experimenter, using the same inbred strain of laboratory rat, defined three environments. In an intermediate environment, poorly equipped with objects for the young rats to manipulate, he placed mothers and their litter of offspring in separate compartments. This was an impoverished environment, which restricted intensity of the social interactions of the young rats. In a second environment consisting of a bare cage he placed individual baby rats. In a third, and superior environment, consisting of large cages, he placed a number of mothers and their offspring. This group was favored with an abundance of "rat toys," whatever these may be. So we have three different levels of environment as the variable, while the genetic factor is held essentially constant by using inbred rat strains. Now it would not be surprising if the investigator found, as indeed he did, considerable variation in the development of social behavior in his rats placed in these different circumstances. But what was surprising and unexpected was that the rats in the superior environment grew significantly larger brains than those in the controlled environment, while brain size was deficient in those raised in the underprivileged environment. Since nutrition was held constant, the investigators concluded that brain growth in young mammals is directly influenced by the number of objects they can manipulate in early age, and perhaps even more by the frequency and intensity of their social interactions. Out of experiments like this has grown the idea that *the brain is something like a muscle;* it benefits and grows larger with exercise of the proper sort and diminishes in volume and loses functional tone if the proper stimuli are lacking. The reader will not have far to project these findings, for Operation Head Start was founded in response to such findings.

Another group of experimenters has found an alarmingly direct relationship between growth of brain in the unborn infant and the mother's diet. In rats in which the pregnant mothers are fed a protein deficient diet, brain growth is likewise deficient. If mothers of the same strain are fed an enriched protein diet, the growth of the brain of the fetal rat exceeds normal expectations. Since the size of the cortical neuron is constant within a species, these experiments suggest that the larger brains are anatomically and neurologically more complex and should function better. If this experiment is projected to human beings, we must come to the appalling conclusion that as long as human mothers carry fetuses while eating a deficient diet, so long will the world find new generations of young whose brains are organically deprived and whose actions behavoristically fall short of their full genetic potential. Since this problem occurs irrespective of skin color, we are again dealing with a problem of the poor, and not one directly involved with race. And yet there remains the undoubted fact that in this country, as well as in many others, minority groups are unduly represented in the lowest economic stratum. To this extent the problem of the underprivileged overlaps the problem of "race." This nation had set as one of its goals the landing of a man on Mars in the 1980s; it could do better to pledge itself to the massive solutions of the problem of the poor. For a look into the biological future see Supplement No. 21.

Bibliography

MEGGITT, M. J.
  1962   Desert People: A Study of the Walpiri Aborigines of Central Australia. Sydney: Angus and Robertson.

The Impact of the New Biology

Man is constantly striving to extend his control over his environment, even though his efforts to control his own social life have not been completely successful. And now in the near future technology promises to allow him to bring into being a *new biology*. The skilled manipulating of biological factors promises to open series of doors leading into long, dark passageways and going who knows where.

The capabilities of the new biology obviously enter into human reproduction in a variety of potential ways. One of the first and simplest promises for the near future is that couples will be able in a high degree of probability to control the sex of their own children. The ability to produce children of the desired sex sounds like a great gift to mankind, bringing immediate satisfaction to millions of couples. But I have repeatedly asked the students in this course to indicate their preferences with regard to the sex of their firstborn. Nearly 80 percent, both men and women, wished their first child to be a boy. Almost an equally high proportion would like their second child to be a boy. In a world in which couples may soon be legally prevented from having more than two children, this initial flood of male children threatens to produce a population in which men outnumber women three or four to one. While this would slow down human reproductive rates, it would also produce an enormous imbalance in human societies. There would be only about one-third as many women as needed in marriage. The world would become flooded with permanent bachelors who could not find a marriage partner. The minority of women would themselves be under great sexual pressures, both from their own husbands and from the unmarried male leftovers. The present fabric of the family has undergone considerable erosion in recent years, and it is hard to see how it could survive with so unbalanced a sex ratio. These and many more repercussions flow from the simple ability of people to choose the sex of their children.

One of the most spectacular acts of scientific social responsibility occurred through announcement in the world press on July 18, 1974, when a committee of scientists asked the National Academy of Science to declare a moratorium on using certain laboratory-built living molecules which could be hazardous if they got loose. Significantly

the group included one Nobel Laureate, Dr. James D. Watson, the co-discoverer of the genetic code. This area of research has developed very rapidly in the past few years and includes such bizarre creations as the hybridizing in vitro (in laboratory test tubes containing the proper solution) permanent crosses between the cells of such diverse animals as mice and rats and more frighteningly mice and men. This type of research has the promise of considerable progress in determining exactly what genetic loci produce what enzymes. In short, it seemed to offer a great increase in our detailed knowledge of the genetics of man, and of course the animals whose cells were hybridized with his own.

But the technique was carried further in another direction. It is chemically possible to cut a long nucleic-acid molecule into shorter pieces. The common colon bacillus, which flourishes in all of us, is the normal organism in such experiments. These pieces can then be recombined and incorporated into bacteria so creating new microorganisms whose potentials for causing disease in man, animals, or plants is both unknown and unpredictable. Such rearranged living molecules might find life on this planet with no immunities developed against its inroads. This of course is a serious threat, and the scientific moratorium is certainly welcome news for most of us.

One of the causes for alarm involved a virus called SV40 which causes cancer in monkeys. It had been considered as possibly useful in hybrid molecule experiments to throw light on its genetic composition. It was discovered after the Salk polio vaccine had been developed in the 1950s that some batches of it had been contaminated by virus SV40. While no human cancer has ever been directly linked to this virus, it does represent a potential hazard. The situation is similar to that portrayed in the movie *The Andromeda Strain*. Thus relatively simple techniques have the capacity for loosening upon us an unknowable number of new forms of living molecules. The fact that these represent recombinations of DNA occurring on this planet instead of coming in from outer space does not diminish the risks involved.

About the same time, three healthy babies were born of mothers normally incapable of producing them. This feat involved surgical assistance to couples who proved infertile because the women suffered from blocked fallopian tubes. The surgery involved removing a matured ovum from the mother, finding the right test tube environment, and in it fertilizing the egg with sperm from the father. The fertilized egg was then replanted surgically in the uterus of the mother at the appropriate time, and presumably the pregnancy proceeded normally.

This kind of manipulation can be extended outside the rigid boundaries of the family. An ovum fertilized with the sperm of a chosen male donor can be implanted in the uterus of another woman. This simple procedure is possible now and will continue to be undertaken in the future, since it only requires the consent of three willing parties, not a difficult situation to find. Its population potential is of some significance, insofar as planned breeding of human beings is to be thought of as becoming more and more commonplace. But it should be noted that in such domestic animals as cattle this technique is used to produce high-bred calves from the uterus of scrub mothers. With cattle the economic justification seems apparent. How this would be translated into elitist ideas about humans and how our reproduction should be controlled is of course quite another matter.

Recent reports indicate that mouse embryos have survived freezing at temperatures ranging from −196° C. to −269° C. From the fair proportion of embryos that survive this treatment a number were transplanted into the uteri of foster mouse mothers. A fair proportion of these gave rise to normal living newborn mice. Whereas the talk of human sperm banks frozen for centuries once seemed startling, there now seems a possibility that human embryos may be held over in deep freezes for long periods of time, finally to reappear in the world at some future date.

Another promise of the new biology arises through vastly improved techniques of tissue culture. It has been possible for a number of decades to keep alive and growing cellular tissues from a variety of parts of the body. Placed in nutrient solutions and maintained at proper temperatures, these severed cells grow in a regular fashion. Some little distance into the future is a promise that tissue culture will be able to take a single cell and grow it into a complete person.

Doctor Shettles, at the College of Physicians and Surgeons of Columbia University, has for some years succeeded in fertilizing a human ovum outside of the mother's body and actually growing it for a period of five days until it has reached the 64-cell stage. At this time it needs implantation in the uterus, and he has already succeeded in achieving this. The artificial incubation of babies in glass uteri will be achieved before the end of this century.

A cell from any part of the body contains the nucleus with its enclosed genetic code. Given the proper set of techniques, reproduction may be possible in this manner. It is proposed that a few cells scraped from the mucous membrane lining of the mouth would be enough to start a clone of descendant persons genetically identical to the person whose mouth was scraped. This is the equivalent of vegetative propagation in plants. When this ability materializes it means that human societies will have the choice of perpetuating future generations consisting totally of proven genotypes. Tissue culture would thus presumably completely oust copulation as the natural expression of bisexual reproduction, since copulation involves enormous genetic assortment and no predictable product.

The production of human beings of specified genetic content by tissue culture is but a part of the problem presented. How shall society determine which individuals are to be propagated in future generatons? Let us assume that the technique had been available during the life of Albert Einstein. Certainly the social pressures would have been great to produce another 10,000 to 100,000 Albert Einsteins. There is no doubt that he was a genius, and intellectual brilliance is a valuable commodity in any complex society. But Einstein was a product not only of his biological inheritance but of his times. There is no assurance that his very great gifts as they were demonstrated in the first half of the twentieth century would be what will be needed in the first half of the twenty-first century. Too many Einsteins might become a drag on the societies of the future.

Turning to the other end of the scale, it is always likely that human muscles will be needed to perform some of the tasks in our complex societies. Likely, too, is the fact that some menial chores will be so boring that only very unimaginative types of human beings can stand them. One can see a great demand for the clonal culturing of big-muscled morons and patient clerks as necessary solutions to labor problems in the stratified societies of the future. Since man was not wise enough to devise a controlled economy in our depression of the early 1930s and now at the threshold of the 1980s cannot seem to outlaw war, we can only question the wisdom of making men to order. We cannot know what will be required for the social betterment of future generations.

Allen's Rule states that in warm-blooded animals living in cold climates, the extremities become reduced in size in order to minimize heat loss through radiation.

Bergmann's Rule declares that in widespread warm-blooded species, population body size increases as the regional temperature decreases.

Cope's Rule states that animals tend to increase in size progressively.

Dollo's Principle claimed that evolution was irreversible. In fact, it is only irrevocable.

Gause's "Law" states that two animals with very similar food requirements cannot coexist for long in the same space.

The *Hardy-Weinberg Law* mathematically defines the phenotypic frequencies which will occur for a given set of gene values in a random mating population, provided no evolutionary forces are operating.

Malthus framed the rule that populations increase faster than do their food supplies.

Mendel's "Laws" declared that:
  1) Inheritance is particulate;
  2) Ratios of offspring occur as small whole numbers;
  3) Traits not visible in the parents may resegregate and appear in their children.

Van't Hoff's Rule is that the rate of chemical activity doubles for every 10°C. rise in temperature.

Williston's "Law" states that in the course of evolution of the vertebrates the number of bones in the skull tend to be reduced in numbers.

GENERAL TRENDS OPERATING IN ORGANIC EVOLUTION

GLOSSARY

absolute dating A series of dating techniques which provide age designations in terms of actual years.

Acheulian industry The stone tool industry of the Middle Pleistocene of Europe and Africa. It is characterized by the hand axe and the cleaver.

adaptive radiation Animals which rapidly diversified structurally and adapted to new niches within the overall habitat are said to have undergone adaptive radiation.

age-graded male group The type of primate group in which some of the young males growing up in it are allowed to remain into their adulthood, as a result of increased tolerance by the dominant male.

agglutination The action of clotting of the red corpuscles, caused by antigen-antibody reaction.

allele (allelomorph) A series of genes which may be found at any given locus. Each variable locus has its own series of allelomorphs.

amino acid dating A relative dating technique which uses the conversion of L-amino acids to D-amino acids as a measure of age to bone specimens. The older the specimen, the greater the ratio of D- to L-amino acids it will contain.

amniotic egg The desiccation-resistant egg developed by the reptiles composed of an amnion, yolk sac, allantois, and chorion.

amphibian A vertebrate class which is a kind of transitional group—adapted at once to land and water, they usually return to water to reproduce.

analogous structures Structures in different animals which have functional, but not true structural, similarities.

anemia A condition in which either the number of red corpuscles is reduced in the body, or the amount of hemoglobin is lessened, or both.

angiosperm Flowering plants.

annual productivity This may be expressed as total productivity, above ground productivity, or below ground productivity, and deals with the total production of biomass in a given area in a given year.

anthropoidal plate In the primates, the ilium grew outward into an enlarged plate of bone.

anthropometry The techniques and instruments used to measure the human body.

anticlinal vertebra In smaller animals, the thoracic vertebrae and the lumbar vertebrae converge toward a single central vertebra, the anticlinal vertebra, which represents the point about which the backbone flexes in running.

antigens Large protein molecules found on red blood cells. They react with antibodies present in the serum, and their action is to cause the clotting of red corpuscles.

arboreal Animals which live in trees are arboreal.

arthropods Spiders, centipedes, and the broad category of insects.

artiodactyls Hoofed animals with an even number of toes.

asterion The asterion is formed where the suture bounding the occipital bone unites with that separating the parietal from the temporal bone.

Aurignacian Upper Paleolithic, Cro-Magnon culture appearing in Europe 35,000 years ago.

axiom A statement which is universally accepted as true.

baboons The largest of the monkeys, an Old World cercopith ranging throughout Africa south of the Sahara.

bacteria The lowest among the truly living organisms—single-celled in organization and parasitic in habit.

balanced polymorphism A stable polymorphic condition in which selective pressure on the heterozygote acts to maintain the frequencies of particular alleles in balanced form.

biasterionic breadth The breadth between the asterions.

binomial nomenclature The system devised by Linnaeus in which two latinized names—the genus and the species names—are assigned to every animal group.

biomass The living weight of the total community, including all stored food, in a given environment.

biosphere   The zone of our planet in which life can occur.

biota   The total mass of living organisms present in an area.

bipedalism   The locomotive pattern involving walking on two legs.

brachiation   A specialized form of locomotion, swinging arm over arm through the trees, characteristic of gibbons.

burin   A specially shaped stone tool used for bone engraving.

canine   The single cusped tooth found just behind the incisors, sometimes called the cuspid or the eyetooth, and prominent in dogs.

canine diastema (-ata, pl.)   The gap left in the dental series to accomodate the long canines of most primates. Hominid forms lack the diastema.

carbon dating   An absolute dating technique based on the decay of carbon 14 to carbon 12. The older an organic substance is, the less radio-carbon will be left in it.

carnivores   Flesh-eaters.

carotene   The pigment which provides skin color with its yellowish tones.

Caucasoid   A term usually applied to light-skinned dwellers of Europe, the circum-Mediterranean area of the Near East and the Middle East.

ceboid   The larger forms of New World monkeys.

centriole   The small bodies in the cytoplasm of cells which act to organize the poles of the cell in division.

cercopith   The Old World monkeys.

cervical vertebrae   The neck vertebrae.

cetaceans   A group of aquatic mammals including whales, porpoises, and dolphins.

chiasma (chiasmata, pl.)   The condition of cross-over in meiosis when an actual cross or an X appears between the chromosome strands.

chimpanzee   The smaller of the African pongids which inhabits rain forests, open and deciduous woodlands, and tree-dotted woodlands.

chopper/chopping tools   A simple, multipurpose tool made by striking a few thick flakes off a core or nodule of stone.

chromosomes   The color-staining bodies in the nucleus of a cell which contain the deoxyribonucleic acid (DNA) responsible for inheritance.

cline   A graphic expression of biologically varying characters in geographical space.

clone   A colony of identical single celled organisms created by simple mitotic reproduction.

codominance   Different alleles are referred to as codominant if they both manifest themselves in the phenotype.

codon   The coding unit, or the message unit, in the DNA strand which consists of a sequence of three base pairs (a triplet). The 64 possible triplets code for the 20 amino acids commonly used in protein manufacture.

coefficient of inbreeding (F)   A symbol to express the consequences predicted in inbreeding. The value of F merely indicates the amount of deviation in the heterozygotes from the value predicted by the Hardy-Weinberg expansion.

coefficient of selection   A numerical expression of the force of evolutionary pressure as measured by numerical differences in terms of differential effective fertility. It essentially reflects the strength of the selective forces operating on a particular phenotype and so effects the frequencies of the genes in the pool.

convergence (convergent evolution)   The development of similar (analogous) forms in separate, unrelated evolutionary lines.

core tool   An instrument made by striking off a series of flakes from a nodule of stone.

cranial capacity   The volume of the cranium (not the size of the brain itself).

Cro-Magnon   Upper Paleolithic, *Homo sapiens* peoples of Europe.

cross-dating   The dating of a site of unknown age by comparing it with a site of known age.

crossing-over   The exchange of homologous blocks of genes between adjacent strands of the synapsed chromosomes during meiotic division.

cynodont tooth   Teeth with a small pulp cavity.

deletion   This occurs when a broken section of chromosome fails to fuse to another chromosome and so simply becomes lost in the cytoplasm of the cell.

dental formula   The formal, numerical designation of the types and number of teeth found in one quadrant of the mouth, as counted from front to back. The typical dental formula in man is 2-1-2-3, designating two incisors, one canine, two premolars, and three molars.

dialectic tribe   A population unit made up of the people within a region who share a common language. It is common to find that such groupings approximate 500 persons among many hunting and gathering people.

differential effective fertility   This concept revolves about the way in which parents contribute their genes to the next generation and whether or not the new generation is effectively raised to full reproductive competence. It is measured by the number of children who reach the midpoint of their own reproductive age.

diploid number One complete set of pairs of chromosomes. The human diploid number is 46 chromosomes.

DNA (deoxyribonucleic acid) The long, double spiral molecule (the double helix) composed of sugars, phosphates, and bases which form the genetic material. The sequence of the bases codes information necessary for self-replication and protein production.

dogma A belief held by most people. A consensus among interested scientists without any necessary implication that it represents a final statement.

dominance In Mendelian genetics, when two alleles are combined in a heterozygote and only one is expressed in the phenotype, it is referred to as being dominant over the unexpressed allele.

dominance hierarchy A sort of group pecking order; a priority system which may vary from one situation to the next.

ecological niche The specific type of environment occupied by an animal species (its habitat) and the way it exploits this habitat.

ecology The relationship between a species of organism and its total environment, organic and inorganic.

effective breeding population The people in a population who are of reproductive age, who have succeeded in having children, and who also succeed in raising them to adulthood. The effective breeding population is usually approximately one-third of the real head count.

endoskeleton An internal skeleton or supporting framework in an animal.

epicanthic fold A skin fold on the eyelid which comes down over the inner corner of the eye and runs on a line with the edge of the upper eyelid. It is characteristic of Mongoloid peoples, but also appears in many people of non-Mongoloid ancestry.

epiphysis (-ses, pl.) The end caps of bone shafts, which fuse to the central portion of the bone at different ages as maturity is approached.

epochs Subdivisions of the periods of the Cenozoic.

era The broadest classification in the geologic time chart.

erythrocytes The red blood corpuscles.

estrus A period of sexual receptivity due to female subhuman mammals coming into heat.

ethnocentrism When a person evaluates another culture solely in terms of his own, making value judgments which would elevate his own position by lowering the position of others.

ethology The study of animal behavior.

eustatic The uniform movements in sea level which occurred as glaciers were built up (thus lowering the level of the seas) or melted off (thus raising sea levels).

evolution Changing characteristics in time. In organic evolution it is a change in the genetic contents of the pool which represents the total inheritable traits of a population.

exogamous The practice of marrying outside of some recognized social group.

exoskeleton A hard supporting or protective structure developed on the outside of the body.

fibula One of the paired bones of the distal portion of the hind limb.

fission track dating An absolute dating technique based on the observation of damage paths, fission tracks, in natural and man-made glasses, caused by the fission of U^{238} nuclei. The more tracks there are, the older the substance.

flake tool A tool made from a flake struck off a nodule of stone.

fleshy-finned fish (lobe-finned fish) An early ancestor of the amphibians and all subsequent terrestrial vertebrates whose paired fins had common fleshy bases instead of spreading out in flexible horny rays.

fluorine dating A relative dating technique which uses the relative fluorine content of the bones in an assemblage to determine whether particular fossils belong to the beds in which they are found.

foramen magnum The hole in the base of the skull through which the spinal cord joins the brain.

galaxy A giant cluster of stars.

gametes The matured sex cells necessary for reproduction which are produced by meiosis.

gene Long considered the basic unit of inheritance, the gene corresponds to that segment of the DNA code responsible for determining the nature and function of some physical or behavioral trait.

gene flow   The hybridization which occurs when the genes from one group are transmitted to another. Since they differ in their initial frequency they produce changes in the gene pool which receives them.

gene pool   The sum of all the different genes contained within a population.

genetic coadaptation   The idea that an evolutionary shift in the frequency of one gene might have impact upon others.

genotype   The actual genetic makeup of an individual, including those alleles which are recessive and therefore not expressed in the phenotype.

glaciation   A period during which glacial advance takes place, usually accompanied by greater cold and an equatorial weather shift, more water locked up in ice caps, and lowering of sea levels.

gorilla   The largest of the hominoids which inhabits the humid forests of tropical Africa.

grade   A level of general organization.

grooming   Grooming behavior, which consists of cleaning the coat by removing lice, detritus, etc., is an important part of primate social life.

G6PD (glucose-6-phosphate dehydrogenase)   A protein found in the red blood cells which acts as an enzyme furthering chemical processes within the cell.

habitat   The place within an environment which is occupied by a particular animal or plant species.

half-life   The length of time it takes for half of a given quantity of radioactive material to decay to a nonradioactive substance.

haploid number   One complete set of chromosomes. The human haploid number is 23 chromosomes.

haptoglobin   Substances in the serum of the blood that combine with free hemoglobin to prevent them from passing through the capillary network of the kidney and there doing damage because of the large size of the molecules involved.

helix (helices, pl.)   A spiral. The DNA molecule is referred to as the double helix because it consists of a pair of equidistant and entwined spirals.

hemoglobin   A complex blood protein whose primary function is to fix oxygen and transport it to the body capillary system and on return to carry out the carbon dioxide to the lungs.

heterozygous   When different alleles are found at corresponding loci of a homologous pair of chromosomes, the condition is referred to as heterozygous.

homeotherm   The so-called warm-blooded animals whose body temperatures are regulated at approximately the same level, regardless of the temperature of their surroundings.

hominid   Man including his ancestors and modern representatives.

hominoids   The manlike forms, including great apes and man.

homologous   Fundamental similarity in structure (regardless of actual function) found in animals because of descent from some common ancestor. A seal's flipper and a horse's foreleg are homologous although quite different in actual form.

homozygous   When the same allele is found at corresponding loci of a homologous pair of chromosomes, the condition is referred to as homozygous.

hybrid   The offspring of two parents of different species (a mule, for example is the hybrid offspring of a horse and a donkey) or of two parents from rather inbred lines (the "mongrel" progeny produced by a Dalmation/Irish setter cross, for example).

hybridization   See **gene flow.**

hypothesis   A proposition formulated to explain certain observed facts, preferably one which refers to the relationships between two or more variable factors.

ilium   The upper blade of the pelvis.

inbreeding   The mating of persons having common ancestors.

incisor   The sharp-edged, chisel-like front teeth of mammals.

industrial melanism   A change in color which has taken place in certain populations of moths in the industrial regions of Europe during the past hundred years.

infanticide   The practice of killing newborn infants when awkwardly spaced or in excess for maintaining a stable population.

inheritance of acquired characteristics   The erroneous idea that characteristics acquired during the lifetime of an organism can be passed on to the offspring of that organism.

innate behavior   Behavior which is considered to be genetically determined as opposed to learned.

inorganic   A nonliving material.

instinct   Behavior patterns which are considered

to be due to genetic programming as opposed to learning.

intergenerational drift A type of sampling error which goes on continuously in populations whose effective breeding size is numerically small. Under such circumstances, children have a gene pool which always differs from that of their parents to some degree.

interglacial The warmer time periods between major glacial advances.

interphase The stage in which the cell is found prior to beginning its division and the state to which it returns after the act is completed.

inversion A type of mutation which consists of a broken fragment of one chromosome healing in an inverted position to the section from which it was torn.

invertebrate Animals without backbones.

ischium The posterior, lower bone of the pelvis.

isodont The dentition of animals which retain simple, cone-shaped undifferentiated teeth.

isogenic Genetically homogeneous.

kill site A place where animals have been trapped (by driving them into a swamp, over a cliff, etc.) and killed.

labrynthodonts An early group of amphibians, named for the fact that their teeth became complicated through infoldings of the enamel in the dentine.

law A sequence of events in nature that has been observed to occur with unvarying uniformity as long as the conditions are constant.

learned behavior Behavior which is conditioned as opposed to being inherited.

lemurs A group of prosimians restricted mainly to Madagascar.

leukocyte The white blood cells.

linked genes Genes are usually referred to as linked if they occur on the same chromosome, since they are inherited as a unit.

living floor A site on which the ancient inhabitants actually lived, camped, etc.

locus (loci, pl.) The physical position of a gene on a chromosome.

longitudinal arch The arch in the foot which extends from the heel to the ball of the foot.

lumbar vertebrae Vertebrae in the small of the back.

macroevolution Long-term changes, primarily manifested as a series of seemingly systematic shifts in time.

Magdalenian Upper Paleolithic tool industry appearing 17,000 years ago in Europe and characterized by bone harpoons and cave art.

mammals The class of advanced suckling vertebrates to which the primates and man belong.

marmosets The smaller, furry-tailed forms of New World monkeys.

marsupial The pouched mammals, which give birth to tiny embryonic young who must find their way instinctively to their mother's pouch where they attach themselves to a nipple and continue to develop.

meiosis The type of cell division which occurs in the sexual cells of physiologically adult individuals and functions to produce the gametes necessary for reproduction. Meiosis is a process of division which produces haploid cells.

melanin The brownish black pigment which provides most of the color in our skin, hair, eyes.

metazoans The multicellular animals which include all the higher forms of life.

microevolution Small scale evolution involving short time periods. It is observed primarily in living populations.

mitosis The basic type of cell division which occurs in all of the body cells to produce growth, maintain function, and repair damage. Mitosis is a process of division which produces diploid daughter cells.

model Something constructed to simulate complex processes in the world around us, designed to aid the investigator in comprehending what goes on in very complicated situations.

modernization of growth The trend, seen recently in western Europe, the United States, and Japan, for children to average almost an inch taller at maturity than their parents did.

molarization The trend, seen in some early hominids, for the premolars to change from their original bicuspid shearing function to become flattened, functional members of the grinding cheek teeth.

molars The large cheek teeth found behind the premolars which have multiple cusps for grinding the animal's food.

Mongoloid A term applied to people of the Far East and the American Indians. They share characteristically straight, coarse hair, epicanthic eyefolds, and moderately pigmented skin which usually shows some yellowish tones.

monotreme The egg-laying mammals, duckbilled platypus, and spiny anteater, which nurse their young after the eggs hatch.

Mousterian industry The tool industry characteristic of Neanderthal man. It is basically a flake industry characterized by scrapers, denticulate tools, and numerous distinguishable "tool kits."

multi-male group Among primates, a group characterized by an oligarchy of adult males who are roughly equivalent in age.

mutation Abrupt changes in inheritance, probably brought about by a change in the base pair sequences within the DNA molecule. These changes can be passed on to subsequent generations if they are not so extensive as to damage the organism bearing them.

negative assortative mating If patterns of mating result in encouraging marriage between people who are physically unalike, then the situation is a case of negative assortative mating.

Negroid A term usually applied to the darkly pigmented inhabitants of Africa south of the Sahara.

nonvariable loci Genetic loci at which no genetic variants (alleles) are found. These loci are probably the portion of the life code in which no variation is tolerable to the proper development and life of the organism.

notochord An unsegmented, cartilage stiffening rod found in lower vertebrate forms (and at certain stages of embryonic development in higher forms).

nuchal crest The ridge of bone on the occipital bone where the neck muscles are attached.

nucleus That portion of the cell which within its own containing membrane encloses the chromosomes.

occipital The bone of the skull which forms the back and part of the base of the brain case.

Oldowan industry The stone tool industry of the Lower Pleistocene, characterized by choppers and flake tools.

order Taxonomically, groups of families.

organic A living substance.

orthogenetic evolution A series of changes proceeding in so direct a fashion as to appear as if the animals were straightforwardly seeking a goal.

ovoviviparity Giving birth to live young through the process of retaining the eggs internally.

ovum (ova, pl.) Female sex cells, each of which carries the haploid number of chromosomes.

parallel evolution The development of similar (homologous) forms in two separate, but related, evolutionary lines.

patrilineal band A group found among many generalized hunter/gatherer groups. It is essentially a male line of descent—grandfathers, fathers, and sons who own and inherit the food rights to a tract of land whose boundaries are well defined. Women who marry into the group and daughters not yet married are also a part of it.

Perigordian Upper Paleolithic, European culture appearing 35,000 years ago.

periods Subdivisions of geological eras.

perissodactyls Hoofed mammals with an odd number of toes.

phenotype The visible, measurable characteristics of an individual as determined by his genetic makeup (genotype) and the environment.

placental Those mammals which possess a placenta, the organ which functions to provide the growing fetus with food and oxygen and remove its waste material.

plantigrade Those animals that walk on the entire, flat of the foot.

platyrrhine The monkeys of the New World, so called because of the wide nasal cartilage between their nostrils.

pleiotropy Genes which influence more than one character.

poikilotherm The so-called cold-blooded animals whose body temperatures are dependent upon their surroundings. They cannot produce and maintain a constant heat level.

polygenic characteristics (multifactorial characters) Genetically complicated characteristics which are the expression of numbers of genes located at more than one locus.

polymorphic A species containing many different forms owing to the variation among its individuals.

polymorphism A trait which is controlled by two or more common alleles.

polytypic A species which consists of regional populations which vary in type.

pongids The four living apes—chimpanzee, gorilla, gibbon, and orangutan.

population density Density is the ratio which expresses the number of individuals per unit of land area. May be described in terms of the number of individuals per square mile or in terms of the number of square miles needed to support one individual.

position effect The activity of each section of

coded genetic instructions is dependent to some degree upon the actions of the neighboring sections of code and the sequence in which the instructions are followed.

positive assortative mating If patterns of mating either allow or encourage marriage between physically like partners, the system involves positive assortative mating.

potassium-argon dating An absolute dating technique based on the decay of radioactive argon 40. The more argon which has built up, the older the specimen.

power grip The prehensile fingers wrap around an object as when one grasps a hammer, allowing the tools to be used with rapid powerful motions involving arm and wrist when needed.

precision grip The use of the opposable thumb and one or more fingers to pick up fine objects.

premolar The two cusped teeth found behind the canine, sometimes called the bicuspids.

primate The order of mammals to which the prosimians, monkeys, apes, and man belong.

principle A fundamental law upon which other laws are based.

proboscidians Elephants and mastodons with long prehensile noses or trunks.

prolactin The hormone which apparently acts to intensify mother love in animals which have just given birth to young.

proposition A proposal, assertion, or conjecture which needs confirmation.

prosimians The simplest and earliest evolved level of primates, including lemurs, lorises, tarsiers.

pubis The inferior, anterior bone of the pelvis.

quadrupedal plate In the quadrupeds a plate of bone which developed from the top of the generalized ilium.

race A subspecies, or a regional variation in populations. Members of a subspecies usually resemble each other more than they resemble members of another subspecies, but they are still capable of interbreeding with members of other subspecies within the same species.

racism A complex of attitudes which involves hatred of members of other groups.

random assortment The random way in which the chromosomal pairs align themselves on the equatorial plane during meiosis.

random genetic drift Where effective breeding populations are small, chance factors may result in changes in the genetic structure of a population. It is also known as the Sewall Wright effect and sampling error.

random mating Mating patterns are random when every individual of marriageable age has an equal opportunity of marrying any other mature individual of the opposite sex.

ray-finned fish The modern dominant types of fish are ray-finned, so called because their fins spread out in flexible, horny rays.

recessive In Mendelian genetics, when two alleles are combined in a heterozygote and one does not express itself in the phenotype, it is referred to as being recessive to the one that is so expressed. Recessives are only observable in the phenotype when they occur in the homozygous condition.

relative dating A series of techniques which allow strata, fossils, and human artifacts to be dated relative to each other.

reptile A vertebrate class made up of animals which have completely broken their dependency on water and can reproduce on dry land by means of amniotic egg.

rule A statement of what normally or usually happens.

ruminants Those mammals which are cud-chewers.

scapula The shoulder blade.

selection The process by which the environment imposes pressures on a population which can be handled better by some members than by others. The more "fit" will live longer and thus will be able to leave more offspring.

Sewall Wright effect See **random genetic drift.**

sex ratio A calculation made by dividing the number of male children by the number of females. In ordinary populations the value obtained will run close to 100 percent. Where female infanticide is practiced the values will be anywhere from 130 to 260 percent.

sicklemia (sickle cell anemia) Occurs in people who have an abnormality of the hemoglobin caused by the gene S. In those who are heterozygous for the trait, the red blood cells assume a sickle or crescent shape under conditions of oxygen reduction, causing failures in the blood supply, heart failure, kidney failure, etc.

solar system The sun and its retinue of planets. A galaxy is made up of millions of solar systems, hundreds of thousands of which may be suitable for life.

Solutrean Upper Paleolithic tool industry, characterized by extremely skilled pressure flaked tools, appearing in Europe about 20,000 years ago.

spear thrower Implement used in essence to add an extra joint to the human arm. Use of spear throwers introduced a new and more efficient projectile system for the hunter, probably in Magdalenian.

sperm Male sex cells, each of which carries the haploid number of chromosomes.

spindle threads (spindle fibers) The thin protein fibers which emerge from each polar centriole to connect the centriole to midpoints of each of the reduplicated chromosomes. The spindle threads shorten, pulling the replicated chromosomes towards the poles established by the centrioles.

steatopygia A curious condition which consists of enormously protruding buttocks, which are made up of masses of fat, and some additional slant in the sacrum.

stratigraphic dating The establishment of a relative dating sequence dependent upon the identification of different strata of rock (or site deposit) in terms of their positions relative to other beds. Involves the assumption that those layers on the bottom are older than those which overlie them.

stratigraphy The study of the Earth's strata, or layers.

subspecies See **race.**

super genes Blocks of genes well adapted in functioning together. In the fruit fly, inverted portions of chromosomes, if heterozygous, inhibit crossing over which serves to keep together blocks of genes.

synapsis The stage in meiosis during which the members of each homologous chromosome pair fuse together along the equatorial plane of the dividing cell. This pairing is such that the sequence of the DNA code on each of the fused chromosomes is the same from one end to the other.

taurodontism The tendency toward an enlarged pulp cavity in the molars.

territoriality The establishment of rights to utilize the food products of a given area.

thalassemia Cooley's anemia, or the Mediterranean anemia. Thalassemia is caused by a block in the synthesis of normal hemoglobin.

theory A formulation of apparent relationships or underlying principles of certain observed phenomena. The use of the word usually implies some verification has already been achieved.

thoracic vertebrae Vertebrae of the chest region.

tibia One of the bones of the hind limb.

tool kit A mixture of tools used at a particular time for a specific activity (i.e., butchering kit, food processing kit, tool-making kit, etc.).

totems Spirits which, among the Australian aborigines are associated with some features of the landscape. The spiritual essence of the totem is situated at a focal point, and it is to this point that the spirits of the dead return and from it that the spirits of the newborn emerge to enter the body of the women who are to bear children.

transect A transect is a cross-section through a cline, essentially a two-dimensional curve representing the slope gradient conformation along that line.

transferrins Substances in the serum of the blood which carry iron ions to and from the red marrow.

transformation The idea that changes can be made more quickly in the code of life by instructions which modify existing organs rather than in attempting to create totally new ones.

transient polymorphism A phenomenon which occurs relatively rapidly and involves a shift in selective pressure so that one allele, which was once predominant in its frequency, is in time replaced by its alternate form, another allele.

translocation A portion of one chromosome breaks off and attaches itself to the other chromosome of the pair.

transverse arch The arch in the foot which runs from the large toe to the fifth through the ball of the foot.

trophic pyramid The energy pyramid which graphically illustrates the energy levels which take place in nature—from plants to decomposers.

tupaia The living form of tree shrew which is considered by many to be a reasonable living representative of the placental ancestor of mammals.

uni-male group A primate group consisting of a male and a number of females and infants.

universe The totality of all things that exist; the cosmos.

uranium dating An absolute dating technique

based on the decay of the radioactive isotopes U^{235}, U^{238}, and U^{239}.

variable loci   Genetic loci which contain two or more variant genes (alleles). These loci are probably concerned with less vital organic functions and thus can tolerate some variations.

Veddoid   The Vedda, a small hunting and gathering group in Ceylon, seem to be a remnant of the original inhabitants of India. They are dark skinned, linear in body build, but smaller than Africans and with fine-textured hair that is straight or wavy. They are usually accepted as a fourth major human subspecies.

vertebrate   A rather comprehensive category of animals (a subphylum) including all those which possess a segmented spinal column or an internal stiffening rod (notochord).

virus   Organisms on the borderline between living and nonliving things, since they cannot reproduce themselves in the way typical of other living forms.

Y-5   The molar cusp pattern characteristic of the dryopithecines and later men. The five cusps are separated by grooves which vaguely resemble the letter Y.

zygote   An ovum, fertilized when a sperm penetrates its membrane, thereupon referred to as a zygote.

INDEX

Cover and end paper illustration by Ken Schaber